Rapidly Evolving Genes and Genetic Systems

Rapidly Evolving Genes and Genetic Systems

EDITED BY

Rama S. Singh

McMaster University, Canada

Jianping Xu

McMaster University, Canada

and

Rob J. Kulathinal

Temple University, USA

Rapidly Evolving Genes and Genetic Systems. First Edition. Edited by Rama S. Singh, Jianping Xu, and Rob J. Kulathinal. © 2012 Oxford University Press. Published 2012 by Oxford University Press.

Great Clarendon Street, Oxford, OX2 6DP,
United Kingdom

Oxford University Press is a department of the University of Oxford. It furthers the University's objective of excellence in research, scholarship, and education by publishing worldwide. Oxford is a registered trade mark of Oxford University Press in the UK and in certain other countries

First Edition published in 2012

Impression: 1

British Library Cataloguing in Publication Data
Data available

Library of Congress Cataloging in Publication Data
Library of Congress Control Number: 2012937854

ISBN 978–0–19–964227–4 (hbk)
978–0–19–964228–1 (pbk)

Printed and bound by
CPI Group (UK) Ltd, Croydon, CR0 4YY

Contents

Foreword

Richard Lewontin

Our understanding of the dynamics of the evolutionary process has long been dominated by a simplified picture built by generations of biologists who followed Darwin, a process of the slow and steady change resulting from the constant accumulation of small inherited changes over very long periods of time. This book, *Rapidly Evolving Genes and Genetic Systems*, has as its purpose to promote our understanding of how developing scientific knowledge of the complexity of the processes of inheritance and development has greatly augmented what was, for a long time, a standardized view based on an overly simple understanding of those processes.

In the ambition to build a new usable theory of some important aspect of the natural world, scientists have to cope with two disturbing factors. First, given the heterogeneity of the real world of the relevant phenomena, the theory that they build must abstract the description of that world, putting aside as minor deviations certain observed details, sometimes even to the extent of ignoring well-known observations. Second, there exist important relevant causal mechanisms which, at a particular time in the history of the science, have yet to be discovered.

While every student of nature recognizes these limitations, the socially created drive to make general theories often leads us to ignore, or at least minimize, some of the complexities of the real world.

From a literature on experiments in plant breeding, Mendel was well aware of the usual outcome of crossing two strains of plants that differed somewhat in height. Plant height, except for the unusual mutant forms like the one used by Mendel, is a typical example of a continuously varying character and plant breeders of his time were cognizant of the fact. Moreover, to get a neat 9:3:3:1 ratio in the F2 of a cross differing in two simple genetic characters, the genes involved have to be on different chromosomes. So Mendel must have been quite careful to choose the mutant forms he worked with, and there were, no doubt, some false starts. Nor was Darwin unaware of cases that did not seem to fit his general theory if evolution by natural selection. Indeed, he devoted an entire chapter of *The Origin of Species* to 'Difficulties of the Theory' although assuring us that 'the greater number are only apparent, and those that are real are not, I think, fatal to the theory.'

If we take 1900, with the rediscovery of Mendel's work as the beginning of modern genetics, the mechanistic apparatus of evolutionary genetics and of selection under domestication remained more or less unchanged for 50 years. R.A. Fisher, Sewall Wright, J.B.S. Haldane, and their respective schools of population genetics built a theoretical apparatus from a model of large numbers of unlinked Mendelian gene loci. That apparatus was used to understand the variation of traits in nature and to design schemes of artificial selection in agriculture. It is the apparatus that remains to this day as the usual skeleton structure for the training of future population geneticists and plant and animal breeders. This theoretical structure fits the general nature of paleontological theory and observations that emphasized the slow and steady evolution that was said to characterize the fossil record. But all that began to change in the 1950s and remains a major activity of genetics to the present time, with the molecularization of genetics, the discovery of the mobility of genetic elements within the genome, the mechanisms of mobility of DNA sequence information between genomes, and the variation in the rules of passage of gene copies from one organism to another and from one generation to another. It is now clear that if we want to understand evolution both in nature and under domestication we cannot model genetic evolution simply by using the classical models built of simple discrete Mendelizing factors. So, a work like my own 'The Genetic Basis of Evolutionary Change,' which depended entirely on the simple observations of allelic variation within populations, is incomplete in the program implied in its title.

It is not only the extraordinary complexity of the underlying genetic basis of evolution that has changed our understanding of evolution. We must also rid ourselves of the classical notion of a causal division between organisms and the environments which they 'inhabit.' We have inherited a notion of a distinction between an organism and the 'ecological niche' into which it 'fits.' Organisms do not 'fit' into preexistent niches. They construct their niches by their life activities. Of the infinity of ways we can describe combinations of physical and biotic factors, only those corresponding the actual ways that organisms make a living are realized niches. As organisms evolve, so their niches evolve with them, but the possibility of evolving a new niche as the organisms change is itself limited and directed by the situation in the world external to the organism. An organism cannot evolve in a particular way unless there is the possibility of remaking its niche in a corresponding fashion. But that means that there are both internal and external factors that limit and, in part, determine the rate of evolution of a species. Given that complex dependence there is no reason to expect that evolutionary rates will be constant over time for any particular phyletic line. What S.J. Gould labeled 'punctuated evolution' is not a simple start and stop process in which, for a long time, nothing at all happens and then suddenly there is a burst of evolutionary change. As recognized by G.G. Simpson in his 'Tempo and Mode in Evolution,' differential rates of change within any evolutionary line and differences in rate between related branches of an evolutionary 'tree' are a universal feature of all evolution, but evolution never stops until the inevitable extinction of every particular branch and, finally, the inevitable extinction of life on Earth.

Preface

Evolution's tempo and mode shapes our understanding of life's fundamental processes and systems. From early discussions between Charles Darwin and Alfred Wallace about the pace of species change to present-day concerns about the transmission of drug-resistant superbugs, biologists have tried to characterize the evolutionary dynamics of genetic systems. A range of theories on rates of evolution—from static to gradual to punctuated to quantum—have mostly been drawn by comparing morphological changes over geological timescales as described in the fossil record. However, new paleontological, experimental, molecular, and genomic investigations have injected a tremendous amount of new data, fresh perspectives, and excitement, offering valuable insights on the rates of evolutionary change, particularly in those fast evolving genetic systems.

This book attempts to capture these recent exciting developments by bringing together information from morphological, molecular, cellular, developmental, and genetic investigations of both natural and experimental populations across a diversity of life forms. An emerging theme among the 27 chapters is that while the rate of evolution can vary significantly, rapid changes are commonly observed. Furthermore, they play extremely important roles in adaptation, speciation, and the generation and maintenance of a diversity of biological traits and properties.

The aim of this book is to not only provide case studies that exemplify rapid evolution but to also showcase the diversity of rapidly evolving genes and genetic systems. Together, these chapters shed light on the rapid rate of evolution at the genetic, genomic, and phenotypic levels that span a diversity of timescales: from a few days in laboratory experimental populations to tens of millions of years on geological timescales. While rates of evolutionary change have been treated in various books and chapters in the post-Modern Synthesis era, to the best of our knowledge this is the first book of its kind in terms of its comprehensive coverage, breadth of genetic mechanisms, and relevant evolutionary processes and case histories. We believe that this book is timely and we hope it will continue to evolve as more and more data and theory gather.

Rama S. Singh
Jianping Xu
Rob J. Kulathinal

List of Contributors

Mihai Albu, Banting and Best Department of Medical Research, University of Toronto, Canada
mihai.balai@googlemail.com

Michael L. Arnold, Department of Genetics, Fred C. Davison Life Sciences Complex, University of Georgia, Athens, GA, USA
arnold@plantbio.uga.edu

Carlo G. Artieri, Department of Biology, Stanford University, Stanford, CA, USA
carlo.artieri@gmail.com

Evangeline S. Ballerini, Department of Genetics, Fred C. Davison Life Sciences Complex, University of Georgia, Athens, GA, USA
anji@uga.edu

Amanda N. Brothers, Department of Genetics, Fred C. Davison Life Sciences Complex, University of Georgia, Athens, GA, USA
abrother@uga.edu

Alberto Civetta, Department of Biology, University of Winnipeg, Winnipeg, Manitoba, Canada
a.civetta@uwinnipeg.ca

Andrew G. Clark, Department of Molecular Biology and Genetics, Cornell University, Ithaca, NY, USA
ac347@cornell.edu

Wendy L. Clement, Department of Ecology and Evolutionary Biology, Yale University, New Haven, CN, USA
wendy.clement@yale.edu

Yujun Cui, Beijing Institute of Microbiology and Epidemiology, Beijing, China
cuiyujun.lam@gmail.com

Steve Dorus, Department of Biology and Biochemistry, University of Bath, Bath, UK
sd236@bath.ac.uk

G. Brian Golding, Biology Department, McMaster University, Hamilton, Ontario, Canada
golding@mcmaster.ca

Felicia Gomez, Department of Genetics and Biology, School of Medicine and School of Arts and Sciences, University of Pennsylvania, Philadelphia, PA, USA; and Department of Anthropology, Center for the Advanced Study of Hominid Paleobiology, The George Washington University, Washington, DC, USA
fgomez@mail.med.upenn.edu

Wilfried Haerty, McMaster University, Hamilton, Ontario, Canada
wilfried@evol.biology.mcmaster.ca

Jennafer A.P. Hamlin, Department of Genetics, Fred C. Davison Life Sciences Complex, University of Georgia, Athens, GA, USA
jennahamlin@gmail.com

Weilong Hao, Department of Biological Sciences, Wayne State University, Detroit, MI, USA
haow@wayne.edu

Daniel L. Hartl, Department of Organismic and Evolutionary Biology, Harvard University, Cambridge, MA, USA
dhartl@oeb.harvard.edu

Ryan D. Hernandez, Department of Bioengineering and Therapeutic Sciences, University of California San Francisco, San Francisco, CA, USA
ryan.hernandez@ucsf.edu

Donal A. Hickey, Department of Biology, Concordia University, Montreal, Quebec, Canada
dhickey@alcor.concordia.ca

Timothy Y. James, Department of Ecology and Evolutionary Biology, University of Michigan, Ann Arbor, MI, USA
tyjames@umich.edu

Timothy L. Karr, Center for Infectious Diseases and Center for Vaccinology and Evolutionary Medicine and Informatics, Biodesign Institute, Arizona State University, Tempe, AZ, USA
tkarr@asu.edu

Amir R. Kermany, Department of Human Genetics, The University of Chicago, Chicago, IL, USA
amir.raji.kermany@gmail.com

Wen-Ya Ko, Department of Genetics and Biology, School of Medicine and School of Arts and Sciences, University of Pennsylvania, Philadelphia, PA, USA
wenyako@gmail.com

Artyom Kopp, Department of Evolution and Ecology, University of California Davis, Davis, CA, USA
akopp@ucdavis.edu

Rob J. Kulathinal, Department of Biology, Temple University, Philadelphia, PA, USA
robkulathinal@temple.edu

Seiji Kumagai, Department of Biology, Duke University, Durham, NC, USA
seiji.kumagai@gmail.com

Brian P. Lazzaro, Cornell University, Ithaca, NY, USA
bl89@cornell.edu

H.A. Lessios, Smithsonian Tropical Research Institute, Balboa, Panama
Lessiosh@post.harvard.edu

Manyuan Long, Department of Ecology and Evolution, The University of Chicago, Chicago, IL, USA
mlong@uchicago.edu

Harmit S. Malik, Division of Basic Sciences and Howard Hughes Medical Institute, Fred Hutchinson Cancer Research Center, Seattle, WA, USA
hsmalik@fhcrc.org

Therese Ann Markow, Section of Cell and Developmental Biology, University of California San Diego, San Diego, CA, USA
tmarkow@ucsd.edu

Annika M. Moe, Syracuse University, Syracuse, NY, USA
ammoe@syr.edu

Patrick M. O'Grady, Department of Environmental Science, Policy and Management, University of California, Berkeley, CA, USA
ogrady@drosophilaevolution.com

Sarah P. Otto, Department of Zoology & Biodiversity Research Centre, University of British Columbia, Vancouver, Canada
otto@zoology.ubc.ca

Melody R. Palmer, Department of Genome Sciences, University of Washington, Seattle, WA, USA
mryner@uw.edu

Kevin C. Roach, Department of Genome Sciences, University of Washington and Division of Basic Sciences, Fred Hutchinson Cancer Research Center, Seattle, WA, USA
kroach@fhcrc.org

Rebekah L. Rogers, Department of Ecology and Evolutionary Biology, University of California Irvine, Irvine, CA, USA
rogersrl@uci.edu

Benjamin D. Ross, Molecular and Cellular Biology Program, University of Washington and Division of Basic Sciences, Fred Hutchinson Cancer Research Center, Seattle, WA, USA
bdross@uw.edu

Rama S. Singh, Biology Department, McMaster University, Hamilton, Ontario, Canada
singh@mcmaster.ca

Willie J. Swanson, Department of Genome Sciences, University of Washington, Seattle, WA, USA
wjs18@u.washington.edu

Sarah A. Tishkoff, Department of Genetics and Biology, School of Medicine, School of Arts and Sciences, University of Pennsylvania, Philadelphia, PA, USA
tishkoff@mail.med.upenn.edu

Dara G. Torgerson, Department of Medicine, University of California San Francisco, San Francisco, CA, USA
dara.torgerson@ucsf.edu

Marcy K. Uyenoyama, Department of Biology, Duke University, Durham, NC, USA
marcy@duke.edu

Maria D. Vibranovski, Department of Ecology and Evolution, The University of Chicago, Chicago, IL, USA
mdv@uchicago.edu

George D. Weiblen, University of Minnesota, St Paul, MN, USA
gweiblen@umn.edu

Mariana F. Wolfner, Department of Molecular Biology and Genetics, Cornell University, Ithaca, NY, USA
mfw5@cornell.edu

Alex Wong, Department of Biology, Carleton University, Ottawa, Ontario, Canada
Alex_Wong@carleton.ca

Xuhua Xia, Department of Biology, University of Ottawa, Ottawa, Ontario, Canada
xxia@uottawa.ca

Jianping Xu, Biology Department, McMaster University, Hamilton, Ontario, Canada
jpxu@mcmaster.ca

Ruifu Yang, Beijing Institute of Microbiology and Epidemiology, Fengtai, Beijing, China
ruifuyang@gmail.com

Yong E. Zhang, Department of Ecology and Evolution, The University of Chicago, Chicago, IL, USA
ZhangLabIOZ@gmail.com

Dongsheng Zhou, Beijing Institute of Microbiology and Epidemiology, Fengtai, Beijing, China
dongshengzhou1977@gmail.com

Kirk S. Zigler, Biology Department, Sewanee: The University of the South, Sewanee, TN, USA
kzigler@sewanee.edu

CHAPTER 1

Introduction

Rama S. Singh, Jianping Xu, and Rob J. Kulathinal

1.1 A gradualist history

Our knowledge of evolutionary rates is as old as evolutionary theory itself. Gradualism—or slow and steady change—has been one of the most important defining characteristics of the theory of evolution since its inception (Darwin 1859). Along with variation and adaptation, gradual evolution quickly became integrated into evolutionary theory for a variety of reasons. First, we hardly observe discernable evolutionary changes during our own lifetimes and we do not expect to since the environment, while fluctuating wildly back and forth, does not appear to change in any net (i.e. predictable) direction in any significant way. Second, evolutionary change requires excess death (of less fit individuals) and/or excess fertility. While some species possess high fertility and hence can afford strong selection on short timescales, it is not true for all organisms, especially those with low reproductive rates, high parental investments, and/or small population sizes such as those found in many large mammals. Third, the immensity of the evolutionary timescale as perceived from the geological and fossil records seemed in favor of slow and constant evolutionary change. Finally, some argue that Darwin promoted the idea of gradual evolution in order to make the theory more palatable to a conservative Victorian society that believed in social stability and had little room for fluctuations of any kind—revolutionary, evolutionary, or otherwise.

Although this paradigm of slow and steady change became generally accepted, an active debate on rates prevailed after Darwin's Origins. Alfred Russell Wallace wrote:

> Mr. Darwin was rather inclined to exaggerate the necessary slowness of the action of natural selection; but with the knowledge we now possess of the great amount and range of individual variation, there seems no difficulty in an amount of change, quite equivalent to that which usually distinguishes allied species, sometimes taking place in less than a century, should any rapid change of conditions necessitate an equally rapid adaptation (Wallace 1889, p. 125).

Even Thomas Huxley, a strong advocate of Darwin's new theory, warned Darwin about his insistence on gradual evolution when he wrote 'you have loaded yourself with an unnecessary difficulty in adopting *natura non facit saltum* so unreservedly' (quoted in Gould 2002, p. 151).

In the early years of the 20th century, gradualism found its way into the mechanistic underpinnings of evolutionary models, thus making the bold leap from phenotypes to their underlying genotypes. Ronald Fisher provided theoretical support for gradual evolution on the basis that micromutations would allow evolutionary changes to occur smoothly without destroying existing adaptations through major mutations (Fisher 1918). Much of this was a response to the saltationist mutational theories of Hugo DeVries and William Bateson at the beginning of the century and later espoused by Richard Goldschmidt (1940). Early population geneticists built on a paradigm of small gradual change and evolution by micromutations and this paradigm has become the cornerstone of population genetics theory (Fisher 1930; Wright 1931; Haldane 1932; Mayr and Provine 1980).

However, there remained an obvious gap between the continuous microevolutionary patterns found in population genetic models and the episodic macroevolutionary patterns observed in systematics and paleontology. The Modern Evolutionary Synthesis attempted to fill this void by integrating population genetic theory with systematics

Rapidly Evolving Genes and Genetic Systems. First Edition. Edited by Rama S. Singh, Jianping Xu, and Rob J. Kulathinal.

and paleontological views (Dobzhansky 1937/1951; Simpson 1944). Using gradualist mechanisms of microevolutionary change, it became possible to explain the discontinuities found in the fossil record. Thus, within this framework, gradualism conforms, on one hand, to the immensity of the geological timescale and, on the other hand, to the imperceptible changes in the environment requiring adaptive modifications. Under this model and in the short term, most changes in nature would be of a stabilizing nature and the changes in the environment, biotic or abiotic, would generally be of a very slow and gradual nature. Thus, evolutionary rates were seen in concordance with the geological timescale.

In addition, population genetic theory suggested that most lineages might not be able to sustain rapid evolution for long, continuous periods of time. Sooner or later, the lack of adequate genetic variation and/or the lack of opportunity would eventually slow the pace of change. Over longer periods of time, evolutionary rates would likely average out to increasingly slower rates. Therefore, it is expected that any rapid evolution will be of an intermittent nature—few and far between geologic time points, much like Gould's punctuated equilibrium (see Section 1.3)—and, therefore, in terms of overall long-term evolution, the law of diminishing average returns will prevail. Thus, gradual evolution would appear to be a geologic reality.

Gradualism found its greatest ally in the grand neutral theory of molecular evolution, formalized during the latter part of the last century. Motoo Kimura's neutral theory (1968) posits the regularity of evolutionary change, observed previously in proteins by biochemists such as Emile Zuckerkandl and Linus Pauling (1962) and Emanuel Margoliash (1963), as the basic product of fixed substitutions on neutral mutations. Neutral theory made three important contributions. First, and most importantly to this discussion, neutral theory suggested a clock-like mechanism of evolutionary change at the molecular level. Using this clock, neutral theory offers us the ability to infer historical patterns and estimate rates of molecular evolution on varying timescales across a range of genes and species. Second, it provided a null hypothesis to measure selection and adaptation at the gene level. Without the null model, it would be impossible to evaluate whether an allele is directly under selection, is being hitch-hiked, or has low functional constraints. Finally, neutral theory showed that, even in the face of adaptive evolution, much existing genetic variation can remain standing because of neutral fitness effects, and such mutations can drift to fixation over time. Together, these three contributions made us realize, for the first time, that while populations may be under selective pressures (stabilizing or directional) on a regular basis, this does not necessarily mean that surviving organisms are genetically optimized. It also reinforced the idea that rates of change—whether slow, moderate, or fast—are a *constant* force of nature and provide a seemingly linear rate of evolution over long periods of time.

1.2 Mechanisms of rapid and episodic change

Yet, this tradition of evolutionary regularity does not mean that all evolutionary biologists have always accepted gradual evolution as dogma. In fact, far from it. Debates of one kind or another and arguments have been made in support of rapid, and often episodic, modes of evolutionary change. These arguments have been largely based on non-equilibrium situations such as changes in the neutral landscape, sudden demographic shifts in populations, and adaptive radiations. In the following sections, we highlight representative mechanisms in which non-gradual, and in particular, rapid evolution, occurs from molecular to population level perspectives.

1.2.1 Unconstrained neutral space

Marathon runners dream of running free, unencumbered by other runners or by any other factors. At the start, their pace may be constrained due to an abundance of fellow runners, but soon the crowds thin out and they run in relative autonomy. Neutral mutations are like marathon runners—their evolution may also be constrained at the start line due to neighboring mutations that are armed with selective advantages or disadvantages. However, they may ultimately evolve at rates free of constraint and that are solely governed by neutral mutation rates (Kimura 1968). Depending on the degree of functional constraints, different genes or genetic elements will also evolve at different rates. Third codon positions, intergenic regions, gene duplications, multigene families, satellite DNA, and pseudogenes are examples of genetic elements that evolve at faster rates due to lower selective constraints (Kreitman 1983; Lynch 2007).

Yet a model of strict neutrality does not fit most genetic data. For example, Gillespie (1984) observed that molecular evolution generally did not follow a Poisson model of constant rate change. He used the term 'episodic clock' to directly contrast its behavior to the regularity of the molecular clock, as originally proposed by Kimura (1968). Context-dependent fitness effects offer a mechanism to explain how molecules, which are usually evolving neutrally, can possess different evolutionary rates over time. In particular, landscape models that were originally developed by Sewall Wright have recently begun to be used to incorporate the role of epistasis in the evolution of proteins (e.g. Kulathinal et al. 2004; Weinreich et al. 2006), providing a mechanism to rapidly change the pace of evolution.

1.2.2 Horizontal gene transfer

In complex eukaryotes, mating and sexual reproduction between individuals of the same species is a common feature of reproduction. This mode of reproduction allows genetic exchange and recombination to take place but also at the same time provides a means to safeguard the genome integrities of populations and species from invasions by foreign genetic elements. However, such a mechanism is not present in prokaryotes. Instead, genetic exchange can occur among distantly related organisms through transformation, transduction, and conjugation. Indeed, genomic sequence analyses of bacteria and archaea over the last decade have shown that horizontal gene transfer is a major force in shaping prokaryotic genome size, and gene content, and plays an important role in the adaptation and long-term survival of these organisms. Horizontal gene transfer allows for the fast acquisition of novel genes: if these genes enhance the host cell's survival and reproduction, they will be retained by the host genome. However, if the newly acquired genetic elements are not beneficial, they degenerate quickly and are often rapidly removed from the genomes (Hao and Golding 2006). Different from genetic exchanges in sexual eukaryotes, horizontal gene transfer is generally localized to a small genomic region, and does not involve whole genome shuffling typical of sexual recombination in eukaryotes.

1.2.3 Developmental macromutations

Developmental biology, more often than not, has been at odds with population genetics particularly during their early histories. The most famous example is that of Richard Goldschmidt who proposed the idea of rate genes, i.e. genes controlling molecular and physiological steps, which he envisioned could accelerate rates of evolution. In simple terms, rate genes were macromutations with the capacity to make major phenotypic changes (Goldschmidt 1940). As a first-rate experimental biologist who studied variation and evolution in moths, Goldschmidt accepted microevolution within species. However, he did not believe that microevolution gave rise, or was the natural precursor, to macroevolution, i.e. speciation. As a result, he supported the role of macromutations in speciation. It is important to note that the argument has never been whether macromutations can create rapid change but whether such mutations play a major role in long-term evolution (Singh 2003). The argument, however, has been more about the mode rather than the tempo of evolution. The occasional role of macromutations, especially in a new environment, has been an accepted component of modern population genetics theory (Wright 1982).

The field of evolutionary developmental biology ('evo-devo') has provided many examples of such large evolutionary jumps caused by mutations in developmental genes, gene clusters, and sometimes even whole genomes (e.g. genome duplication and polyploidization events). These events can place these individuals at a selective advantage resulting in saltational leaps in both genotype and phenotype.

1.2.4 Evolution by gene regulation

Changes in gene regulation—the variation in amount and time of gene expression, and in tissue distribution—have always been seen as a source of rapid phenotypic change. A landmark study by Mary-Claire King and Allan Wilson (King and Wilson 1975) based on comparative molecular studies of genetic differences between humans and chimpanzees, highlighted the importance of gene regulation evolution in species differences. King and Wilson observed that the level of protein electrophoretic divergence between humans and chimps was no greater than that found between indistinguishable sibling species of *Drosophila*. However, based on their results, what they proposed was truly shocking at that time: that humans and chimpanzees diverged from a common ancestor 5–7 million years ago (mya), and not 30 mya, as was commonly believed at the time based on anthropological studies. How can humans and chimps be so different, yet share a very recent common ancestor? Thus, the great debate on evolution by gene regulation began.

Gene regulatory changes have the potential to make rapid phenotypic changes but only if the organism-environment circumstances allow for it to happen. For example, it is generally believed that the hominid lineage has evolved faster in cognitive, behavioral, and social characters but it is not evident if gene regulation has played a greater role. Considering that the supply of extant genetic variation is a universal feature of all sexually reproducing organisms, a relatively rapid rate of evolution can be sustained by changes in gene regulation as well (Hoekstra and Coyne 2007). Nevertheless, rapid gene-regulation-driven evolution, especially through cis-acting elements, has been taken as not just a possibility, but the norm, by some authors (e.g. Carroll 2005).

1.2.5 Coevolutionary forces

Aside from abiotic factors, biotic factors such as the activities of other species can exert significant influence on rates of evolution. The influence is stronger if the organisms are more closely linked to each other in their respective life cycles. This is precisely what occurs in coevolution involving interacting partners such as host–parasite, predator–prey, and pollinator–plants. In coupled systems such as these, rates of evolution can, in principle, be slower or faster. And if faster, these rates can accelerate extremely rapidly. For example, the generation and maintenance of disease resistance and immune-response genes in humans are direct reflections of the hosts' responses to the diversity of infectious agents. Similarly, interactions between predators and prey (e.g. Brodie et al. 2002) and between pollinators and flowering plants (Farrell 1998) have been hypothesized as the main mechanism responsible for the rapid evolution of genes and genetic systems in such biological communities.

1.2.6 Sexual selection and sexual arms races

Secondary sexual traits provide some of the most spectacular displays of diversity, especially in birds, insects, and mammals (Andersson 1994). Darwin proposed his sexual selection theory to explain the evolution of secondary sexual traits. These traits are more common in males and are often exaggerated in appearance and look maladaptive from a survival point of view. Darwin surmised that the evolution of these traits is driven by mate choice as females use these traits as a basis for choosing mates. As demonstrated by Fisher's 'runaway selection' model of sexual selection (Fisher 1930), secondary sexual traits, including genes affecting secondary sexual traits in the male and choosiness in females, are expected to evolve at faster rates: the mutual reinforcement between increasing female choice and male traits can lead to a cycle of accelerated evolution of secondary sexual traits in males and rapid speciation (Fisher 1930; Lande 1981).

Other sexual selection mechanisms also incorporate rapid evolution into their models. Although female choice is primarily female-driven, we can imagine the two sexes interacting in many ways to increase their own fitness by what is now called a sexual arms race (Arnqvist and Rowe 2005; Rice 1996). Sexual arms races can quickly lead to the rapid evolution of traits associated with sex and reproduction. Such arms races resemble predator–prey or host–parasite models except that in the case of the former, the effects of response and its counter response is felt by traits affecting sex and reproduction of interacting partners within the same species. Sexual selection has been a driving force in the evolution of sex- and reproduction-related genes in general (Civetta and Singh 1998) and mating system and egg–sperm fertilization proteins in particular (Swanson and Vacquier 1998).

1.2.7 Population demography and genetic revolutions

Major demographic and geographical shifts in populations can also lead to very rapid, and episodic, evolutionary change. Peripatric speciation (Mayr 1954) is a popular extension of the allopatric model of speciation, involving smaller subpopulations located at the periphery of its species' distribution. The combination of small population size and more extreme environmental variation makes rapid genetic changes more likely in what Mayr called speciation by genetic revolution (Mayr 1963). A variation on the same theme was proposed by Hampton Carson when studying Hawaiian *Drosophila* (Carson 1971). For their size and age (molecular evidence demonstrates that the Hawaiian *Drosophila* lineage is quite old), Hawaiian Islands have a disproportionately high number of *Drosophila* species that are morphologically and behaviorally very diverse. Geologic evidence reveals that the Hawaiian Islands form a chain of islands of which the newest ones have arisen through the most recent volcanic eruptions, giving rise to opportunities for new species, while the oldest ones are re-submerged. The founder-effect speciation model was proposed to fit the observation of rapid and diversified speciation in Hawaiian Drosophila. Carson (1971) extended the founder-effect speciation model to involve populations going through repeated rounds of increase (flush) and decrease (crash) in size giving rise to rapid evolution. Carson envisioned these population cycles as a result of volcanic eruptions and decimation of populations followed again by exponential population growth between eruptions with dramatic effects on the constancy of evolutionary rates.

1.2.8 Adaptive radiation

For decades, evolutionary biologists have been describing the widespread variation found in rates of evolution and in levels of diversity among different taxonomic groups. Niche expansion, or the availability of new niches, has the potential to initiate rapid evolution. For example, Gaylord Simpson (1949) described the relatively rapid evolution of mammals after the demise of the dinosaurs: within a span of just 60 million years, a remarkable diversity of mammals evolved. Other recent examples of adaptive radiations include anole lizards, cichlids, and the Hawaiian *Drosophila*.

Since Darwinian evolution is usually taken to mean both adaptive and gradual, do radiations such as the ones listed here fit Darwin's notion of gradualism? Non-gradual or rapid evolution is not necessarily anti-Darwinian. The nature of genetic variation in combination with various evolutionary forces has the potential to produce gradual or non-gradual response. Biologists expounding rapid evolution have historically raised an eyebrow or two among their peers. However, they have traditionally been focusing on the role of either unusual genetic variation (such as macromutations) or dramatic demographic shifts. Niche expansion-associated rapid evolution works fully within the framework of Darwinian evolution. In fact, one of Darwin's favorite evolutionary examples is the rapid radiation of finches found on different island habitats in the Galapagos (Grant 1999).

1.3 Punctuated equilibrium within a microevolution framework

As shown earlier, a wide range of mechanisms, from molecular to population and community, can generate rapid episodic rates of change. All these mechanisms fit within a population genetics framework that can be explained by mutation, selection, drift, and gene flow. In contrast, paleontology examines the changes in fauna and flora as well as their anatomical and morphological features through the geological record. The remarkable breadth and depth of fossil record evidence have provided some of the most direct support for the theory of evolution. With respect to mechanisms of evolution, paleontology has traditionally dealt with macroevolution while population genetics stays within the domain of microevolution. The combinations of the two fields were essential in forging the 'Modern Synthesis' in the 1940s (Simpson 1944). However, these same two fields have often shown divergent views on the mechanisms of macroevolution. Stephen J. Gould and Nile Eldredge's punctuated evolution is a prominent case in point (Gould and Eldredge 1977).

There are many instances of the fossil series where geological records provide distinct and unequivocal evidence of non-gradual evolution, i.e. long periods of stasis followed by brief bursts of diversity and speciation. For example, new species often appear suddenly in the fossil record. While it has been argued that such findings are fortuitous, e.g. that these new species may have originated elsewhere and migrated to the fossilized locale, the idea that speciation can occur through physiological and behavioral changes, with morphological differentiation following later, did not hold ground against Gould's persistent and eloquent arguments to the contrary. Since large populations are more likely to leave fossils than small populations, Gould appealed to Mayr's peripatric model of speciation. This appeal was later extended to sympatric models of speciation, to Goldschmidt's rate genes, and to any theory that would make speciation go faster. While peripatric speciation fully fits in the realms of population genetics theory, as time went on, Gould became more convinced that the rapid rate of speciation was the property of the speciating lineage itself rather than the property of the speciation process.

A careful reading of Gould's writing on punctuated speciation reveals that he was using the term, speciation, in a geological sense and not in a population genetics sense. Based on fossil records, Gould suggested that the punctuation or evolutionary burst period accounted for only about 1–2 % of the length of the stasis period, implying a punctuation period of about 1–2 million years (my) in a lineage undergoing a stasis period of ~100 my. In population genetics terms, 1–2 my is an exceedingly long time for gradual evolution, even though it represents a very short time on the geological scale. One would think that there shouldn't be a disagreement between geologically-inspired punctuated speciation and gradual speciation in population genetics (for a discussion, see Jagdeeshan et al. 2011) but Gould, looking for a mechanistic explanation, kept the two views connected and interchanged them back and forth. Gould treated species as individuals with a propensity to speciate—a characteristic that could not be reduced to the characteristics of the comprising individual organisms and thus not to be explained by natural selection. In fact, Gould went so far as to initially maintain that all significant evolutionary changes occur during speciation (Gould 1982) but later relented. As seen in the previous sections (1.2.1–1.2.8), it is now clear that there are ample microevolutionary mechanisms that could generate such extreme patterns of variable evolutionary rate.

1.4 Tempo, mode, and the genomic landscape

In *Tempo and Mode in Evolution*, Gaylord Simpson (1944) promoted the Modern Evolutionary Synthesis by interconnecting the disparate fields of genetics, systematics, and paleontology within a gradualist framework. Since then, researchers have been finding more and more examples of non-gradual change, as either an episodic phase or a rapid and continual process in a species' evolutionary trajectory. All known cases of rapid evolution, whether in the wild or in the lab, are directly or indirectly associated with high selective pressure or sudden changes in the direction of selection. For example, the high diversity of Hawaiian *Drosophila* and island fauna and flora in general are associated with geographic isolation, environmental variation, and population dynamics. The spectacular diversity of cultivated plants and animals during the last few thousand years is the result of not new genetic variation but strong man-made selection pressure. The famous long-term University of Illinois selection experiment for increased oil content in corn from 4% to nearly 20% over 90 generations is a classic case of selection and human-driven rapid change (Dudley and Lambert 1992). More recent cases of very rapid evolution include sexual selection-driven changes in color spots in male guppies within months (Endler 1986), the evolution of new carbon sources in bacteria (Lenski and Travisano 1994), the domestication of silver foxes within 35 generations (Trut 1999), and the evolution of diet (from insectivore to herbivore) and associated changes in head shape and muscles in the lizard, *Podarcis sicula*, on two Adriatic islands off the coast of Croatia within a span of a mere 30 years (Herrel et al. 2008).

The field of evolutionary biology has also expanded in tools, resources, and scope, and we now can readily evaluate the tempo and mode of molecular evolution at the genomic level, and to relate it to the tempo and mode observed at the phenotypic level. The super-exponential increase of genomic information has propelled the field into exciting new domains. Evolutionary genomics allows large-scale comparisons of sequence and gene expression changes in both closely and widely separated taxa and thus provides relevant data for comparing rates of evolution between different genetic elements as well as in different evolutionary lineages over any desired length of time. In a very short period of time, evolutionary genomics has revised our view of almost all aspects of genetic variation: the nature of point mutation, nucleotide repetition, insertion-deletions, copy number polymorphism and gene duplication, retrotransposition, gene families, and structural and functional redundancies (Lynch 2007). The rate of progress in sequencing technologies and downstream informatics is indeed breathtaking. Comparative genomics is allowing researchers to study rates of evolution from single nucleotides to whole genomes and is thus providing in-depth views on how the various components of the genome have changed over time.

In microorganisms, unique genetic mechanisms (e.g. horizontal gene transfer) and/or rapid changes in the environment, such as the use of antibiotics, provides opportunities for rapid change. In prokaryotes, rapid changes as seen in eukaryotic sexual system genes and immune response genes are more likely due to responses to changes in the environment as has likely been the case in the Hawaiian *Drosophila* and the cichlids of Lake Victoria. Common sense dictates that in the long term, genetic and phenotypic evolution would necessarily be coupled and the law of gradualism would apply as a result of a rolling average of rates. However, in the short term, evolutionarily speaking, and in different organisms and at different times, this need not be the case: different genetic elements and lineages can evolve faster or slower than the average rates.

During much of the last 150 years, evolutionary biology has preoccupied itself with the study of phenotypic evolution. Such studies have greatly enriched our perception and appreciation of organismal diversity from molecules to humans. Now, the arrival of the genomics era promises not only to unravel the structure of molecular machinery, but also to provide us with an unprecedented knowledge of the rates and limits of evolutionary change. Such knowledge would be needed as we embark on manipulating genomes for food, medicinal, and commercial purposes. Just as comparative genomics has helped settle the debate between neutralist and adaptationist views of molecular variation by supplying evidence that supports both views (Kimura 1983; Kreitman 1983; Andolfatto 2005), functional and developmental genomics will progressively reveal the diverse rates of evolution, unravel the complex relationships between genotypes and phenotypes (Lewontin 1974; Artieri et al. 2009; Artieri and Singh 2010) and provide the material basis for understanding the tempo and mode of both molecular and phenotypic evolution.

1.5 'Rapidly evolving genes and genetic systems': a brief overview

The objective of this book is to provide an advanced, comprehensive, and topical overview of evolutionary rates in biological systems by drawing upon evidence for the rapid evolution of genes and genetic systems from diverse perspectives. The 27 chapters, together, describe a fantastic array of rapidly evolving systems: from individual phenotypes to sets of related traits and developmental pathways; from natural populations to experimental evolutionary studies; from bacteria to fungi; from plants to animals; from simple genetic elements to a complex of species. These chapters are placed into the following five parts.

The first part, 'From Theory to Experiment', contains four chapters and provides readers with a theoretical, and experimental foundation on the rate of evolution. Otto introduces the concept in Chapter 2 from a historical and theory-driven perspective. In Chapter 3, Albu et al. model the effects of recombination on the rate of selective change while Kumagai and Uyenoyama, in Chapter 4, explore how sex-specific hybrid incompatibilities can drive the distribution of evolutionary rates across the genome. Xu (Chapter 5) provides a review of the experimental evolution field. Together, this first part shows how such parameters as population size, recombination, and intensity of selection, impact the rate of evolution for a variety of genotypic and phenotypic traits in natural, experimental, and *in silico* biological systems.

The second part, 'Rapidly Evolving Genetic Elements', contains seven chapters that showcase a variety of genetic and genomic examples for variable and often rapid rates of DNA sequence change. Chapter 6 by Haerty and Golding, focuses on the rapid evolution of a specific category of coding sequences, the low complexity sequences, and amino acid repeat regions. Hao (Chapter 7) describes the ubiquitous signature of horizontal gene transfer in bacteria and how these transfers impact bacterial genome evolution, including those related to virulence properties. Xia (Chapter 8) describes the patterns of animal mitochondrial genome evolution and discusses the mechanisms for its rapid evolution. Roach et al. highlight the rapid evolution of centromere and centromeric/kinetochores in Chapter 9. Rogers and Hartl (Chapter 10) discuss the rapid origin and evolution of novel genes and gene functions through the formation of chimeric genes. In Chapter 11, Long et al. highlight the interactions between sex chromosomes and autosomes. Torgerson and Hernandez discuss, in the last chapter of this part, the general patterns of non-coding DNA sequence variation, with a special focus on the human genome (Chapter 12).

The third part, 'Sex- and Reproduction-Related Genetic Systems', presents seven chapters highlighting the rapid evolution of sex and reproduction-related genes. The chapter topics include Palmer and Swanson's review on sperm–egg interactions (Chapter 13), Lessios and Zigler's analysis of evolutionary rates in the sea urchin bindin protein (Chapter 14), fast evolving *Drosophila* seminal proteins and their networks by Wong and Wolfner (Chapter 15), the evolution of the sperm proteome in *Drosophila* by Karr and Dorus (Chapter 16), Civetta's essay on natural selection versus sexual selection of reproductive systems (Chapter 17), O'Grady and Markow's chapter on behavioral traits involved in mating and host use (Chapter 18) and, finally, fungal mating systems and mating type genes by James (Chapter 19).

The fourth part, 'Pathogens and their Hosts', contains three chapters that highlight the evolution of pathogens and the impacts of pathogen genomic changes on their hosts, from *Drosophila* to humans. Lazzaro and Clark (Chapter 20) provide a review of the evolution of the *Drosophila* innate immunity system. Yang et al. (Chapter 21) focus on the rapid evolution of the human plague pathogen, *Yersinia pestis*. Lastly, an excellent example of how the human genome responds to the evolution of the malaria parasite is covered by Ko et al. (Chapter 22).

The fifth and final part, 'From Gene Expression to Development to Speciation' comprises five chapters that highlight above species-level consequences of rapidly evolving processes. Artieri (Chapter 23) discusses the rapid changes in gene expression across closely related species and focuses on the roles that these changes often bring to dramatic phenotypic divergence between these species. Kopp (Chapter 24) utilizes an evo-devo approach to understanding the consequences of rapidly evolving genes. Arnold et al. (Chapter 25) and Moe et al. (Chapter 26) explore how, respectively, species hybridization and coevolutionary processes, can drive the rapid evolution of traits and lineages. Finally, Kulathinal and Singh (Chapter 27) provide a synthesis in speciation theory, based on the consequences of rapidly evolving reproductive genetic systems.

1.6 Future prospects

Exciting new developments in evolutionary research are providing an opportunity to study genetic systems in unprecedented detail with the promise of learning not only about genic rates of evolution but about the mechanisms as well. A comprehensive understanding of evolutionary rates across lineages would allow us, for the first time, to infer from rates of genic changes, the rates of environmental changes—both abiotic and biotic—that have generated the diversity of life on our planet. While a post-genomic synthesis awaits us, the exceptional chapters in this book, together, provide an exciting step to understanding the important evolutionary forces that have shaped our world.

References

Andersson, M. (1994) *Sexual selection*. Princeton, NJ: Princeton University Press.

Andolfatto, P. (2005) Adaptive evolution in non-coding DNA in Drosophila. *Nature* **437**: 1149–52.

Arnqvist, G. and Rowe, L. (2005) *Sexual conflict*. Princeton, NJ: Princeton University Press.

Artieri, C.G. and Singh, R.S. (2010) Demystifying phenotypes: The comparative genomics of evo-devo. *Fly* **4**: 18–20.

Artieri, C.G., Haerty, W., and Singh, R.S. (2009) Ontogeny and phylogeny: Molecular signatures of selection, constraint, and temporal pleiotropy in the development of Drosophila. *BMC Biology* **7**: 4.

Brodie, E.D., Jr., B.J. Ridenhour, and E.D. Brodie III. (2002) The evolutionary response of predators to dangerous prey: Hotspots and coldspots in the geographic mosaic of coevolution between garter snakes and newts. *Evolution*: **56**: 2067–82.

Carroll, S.B. (2005) *Endless forms most beautiful*. London: W.W. Norton & Company Ltd.

Carson, H.L. (1971) Speciation and the founder principle. *Stadler Symp* **3**: 51–70.

Civetta, A. and Singh, R.S. (1998) Sex-related genes, directional sexual selection and speciation. *Mol Biol Evol* **15**: 901–9.

Darwin, C. (1859) *On the origin of species*. London: Murray.

Dobzhansky, Th. (1937/1951) *Genetics and the origin of species*. New York: Columbia University Press.

Dudley, J.W. and Lambert, R.J. (1992) Ninety generations of selection for oil and protein in maize. *Maydica* **37**: 81–7.

Endler, J.A. (1986) *Natural selection in the wild*. Princeton, NJ: Princeton University Press.

Farrell, B. D. (1998) "Inordinate fondness" explained: Why are there so many beetles? *Science* **281**: 555–9.

Fisher, R.A. (1918) The correlation between relatives under the supposition of Mendelian inheritance. *Trans R Soc Edinb* **52**: 399–433.

Fisher, R.A. (1930) *The genetical theory of natural selection*. Oxford: Clarendon Press.

Gillespie, J.H. (1984) The status of the neutral theory. *Science* **224**: 732–33.

Goldschmidt, R.B. (1940) *The material basis of evolution*. New Haven, CT: Yale University Press.

Gould, S.J. (1982) Darwinism and the expansion of evolutionary theory. *Science* **216**: 380–7.

Gould, S.J. (2002) *The structure of evolutionary theory*. Cambridge, MA: Harvard University Press.

Gould, S.J. and Eldredge, N. (1977) Punctuated equilibria: the tempo and mode of evolution reconsidered. *Paleobiology* **3**: 115–51.

Grant, P. (1999) *Ecology and evolution of Darwin's Finches*. Princeton, NJ: Princeton University Press.

Haldane, J.B.S. (1932) *The causes of evolution*. Ithaca, NY: Cornell University Press.

Hao, W., and Golding, G.B. (2006) The fate of laterally transferred genes: Life in the fast lane to adaptation or death. *Genome Res* **16**: 636–43.

Herrel, A., Huyghe, K., Vanhooydonck, B., Backelju, T., Breugelmans, K., Grbac, I., et al. (2008) Rapid large scale evolutionary divergence in morphology and performance associated with exploitation of a different dietary resource. *Proc Natl Acad Sci U S A* **105**: 4792–5.

Hoekstra, H.E. and Coyne, J.A. (2007) The locus of evolution: Evo Devo and the genetics of adaptation. *Evolution* **61**: 995–1016.

Jagadeeshan, S., Haerty,W., and Singh, R.S. (2011) Is speciation accompanied by rapid evolution? Insights from comparing reproductive and non-reproductive transcriptomes in Drosophila. *Int J Evol Biol* (Published online August 22. doi: 10.4061/2011/595121).

Kimura, M. (1968) Evolutionary rate at the molecular level. *Nature* **217**: 624–6.

Kimura, M. (1983) *The neutral theory of molecular evolution*. Cambridge, MA: Cambridge University Press.

King, M.C. and Wilson, A.C. (1975) Evolution at two levels in humans and chimpanzees. *Science* **188**: 107–16.

Kreitman, M. (1983) Nucleotide polymorphism at the alcohol dehydrogenase locus of Drosophila melanogaster. *Nature* **304**: 412–17.

Kulathinal, R.J., Bettencourt, B.R., and Hartl, D.L. (2004) Compensated deleterious mutations in insect genomes. *Science* **306**: 1553–4.

Lande, R. (1981) Models of speciation by sexual selection on polygenic traits. *Proc Natl Acad Sci U S A* **78**: 3721-5.

Lenski, R.E. and Travisano, M. (1994) Dynamics of adaptation and diversification: A 10,000-generation experiment with bacterial populations. *Proc Natl Acad Sci U S A* **91**: 6808–14

Lewontin, R.C. (1974) *The genetic basis of evolutionary change*. New York: Columbia University Press.

Lynch, M. (2007) *The origins of genome architecture*. Sunderland, MA: Sinauer Associates, Inc.

Margoliash, E. (1963) Primary structure and evolution of cytochrome C. *Proc Natl Acad Sci U S A* **50**(4): 672–9.

Mayr, E. (1954) Change of genetic environment and evolution. In J. Huxley, A.C. Hardy, and E.B. Ford (Eds) *Evolution as a Process*, pp. 157–80. London: Allen and Unwin.

Mayr, E. (1963) *Animal species and evolution*. Cambridge, MA: Harvard University Press.

Mayr, E. and Provine, W.B. (1980) *The evolutionary synthesis: Perspectives on the unification of biology*. Cambridge, MA: Harvard University Press.

Rice, W.R. (1996) Sexually antagonistic male adaptation triggered by experimental arrest of female evolution. *Nature* **381**: 232–4.

Simpson, G.G. (1944) *Tempo and mode in evolution*. New York: Columbia University Press.

Simpson, G.G. (1949) *The meaning of evolution*. New Haven, CT: Yale University Press.

Singh, R.S. (2003) Comment on "Epigenetics and the renaissance of heresy". *Genome* **46**(6): 968–72.

Swanson, W.J. and Vacquier, V.D. (1998) Correlated evolution in an egg receptor from a rapidly evolving Abalone sperm protein. *Science* **281**: 710–12.

Trut, L.N. (1999). Early canid domestication: the farm-fox experiment. *Am Sci* **87**: 160–9.

Wallace, A.R. (1889) *Darwinism*. London: Macmillan.

Weinreich, D.M., Delaney, N., DePristo, M.A., and Hartl, D.L. (2006) Darwinian evolution can follow only very few mutational paths to fitter proteins. *Science* **312**: 111–14.

Wright, S. (1931) Evolution in Mendelian populations. *Genetics* **16**: 97–159.

Wright, S. (1982) Character change, speciation and higher taxa. *Evolution* **36**: 427–34.

Zuckerkandl, E. and Pauling, L. (1962) Molecular disease, evolution, and genic heterogeneity. In M. Kasha and B. Pullman Horizons (Eds) *Biochemistry*, pp. 189–225. New York: Academic Press.

PART I

From Theory to Experiment

CHAPTER 2

Theoretical perspectives on rapid evolutionary change

Sarah P. Otto

2.1 Introduction

Rapid evolutionary change is a common outcome of strong phenotypic selection—from exposure to unusual temperatures to attack by unfamiliar parasites to growth on new food resources—organisms evolve rapidly under challenging circumstances as long as the requisite genetic variation is present. As recounted throughout this book, revolutionary changes in sequencing technology and bioinformatics have allowed biologists to develop a detailed understanding of the genetic basis of rapid evolutionary change from a variety of case studies, from antibiotic resistance to evolutionary responses to sexual conflict.

This chapter explores the theoretical implications of strong selection. Much of basic evolutionary theory assumes weak selection acting on a gene; here we will ask when and to what extent strong selection would alter our theoretical predictions. To begin, we must address the issue that strong selection at the phenotypic level need not generate strong selection at the genic level. We then explore the implications of selection when it is strong at the genic level.

2.2 When is strong selection strong?

A rapid response to selection is possible by accumulating many small changes or a handful of larger ones. If variation is due to many alleles of weak effect, even very strong selection at the phenotypic level can result in weak selection at any one site. For example, if a trait were equally affected by 1000 single nucleotide polymorphisms (SNPs), selection could shift the value of the trait by 0.1 haldane (1/10th of a standard deviation) in a single generation (faster than most observed cases of natural selection; Hendry and Kinnison 1999), yet generate weak selection on any one SNP (e.g. $s = 0.013$ assuming alleles at frequency 1/2, additive diploid selection, independent selective effects on each locus, no initial disequilibrium, and a heritability of 1/2). Indeed, in the infinitesimal limit, there is no appreciable selection or change in allele frequency at any one of the infinitely many loci underlying the trait (Crow and Kimura 1970; Bulmer 1971). Consequently, for strong selection at the phenotypic level to generate strong selection at the genic level requires that one or a few loci contribute disproportionately to the genetic variation present in a population.

Up until the late 1900s, many evolutionary biologists thought that major effect loci, while they underlie traits exhibiting Mendelian inheritance, would explain little of the response to selection on quantitative traits. Instead, it was widely regarded that quantitative traits evolved via slight changes in gene frequency at a great number of minor sites. To understand why this view was so commonplace and why it shifted, we must look back to the early days of population genetic theory.

In his groundbreaking reconciliation of Mendelian genetics and biometry, Ronald A. Fisher (1918) demonstrated that a normal distribution would emerge for a quantitative character influenced by numerous genetic, developmental, and environmental factors. Mathematically, this is the naturel outcome of the central limit theorem: given a sufficiently large number of contributing factors, each of which is independent and identically distributed (or roughly so), the resulting distribution is

Rapidly Evolving Genes and Genetic Systems. First Edition. Edited by Rama S. Singh, Jianping Xu, and Rob J. Kulathinal.

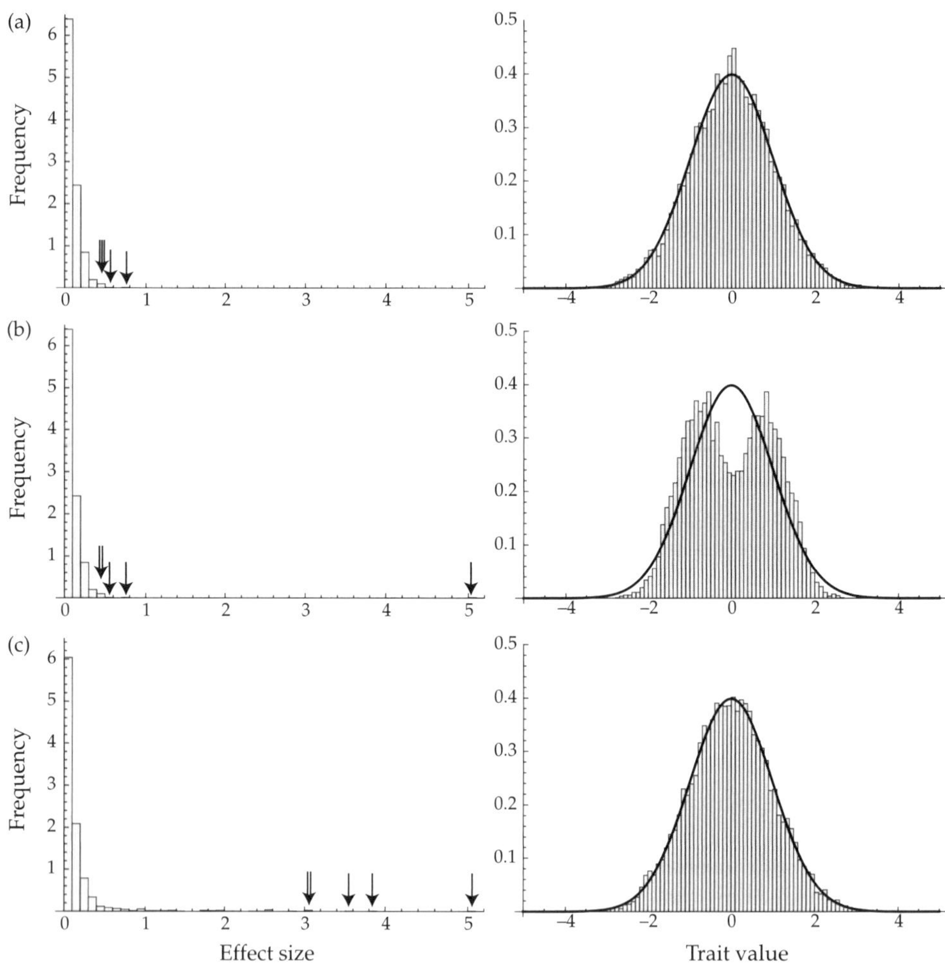

Figure 2.1 Phenotypic distributions with major effect alleles. (a) 1000 loci affecting a trait were randomly assigned an allele with frequency drawn uniformly between 0 and 1 and effect size drawn from an exponential distribution (left panel). 10,000 individuals were then drawn from this population to obtain a phenotypic distribution, which was standardized to have a mean of 0 and a standard deviation of 1 (right panel; solid curve shows a normal distribution for comparison; panel assumes no environmental variation and no linkage disequilibrium). (b) Same as (a) but one allele was replaced with one having ~50 times larger effect, resulting in a decidedly non-normal phenotypic distribution (right). (c) 1000 loci were redrawn from a long-tailed distribution (90% chance of being drawn from the same exponential as in (a) and a 10% chance of being drawn from a second exponential with a tenfold higher mean). Arrows show the five alleles with greatest effect.

normal (Fig. 2.1a). If a single gene contributes substantially to variation in a trait, however, the result will often be decidedly non-normal (Fig. 2.1b). Consequently, the widespread occurrence of normally distributed characters suggested that major effect alleles rarely contribute to variation in quantitative traits.

In his 1930 book, Fisher introduced a second argument against major effect alleles. He argued by metaphor that mutations were akin to random turns of the focus knobs of a microscope. If the image was slightly out of focus, a large random turn to a knob would be unlikely to bring the image closer in focus. A small turn in either direction, however, would have a 50:50 chance of sharpening the image. Geometrically, with n knobs, the degree of focus can be represented as a point on an n-dimensional sphere, whose center represents

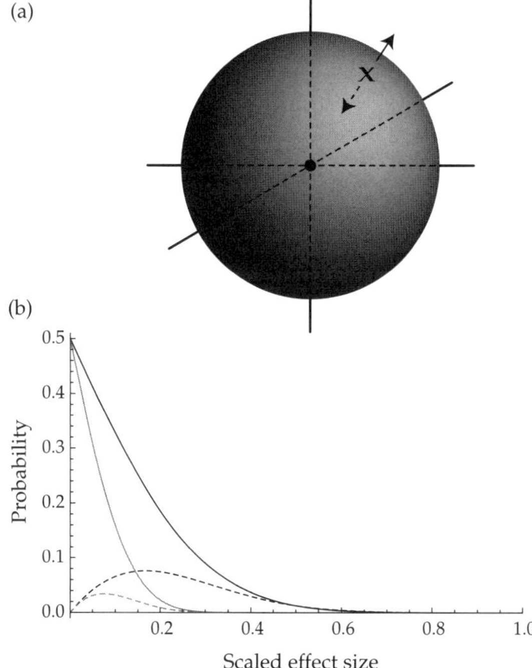

Figure 2.2 Fisher's geometric model. (a) Shown is an imaginary three-dimensional trait space, where the current population lies at a point, x, on the surface of a sphere, whose centre represents the optimal trait combination (black circle). All trait combinations that are equally distant from the optimum (on the sphere's surface) have the same fitness. Mutations pointing out of the sphere are thus disadvantageous (solid arrow), while mutations pointing inward are beneficial (dashed arrow). (b) Fisher (1930) approximated the probability (solid curves) that a mutation pointing in a random direction would be beneficial given a particular effect size (x-axis) in n-dimensional trait space (black: $n = 20$; grey: $n = 100$), where the scaled effect size of a mutation is its magnitude times $\sqrt{n}/d$ for a sphere of diameter d. Kimura (1983) noted, however, that only a fraction of these (roughly proportional to the effect size) would survive loss while rare (dashed curves), shifting the modes of the curves toward larger effect mutations.

the optimal focus. By analogy, we can imagine an organism positioned on a sphere some distance from its optimum in n-dimensional trait space (Fig. 2.2a). The beauty of this metaphor is that it can be used to determine the probability that a mutation, randomly pointing in any direction, falls inside the sphere and so is advantageous (Fig. 2.2b). Fisher noted that this probability falls sharply as mutations increase in their effect size. This argument bolstered the view that small-effect mutations would contribute to the bulk of evolutionary change.

The opposing view, that large-effect mutations matter, has gained prominence in recent decades, with a combination of theoretical and empirical support. Building on Fisher's microscope metaphor, Kimura (1983) pointed out that small-effect mutations, even if more numerous, are more likely to be lost by random genetic drift while rare. Consequently, the distribution of mutations that are both beneficial and survive loss while rare is shifted, with a mode no longer at zero (Fig. 2.2b). Orr (1998) further showed that if we consider not just the first step, but a series of steps toward the optimum, the largest step taken need not be the first step and so is slightly larger, on average, than the first step. A number of other theoretical studies also predict that large-effect mutations contribute substantially more often to adaptation than one would expect based on the frequency of such mutations. These include models that consider the fitness distribution of all possible sequences reachable by single mutations (Gillespie 1984; Orr, 2002; Joyce et al. 2008), that track a changing optimum (Otto and Jones 2000; Griswold and Whitlock 2003), that account for migration swamping small-effect mutations (Griswold 2006; Yeaman and Whitlock 2011), and that incorporate clonal interference in asexual populations (Rozen et al. 2002).

Returning to the argument that quantitative characters are typically normally distributed, it turns out that the central limit theorem is remarkably robust; even if factors are drawn from a distribution with a long tail, such that large-effect alleles are reasonably common, a normal distribution still emerges with enough underlying factors. In Fig. 2.1a, for example, we drew factors from an exponential distribution, the tail of which included two major alleles whose effects were 5.9 and 7.9 times larger than the average. If allele effects are drawn from even more long-tailed distributions, the resulting genotypic distribution may still appear normal, despite the presence of major effect alleles (as in Fig. 2.1c, where the two largest alleles have effects 21.0 and 27.5 times the average). Thus, the observation of a normal phenotypic distribution does not exclude the possibility of large-effect alleles, and such alleles could rise rapidly in frequency in response to strong phenotypic selection. The question is fundamentally an empirical one: how often do major alleles underlie variation in quantitative traits?

Empirically, two forms of data have provided evidence that large-effect alleles can contribute to phenotypic differences in quantitative traits. First, studies searching for quantitative trait loci (QTL) typically find some genomic regions that have a major effect on the phenotype. For example, among the QTL studies summarized by Lynch and Walsh (1998, supplementary table available online), the maximum QTL accounted for an average of 21.6% of the phenotypic variance across 201 different trait and species combinations (SE = 1.0%, range 2–86%). Granted, the effect sizes are somewhat overestimated due to the 'Beavis effect' (Beavis 1994), and many smaller QTL are typically found as well (average minimum QTL: 7.0 ± 0.47% of the phenotypic variance; average number of QTL: 6.1 ± 0.3). Moreover, many small-effect QTL certainly remain undetected. Indeed, in the earlier mentioned studies, the total percentage of phenotypic variation explained by all of the detected QTL was only 42.7% ± 1.5%, on average. Furthermore, even those major QTL detected might be comprised of linked loci of smaller effect (e.g. Perez and Wu 1995). More powerful QTL studies based on genome-wide SNPs have revealed hundreds of potential underlying factors, but even then only a handful are often responsible for the bulk of the variation. For example, in a recent genomic analysis of trait variation in *Drosophila melanogaster*, 6–10 SNPs accounted for 65–90% of the genetic variation in all three traits considered (Mackay et al. 2012).

Overall, it is not uncommon to find one or more QTL contributing a substantial fraction of the phenotypic variation in a quantitative trait, substantial enough that we can expect strong selection at the phenotypic level to translate into strong selection at the genetic level. This is an essential issue if we are to have any hope of observing signs of past selection from sequence data. If numerous very small-effect loci underlie a phenotypic trait, selection on the phenotype might lead to only small shifts in allele frequency, occurring so slowly over time that recombination would destroy the signals of selection. Nevertheless, as Roff (2003) cautions, we should not throw out the baby with the bathwater: the major-effect loci often co-occur with enough small-effect loci that the Gaussian assumption underlying quantitative genetic theory is reasonable, and genetic variation may remain roughly constant even as the major-effect alleles spread, especially in the presence of new mutations (Turelli and Barton 1994; Barton and Keightley 2002).

A second type of evidence for large-effect alleles has come from genomic scans revealing signs of selection at a number of sites. For example, Sabeti et al. (2007) report finding ~300 candidate regions showing longer than expected haplotypes in a cross-population analysis of human genomes (XP-EHH: cross-population extended haplotype homozygosity). Several of these sites are implicated in disease resistance, diet (e.g. lactose tolerance), or skin and hair variation. Very weak selection would not generate such extended haplotypes, because recombination would have time to break down the extended haplotype during the slow spread of a weakly favored allele. A plethora of such genomic scans have now been published (Akey (2009) reviews 21 genome-wide scans), based on a variety of different metrics, including allele frequency distributions, linkage disequilibria (including extended haplotype methods), and population differentiation (e.g. studies based on F_{st}). Regardless of the metric, however, weak selection at a single site is almost certain to remain undetected. Indeed, it has been estimated that selection on a site must be stronger than $\sim 100/N_e$, where N_e represents the effective population size, for there to be a reasonable chance of detection (Akey 2009). Of course, humans might be a poor species in which to study selective differences among populations (except for the obvious reason of self-interest) because we are all so closely related, with recent common ancestry within Africa and a relatively small effective population size. Genomic scans in non-human species promise to provide fascinating information about the prevalence and nature of major effect alleles.

2.3 Does strong selection differ in kind from weak selection?

Strong selection obviously differs in degree from weak selection, but does it ever differ in kind? Does it ever lead to dramatically different evolutionary outcomes than weak selection? Theoreticians often resort to assuming weak selection in order to obtain analytical solutions; if selection is strong, do the results differ in any substantial way? As we review here, the answer to this set of questions is mixed, depending on the phenomenon being modeled. Often, the expected outcome of evolution under strong selection exhibits only minor quantitative discrepancies from what we would predict by increasing the strength of selection in weak selection approximations. In other cases, however, predictions are fundamentally different when selection is strong rather than weak.

To begin, consider one of the core equations in evolutionary biology, which describes the change in frequency, $p(t)$, of an allele A over time t. To keep things simple, we consider a haploid population subject to discrete and non-overlapping generations, where allele A causes its carriers to have fitness is $1 + s$ times that of an alternate allele, a. After one generation of selection, the A allele changes in frequency to:

$$p(t+1) = \frac{(1+s)\,p(t)}{(1+s)\,p(t) + (1-p(t))} \tag{2.1}$$

(see Otto and Day (2007) for derivations of the results presented in this chapter). This equation can be solved exactly for any strength of selection to give the allele frequency in any future generation:

$$p(t) = \frac{(1+s)^t\,p(0)}{(1+s)^t\,p(0) + (1-p(0))}. \tag{2.2}$$

In most models of evolutionary change, where additional complications are incorporated (e.g. dominance or frequency dependence), an exact solution is not possible. Theoreticians then often invoke a weak-selection approximation, for example, replacing the recursion equation (2.1) with the analogous differential equation:

$$\frac{dp}{dt} = s\,p(t)\,(1-p(t)). \tag{2.3}$$

A wider array of differential equations can be solved, which is why a weak-selection approximation is often invoked. For example, the solution to equation (2.3) is:

$$p(t) = \frac{e^{st}\,p(0)}{e^{st}\,p(0) + (1-p(0))}. \tag{2.4}$$

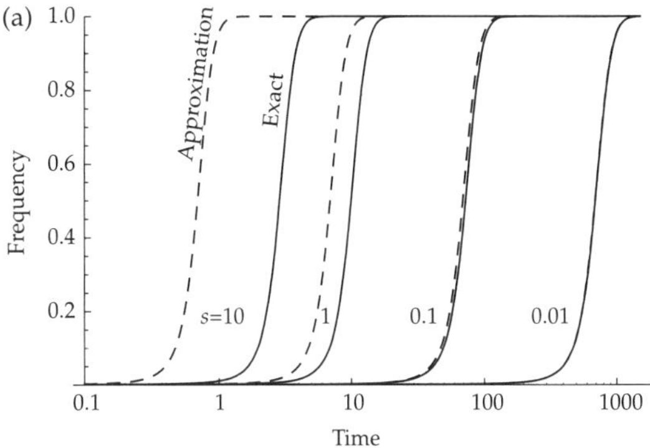

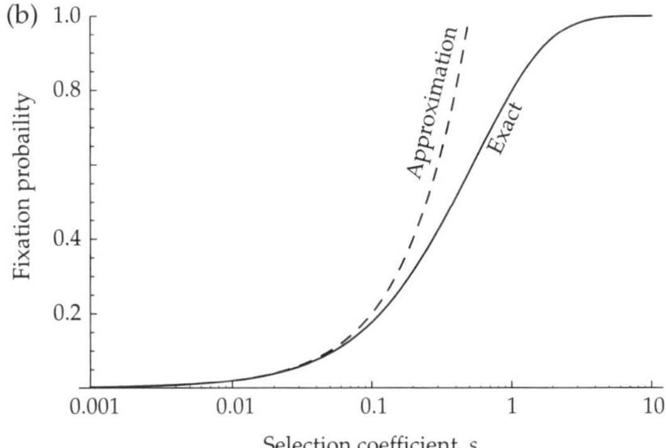

Figure 2.3 Weak versus strong selection. (a) The frequency of a beneficial allele with relative fitness $1 + s$ is shown over time in a haploid population, starting at frequency 0.001. The exact frequency in a haploid population (equation 2.2; solid) predicts a slower response to selection than the weak selection approximation (equation 2.4; dashed). (b) The exact fixation probability of a beneficial allele in an infinitely large population (equation 2.6; solid) is lower than the weak selection approximation, $2s$ (equation 2.7; dashed).

While different in form from equation (2.2), the qualitative behavior is very similar (Fig. 2.3a), and visible differences in the speed of the trajectories only appear for s greater than about 0.1. For example, the time to spread from any initial frequency to any final frequency is 4.7% faster under the approximation (2.4) than with the exact solution (2.2) when $s = 0.1$. Even then, the trajectories remain S-shaped and differ solely in timing. Only if we needed precise predictions about the timing of the spread of a favorable allele would strong selection violate the weak-selection approximation (2.4).

A second core equation in evolutionary theory concerns the fixation probability of an allele. Even beneficial alleles can be lost, by chance, after they first arise, simply because their carriers fail to reproduce. In very large populations, Haldane (1927) argued that the probability that a newly arisen allele will ultimately be lost from the population must equal the probability that all j offspring carrying the allele are ultimately lost. Letting $1 - P$ equal the probability of loss and assuming that the fate of each allele is independent, this logic results in the equation:

$$1 - P = \sum_j \text{prob}\,(j \text{ offspring})\,(1 - P)^j, \qquad (2.5)$$

where the sum is taken over the probability distribution of having j offspring. If this offspring distribution is Poisson with mean $1 + s$ for individuals carrying the beneficial mutation, then the sum in equation (2.5) can be evaluated, leading to:

$$1 - P = e^{-(1+s)P}. \qquad (2.6)$$

The fixation probability, P, is implicitly given by equation (2.6), but this is still a complicated equation to evaluate. Theoreticians often simplify such equations by assuming weak selection, obtaining answers that keep the leading order terms (e.g. of order s) but drop terms of smaller order (e.g. s^2 or smaller). In this case, such a 'Taylor series' approximation gives us the oft-cited probability of fixation for a beneficial allele:

$$P \approx 2s. \qquad (2.7)$$

Fig. 2.3b shows that the weak-selection approximation (2.7) performs well for selection coefficients up to about 0.1 (the $2s$ approximation then overestimates the fixation probability by 13.6%).

These considerations suggest that, roughly speaking, the boundary between weak and strong selection occurs at 0.1. This is a reasonable rule of thumb when it comes to theory considering only selection acting at a single locus. In general, however, what constitutes strong versus weak selection depends on the context. In reality, selection is never the only process acting on a population to affect allele frequencies. Thus, whether selection is strong is relative, depending on the magnitude of selection in comparison to the impact of other processes, such as drift, mutation, migration, and recombination.

Selection may be weak relative to 0.1, for example, but strong relative to random genetic drift. Selection overwhelms drift when the best predictor of the future distribution of allele frequencies is s rather than the population size, N. With respect to the fate of an allele, we say that selection is strong relative to drift when the fixation probability of an allele is predicted to be higher as a result of selection alone ($\sim 2s$) than drift alone (fixation probability is $1/(cN)$, for an organism whose ploidy level is c). For diploids ($c = 2$), this leads to the oft-stated rule that selection overwhelms drift whenever $2s >> 1/(2N)$, commonly written as $\theta = 4Ns >> 1$.

Selection can be strong enough to overwhelm drift but not strong enough to overwhelm other processes, such as migration. Consider the simplest haploid case where an allele A is favored in a patch (with fitness 1 relative to the alternate allele with fitness $1 - s$) but is reduced in frequency by migration from a source population (at rate m). If selection is weak relative to migration ($s < m$), the A allele fails to establish and the population in the patch remains locally maladapted. With stronger selection ($s > m$), however, population differentiation can occur, with the A allele approaching a frequency of $1 - m/s$ (Haldane (1930), who also considered diploidy and sex linkage; see also Yeaman and Otto (2011) for cases with bidirectional migration and drift). Empirically, these conditions are important because they indicate that population differentiation in the face of ongoing migration will typically not be comprised of small-effect alleles. Here, the strength of selection is critical; only alleles experiencing strong selection relative to migration are capable of being maintained and contributing to local adaptation in the face of gene flow (Griswold 2006; Yeaman and Whitlock 2011).

Another situation in which strong selection can behave qualitatively differently than weak selection is when the fitness surface is multipeaked. The ability to traverse fitness valleys then depends critically on the strength of selection relative to other forces, especially the rate of recombination. For example, consider a population currently fixed for one gene combination (ab) at two loci, where each single mutant is selected against (creating the 'fitness valley'), but where the double mutant (AB) has highest fitness ($W_{AB} > W_{ab,} > W_{Ab}, W_{aB}$). If selection is weak relative to the recombination rate between the two loci, r, the fitness valley forms an insurmountable barrier, and the population remains fixed for *ab*. If selection is strong relative to recombination, such that W_{AB} (1 - r) > W_{ab} (or, in terms of selection coefficients, $(s_{AB} - s_{ab})/(1 + s_{AB}) > r$), then the favorable gene combination, *AB*, spreads (Bodmer and Felsenstein 1967). Thus, whether or not evolution is prevented from traversing a multipeaked fitness surface depends on just how strong selection is, with strong selection being able to drive beneficial gene combinations through to fixation, despite the fact that these combinations are broken apart by recombination.

A related context in which the strength of selection matters involves hitchhiking of alleles linked to a site under selection. Maynard Smith and Haigh (1974) first tackled this question, assuming that a favorable allele spreads deterministically, having arisen on one particular genetic background. Considering a specific allele at a linked neutral site that initially co-occurs with the beneficial allele, they showed that hitchhiking in a haploid population causes a proportional reduction in the frequency of any other allele at this site by approximately:

$$\frac{r}{s}\log\left(\frac{1}{p_0}\right) \tag{2.7}$$

where r is the recombination rate between the selected and neutral site and p_0 is the initial frequency of the selected allele in a haploid population ($p_0 = 1/N$). This approximation assumes that selection is weak (again, s not much greater than 0.1) but greater than the recombination rate ($r << s$). Importantly, equation (2.7) tells us that variation at linked neutral sites will be impacted whenever the recombination distance is small relative to the selection coefficient, s. Thus, what we might consider to be weak selection in an absolute sense ($s <<$ 0.1) may still be strong selection when it comes to hitchhiking, as long as recombination rates are less than the selection coefficients. More sophisticated stochastic models have similarly shown that hitchhiking has a major impact on genetic variation and coalescence times only when selection is strong relative to recombination (Kaplan et al. 1989; Barton 1998). This is the underlying reason why genomic scans of selection based on unusual patterns of variation surrounding a site can only detect alleles of moderately strong effect.

Strong selection also affects the type of hitchhiking event that we should expect within a genomic region. The scenario considered by Maynard Smith and Haigh (1974) was of a beneficial allele that arose in only a single copy; the reduction in genetic variation at surrounding sites due to the spread of this allele is known as a 'hard selective sweep.' But if the environment changes and there are multiple copies of the beneficial allele when it first becomes favorable (i.e., there is standing genetic variation), then strong selection makes it more likely that more than one copy of the allele will survive loss while rare and contribute to adaptation (Hermisson and Pennings 2005). The cofixation of distinct haplotypes bearing the favorable allele causes a so-called 'soft selective sweep.' Compared to hard selective sweeps, soft selective sweeps preserve more variation at neighboring sites, but they can also generate stronger patterns of linkage disequilibrium, extending further into surrounding regions, whenever the haplotypes are distantly related. Interestingly, one might also predict that strong selection would make it more likely to fix the same mutation if it recurred by chance after the sweep was underway, but the increased chance that such recurrent mutations would fix is almost entirely counterbalanced by the decreased amount of time taken until the sweep is over (Pennings and Hermisson 2006). Thus, strong selection has a large impact on how many alleles survive from the pool of standing genetic variation when the environment changes, but less impact on how many recurrent mutations get captured along the way.

A particularly negative consequence of hitchhiking with a favored allele is that less fit alleles at neighboring sites can rise in frequency and even fix (Barton 1995; Yu and Etheridge 2010; Hartfield and Otto 2011). Hitchhiking to fixation of deleterious alleles requires that selection for the favorable allele is even stronger relative to the recombination rate (compared to the case with one neutral and one selected site), because recombinant lineages that combine together the fittest alleles are especially likely to persist (Hartfield and Otto 2011). At the extreme of no recombination, deleterious alleles hitchhiking with favored alleles can cause a ratcheting decline in the fitness of an asexual lineage, which lacks recombination entirely, relative to a sexual population in the face of a changing environment, a decline that may eventually drive asexual lineages to extinction (Hadany and Feldman 2005).

In a related fashion, alleles subject to strong selection because of their effect along one particular phenotypic axis can drag along stronger pleiotropic side consequences with respect to other traits. For example, consider an environmental change selecting for a particular trait, such as increased temperature tolerance. An allele that slightly increased heat tolerance but had numerous pleiotropic effects on other characters is more likely to have a net deleterious effect on fitness and fail to spread. By contrast, an allele that strongly increased heat tolerance, and experienced a strong selective advantage in doing so, is more likely to spread, despite disrupting other features. Otto (2004) estimated that the selective advantage of such favorable alleles would, on average, be halved by negative pleiotropic side-consequences. Because organisms having experienced strong selection in the past may have changed along multiple dimensions, it might be difficult to determine the precise trait(s) that was positively selected. We see this with domesticated corn, for example, where the *tb1* allele that differentiates corn from teosinte accounts for phenotypic differences in a large number of traits, including seed shattering, cob size, and apical dominance (Doebley et al. 1995), making it difficult to know which character(s) was most directly under selection.

We end with one final example where strong selection differs in kind from weak selection, involving the Red Queen explanation for sex and recombination. With strong selection acting on genes mediating interactions between hosts and parasites, gene combinations that have been favored in the recent past can be currently prevalent but disadvantageous, selecting for sex and recombination to break apart these combinations. Such a mismatch between which gene combinations are prevalent and which gene combinations are fittest only really occurs, however, if selection is strong relative to the forces breaking apart genetic associations, such as segregation, recombination, and mutation (Gandon and Otto 2007). With weak selection, host–parasite cycles are relatively slow, allowing enough time for genetic associations to re-equilibrate and erasing previously favored gene combinations (Otto and Nuismer 2004). Thus, an important open empirical question facing the Red Queen hypothesis is how often strong selection acts on the genes mediating host–parasite interactions.

2.4 Concluding thoughts

Evolutionary biologists live fairly comfortably with using the terms 'weak' and 'strong' selection without strictly defining what we mean. The reason for this vagueness is that the dividing line between strong and weak depends so much on the context. When we are concerned with whether the change in allele frequency due to selection is overwhelmed by drift or by mutation, what counts as 'strong' selection may still be exceedingly weak (e.g., $2s >> 1/(2N)$ or $2s >> \mu$ with mutation rate μ). In contrast, when we approximate complicated equations in terms of simpler functions of s (e.g. using a Taylor series, as in equation (2.7)), our standards shift; now, selection only starts to wreak havoc with our approximations once the strength rises above, approximately, 0.1 (essentially because terms involving s^2 may no longer be small relative to terms involving only s). Finally, the strength of selection can qualitatively impact the outcome of an evolutionary process when selection is opposed by other processes, such as migration of a disfavored allele or recombination breaking apart favorable gene combinations. In these contexts, what defines strong selection is, by necessity, relative (e.g. relative to m or to r).

Finally, regardless of where we might place the dividing line between strong and weak selection at the genic level, the correspondence to strong and weak selection at the phenotypic level is fuzzy. Strong selection at the phenotypic level might generate only weak selection on the underlying loci, if there are many alleles of similar effect and/or substantial environmental variation contributing to the phenotypic variation. Conversely, relatively modest selection at the phenotypic level could cause strong selection at the genic level, if a few underlying loci contribute the bulk of phenotypic variation.

Although applying a more stringent and consistent definition for the term 'strong selection' might seem appealing, Aldous Huxley was almost certainly right when he cautioned that 'Consistency is contrary to nature, contrary to life. The only completely consistent people are dead.'

References

Akey, J.M. (2009) Constructing genomic maps of positive selection in humans: where do we go from here? *Genome Res* **19**: 711–22.

Barton, N.H. (1995) Linkage and the limits to natural selection. *Genetics* **140**: 821–41.

Barton, N.H. (1998) The effect of hitch-hiking on neutral genealogies. *Genet Res* **72**: 123–33.

Barton, N.H. and Keightley, P.D. (2002) Understanding quantitative genetic variation. *Nat Rev Genet* **3**: 11–21.

Beavis, W.D. (1994) The power and deceit of QTL experiments: Lessons from comparative QTL studies. In *Proceedings of the Forty-ninth Annual Corn and Sorghum Research Conference*. Washington, DC: American Seed Trade Association, pp. 250–66.

Bodmer, W.F. and Felsenstein, J. (1967) Linkage and selection: theoretical analysis of the deterministic two locus random mating model. *Genetics* **57**: 237–65.

Bulmer, M.G. (1971) The effect of selection on genetic variability. *Amer Nat* **105**: 201–11.

Crow, J.F. and Kimura, M. (1970) *An Introduction to Population Genetic Theory*. New York: Harper & Row.

Doebley, J., Stec, A., and Gustus, C. (1995) *teosinte branched1* and the origin of maize: evidence for epistasis and the evolution of dominance. *Genetics* 141: 333–46.

Fisher, R.A. (1918) The correlation between relatives under the supposition of Mendelian inheritance. *Trans R Soc Edinb* **52**: 399–433

Fisher, R.A. (1930) *The Genetical Theory of Natural Selection*. Oxford: Oxford University Press.

Gandon, S. and Otto, S.P. (2007) The evolution of sex and recombination in response to abiotic or coevolutionary fluctuations in epistasis. *Genetics* **175**: 1835–53.

Gillespie, J.H. (1984) Molecular evolution over the mutational landscape. *Evolution* **38**: 1116–29.

Griswold, C.K. (2006) Gene flow's effect on the genetic architecture of a local adaptation and its consequences for QTL analyses. *Heredity* **96**: 445–53.

Griswold, C.K. and Whitlock, M.C. (2003) The genetics of adaptation: The roles of pleiotropy, stabilizing selection and drift in shaping the distribution of bidirectional fixed mutational effects. *Genetics* **165**: 2181–92.

Hadany, L. and Feldman, M.W. (2005) Evolutionary traction: the cost of adaptation and the evolution of sex. *J Evolution Biol* **18**: 309–14.

Haldane, J.B.S. (1927) A mathematical theory of natural and artificial selection, part V: selection and mutation. *Math Proc Camb Philos Soc* **23**: 838–44.

Haldane, J.B.S. (1930) A mathematical theory of natural and artificial selection. Part VI. Isolation. *Math Proc Camb Philos Soc* **26**: 220–30.

Hartfield, M. and Otto, S.P. (2011) Recombination and hitchhiking of deleterious alleles. *Evolution* **65**: 2421–34.

Hendry, A.P. and Kinnison, M.T. (1999) Perspective: The pace of modern life: Measuring rates of contemporary microevolution. Evolution 53: 1637–1653.

Hermisson, J. and Pennings, P.S. (2005) Soft sweeps: molecular population genetics of adaptation from standing genetic variation. *Genetics* **169**: 2335–52.

Joyce, P., Rokyta, D.R., Beisel, C.J., and Orr, H.A. (2008) A general extreme value theory model for the adaptation of DNA sequences under strong selection and weak mutation. *Genetics* **180**: 1627–43.

Kaplan, N.L., Hudson, R.R., and Langley, C.H. (1989) The "hitchhiking effect" revisited. *Genetics* **123**: 887–99.

Kimura, M. (1983) *The Neutral Theory of Molecular Evolution*. Cambridge: Cambridge University Press.

Lynch, M. and Walsh, B. (1998) *Genetics and Analysis of Quantitative Traits*. Sunderland, MA: Sinauer. [Supplementary table available at: http://nitro.biosci.arizona.edu/zdownload/QTLtable.pdf]

Mackay, T.F.C., Richards, S., Stone, E.A., Barbadilla, A., Ayroles, J.F., Zhu, D., Casillas, S., et al. (2012). The *Drosophila melanogaster* Genetic Reference Panel. *Nature*, **482**: 173–178.

Maynard Smith, J. and Haigh, J. (1974) The hitch-hiking effect of a favourable gene. *Genet Res* **23**: 23–35.

Orr, H.A. (1998) The population genetics of adaptation: The distribution of factors fixed during adaptive evolution. *Evolution* **52**: 935–49.

Orr, H.A. (2002) The population genetics of adaptation: the adaptation of DNA sequences. *Evolution* **56**: 1317–30.

Otto, S.P. (2004) Two steps forward, one step back: the pleiotropic effects of favoured alleles. *Proc Roy Soc Lond B* **271**: 705–14.

Otto, S.P. and Day, T. (2007) *A Biologist's Guide to Mathematical Modeling in Ecology and Evolution*. Princeton, NJ: Princeton University Press.

Otto, S.P. and Jones, C.D. (2000) Detecting the undetected: estimating the total number of loci underlying a quantitative trait. *Genetics* **156**: 2093–107.

Otto, S.P. and Nuismer, S.L. (2004) Species interactions and the evolution of sex. *Science* **304**: 1018–20.

Pennings, P.S. and Hermisson, J. (2006) Soft sweeps II – molecular population genetics of adaptation from recurrent mutation or migration. *Mol BiolEvol* **23**: 1076–84.

Perez, D.E. and Wu, C.I. (1995) Further characterization of the *Odysseus* locus of hybrid sterility in Drosophila: one gene is not enough. *Genetics* **140**: 201–6.

Roff, D. (2003) Evolutionary quantitative genetics: Are we in danger of throwing out the baby with the bathwater? *Ann Zool Fennici* **40**: 315–20.

Rozen, D.E., de Visser, J.A., and Gerrish, P.J. (2002) Fitness effects of fixed beneficial mutations in microbial populations. *Curr Biol* **12**: 1040–5.

Sabeti, P.C, Varilly, P., Fry, B., Lohmueller, J., Hostetter, E., Cotsapas, C., et al. (2007) Genome-wide detection and characterization of positive selection in human populations. *Nature* **449**: 913–18.

Turelli, M. and Barton, N.H. (1994) Genetic and statistical analyses of strong selection on polygenic traits: what, me normal? *Genetics* **138**: 913–41.

Yeaman, S. and Otto, S.P. (2011) establishment and maintenance of adaptive genetic divergence under migration, selection, and drift. *Evolution* **65**: 2123–9.

Yeaman, S. and Whitlock, M.C. (2011) The genetic architecture of adaptation under migration-selection balance. *Evolution* **65**: 1897–911.

Yu. F. and Etheridge, A. (2010) The fixation probability of two competing beneficial mutations. *Theor Pop Biol* **78**: 36–45.

CHAPTER 3

Recombination reshuffles the genotypic deck, thus accelerating the rate of evolution

Mihai Albu, Amir R. Kermany, and Donal A. Hickey

3.1 Introduction

The ubiquity of sexual reproduction, especially among multicellular plants and animals, gives strong support for the belief that sex provides an important biological function. It has proved surprisingly difficult, however, to pinpoint what exactly this function might be. Sex is not a necessity of life, nor of reproduction, as is clearly illustrated by the many instances of asexual reproduction that exist, especially among microbes and plants. But, because of its wide distribution in nature, we can conclude that sexual reproduction—and the resulting genetic recombination—are necessities, or near-necessities of adaptive evolution. Eighty years ago Ronald Fisher wrote that sexual reproduction 'is a development of some special value to the organisms which employ it' (Fisher 1930). Eight decades later, we are still trying to decipher what that 'special value' might be. As Sarah Otto remarked (see Otto 2009), we are still trying to 'solve the paradox of sex.'

Theories about the evolution of sex and recombination can be divided into two broad categories: those that deal with the origin of sex and those that deal with its maintenance. For instance, the enzymatic processes involved in genetic recombination may have evolved due to prior selection for DNA repair (Bernstein et al. 1985; Long and Michod 1995), while the process of conjugation could have originated due to selection for the spread of transposable elements (Hickey 1982). Regardless of how sex originated, however, it is maintained despite its obvious biological costs. The common thread among the several theories that have been advanced to explain the maintenance of sexual reproduction is that they focus on the potential selective benefits of homologous recombination. But these theories about the potential advantage of recombination may be subdivided further according to whether they deal with: (i) recombination of new favorable mutations (Fisher 1930; Muller 1932; Crow and Kimura 1965); or (ii) recombination of new deleterious mutations (Kondrashov 1982); or (iii) recombination of the standing genetic variation (Weismann 1887; Goddard et al. 2005; Teotónio et al. 2009).

The proposal that the benefit of recombination lies in the fact that it combines different beneficial mutations into a single genotype is intuitively appealing to biologists and it was first explored many decades ago (Muller 1932, 1964). Later work, however, showed that the benefits of combining advantageous mutations are offset by the fact that recombination also breaks up these favorable combinations once they have been formed (Maynard Smith 1968, 1978). In other words, recombination simply randomizes genotypes, without regard to the fitness of the alleles being recombined. This led to a shift of focus from the consideration of favorable mutations to a consideration of the more frequently-occurring deleterious mutations, and it was shown that recombination could provide a selective advantage if the number of mutant loci in the genome were sufficiently large (Kondrashov 1982). This advantage would be even greater if there were negative epistatic interactions between the new deleterious mutations (for a review, see

Rapidly Evolving Genes and Genetic Systems. First Edition. Edited by Rama S. Singh, Jianping Xu, and Rob J. Kulathinal.

Otto and Feldman 1997). It is not obvious, however, that such negatively epistatic interactions are the norm in nature (Kondrashov and Kondrashov 2001). Whether one is dealing with positive or negative mutations, it is difficult to test these ideas experimentally because, first, mutations are rare and, secondly, the time from the occurrence of the initial mutation until its eventual fixation by natural selection is usually very long. In contrast, theories about the effects of recombination on the standing genetic variation are more amendable to experimental verification and, indeed, there is some experimental evidence that recombination can provide a significant advantage when selecting on this standing genetic variation (Goddard et al. 2005; Teotónio et al. 2009). It should be noted that theories of fluctuating selection based on host–parasite cycles are also implicitly theories about the effects of recombination on the standing genotypic composition (Hamilton et al. 1990; Peters and Lively 1999).

3.2 Simulating selection on multilocus genotypes

In this chapter it is not our goal to choose between these theories but rather to consider the effects of recombination from a different perspective. Specifically, we will ask if the effect of recombination changes as the number of genetic loci under consideration increases from fewer than ten loci to hundreds or thousands of loci. Although we all agree that eukaryotic genomes contain several thousand genes, most of the past work has focused on models describing very few loci, usually fewer than five. If the benefits of recombination are due to its effects on the genome as a whole, then many of these models may be missing an essential element of the question. As explained in this section, we suggest that the benefit of recombination could be that it helps to alleviate the 'curse of dimensionality' (Bellman 1957) that is inherent in multilocus genotypes.

We reasoned as follows. Let's assume a population of 100,000 individuals and one bi-allelic locus where the two alleles are equally frequent. If we assume, to simplify matters further, that the individuals are haploid then we will expect to find 50,000 individuals of each of the two types, and there will be minimal genetic drift. Now, if we assume two such loci in the same haploid population, we will have four possible genotypes with 25,000 individuals of each type. We can continue increasing the number of loci in this manner; for example, 10 loci will give us 1024 genotypes and we expect to see approximately 100 individuals of each type. As soon as we assume 17 loci, however, this simple calculation begins to run into difficulty. Now there are more than 130,000 genotypic combinations for 17 bi-allelic loci, and this means that the expected number of individuals with any particular genotypic combination is less than one. At 20 loci, we have more than a million possible allelic combinations, making the number of genotypes an order of magnitude larger than the population size. At 100 loci—which is still a tiny genome—the number of genotypes is vastly greater than the number of individuals in the population. This means that in any real finite population, the existing array of multilocus genotypes is only a miniscule sample of all possible combinations of the segregating alleles. To illustrate how this affects the interaction between recombination and selection, we developed a simple, individual-based simulation, and we show some of the results in the following paragraphs. Our conclusion is that the beneficial effect of recombination becomes obvious as the number of selected loci increases.

Our model was as follows. The initial allele frequencies were 0.5 at all loci. For the initial population, we generated each individual by creating an allelic string equal to the number of loci. For example, if there were 100 loci, the string length was 100 and the chance of an allele being designated as '+' was 0.5 at each locus. We assigned a fitness value of 0.01 to each '+' allele. Thus the initial population had a binomial distribution of fitness centered on a mean fitness of 0.5. Each individual produced offspring from a Poisson distribution with a mean of 2, and the probability of survival of these offspring was based on their genotypic fitness. The population size was maintained as constant by choosing N individuals at random each generation among the selected offspring. Recombination was simulated by randomly pairing genotypes and generating a new pair of genotypes as a result of a single, randomly-placed recombination event between the two strings. The simulation could be performed either with or without recombination. The graphical user interface allows one to input various combinations of parameters and to compare the results between recombining and non-recombining populations.

A sampling of the results from one such simulation is shown in Fig. 3.1. As can been seen from the figure, both the recombining and the non-recombining populations have initial fitness distributions that are identical (Fig. 3.1a). Also, as is shown in the second panel of the figure, the mean fitness of both populations increases in response to 60 generations of selection. But we can also clearly see that the non-recombining population is

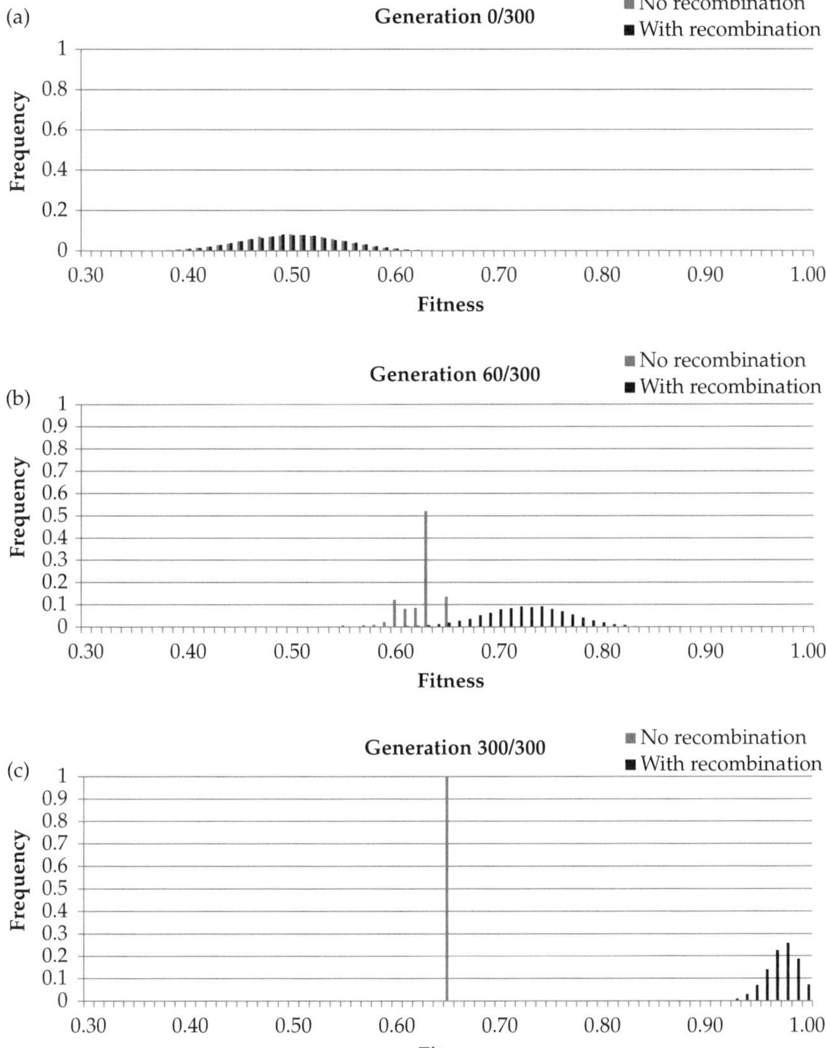

Figure 3.1 Individual-based simulations of multilocus selection, in the presence or absence of recombination. Each individual had a linear array of 100 loci and the initial frequency of the selected alleles was 0.5 at all loci. The recombination rate between adjacent loci was 0.01. The fitness effect of each selected allele was 0.01. At the beginning of the simulation alleles were randomly associated between loci, i.e. in linkage equilibrium. The population size was 10,000. Panel (a) shows the initial frequency distributions of genotypic fitnesses in the recombinant and non-recombinant populations. Panel (b) shows these distributions after 60 generations of selection, and panel (c) shows the distributions after 300 generations of selection.

limited by the fittest type that was already present in the original population, whereas the recombining population can generate new genotypes distributed around the increasing mean fitness. As the simulation progresses (see Fig. 3.1c) this limitation of the non-recombining population becomes more and more severe. We can summarize the results of the simulation by plotting the mean fitness of both populations over the whole 300 generations of the simulation (see Fig. 3.2a). We see that initially, and indeed for the first 30 generations of selection, there is virtually no effect of recombination on the mean fitness. That is because both populations are responding to selection primarily by adjusting the frequencies of existing genotypic classes. As the selection proceeds, however, the mean begins to approach the limit of the initial genotypic distribution and it is at this point that the non-recombining population is at a serious disadvantage, whereas the recombining population is not subject to this limit. After 300 generations of selection, the recombining population has become almost fixed for '+' alleles at all 100 loci, while the non-recombining population is 'stuck' at a maximum fitness just under 0.7 i.e., '+' alleles at only 70 out of the 100 loci.

The particular simulation described here was performed using 100 loci and 10,000 individuals. We then explored the results using different population sizes, with 100 loci in all cases. The results are summarized in Fig. 3.2b. The figure, which is based on 10 replicate simulations for each population size, shows the final average fitness after 500

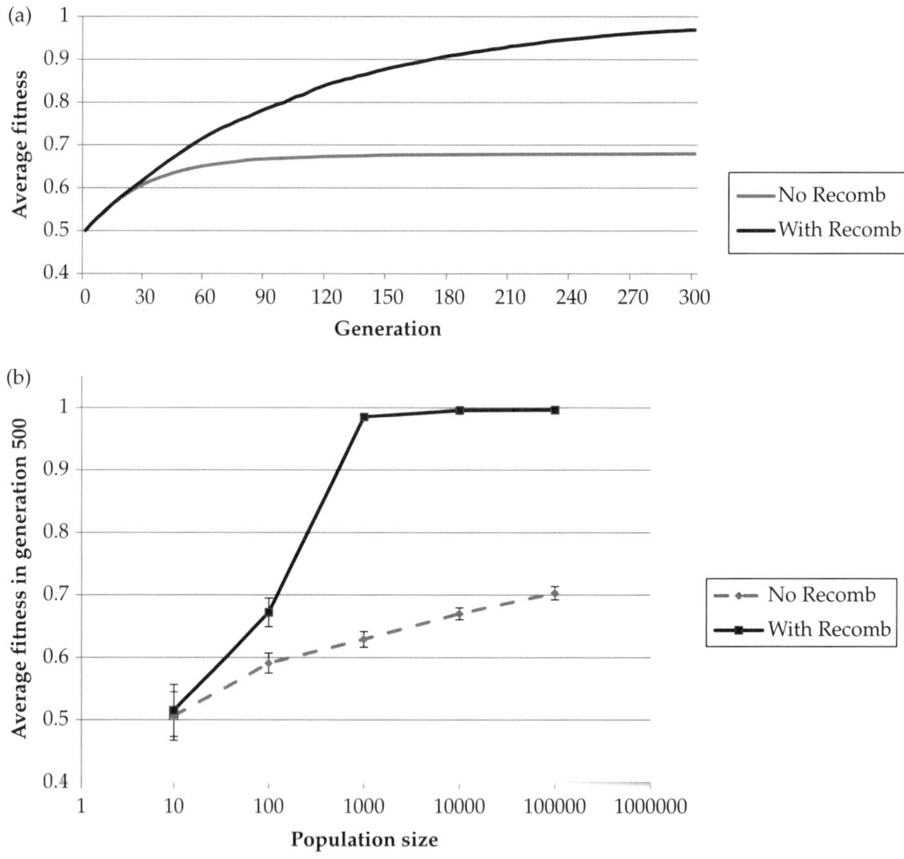

Figure 3.2 Change in average fitness, in the presence or absence of recombination. Panel (a) shows the change in the average fitness for the recombinant and non-recombinant populations that were simulated as described in Fig. 3.1. Panel (b) shows the relationship between the final average fitness (after 500 generations of selection) and the population size. These data are based on 10 replicate simulations; 95% confidence intervals are shown.

generations of selection. From the figure, we see that there is little effect of recombination at very small population sizes. This is because genetic drift erodes the allelic variation at each locus, and there is no advantage to recombining monomorphic loci. When the population size reaches 1000, however, there is little loss of alleles due to genetic drift at individual loci, and this allows recombination to exploit the allelic variation to produce new fitter genotypes. As the population size increases, we see that there is also a predictable increase in the fitness value of the fittest genotype in the non-recombining population, but this increase is relatively small. Thus, for any 'reasonable' population size, i.e. 1000 individuals or more, recombination allows selection to proceed far beyond the limits of the initial population, while still using the allelic variation that was already present in that initial population.

One could say that these results do not so much show an advantage for recombination as that they show a disadvantage or limitation for the lack of recombination. This limitation relates to the fact that if we consider 100 loci, there is necessarily a very limited sampling of the astronomically large number of genotypic combinations in any finite population. We have already explained that this limitation gets more severe as the number of loci increases, but we can ask if it can be compensated for by a corresponding increase in population size. We explored this question and the results are shown in Fig. 3.3. We generated initial populations, of the type described for Fig. 3.1, using a range of population sizes and a range of numbers of loci. From the results, we can see that for a genome with only two loci (which corresponds to the implicit assumption of several theoretical studies) a population size of 100 is sufficient to produce some individuals with the fittest genotype (indeed, we would expect 25 such individuals in a population of 100). When we consider 10 loci, however, we need 10,000 individuals to have a reasonable chance of getting individuals with the highest possible fitness. At 20 loci, even 100,000 individuals is insufficient, and at a 100 loci, the highest fitness will be in the region of 0.7 even in a population of 100,000 individuals. To put this in perspective, if we consider that if even only 1% of the 20,000 human genes were subject to selection,

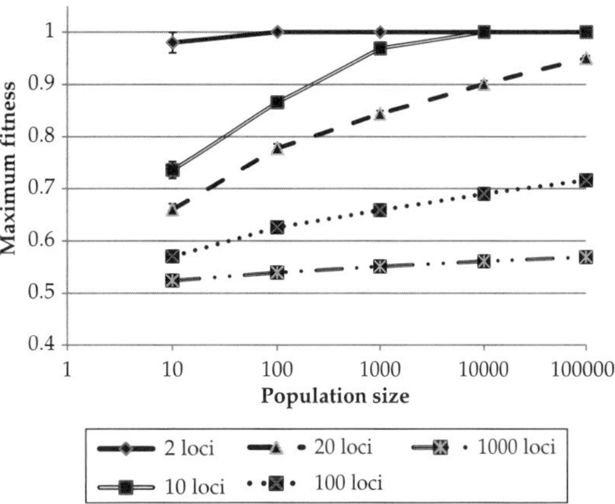

Figure 3.3 The relationship between maximum genotypic fitness and population size in an asexual population. Genotype distributions, of the type shown in Fig. 3.1a, were generated using different numbers of loci and a range of population sizes. Then the genotype with the maximum fitness in that population was scored and its fitness expressed relative to the fitness of an individual with selectively favorable alleles at all selected loci. The data are based on ten replicate simulations and the 95% confidence intervals are shown.

there is no imaginable human population size that would be sufficient to provide even a tiny fraction of all the possible genotypic combinations at the 200 selected loci.

3.3 Discussion

Our results show that, in a multilocus context, recombination can be a very powerful force for translating existing allelic variation into new genotypic variation that has a fitness distribution that goes far beyond the highest fitness in the original population. This is *not* because recombination can preferentially produce fitter genotypes. It is because selection increases the frequency of the fitter alleles and then recombination generates random genotypic combinations based on these new allelic frequencies. In other words, recombination gives no fitness benefit by itself but it acts in a synergistic way with the directional selection. In contrast to this, the non-recombining population is limited to the genotypes that were originally present in the non-selected population. And as our numerical calculations have shown, in a multilocus genome the original suite of genotypes represents an extremely sparse sampling of the entire range of possible genotypes. We could say that the function of recombination is that it continually 'resets' the genotypic frequencies to reflect the current allelic frequencies. This explains why recombination shows little effect in the early stages of the simulations; its benefit only becomes evident when the genotypic frequencies have changed significantly in response to selection. As stated by Crow (1992), directional selection in an asexual population monotonically decreases the genotypic variance, whereas the sexual population is relatively immune to this erosion of genotypic variance. This effect was also noted by Charlesworth (1993) in the case of directional selection on a quantitative trait. Our simulation shows that not only is the genotypic variance maintained in the recombining population, but that the initial distribution of genotypes is replaced by an equally variable, but fitter range of genotypes.

Our simulation provides a specific illustration of the general claim that recombination can evolve in large finite populations given that there is selection occurring on a sufficient number of genetic loci (Iles et al. 2003). What is striking about our results is that the 'sufficient number' can be a very small fraction of all the loci in the eukaryotic genome, and that the 'large finite populations' can be astronomically large. In other words, if even 1% of the loci in a typical eukaryotic genome are subject to selection, then recombination will provide an advantage even if the population size is in the billions. This statement may seem to be in contradiction to some of the previous work on the relationship between recombination and population size (Hill and Robertson 1966; Felsenstein 1974: Barton and Otto 2005) which concludes that population sizes have to be small in order for recombination to be favored. But this seeming contradiction disappears when we remember that much of this work only considered a few selected loci. Indeed, the previous studies of Otto and Barton (2001) and Iles et al. (2003) are consistent with our finding that the restriction on population size tends to disappear as the number of selected loci increases. This is because the number of possible genotypes quickly outstrips any realistic population size. Thus, while the loss of alleles at individual loci through sampling drift only happens in small populations, the absence of many (or most) multilocus combinations is inevitable even in very large populations.

We have focused on the effects of recombination on the standing genetic variation. One could ask if the same considerations apply to new genetic variation that is introduced into the population by positive or negative mutations. Some recent studies indicate that this is indeed the case. For example, it has been shown that new favorable mutations in yeast usually occur on suboptimal genetic backgrounds (Lang et al. 2011) and Hartfield and Otto (2011) have shown that many deleterious mutations could hitchhike to higher frequencies on high-fitness genetic backgrounds. By randomizing the mutations with respect to their initial genetic backgrounds, recombination allows the beneficial mutations to rise in frequency and the deleterious mutations to be selected out of the population. Both of these studies are assuming, quite realistically, that new mutations occur in an already genetically variable population. Previous studies that focused on two loci only were implicitly assuming monomorphism at all of the other loci in the genome. A related point is that the calculation of the time until the occurrence of a second mutation at a second locus considered possible beneficial mutations only at a given second locus, whereas it would be more reasonable to think of the chance of a second mutation at any one of thousands of other loci. This would increase by several orders of magnitude the probability that beneficial mutations would be selected concurrently, making the conditions for an advantage of recombination much less stringent than is often assumed (Christiansen et al. 1998).

Our results, along with the studies referred to in the preceding paragraph seem to be at odds with the view of Maynard Smith who stated that generation of new beneficial combinations by recombination will be equalized with their breakdown by recombination (see Maynard Smith 1968, 1978). This seeming paradox can be solved when we remember that Maynard Smith was thinking primarily of two loci which would yield only four possible gametic haplotypes. In any reasonably-sized population, we would expect all four types to be represented. As the number of loci increases, however, most of the myriad of possible combination of alleles will be missing. What recombination does in this situation is to allow the population to explore the genotypic space far beyond the initial range of genotypes. Moreover, this exploration is 'guided' by the increasing frequency of the selectively favored alleles, thus increasing the probability of generating genotypes with increased fitness.

We were gratified, but somewhat surprised to see that a single recombination event per chromosome per generation was sufficient to allow the population to respond to selection while maintaining a smooth binomial distribution of fitnesses. The gratification comes from the fact that this is approximately the real recombination rate per chromosome arm in mammals. We were surprised because, intuitively this seems like a very low rate of recombination. The explanation is that the recombination rate is indeed low if we consider a specific lineage within the population, but selection is acting on all lineages simultaneously. Thus while some chromosomes may have a single recombination event between locus 34 and locus 35, for example, another chromosome within the population will have the event occurring between locus 88 and 89, and so on, for 100,000 different events per generation over the entire population. Thus the process is reminiscent of the parallel approach to computing: the population is simultaneously 'searching' for new genotypic combinations in 100,000 independent ways.

3.4 Conclusions

Our results imply that the main function of recombination may *not* be to provide some biological advantage in the strict sense of the word. Instead, it functions to prevent a potential problem and, in so doing, it extends the 'genotypic range' of the population during the course of selection. To relate our findings to the theme of this volume, we could say that while recombination does not actively accelerate evolution, it does allow evolution to proceed past the limitations that the population size puts on the range of genotypic combinations. We could think of recombination as being analogous to the action of topoisomerase during DNA replication. Topoisomerase does not actively accelerate the rate of DNA replication but, in its absence replication would quickly grind to a halt due to the supercoiling of DNA ahead of the replication fork. Another way to state this point is to say that recombination does not increase the genotypic variance but, rather, that it prevents the genotypic variance from being quickly eroded by selection (or indeed by genetic drift).

Finally, our results are reminiscent of Weismann's original suggestion for the function of sex (Weismann 1887). He stated:

> We must attempt to explain the reason why Nature has insisted upon the rise and progress of sexual propagation. If we bear in mind that in sexual propagation twice as many individuals are required in order to produce any number of descendants, and if we further remember the important morphological differentiations which must take place in order to render sexual propagation possible, we are led to the conviction that sexual propagation must confer immense benefits upon organic life. I believe that such beneficial results will be found in the fact that sexual propagation may be regarded as a source of individual variability, furnishing material for the operation of natural selection.

We now understand that not recombination is *not* the source of genic (i.e. allelic) variation. But we have shown that it can interact with selection-based changes in allelic frequencies to produce new adaptive *genotypic* variation. So we could argue that Weismann was mistaken if we interpret 'individual variability' to mean allelic variation, but he was not mistaken it we interpret variability to mean genotypic variation. As we have shown here, recombination does not generate new favorable alleles but it can generate new favorable genotypic combinations as the frequency of certain alleles increases in response to selection.

References

Barton, N.H. and Otto, S.P. (2005) Evolution of recombination due to random drift. *Genetics* **169**: 2353–70.

Bellman, R.E. (1957) *Dynamic programming*. Princeton, NJ: Princeton University Press.

Bernstein, H., Byerly, H.C., Hopf, F.A., and Michod, R.E. (1985) Genetic damage, mutation, and the evolution of sex. *Science* **229**: 1277–81.

Charlesworth, B. (1993) Directional selection and the evolution of sex and recombination. *Genet Res* **61**: 205–24.

Christiansen, F.B., Otto, S.P., Bergman, A., and Feldman, M.W. (1998) Waiting with and without recombination: the time to production of a double mutant. *Theoret Pop Biol* **53**: 199–15.

Crow, J.F. and Kimura, M. (1965) Evolution in sexual and asexual populations. *Amer Natur* **99**: 439–50.

Crow, J.F. (1992) An advantage of sexual reproduction in a rapidly changing environment. *J Hered* **83**: 169–73.

Felsenstein, J. (1974) The evolutionary advantage of recombination. *Genetics* **78**: 737–56.

Fisher, R.A. (1930) *The Genetical Theory of Natural Selection.* Oxford: Oxford University Press.

Goddard, M.R., Godfray, H.C., and Burt, A. (2005) Sex increases the efficacy of natural selection in experimental yeast populations. *Nature* **434**: 636–40.

Hamilton, W.D., Axelrod, R., and Tanese, R. (1990) Sexual reproduction as an adaptation to resist parasites. *Proc Natl Acad Sci USA* **87**: 3566–73.

Hartfield, M. and Otto, S.P. (2011) Recombination and hitchhiking of deleterious alleles. *Evolution* **65**: 2421–34.

Hickey, D.A. (1982) Selfish DNA: a sexually transmitted nuclear parasite. *Genetics* **101**: 519–31.

Hill, W. and Robertson, A. (1966) The effect of linkage on limits to artificial selection. *Genet Res* **8**: 269–94.

Iles, M.M., Walters, K., and Cannings, C. (2003) Recombination can evolve in large finite populations given selection on sufficient loci. *Genetics* **165**: 2249–58.

Kondrashov, A.S. (1982) Selection against harmful mutations in large sexual and asexual populations. *Genet Res* **40**: 325–32.

Kondrashov, F.A. and Kondrashov, A.S. (2001) Multidimensional epistasis and the disadvantage of sex. *Proc Natl Acad Sci U S A* **98**: 12089–92.

Lang, G.I., Botstein, D., and Desai, M.M. (2011) Genetic variation and the fate of beneficial mutations in asexual populations. *Genetics* **188**: 647–61.

Long, A. and Michod, R.E. (1995) Origin of sex for error repair. I. Sex, diploidy, and haploidy. *Theor Popul Biol* **47**: 18–55.

Maynard Smith, J. (1968) Evolution in sexual and asexual populations. *Amer Nat* **102**: 469–73.

Maynard Smith, J. (1978) *The Evolution of Sex.* London: Cambridge University Press.

Muller, H.J. (1932) Some genetic aspects of sex. *Amer Nat* **66**: 118–38.

Muller, H.J. (1964) The relation of recombination to mutational advance. *Mut Res* **1**: 2–9.

Otto, S.P. (2009) The evolutionary enigma of sex. *Am Nat.* **174**(Suppl 1): S1–S14.

Otto, S.P. and Barton, N.H. (2001) Selection for recombination in small populations. *Evolution* **55**: 1921–31.

Otto, S.P. and Feldman, M.W. (1997) Deleterious mutations, variable epistatic interactions, and the evolution of recombination. *Theor Popul Biol* **51**: 134–47.

Peters, A.D. and Lively, C.M. (1999) The Red Queen and fluctuating epistasis: a population genetic analysis of antagonistic coevolution. *Amer Nat* **154**: 393–405.

Teotónio, H., Chelo, I.M., Bradiæ, M., Rose, M.R., and Long, A.D. (2009) Experimental evolution reveals natural selection on standing genetic variation. *Nat Genet* **41**: 251–7.

Weismann, A. (1887) On the significance of the polar globules. *Nature* **36**: 607–9.

CHAPTER 4

Heterogeneity in neutral divergence across genomic regions induced by sex-specific hybrid incompatibility

Seiji Kumagai and Marcy K. Uyenoyama

4.1 Introduction

Genes introduced into a genomic or environmental context different from the context in which they evolved may induce deleterious effects. Among the best-documented cases are hybrid incompatibility factors, which cause severe disruptions in viability, fertility, morphology, and behavior in interspecific hybrids (Coyne and Orr, 2004; Nosil and Schluter, 2011). Further, adaptation to local ecological conditions may engender divergent selection across environments (Charlesworth et al., 1997; Schluter, 2001). Here, we refer to genes that are neutral in their home context and deleterious in other contexts as incompatibility factors.

4.1.1 Detecting incompatibility factors

Many studies of genomic patterns of variation treat locus-specificity as a hallmark of selection, assuming that demographic history and population structure affect all regions of the genome uniformly (e.g. Akey et al., 2002; Innan et al., 2003). Neutral markers tightly linked to targets of incompatibility selection are expected to show low introgression relative to unlinked markers (Bengtsson, 1985; Barton and Bengtsson, 1986; Navarro and Barton, 2003). Kulathinal et al. (2009) examined genome-wide pairwise nucleotide differences in intraspecific comparisons (different strains of *Drosophila pseudoobscura*) and interspecific comparisons (*D. pseudoobscura* to sister species *D. persimilis* or to outgroup *D. miranda*). This exploration appeared to indicate greater excess of interspecific over intraspecific differences in genomic regions adjacent to known incompatibility factors.

A departure from earlier theoretical expectations is that sex-specific incompatibility can differentially impede introgression of neutral markers even in the absence of physical linkage (Fusco and Uyenoyama, 2011b). This effect derives from associations of genetic regions with sex, in the absence of functional epistasis between targets of selection and neutral markers. For example, deleterious factors tend to be eliminated more slowly from the sex in which they are more benign and neutral markers transmitted primarily or exclusively through one sex experience incompatibility factors predominantly in the context of that sex.

4.1.2 Within-species polymorphisms for incompatibility factors with sex-limited transmission

Chippindale and Rice (2001) detected remarkably strong effects on male fitness of Y chromosomes segregating within a laboratory strain of *Drosophila melanogaster*. Recent analyses have demonstrated pervasive disruptions in expression of genes throughout the genome upon introgression of Y chromosomes between laboratory strains or natural populations (Lemos et al., 2008, 2010; Jiang et al., 2010). Earlier work had documented greater divergence between *D. melanogaster* and *D. simulans* of genes with male-biased expression (Ranz et al., 2003). Male-biased genes contributed disproportionately to those subject to Y-linked regulatory variation (YRV), and YRV genes show greater

Rapidly Evolving Genes and Genetic Systems. First Edition. Edited by Rama S. Singh, Jianping Xu, and Rob J. Kulathinal.

divergence between *D. melanogaster* and *D. simulans* (Lemos et al., 2008).

Substitution of mitochondrial genomes among strains also induces pervasive changes in expression throughout the *D. melanogaster* genome (Innocenti et al., 2011), with genes with male-biased expression again over-represented among affected genes. The authors suggested that near-complete restriction of mitochondrial transmission through the female line indicates that selection on mitochondrial genomes, driven primarily by their effects on female but not male fitness, may be strongly sexually antagonistic (reviewed by Rice, 1998).

As extensive disruption of expression is likely to be highly deleterious, these examples suggest that polymorphisms for incompatibility factors with sex-limited transmission may induce substantial barriers to introgression among populations of conspecifics. Further, that genes with sex-biased expression diverge at accelerated rates (Ranz et al., 2003) suggests that incompatibility at the intraspecific level as well as the interspecific level may tend to be sex-specific.

In this chapter, we summarize the nature of differential rates of introgression across the genome generated by sex-specific incompatibility. We explore the implications of this process for the inference of population structure.

4.2 Genealogical migration rate

4.2.1 Definition

A number of indices have been proposed to quantify rates of neutral introgression induced by incompatibility factors. The index most widely used in the context of interspecific hybridization (e.g. Barton and Bengtsson, 1986; Navarro and Barton, 2003) is Bengtsson's (1985) 'gene flow factor,' which corresponds to the probability that a foreign marker allele will succeed in transferring to a genomic background free of incompatibility factors. Gavrilets (1997) used the inverse of the equilibrium frequency of a marker allele in a deme in which its existence requires gene flow in opposition to selection. Takahata and Slatkin (1984) studied the rate of replacement of the local neutral marker allele by a foreign allele repeatedly introduced together with an incompatibility factor by one-way migration. Kobayashi et al. (2008) defined 'neutral effective migration rate' in terms of the change in frequency of a foreign allele relative to the difference in frequency of the allele within the deme and among migrants.

Our objective is to characterize the pattern of variation at neutral markers in a genome containing incompatibility factors. Fusco and Uyenoyama (2011a,b) approximated the full, complex process using a structured population model of neutral variation with a migration rate scaled to account for selection targeted to incompatibility factors and crossing-over between the marker and incompatibility loci. The genealogical migration rate (g) corresponds to the parameter of an exponential waiting time density of migration events along a random lineage traced backwards in time:

$$g = m\omega, \tag{4.1}$$

for m the backward migration rate (proportion of the local gamete pool derived from migrants) and ω the relative reproductive rate. For Ω indicating the locations of incompatibility factors, relative reproductive rate $\omega_{\Omega}^{l,s}$ represents the expected number of descendants in a generation far into the future of a neutral marker gene at genomic location l on a gamete (v_m) transmitted by a migrant of sex s, relative to a marker gene on a resident gamete (v_r). It corresponds to the limit

$$\omega_{\Omega}^{l,s} = \lim_{t\to\infty} \frac{v_m(\mathbf{ZSG})^t \mathbf{e}}{v_r(\mathbf{ZSG})^t \mathbf{e}}, \tag{4.2}$$

for $\mathbf{Z}$ describing the generation of zygotes from gametes; $\mathbf{S}$ selection on zygotes; $\mathbf{G}$ the transmission by reproducing zygotes of gametes bearing neutral marker genes; t the number of generations since the focal migration event; and $\mathbf{e}$ the vector with all elements equal to 1 (Fusco and Uyenoyama, 2011a). Results of simulations (Fusco and Uyenoyama, 2011b) suggest that for backward migration rates (m) sufficiently low so that the interval between migration events is long relative to the time to convergence of the limit in (4.2), the waiting time to the most recent migration event along a random lineage at the neutral marker locus is indeed well-approximated by an exponential distribution with parameter given by the genealogical migration rate (4.1).

For c the proportion of females among reproducing migrants, incompatibility loci at genomic locations indicated by Ω induce an overall relative reproductive rate at a neutral marker locus at location l of

$$\bar{\omega}_{\Omega}^{l} = c\omega_{\Omega}^{l,f} + (1-c)\omega_{\Omega}^{l,m}. \tag{4.3}$$

4.2.2 Non-sex-specific incompatibility

While previous analyses of barriers to interspecific introgression have assumed the absence of disfavored incompatibility alleles in pure-species populations, polymorphisms appear to be common in plants (Rieseberg and Blackman, 2010) and are expected to arise among conspecific demes adapted to local ecological conditions. Relative reproductive rate as defined in (4.2) easily accommodates polymorphisms maintained by a balance between selection and migration (Fusco and Uyenoyama, 2011a).

Under purifying or disruptive selection, regimes that promote within-deme monomorphism, genealogical migration rate g (4.1) declines with increasing difference in incompatibility allele frequency between the local gamete pool and gametes transmitted by migrants. In contrast, overdominant selection within demes engenders almost no barrier to introgression, even for very large between-deme differences in equilibrium frequencies of incompatibility alleles. Under meiotic drive opposing purifying selection within demes, relative reproductive rate $\omega_{\Omega}^{l,s}$ (4.2) can exceed unity, signifying that migrants have greater expected numbers of descendants than residents.

4.2.3 Sex-specific incompatibility

Sex-specific incompatibility may reflect differential impairment of the sexes by foreign alleles, linkage to genomic regions transmitted primarily or exclusively through one sex, or differences between the sexes in rates of crossing-over between neutral marker loci and targets of incompatibility selection. For example, incompatibility factors borne on Y chromosomes or mitochondrial genomes (e.g. Lemos et al., 2008; Innocenti et al., 2011), which show maximal associations with sex, differentially impede introgression of neutral markers on autosomes, sex chromosomes, or mitochondria (table 1 of Fusco and Uyenoyama, 2011b).

In general, barriers to neutral introgression engendered by sex-specific incompatibility depend on the locations within the genome of incompatibility loci and neutral marker loci and on the sex of migrants. Table 4.1 provides expressions for relative reproductive rate (4.2) in ZW sex determination systems.

Sex-specificity also causes barriers generated by multiple sex-specific incompatibility factors to depart from earlier expectations. In the absence of sex-specificity, the total barrier to introgression induced by incompatibility factors showing no functional epistasis corresponds to the product of the barriers induced by the factors individually (Barton and Bengtsson, 1986; Fusco and Uyenoyama, 2011b). In contrast, associations with sex developed by multiple sex-specific incompatibility factor influence their joint distribution. Non-epistatic factors with concordant effects on the sexes (e.g. impairing males more than females or higher rates of crossing-over with the neutral marker in females) generate a submultiplicative total barrier (below the multiplicative expectation) and factors with discordant effects a supermultiplicative total barrier (Fusco and Uyenoyama, 2011b). This effect reflects the greater efficiency of selection acting to purge incompatibility factors associated with the same sex (compare Hill and Robertson, 1966; Barton, 1995).

4.3 Applications

We explore some implications of sex-specific incompatibility for patterns of variation across genomic regions.

4.3.1 Mitochondrial introgression

Petit and Excoffier (2009) conducted a literature survey of patterns of interspecific introgression in 37 mammal, bird, and insect species known to have sex-biased dispersal. Their results appeared to indicate a trend opposite to prediction based on sex-biased migration alone: all 16 organisms with

Table 4.1 Relative reproductive rates under ZW sex determination

Position[a]		Relative reproductive rate	
Factor	Marker	Female migrant	Male migrant
A	A	$\frac{\sigma_f r_f + \sigma_m r_m}{1 - \sigma_f(1 - r_f) + 1 - \sigma_m(1 - r_m)}$	$\frac{\sigma_f r_f + \sigma_m r_m}{1 - \sigma_f(1 - r_f) + 1 - \sigma_m(1 - r_m)}$
A	Z	$\frac{\sigma_m(4 + \sigma_f)}{8 - \sigma_m(2 + \sigma_f)}$	$\frac{2(\sigma_f + \sigma_m) + \sigma_f \sigma_m}{8 - \sigma_m(2 + \sigma_f)}$
A	W	$\frac{\sigma_f}{2 - \sigma_f}$	0
A	mt	$\frac{\sigma_f}{2 - \sigma_f}$	0
Z	A	$\frac{4 + \sigma_m}{8 - \sigma_m(2 + \sigma_f)}$	$\frac{2(\sigma_f + \sigma_m) + \sigma_f \sigma_m}{8 - \sigma_m(2 + \sigma_f)}$
Z	Z	$\frac{2\sigma_m r_m}{2 - \sigma_m(1 + \sigma_f)(1 - r_m)}$	$\frac{\sigma_m(1 + \sigma_f)r_m}{2 - \sigma_m(1 + \sigma_f)(1 - r_m)}$
Z	W	1	0
Z	mt	1	0
W	A	$\frac{1}{2 - \sigma_f}$	1
W	Z	1	1
W	W	0	0
W	mt	0	0
mt	A	$\frac{\sigma_m}{2 - \sigma_f}$	1
mt	Z	σ_m	1
mt	W	0	0
mt	mt	0	0

[a] Genomic location: autosomal (A), Z-linked (Z), W-linked (W), mitochondrial (mt)

female-biased dispersal showed lower introgression of mitochondrial than nuclear markers and all but two of 21 organisms with male-based dispersal showed higher introgression of mitochondrial markers. Most of the species designated as having female-biased dispersal also had a ZW sex determination system with homogametic (ZZ) males and most of the species with male-biased dispersal had heterogametic (XY or XO) males.

We compare XY and ZW sex determination systems with respect to barriers to mitochondrial introgression induced by incompatibility factors on the Y or W chromosome. A major difference between XY and ZW sex determination systems is the nature of cosegregation of mitochondria with sex. In descendants of a female migrant, the foreign mitochondrial genome, foreign W chromosome, and femaleness show complete cosegregation. In particular, Table 4.1 indicates that a W-linked incompatibility factor completely blocks mitochondrial introgression ($\omega_W^{mt,f} = 0$), while a Y-linked factor presents no barrier at all ($\omega_Y^{mt,f} = 1$). To the extent that incompatibility factors occur on the chromosome held exclusively by the heterogametic sex (Y or W), this comparison suggests greater mitochondrial than nuclear introgression in natural populations with XY but not ZW sex determination systems.

This prediction appears to be consistent with Petit and Excoffier's (2009) finding of higher mitochondrial than nuclear introgression in 13 of the 17 organisms with XY sex determination and in four of the 17 organisms with ZW sex determination. Further, in one of the first empirical observations documenting differential interspecific divergence among genomic regions, Powell (1983) reported that sympatric but not allopatric populations of *Drosophila pseudoobscura* and *D. persilimis* (XY) share mitochondrial genomes despite abundant evidence of nuclear divergence in sympatry as well as allopatry.

4.3.2 Interpreting region-specific F_{ST}

A seminal analysis by Seielstad et al. (1998) revealed striking differences in F_{ST} estimated from autosomal, mitochondrial, and Y-linked variation at a local scale (hundreds of kilometers) among human European populations. They suggested that the large (eightfold) difference in numbers of migrants inferred from mitochondrial and Y-linked variation primarily reflects higher migration rates in females. A number of subsequent studies have based inferences about sex-bias in effective number or migration rate on comparisons of F_{ST} (reviewed by Wilkins, 2006). Exploring sex-bias at the genome scale in global samples has proven to be more complex than anticipated, with some data sets failing to show expected patterns (Wilder et al., 2004), showing conflicting patterns across data sets (Hammer et al., 2008; Keinan et al., 2009; Bustamante and Ramachandran, 2009), or showing no consistent pattern across loci (Garrigan et al., 2007).

Sex-bias in migration or effective numbers: A number of methods have been proposed to estimate sex-bias in migration rate or effective number based on genomic patterns of variation (e.g. Hamilton et al., 2005; Hammer et al., 2008; Ramachandran et al., 2008; Ségurel et al., 2008). In particular, low F_{ST} estimated from X-linked (F_{ST}^X) relative to autosomal (F_{ST}^A) markers has been interpreted as indicative of greater numbers of female migrants. For the island model with small rates of mutation relative to migration,

$$F_{ST} = \frac{1}{1 + 4N_e m_e[d/(d-1)]^2}, \tag{4.4}$$

for N_e the effective number of zygotes, m_e the effective rate of migration, and d the number of demes (e.g. Hudson, 1990; Slatkin, 1991). Under sex-biased dispersal or effective numbers,

$$N_e = \frac{4N_f N_m}{N_f + N_m}$$

$$m_e = (m_f + m_m)/2$$

for autosomal marker loci and

$$N_e = \frac{9N_f N_m}{2N_f + 4N_m}$$

$$m_e = (2m_f + m_m)/3$$

for X-linked marker loci, in which N_f and N_m denote effective numbers of female and male reproductives within demes and m_f and m_m female and male backward migration rates.

Ramachandran et al. (2008) addressed the relationship between the ratio of migrant numbers ($M_f/M_m = N_f m_f/(N_m m_m)$) and the proportion of females among reproductives within a deme ($r = N_f/(N_f + N_m)$):

$$\frac{M_f}{M_m} = \frac{r\left[-3\left(\frac{1}{F_{ST}^A} - \frac{1}{F_{ST}^X}\right) + (5-4r)\left(\frac{1}{F_{ST}^X} - 1\right)\right]}{2(1-r)\left[3\left(\frac{1}{F_{ST}^A} - \frac{1}{F_{ST}^X}\right) - (1-2r)\left(\frac{1}{F_{ST}^X} - 1\right)\right]}. \tag{4.5}$$

Sex-specific incompatibility: Here, we explore inferences about sex-biased dispersal or effective numbers that might be drawn from F_{ST} generated under a distinct model: sex-specific incompatibility with sex-independent effective numbers ($N_f = N_m = N/2$) and an overall backward migration rate of m. Under our model,

$$F_{ST}^A = \frac{1}{1 + 4Nm\bar{\omega}_\Omega^A[d/(d-1)]^2}$$
$$F_{ST}^X = \frac{1}{1 + 3Nm\bar{\omega}_\Omega^X[d/(d-1)]^2}, \tag{4.6}$$

for $\bar{\omega}_\Omega^A$ and $\bar{\omega}_\Omega^X$ defined in (4.3). Substitution of these expected values into (4.5) produces

$$\frac{M_f}{M_m} = \frac{r\left[(2-r)\bar{\omega}_\Omega^X - \bar{\omega}_\Omega^A\right]}{(1-r)\left[2\bar{\omega}_\Omega^A - (2-r)\bar{\omega}_\Omega^X\right]}. \tag{4.7}$$

Positivity of the inferred ratio of female to male migrants (M_f/M_m) requires

$$2\bar{\omega}_\Omega^A/(2-r) > \bar{\omega}_\Omega^X > \bar{\omega}_\Omega^A/(2-r),$$

which reduces, under equal effective numbers of males and females ($r = 1/2$), to

$$\frac{4}{3}\bar{\omega}_{\Omega}^{A} > \bar{\omega}_{\Omega}^{X} > \frac{2}{3}\bar{\omega}_{\Omega}^{A}. \tag{4.8}$$

Satisfaction of the inequality on the left ensures $F_{ST}^{X} > F_{ST}^{A}$, consistent with lower effective numbers of X-linked than autosomal genes. However, sex-specific incompatibility can induce sufficient greater barriers to neutral introgression of autosomal than X-linked markers to cause F_{ST}^{A} to exceed F_{ST}^{X}.

Fig. 4.1 shows the ratio of female to male migrants (4.7) inferred from values of F_{ST}^{A} and F_{ST}^{X} expected to arise in response to a sex-specific Y-linked incompatibility factor that reduces the viability of its carriers to $\sigma_m = 0.5$ (solid) or to $\sigma_m = 0.8$ (dashed). For increasing proportions of females among reproducing migrants (c), the inferred M_f/M_m uniformly *decreases*. Because X-linked marker genes borne by a male migrant are transmitted only to factor-free daughters, Y-linked incompatibility factors induce no barrier to their introgression ($\bar{\omega}_{Y}^{X} = 1$). In contrast, autosomal markers transmitted to sons of a male migrant suffer reductions in fitness induced by the Y-linked factor, implying

$$\bar{\omega}_{Y}^{A} = c + (1 - c)/(2 - \sigma_m)$$

(from table 1 of Fusco and Uyenoyama, 2011b). Strong incompatibility ($\sigma_m < 2/3$) can cause F_{ST}^{A} to exceed F_{ST}^{X} (violation of the left inequality of (4.8)) for sufficiently low c, in spite of the higher effective number of autosomal genes. This effect generates a discontinuity in the inferred M_f/M_m ratio, as exemplified by the hyperbola in Fig. 4.1 ($\sigma_m = 0.5$).

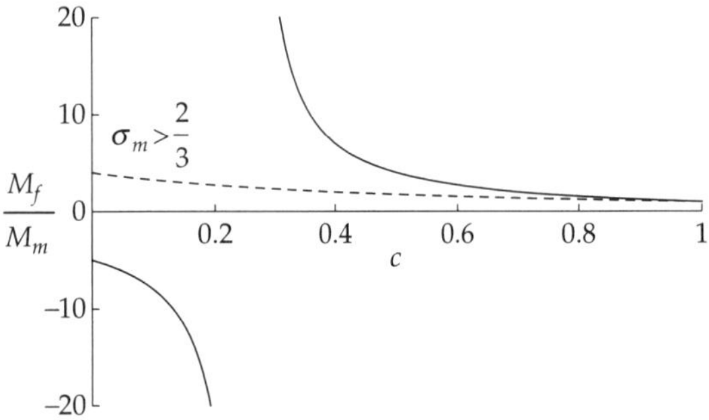

Figure 4.1 Ratio of numbers of female to male migrants (M_f/M_m) inferred from values of F_{ST}^{A} and F_{ST}^{X} expected under sex-specific incompatibility induced by Y-linked incompatibility factors as a function of the proportion of females among migrants (c). The solid lines represent the effects of a factor that reduces the viability of its carriers to $\sigma_m = 0.5$ and the dashed line to $\sigma_m = 0.8$.

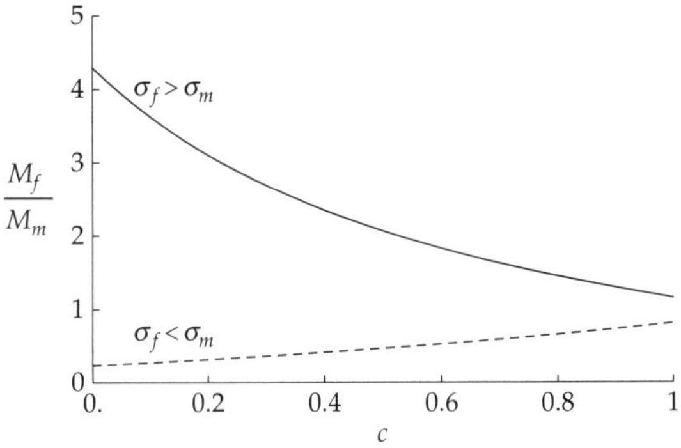

Figure 4.2 Ratio of numbers of female to male migrants (M_f/M_m) inferred from values of F_{ST}^{A} and F_{ST}^{X} expected under sex-specific incompatibility induced by an autosomal factor that reduces the viability of its female and male carriers to σ_f and σ_m as a function of the proportion of females among migrants (c). The solid line corresponds to $\sigma_f = 0.6$ and the dashed line to $\sigma_f = 0.4$, under the constraint $\sigma_f + \sigma_m = 1$.

Fig. 4.2 illustrates trends in the inferred M_f/M_m (4.7) under sex-specific incompatibility due to an autosomal factor that reduces the viability of its female (σ_f) and male (σ_m) carriers relative to factor-free individuals ($1 > \sigma_f, \sigma_m$). To maintain a fixed total amount of selection across the comparison, we imposed the constraint $\sigma_f + \sigma_m = 1$. Fig. 4.2 indicates that the inferred M_f/M_m increases with the proportion of females among reproducing migrants (c) only for factors that impair females more than males ($\sigma_m > \sigma_f$).

An autosomal factor induces relative reproductive rates at markers on a separate autosome and on the X-chromosome of

$$\omega_{A}^{A,f} = \omega_{A}^{A,m} = \frac{\sigma_f + \sigma_m}{4 - \sigma_f - \sigma_m}$$

$$\omega_{A}^{X,f} = \frac{2(\sigma_f + \sigma_m) + \sigma_f\sigma_m}{8 - \sigma_f(2 + \sigma_m)}$$

$$\omega_{A}^{X,m} = \frac{\sigma_f(4 + \sigma_m)}{8 - \sigma_f(2 + \sigma_m)}$$

(from table 1 of Fusco and Uyenoyama, 2011b). Under the constraint $\sigma_f + \sigma_m = 1$, the relative reproductive rate at the unlinked autosomal marker ω_{A}^{A}

reduces to 1/3. For incompatibility factors causing greater impairment of females than males ($\sigma_f < \sigma_m$), X-linked markers introgress at lower rates than autosomal markers, with $\omega_A^{X,m} < \omega_A^{X,f} < \omega_A^A$. This relation reflects a stronger association of the X-chromosome with females, the sex more impaired by incompatibility. The association between femaleness and the X-linked marker strengthens with higher proportions of male migrants, which transmit the X-linked marker exclusively to daughters. The inferred ratio of numbers of female to male migrants (4.7) increases with increasing c (decreasing proportion of male migrants), although it indicates a male-bias in migrant number ($M_f/M_m < 1$) even for c exceeding 1/2.

For incompatibility factors causing greater impairment of males than females ($\sigma_f > \sigma_m$), however, the inferred ratio M_f/M_m *declines* with increasing proportions of female migrants (solid line in Fig. 4.2). In this case, X-linked markers introgress at higher rates than autosomal markers ($\omega_A^{X,m} > \omega_A^{X,f} > \omega_A^A$), with the dependence on c reflecting the strengthening of the association of the X-linked marker with femaleness as the proportion of male migrants increases. Inferences based on (4.7) would indicate uniformly higher numbers of female migrants ($M_f/M_m > 1$), even for c less than 1/2.

These examples illustrate that sex-specific incompatibility can induce differential rates of introgression across the genome, especially among regions that differ in their cosegregation with sex. Associations between incompatibility factors and neutral markers derive not from functional epistasis but rather from an association of each class of loci with sex. Ignoring sex-specific incompatibility may cause patterns in F_{ST} across the genome to be misinterpreted with respect to the existence or direction of sex-bias in effective number or dispersal.

4.4 Conclusions

In *Drosophila*, the model system for speciation in which key genetic and epigenetic mechanisms have been best-characterized, various lines of evidence have now well established that genes with male-limited or male-biased expression diverge at accelerated rates (e.g. Civetta and Singh, 1998; Ranz et al., 2003). Even at the intraspecific level, transfer between populations of genomic regions with sex-limited transmission can cause pervasive disruptions of expression of genes throughout the genome (Lemos et al., 2008; Innocenti et al., 2011).

Rapidly evolving sex-specific hybrid incompatibility can generate heterogeneity in neutral divergence across genomic regions. The locus-specific nature of the induced barriers to introgression derive not from physical linkage to incompatibility factors but rather from associations between sex and neutral markers and between sex and targets of sex-specific selection (Fusco and Uyenoyama, 2011b). As such associations arise even in the absence of linkage, a single sex-specific incompatibility factor can induce locus-specific patterns of neutral divergence across genomic regions. We suggest that the locus-specific nature of barriers to introgression both between species and between conspecific populations induced by sex-specific incompatibility may affect inferences regarding sex-biased dispersal and selection.

Acknowledgments

Public Health Service grant GM 37841 (MKU) provided partial funding for this research. We gratefully acknowledge support for the National Evolutionary Synthesis Center (NESCent) working group on Genomic Introgression.

References

Akey, J.M., Zhang, G., Zhang, K., Jin, L., and Shriver, M.D. (2002) Interrogating a high-density SNP map for signatures of natural selection. *Genome Res* **12**: 1805–14.

Barton, N.H. (1995) Linkage and the limits to natural selection. *Genetics* **140**: 821–41.

Barton, N.H. and Bengtsson, B.O. (1986) The barrier to genetic exchange between hybridising populations. *Heredity* **57**: 357–76.

Bengtsson, B.O. (1985) The flow of genes through a genetic barrier. In P. J. Greenwood, P. H. Harvey, and M. Slatkin (Eds) *Evolution: Essays in Honor of John Maynard Smith*, 31–42. New York: Cambridge University Press.

Bustamante, C.D. and Ramachandran, S. (2009) Evaluating signatures of sex-specific processes in the human genome. *Nat Genet* **41**: 8–10.

Charlesworth, B., Nordborg, M., and Charlesworth, D. (1997) The effects of local selection, balanced polymorphism and background selection on equilibrium patterns of genetic diversity in subdivided populations. *Genet Res* **70**: 155–74.

Chippindale, A.K. and Rice, W.R. (2001) Y chromosome polymorphism is a strong determinant of male fitness in *Drosophila melanogaster*. *Proc Natl Acad Sci (USA)* **98**: 5677–82.

Civetta, A. and Singh, R.S. (1998) Sex-related genes, directional sexual selection, and speciation. *Mol Biol Evol* **15**: 901–9.

Coyne, J.A. and Orr, H.A. (2004) *Speciation*. Sunderland, MA. Sinauer Associates, Inc.

Fusco, D. and Uyenoyama, M.K. (2011a) Effects of polymorphism in locally-adapted genes on rates of neutral introgression in structured populations. *Theor Pop Biol* **80**: 121–31. doi: 10.1016/j.tpb.2011.06.003.

Fusco, D. and Uyenoyama, M.K. (2011b) Sex-specific incompatibility selection generates locus-specific rates of introgression. *Genetics* **189**: 267–88.

Garrigan, D., Kingan, S.B., Pilkington, M.M., Wilder, J.A., Cox, M.P., Soodyall, H., et al. (2007) Inferring human population sizes, divergence times and rates of gene flow from mitochondrial, x and y chromosome resequencing data. *Genetics* **177**: 2195–207.

Gavrilets, S., 1997 Hybrid zones with Dobzhansky-type epistatic selection. *Evolution* **51**: 1027–35.

Hamilton, G., Currat, M., Ray, N., Heckel, G., Beaumont, M. A., and Excoffier, L. (2005) Bayesian estimation of recent migration rates after a spatial expansion. *Genetics* **170**: 409–17.

Hammer, M.F., Mendez, F.L., Cox, M.P., Woerner, A.E., and Wall, J.D. (2008) Sex-biased evolutionary forces shape genomic patterns of human diversity. *PLoS Genet* **4**: e1000202.

Hill, W.G. and Robertson, A. (1966) The effect of linkage on limits to artificial selection. *Genet Res* **8**: 269–94.

Hudson, R.R. (1990) Gene genealogies and the coalescent process. In D. Futuyma and J. Antonovics (Eds) *Oxford Surveys in Evolutionary Biology* volume 7, pp. 1–44. New York: Oxford University Press.

Innan, H., Padhukasahasram, B., and Nordborg, M. (2003) The patterns of polymorphism on human chromosome 21. *Genome Res* **13**: 1158–68.

Innocenti, P., Morrow, E.H., and Dowling, D. K. (2011) Experimental evidence supports a sex-specific selective sieve in mitochondrial genome evolution. *Science* **332**: 845–8.

Jiang, P.-P., Hartl, D.L., and Lemos, B. (2010) Y not a dead end: Epistatic interactions between Y-linked regulatory polymorphisms and genetic background affect global gene expression in *Drosophila melanogaster*. *Genetics* **186**: 109–18.

Keinan, A., Mullikin, J.C., Patterson, N., and Reich, D. (2009) Accelerated genetic drift on chromosome X during the human dispersal out of Africa. *Nat Genet* **41**: 66–70.

Kobayashi, Y., Hammerstein, P., and Telschow, A., (2008) The neutral effective migration rate in a mainland-island context. *Theor Pop Biol* **74**: 84–92.

Kulathinal, R.J., Stevison, L.S., and Noor, M. A.F. (2009) Genomics of speciation in *Drosophila*: Diversity, divergence, and introgression estimated using low-coverage genome sequencing. *PLoS Genet* **5**: e1000550.

Lemos, B., Araripe, L.O., and Hartl, D.L. (2008) Polymorphic Y chromosomes harbor cryptic variation with manifold functional consequences. *Science* **319**: 91–93.

Lemos, B., Branco, A.T., and Hartl, D.L. (2010) Epigenetic effects of polymorphic Y chromosomes modulate chromatin components, immune response, and sexual conflict. *Proc Natl Acad Sci USA* **107**: 15826–31.

Navarro, A. and Barton, N. H. (2003) Accumulating postzygotic isolation genes in parapatry: A new twist on chromosomal speciation. *Evolution* **57**: 447–59.

Nosil, P. and Schluter, D. (2011) The genes underlying the process of speciation. *Trends Ecol Evol* **26**: 160–67.

Petit, R.J. and Excoffier, L. (2009) Gene flow and species delimitation. *Trends Ecol Evol* **24**: 386–93.

Powell, J.R. (1983) Interspecific cytoplasmic gene flow in the absence of nuclear gene flow: Evidence from *Drosophila*. *Proc Natl Acad Sci USA* **80**: 492–95.

Ramachandran, S., Rosenberg, N. A., Feldman, M.W., and Wakeley, J. (2008) Population differentiation and migration: Coalescence times in a two-sex island model for autosomal and X-linked loci. *Theor Pop Biol* **74**: 291–301.

Ranz, J.M., Castillo-Davis, C.I., Meiklejohn, C.D., and Hartl, D. L. (2003) Sex-dependent gene expression and evolution of the *Drosophila* transcriptome. *Science* **300**: 1742–5.

Rice, W.R. (1998) Intergenomic conflict, interlocus antagonistic coevolution, and the evolution of reproductive isolation. In D. J. Howard and S. H. Berlocher, eds., *Endless Forms: Species and Speciation*, pp. 261–70. New York: Oxford University Press.

Rieseberg, L.H. and Blackman, B.K. (2010) Speciation genes in plants. *Ann Bot (Lond)* **106**: 439–55.

Schluter, D. (2001) Ecology and the origin of species. *Trends Ecol Evol* **16**: 372–80.

Ségurel, L., Martínez-Cruz, B., Quintana-Murci, L., Balaresque, P., Georges, M., Hegay, T., et al. (2008) Sex-specific genetic structure and social organization in Central Asia: Insights from a multi-locus study. *PLoS Genet* **4**: e1000200.

Seielstad, M.T., Minch, E., and Cavalli-Sforza, L.L. (1998) Genetic evidence for a higher female migration rate in humans. *Nat Genet* **20**: 279–80.

Slatkin, M. (1991) Inbreeding coefficients and coalescence times. *Genet Res* **58**: 167–75.

Takahata, N. and Slatkin, M. (1984) Mitochondrial gene flow. *Proc Natl Acad Sci (USA)* **81**: 1764–7.

Wilder, J.A., Kingan, S.B., Mobasher, Z., Pilkington, M.M., and Hammer, M. F. (2004) Global patterns of human mitochondrial DNA and Y-chromosome structure are not influenced by higher migration rates of females versus males. *Nat Genet* **10**: 1122–5.

Wilkins, J.F. (2006) Unraveling male and female histories from human genetic data. *Curr Opin Genet Devel* **16**: 611–17.

CHAPTER 5

Rapid evolution in experimental populations of major life forms

Jianping Xu

5.1 Introduction

Questions in evolutionary biology are typically addressed by one of two approaches. In the first and the more common approach, biological samples from natural environments are analyzed and compared with other biotic (e.g. symbionts, predators, pathogens, etc.) and abiotic data (e.g. climatic, geological, geographic, physical/chemical properties of the environment, etc.) in order to infer historical processes, generate novel patterns, and/or test specific hypotheses. The second approach uses experimental evolution: monitoring and analyzing human-controlled long-term selection, adaptation, or mutation accumulation in populations of organisms. While most chapters in this book deal with data from the first approach, this chapter highlights methods and data from the experimental evolution approach. As shown in the following paragraphs, experimental evolutionary studies have provided some of the best and most direct evidence for rapid evolution of genes and genetic systems. And the recent arrival of highthroughput genomics, proteomics, and metabolomics technologies has opened an exciting new chapter in experimental evolution studies.

Experimental evolution studies have roots dating to early human civilization, during the domestication of plants and animals. Over the past few millennia, for some organisms, such domestication and selective breeding experiments have led to a large number of varieties that phenotypically often differ dramatically from each other and from their original wild-type ancestors. Notable examples include dogs, horses, cats, maize, rice, and cabbages. Indeed, the effects of human breeding to create varieties with extreme differences within a single species were recognized by Charles Darwin over 150 years ago, who started his book, *On The Origin of Species*, with a chapter on the widespread variation found in domestic animals. For example, there are now over 400 breeds of dogs that differ tremendously in size, shape, coat color, sensitivity to smell, and behavior (Shearman and Wilton 2011). However, most early selection and breeding experiments were not well documented, and not based on carefully designed and reproducible experiments.

The British microbiologist and microscope expert William Dallinger performed one of the first controlled long-term evolution experiments. From 1880–1886, he continuously cultivated and recorded the phenotypic changes of three flagellated monads, *Tetramitus rostratus*, *Monas dallingeri*, and *Dallingera drysdali*, with one species in each of three custom-built incubators (Dallinger 1887). During the 7-year period, he slowly increased the temperature of the incubators from an initial 16°C to 70°C and continuously monitored their cellular morphology and development using microscopy. The initial cultures grew well at 16°C but were incapable of surviving at 70°C. In contrast, those at the end of the selection experiments all survived well at 70°C but grew poorly or were unable to grow at the initial 16°C. Unfortunately, there was no repetition of his treatments, and only one sample was examined for each of the three species. As a result, the potential variations in the trajectories of adaptation could not be inferred. In addition, none of the ancestral or evolved strains were saved for future investigations, including eliminating the possibility that these results were due to contaminations. And

Rapidly Evolving Genes and Genetic Systems. First Edition. Edited by Rama S. Singh, Jianping Xu, and Rob J. Kulathinal.

lastly, the incubators were accidentally destroyed, leaving no possibility for follow-up experiments for this seemingly remarkable adaptation to high temperature growth for these organisms. Similar to the studies of Dallinger, most experimental evolution studies from the 1880s to the 1960s had no or few repetitions, often lacked stored ancestral and intermediately evolved populations, or were conducted for relatively few generations.

5.2 Features of experimental evolution

Both repetition and a relatively large number of generations are critical in experimental evolutionary investigations. Their importance is linked to the two most fundamental features of biological evolution: (i) the adaptation of organisms to their environments; and (ii) the divergence of populations and species from each other. Without repetition, the roles of stochastic chance events and selection can not be unambiguously distinguished. Similarly, since evolution in nature is a historical process and accumulating notable changes can take time, long-term laboratory evolution (although still much shorter than most evolutionary processes in nature) is often essential in making biologically meaningful inferences. Furthermore, though not essential or even possible for some species, having stored ancestral and evolved populations (both intermediate and final) allow for comparisons of traits at the same time and under the same environmental conditions. The availability of stored materials for comparisons minimizes experimental errors and permits better estimates of the potential contributions of accumulated genetic changes and genotype–environment interactions to the observed phenotypic characteristics. In addition, potential problems such as contamination could also be effectively addressed by comparing the genotypes of the stored strains and populations across the duration of the experiment.

At the beginning of the 1960s, a number of evolutionary biologists working with different groups of organisms began to establish long-term, multiple parallel experimental evolution lines. The organisms include fruit flies, mice, crop plants, viruses, yeasts, and a common mammalian gut bacterium *Escherichia coli*. A topical book titled *Experimental Evolution* (Garland and Rose 2009) was recently published that highlighted the contributions of experimental evolution to biology, agriculture, animal husbandry, and medicine. Although any organism could theoretically be used for experimental evolution studies, for several reasons, microbes have become the organisms of choice. First, microbes typically have short generation times, allowing a large number of generations to be passed in a relatively short period of time. Second, microbes are, by definition, small. Therefore, a large number of cells (billions to trillions) can be easily grown and maintained in the lab. Third, microbes can be readily reproduced clonally and a large number of genetically identical (or nearly identical) cells can be stored permanently as 'living fossils' for comparisons among themselves as well as with their ancestral genotypes, making estimates of various contributors (genotype, environment, and genotype–environment interactions) to overall phenotype variation relatively straightforward. Fourth, most microbes are, or can be made, genetically tractable, allowing specific marker genes to be followed through many generations. Such genetic tractability allows both testing the effects of specific genes and alleles and helping to identify potential contaminations. Finally, certain microbes, e.g. those with RNA-based viruses, have naturally high mutation rates and thus allow a large number of genetic changes to be accumulated rapidly. For microbes with DNA-based genomes, there are genes known to impact mutation rate. As a result, the deletion/alteration of these genes through genetic manipulations can increase mutational rate of the host genome, accelerate overall genetic change, and quickly impact on divergence and adaptation of experimental populations to specific environmental conditions.

In the following sections, I first provide a brief overview of experimental evolution approaches. I will then review a few selected organisms and traits to showcase the rapid rates of adaptation and divergence in these populations. While experimental evolution of plants and animals will be discussed, the focus here is on microbial experimental evolution studies.

5.3 Types of experimental evolution

Based on the type and amount of selective pressure imposed on evolving experimental populations, experimental evolution studies can be broadly classified into three types: directional selection, adaptation, and mutation accumulation.

5.3.1 Directional selection

In directional selection experiments, individuals with specific trait(s) or trait value(s) are directly selected at an appropriate life stage by the experimenter to establish subsequent generations. Typically, this process is repeated over many generations where the selected trait or trait value can change through the selection process. At the end of the selection phase of the experiment, the traits and trait values are analyzed and compared with progenitors and among parallel selection lines. The Danish botanist Wilhelm Johanssen (1903) reported one of the first directional selection experiments on beans where he selected both the largest and smallest beans in self-fertilizing populations. The underlying genetic changes accumulated during selection are often further analyzed by genetic crossing and progeny analyses, and in recent years, whole-genome resequencing, and gene expression profiling.

5.3.2 Adaptation

In adaptation experiments, a selective pressure is typically applied to parallel populations of the experimental organism and only those that survive and reproduce in the specific environment will have a chance to be passed onto the next generation of experimental populations. The selective pressures can be from abiotic physical/chemical factors such as high or low values of temperature (e.g. Dallinger 1887), pH, salt concentrations, and antibiotics or from biotic factors such as pathogens, predators, and symbionts. Similar to those in directional selection experiments, the intensity of selective pressure can change over time in adaptation experiments during the long-term adaptation process. However, unlike in directional selection experiments where organisms for the next generations are individually chosen, the organisms to establish future generations in adaptation experiments are not individually picked but, rather, are randomly drawn from all those that survived the previous generation. Each individual that survived the previous generation has the potential to be a founder for the next round of adaptation. In addition, the founder population for future generations in adaptation experiments is often larger than those in directional selection experiments.

5.3.3 Mutation accumulation

The third type of experimental evolution is mutation accumulation (Mukai 1964). Different from directional selection and adaptation experiments, selective pressure is typically not applied to the study populations in mutation accumulation experiments. As a result, aside from mutations that are lethal in the specific mutation accumulation environment, all other mutations (be they beneficial, neutral, or deleterious) can be accumulated in these populations at each round of the mutation accumulation phase and different replicate populations can accumulate different types and numbers of mutations. To enhance the chances that all cells (including both mutation-free cells and mutants) have equal probabilities of being passed on to the next generations, a single random individual (or pairs of individuals in obligate sexual organisms) is generally taken at each transfer for each line for the next generation, maximizing genetic drift and enhancing the possibility of continuously accumulating random mutations. Evolving populations at intermediate steps and the final stage of the mutation accumulation phase are typically stored. These populations, along with that of the starting population, are then examined for specific phenotypes. The phenotypic differences are used to calculate the rates of change over time and the degree of divergence among parallel mutation accumulation lines. Such data are often used to infer fundamental organismal parameters such as mutation rates and phenotypic effects per mutation for many traits in a variety of organisms.

In the following sections, I highlight examples of rapid evolution in each of the three experimental evolution approaches. I specifically focus on the rate of change in population mean phenotype and the rate of divergence among parallel experimental evolution populations. Together, these approaches have allowed for the examination of both chance and necessity in the evolution of a wide variety of traits.

5.4 Rapid change and divergence among mutation accumulation population lines

Compared to the total number of taxa that exist in the biosphere, the number and diversity of organisms that have been investigated in mutation accumulation experiments are extremely small. However, representatives from the major groups of organisms belonging to viruses, bacteria, fungi, plants, and animals have been investigated in laboratory mutation accumulation experiments. Most of these organisms are model species that have been used extensively in other types of studies. Table 5.1 shows selected examples of the diversities of organisms and traits that have been examined so far through this approach. Next, I present a brief summary of these studies.

5.4.1 Microbial growth rate

Among microorganisms, a commonly examined trait in mutation accumulation experiments is asexual reproductive rate or vegetative growth rate. As shown in Table 5.1, all examined microbial species reveal noticeable declines in vegetative growth after mutation accumulation, even after only a relatively short period of time (within a few months). The results are consistent with the hypothesis that most mutations are deleterious and that bottlenecking allows effective accumulation of such mutations in laboratory experimental populations. Interestingly, the rates of decline differed greatly among the four microbial species in Table 5.1. For example, the rate of decline in both the bacteriophage Ø6 and the mutation repair-competent strain of the budding yeast, *Saccharomyces cerevisiae*, were over 55–60 times greater than that observed in the common gut bacterium, *Escherichia coli* (Chao 1990; Kibota and Lynch 1996; Zeyl and de Visser 2001). The rate of decline in the human basidiomycete yeast pathogen, *Cryptococcus neoformans*, was intermediate (Xu 2002, 2004), about 25 times of that in *E. coli* and less than half of those in bacteriophage Ø6 and *S. cerevisiae*. The exact reason for the relatively low rate of decline in *E. coli* is unknown but may be related to the extended maintenance of this organism in the laboratory setting, likely adapted better to the lab environment than the other organisms and therefore experiencing lower rates of fitness decline when further cultured in the same lab environment. As expected, the mutation repair-deficient strain of *S. cerevisiae* had a greater growth rate decline per generation, over twice of that in the mutation repair-competent strain of the same species (Table 5.1).

In contrast to the large differences in the rates of fitness decline, rates of divergence among replicate mutation accumulation lines were similar among three of the four organisms (bacteriophage Ø6, *E. coli*, and *C. neoformans*), all within a threefold difference (Table 5.1). However, the rates of divergence among populations of both strains of *S. cerevisiae* were over 20 times higher than those in the other three organisms, with the mutation repair-competent strain about four times greater than that in the mutation repair-deficient strain. The relatively large rates of decline and rapid divergence among replicate lines in *S. cerevisiae* was mainly due to the loss of mitochondrial genes and genomes in these mutation accumulation populations, resulting in the formation of petit colonies (Zeyl and De Visser 2001). *S. cerevisiae* is a facultative aerobic/anaerobic yeast. In the absence of oxygen (i.e. in anaerobic conditions), oxidative phosphorylation by the mitochondria for energy generation is not required and losing mitochondria can be advantageous for the cells in such an environment. Even in the presence of oxygen, some cells are prone to lose their mitochondrial genes, resulting in petit colonies on solid agar. In contrast, *E. coli* and *C. neoformans* have different metabolic needs with both growing much faster under aerobic conditions than anaerobic ones. Indeed, *C. neoformans* is incapable of growing in anaerobic environments. Excluding petit mutants from the analyses yields estimates more similar to those observed in the other three microbes (Zeyl and De Visser 2001).

Table 5.1 Representative studies highlighting the rapid change and divergence in traits among experimental populations of bacteriophage Ø6, bacteria, fungi, plants, and animals that have undergone mutation accumulation and genetic drift

Organism	Phenotypic trait	Number of replicate lines	Number of generations	Rate of change per generation[a] ($\times 10^{-3}$)	Rate of divergence per generation ($\times 10^{-3}$)	Reference
Bacteriophage Ø6	Growth rate	20	1320	−0.167	0.039	Chao et al. 1990
Pseudomonas aeruginosa	Utilization pattern of 95 carbon sources	20	2000	−0.41/95 carbon sources	1.41/95 carbon sources	MacLean and Bell 2003
Escherichia coli	Cell size[b]	12	10,000	0.045	0.015	Barrick et al. 2009
	Growth rate	50	6600	−0.003	0.021	Kibota and Lynch 1996
Cryptococcus neoformans	Mating ability	16	600	−1.222	0.322	Xu 2002, 2004
	Filamentation ability	16	600	−0.643	0.298	
	Growth rate[c]	16	600	−0.075	0.075	
Saccharomyces cerevisiae[c]	Growth rate (wild type strain)	50	600	−0.183	5.874	Zeyl and De Visser 2001
	Growth rate (mutator strain)	50	600	−0.348	1.472	
Drosophila melanogaster	Abdominal bristle number (ancestral base number =15)	25	100	−0.153	8.401	Mackay et al. 1992
	Sternopleural bristle number (ancestral base number ~16)	25	100	−0.094	3.687	
	Viability	47	44	−3.8	0.103	Houle et al. 1992
	Life-time fitness	47	44	−12.8	0.828	
Caenorhabditis elegans	Life-time fitness	50	60	−0.3	1.2	Keightley and Caballero 1997
	Life span	50	60	−0.714	0.4	
Arabidopsis thaliana	Germination rate	924	10	−0.710	0.1	Schultz et al. 1999
	Fruit set	924	10	−0.421	0.01	
	Seeds per fruit	143	10	−0.441	0.083	
	Total fitness	127	10	−5.47	0.137	

[a] The values here refer to change relative to that of the ancestral strain/population. The ancestral strain/population trait values were all scaled to 1

[b] Cells were selected for adaptation to minimum medium, increase in cell size is an accidental side effect

[c] In these studies, the fitness of evolved populations were examined on several different growth media/conditions. However, only result from the growth medium/condition originally used during mutation accumulation is shown here.

5.4.2 Other microbial traits

Aside from vegetative growth rates, several other traits have also been examined using the mutation accumulation approach in microorganisms. These included carbon source utilization abilities in the plant and animal pathogenic bacterium, *Pseudomonas aeruginosa*, cell size in *E. coli*, and the mating and filamentation abilities in *C. neoformans* (Table 5.1). Of these four traits in Table 5.1, cell size in *E. coli* increased in the absence of selection, suggesting that large cell size might be deleterious in long-term evolution (Fig. 5.1). However, the cell size data here were obtained from a long-term adaptation experiment to a minimum medium (Barrick et al. 2009). Therefore, it is entirely possible that the observed cell size increase was due to the pleiotropic effects of adaptation to the minimal medium in these populations, even though cell size was not directly selected.

The remaining fitness-related traits other than the vegetative growth rate listed in Table 5.1 all demonstrated declines in mutation accumulation lines. For example, the loss of carbon source utilization ability in *P. aeruginosa* was rapid, about 1 for every 2000 generations for the 95 examined carbon sources (MacLean and Bell 2003). Similarly, both the mating ability and filamentation ability in *C. neoformans* showed much greater declines than those observed for vegetative growth in the examined microorganisms (Table 5.1; Xu 2002, 2004). The overall results suggest that while mutational targets for vegetative growth rates might be broader than most specialized traits, with rare exceptions (e.g. loss of mitochondrial genomes in *S. cerevisiae*), the effect per mutation for vegetative growth in typical mutation lines are overall smaller than other traits.

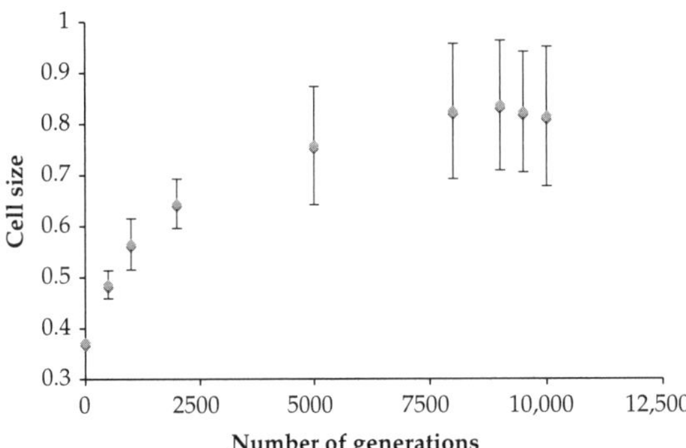

Figure 5.1 Changes in mean cell size in evolving *E. coli* populations over 10,000 generations. The x-axis shows the number of generations. The y-axis shows the mean cell volume ($\times 10^{-15}$ L). Note the relatively rapid change during the first 2000 generations. Vertical lines indicate standard deviations calculated from 12 experimental populations. The original data for this graph were from Lenski (2012; http:/myxo.css.msu.edu/ecoli).

5.4.3 Plants and animals

Similar to declines in fitness-related traits observed in microorganisms, all fitness-related traits examined so far showed declines in mutation accumulation lines in complex multicellular organisms such as the fruit fly, *Drosophila melanogaster*, the nematode, *Caenorhabditis elegans*, and the mustard weed plant, *Arabidopsis thaliana* (Table 5.1). As expected, among the examined traits, the largest declines were seen for lifetime fitness or total fitness where all individual fitness trait values were combined in the analyses (Mackay et al. 1994; Keightley et al. 1997; Schultz et al. 1999). Of the individual traits, their rates of decline varied by about 30-fold, from the lowest in sternopleural bristle number to the highest in viability, both in the fruit fly (Table 5.2). In contrast to the big differences among traits in the fruit fly, the three fitness related traits in *A. thaliana* showed remarkably similar rates of decline, all within a twofold difference and similar to the declines in mating and filamentation abilities in *C. neoformans*. However, the lifetime reproductive fitness in the fruit fly and the mustard weed showed much greater declines than that in the nematode. Akin to the large differences in the rates of decline, the rates of divergence also differed greatly among the organisms and traits. However, as shown for microorganisms, there was no clear correlation between the rate of decline and the rate of divergence. The most rapidly declining trait didn not show the biggest divergence among replicate lines and vice versa. The mechanisms underlying the differences among the species are largely unknown but may be related to differences in their ecology and reproductive biology. Further comparative investigations are urgently needed in this area.

All mutation accumulation populations examined so far have limited or no recombination. Instead, the populations were propagated through

Table 5.2 Examples of responses to adaptation and directional selection in experimental evolution populations. Highlighted in bold are examples of directional selection while others are examples of adaptation

Organism	Phenotypic trait	Number of replicate lines	Number of experimental generations	Rate of change per generation[a] ($\times 10^{-3}$)	Rate of divergence per generation ($\times 10^{-3}$)	Reference
Escherichia coli	Growth rate in an arabinose minimum medium	12	2000	0.192	0.32	Lenski et al. 1991
	Growth rate in an arabinose minimum medium	12	20,000	0.042	0.008	Barrick et al. 2009
	Growth rate at 42°C	6	400	0.18	0.007	Bennett and Lenski 1999
Pseudomonas aeruginosa	Utilization pattern of 95 carbon sources	4 lines each in 23 novel carbon source media	1100	2.003/95 carbon sources	2.183/95 carbon sources	MacLean and Bell 2003
Drosophila melanogaster	**Abdominal bristle number (high selection lines)**	**3 lines**	**125**	**1.76**	**0.613**	**Mackay et al. 1994**
	Abdominal bristle number (low selection lines)	**3 lines**	**125**	−4.96	**1.63**	
	Sternopleural bristle number (high selection lines)	**3 lines**	**125**	**2.65**	**2.02**	
	Sternopleural bristle number (low selection lines)	**3 lines**	**125**	**−1.5**	**0.865**	
	Ethanol resistance (large population size 1600)	2 outbred lines	65	70.553	3.938	Weber and Diggins 1990
	Ethanol resistance (small population size 160)	4 outbred lines	65	44.369	4.738	
Mus domesticus	**Body weight at 6 weeks (high-weight selection line)**	**1 line**	**50**	**1.818**	**N/A**	**Keightley 1998**
	Body weight at 6 weeks (low-weight selection line)	**1 line**	**50**	**−3.909**	**N/A**	
	Wheel-running: total revolution/day in first 2 weeks	**4 lines**	**10**	**50**	**5.68**	**Swallow et al. 1998**
Saccharomyces cerevisiae	Growth rate in test tube-sexual populations	5	200	0.378	0.197	Grimberg and Zeyl 2005
	Growth rate in test tube-asexual populations	5	250	0.221	0.046	
	Growth rate in mouse-sexual populations	4	260	0.197	0.023	
	Growth rate in mouse-asexual populations	4	150	0.659	0.0987	
øX 174	Growth at high temperature (42°C)	5	(11 days, ~3000 genome replications)	1755.6	447.3	Bull et al. 1997

[a] Experimental evolution studies highlighted in **bold** employed the directional selection approach while those in normal font used the adaptation approach.

clonal asexual reproduction (for microorganisms), selfing (e.g. *C. elegans* and *A. thaliana*), or employing balancer chromosomes to prevent recombination (*D. melanogaster*). Limiting recombination in these populations is crucial for understanding the effects of mutation alone on specific phenotypes. Indeed, follow-up analyses of progeny from crosses using mutation accumulation lines as parental strains often showed rapid recovery of population fitness (e.g. Chao et al. 1992). Results from these mutation accumulation experiments clearly demonstrated the importance of sex and recombination in maintaining population fitness.

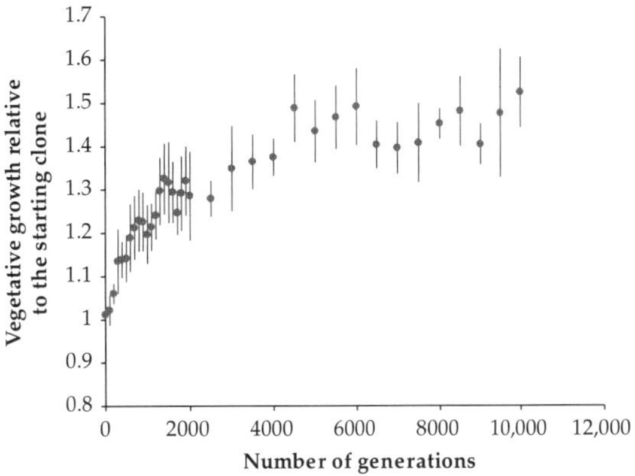

Figure 5.2 Changes in mean relative fitness of evolving *E. coli* populations over 10,000 generations. The x-axis shows the number of generations. The y-axis shows fitness of evolving *E. coli* populations with the fitness of the ancestral population scaled to 1. Note the relatively rapid change during the first 2000 generations. Vertical lines indicate standard deviations calculated from 12 experimental populations. The original data for this graph were from Lenski (2012; http:/myxo.css.msu.edu/ecoli).

5.5 Adaptation and directional selection experiments

In contrast to mutation accumulation studies discussed earlier where no or little selective pressures are applied to the evolving experimental populations, significant selective pressures are usually applied to populations under adaptation and directional selection regimes. There are many examples of experimental evolutionary studies that deal with adaptation and directional selection. Listed in Table 5.2 are a few examples representing both a broad group of organisms and a diverse spectrum of traits. In these studies, noticeable changes in the selected phenotypic traits were evident in the evolved populations. As expected, the overall rate of change in adaptation/directional selection experiments was greater than those seen in mutation accumulation experiments.

5.5.1 Adaptation of *E. coli* populations

In one of the most elegant series of experimental evolutionary studies, Lenski and colleagues performed a series of experiments to examine the relative roles of adaptation, chance, and historical contingency in shaping phenotypic and genotypic diversity among populations of *E. coli*. Twelve populations of *E. coli* were permitted to evolve in a novel sugar (arabinose) environment continuously for over 20 years in batch culture with regular transfers to fresh medium in order to maintain active growth. Interestingly, all 12 populations showed much greater fitness gains at the beginning of the experiment (Lenski et al. 1991), with the rate of fitness increase decreasing over time (Table 5.2, Fig. 5.2). Recent investigations showed that the diminishing returns were largely due to negative epistasis among the newly accumulated beneficial mutations (Khan et al. 2011). Indeed, negative epistasis is commonly observed among beneficial mutations in diverse organisms (Weinreich et al. 2005; Chou et al. 2011).

Despite negative epistasis, fitness gains were still evident in evolving *E. coli* populations even after 40,000 generations under the same growth condition (Barrick et al. 2009). In addition, similar rates and patterns of fitness gain were found for populations of *E. coli* adapting to two different environments, a minimum medium with arabinose as the carbon source and a high temperature environment at 42°C (Table 5.2; Bennett and Lenski 1999 Lenski et al. 1991).

5.5.2 Adaptation of viral populations

Among the other studies shown in Table 5.2, the most remarkable change in phenotype was the adaptation of the bacteriophage øX174 to a high temperature stressful environment at 42°C (Bull et al. 1997). Through 11 days of adaptation that corresponds to about 3000 phage generations, phage growth rates improved by over 5000-fold. In addition, because øX174 has a small genome, Bull and colleagues analyzed the specific changes in DNA sequences in each of the five replicate populations. Sequencing results indicated that approximately half of the changes occurred in more than one of the lineages. This study indicated that the same evolutionary changes could appear repeatedly in populations exposed to the same novel environmental stress conditions.

5.5.3 Adaptation and directional selection in fruit flies

In the fruit fly, rapid responses to both directional selection of bristle numbers and to adaptation to ethanol stress were observed. In the directional selection experiments to both low and high numbers of bristles (for both sternopleural and abdominal bristle numbers), the rates of change were over 10 times higher than those found in mutation accumulation lines (McKay et al. 1992, 1994). The response to ethanol resistance in fruit flies was even greater, with a faster rate of change in larger populations than that in smaller populations (Weber and Diggins 1990), consistent with the hypothesis that mutation supply was an important factor in population adaptation. A larger population provides a greater number of mutation targets and mutations per generation than a smaller population.

5.5.4 Adaptation in yeast

In addition to population size, other factors such as the mode of reproduction and the heterogeneity of environmental conditions can also influence the rate of adaptation and the fixation of sexual and asexual strains. For example, Grimberg and Zeyl (2005) tested the advantages of sex in *S. cerevisiae* strains in two contrasting environments: a homogeneous *in vitro* laboratory environment of minimal medium in test tubes, and a relatively heterogeneous *in vivo* environment of a mouse brain. They established initial populations as equal mixtures of sexual and obligate asexual genotypes. In test tubes where no sex was induced, the sexual genotypes were fixed in all five replicates within 250 mitotic generations, whereas in mouse brain with no sex induced, the asexual genotype was fixed in all four replicates by 170 generations. However, inducing sex altered the outcomes in opposite directions, i.e. decreasing the fixation frequencies of the sexual genotypes in test tubes but increasing them in mice. The rate of change and the rate of divergence among replicate lines followed similar patterns (Table 5.2).

5.5.5 Directional selection in mammals

Similar to mutation accumulation experiments, mammals have also been used as models for experimental evolution in adaptation and directional selection response studies. Two such studies are briefly discussed here. In the first, Swallow et al. (1998) selected populations of mice on the basis of their voluntary running behavior. Mice that ran the most in cage wheels were mated with each other. After 10 generations of selection and breeding, these lineages were compared to controls that were bred at random for their voluntary running behavior. Evan after only 10 generations, the voluntary running ability of the selected mice approximately doubled and their maximum oxygen consumption in the selected mice similarly increased. In the second study, Keightley (1998) divergently selected a long-established inbred strain of mice for both high and low body weight for 50 generations. His selection of new mutations affecting body weight eventually led to a divergence of approximately three phenotypic standard deviations between the high and low lines. Heritability for body weight increased at a rate between 0.23–0.57% per generation from new mutations. Because the response to selection was episodic, a substantial contribution from the selection was likely due to mutations with large effects on the trait. The analyses of data from a cross between the selected high line and an unselected control line indicated that two major loci were involved, with the potential of an additional minor locus. Together, these populations and studies can serve as models to test more general evolutionary theories concerning performance trade-offs in mammals including humans.

5.5.6 Correlated changes between traits

Often, phenotypic changes in one character can impact another character. For example, Rose and colleagues (1990) found trade-offs between early reproduction and longevity in experimental evolution populations. Specifically, they selected some populations of fruit flies for early reproduction and prevented others from reproducing until later in life. As predicted, they found that lifespan increased significantly in those that reproduced late in life. Interestingly, and somewhat unanticipated, the longer-lived populations also evolved greater resistance to some types of environmental stresses, including desiccation and starvation. When selection for delayed reproduction was removed, longevity decreased, as did stress resistance. These studies have greatly influenced our thinking about aging and its evolutionary and physiological bases.

5.5.7 Acquisition of novel phenotypes

The previous examples all examined the quantitative changes of existing phenotypes. However, the origin of novel traits can also be studied using experimental evolution systems. The most common trait studied in this category is probably the gain of resistance to antibiotics by infectious pathogens. This topic has been reviewed extensively since the emergence of antibiotic resistance in the 1950s, soon after the widespread application of these agents in treating infectious diseases in humans, and later for protecting vegetables, crops, forests, and farm animals (Levin and Anderson 1999; Rowe-Magnus and Mazel 2006). Significant variations in the potential for antibiotic resistance have been found among different microbial pathogen species as well as among strains within the same species (e.g. Xu et al. 2001). Aside from antibiotic resistance, the gain of novel carbon source utilizations was also examined in experimental populations of the bacterium, *Pseudomonas aeruginosa* (MacLean and Bell 2003). Their results showed a fourfold higher rate of acquiring the ability to utilize a new carbon source through adaptation than losing one through mutation accumulation (Tables 5.1 and 5.2).

The origination of new phenotypes can be impacted by many factors such as species and strain genetic background, biotic and abiotic environmental factors, the availability of genes coding for such phenotypes in other species and strains in the immediate environments, and the complexity of the phenotypic traits themselves. For example, in the 12 evolving *E. coli* populations maintained by Lenski and colleagues, Blount et al. (2008) identified that none of the 12 populations evolved the capacity to utilize citrate, a substrate in the medium since the beginning of the experiment in 1988, until about 31,500 generations in one of the 12 populations. Two hypotheses were proposed for the long-delayed and unique evolution of this function. In the first, citrate utilization is an uncommon trait and only extremely rare mutations could render the strain capable of utilizing citrate. In the second, the mutation to utilize citrate may be ordinary and similar to many other traits. However, its occurrence or functional expression might be contingent on other prior mutations in the population. Blount et al. (2008) tested these two hypotheses in experiments that 'replayed' evolution from different points in that population's history. They found no citrate-utilizing (Cit+) mutants among 8.4×10^{12} screened ancestral cells, nor among the 9×10^{12} cells from 60 clones sampled in the first 15,000 generations. Interestingly, a significantly greater tendency for later clones to evolve Cit+ was found, consistent with some potentiating mutation(s) that arose by about 20,000 generations. Furthermore, this potentiating change seemed very specific, only increased the mutation rate to Cit+ and did not cause generalized hypermutability. The authors concluded that the evolution of the Cit+ phenotype was contingent on the particular history of the population that gave rise to citrate utilization (Blount et al. 2008).

Overall, these results are consistent with the genomic and metabolic flexibilities of microorganisms in their adaptation to utilizing novel substrates and to coping with environmental stresses, including antibiotics.

5.6 Genomic analysis of experimental evolution populations

Most experimental evolution studies so far have focused on phenotypic changes over time. In the last few years, due to the arrival of high-throughput next-generation sequencing in addition to newer and even more efficient sequencing technologies, genome analyses of experimental evolution populations are beginning to emerge. For example, Barrick et al. (2009) examined the relationship between rates of genomic evolution and phenotypic changes in populations of *E. coli* that have undergone long-term adaptation to the arabinose carbon source. They sequenced genomes sampled through 40,000 generations of adaptation and found that although the rate of fitness gains in the arabinose medium changed and decelerated sharply, genomic evolution was nearly constant. In addition, while such clock-like genome evolution regularity is usually viewed as the signature of neutral evolution, their analyses suggested that almost all of these mutations were beneficial. In contrast, the population that later evolved an elevated mutation rate (due to mutation in DNA repair pathways) and accumulated hundreds of additional mutations was dominated by signatures of neutral mutation. The experimental results thus suggest that the relationship between genome evolution and adaptive phenotype evolution is complex and can be counterintuitive even in a constant environment such as that in the long-term evolving *E. coli* populations. In particular, results from this study indicated that while adaptive phenotypic changes in evolving *E. coli* populations were largely episodic, beneficial substitutions were surprisingly uniform over time (Barrick et al. 2009). In contrast, neutral substitutions were highly variable due to unpredictable mutations in genes involved in DNA-repair pathways.

5.7 Conclusions and perspectives

Over the last 50 years, experimental evolutionary studies have become a mainstream method for testing a variety of hypotheses and addressing many biological questions that have often been difficult to investigate through other approaches. Indeed, these studies have not only complemented our analyses of natural populations and communities, but also provided novel insights and direct empirical observations about evolution by natural selection. The representative studies discussed here showed that even in relative short periods of time, from a few days to weeks and months, rapid changes can be observed in experimental populations and that these populations can diverge significantly from each other.

In situations such as those during the initial introduction of a new species into a novel environment or after a catastrophic event (e.g. hurricane, flooding, wild fire), rapid evolution may occur and be comparable to what we observe in laboratory settings. However, caution should be placed on using results from controlled laboratory experimental evolution studies to directly infer the rates of evolution in natural populations. One caveat is that phenotypic changes in laboratory experimental populations are often unidirectional, regardless whether the experimental populations undergo mutation accumulation (which almost universally result in fitness decline) or adaptation/directional selection (which would change according to the direction of selection). Yet in nature, environmental conditions often fluctuate and the direction of phenotype change may change as well. As a result, some of those phenotypic changes may cancel each other to yield lower rates of changes over the long term in nature than those observed in laboratory experimental populations.

So far in experimental evolution studies, the common traits examined have been those relatively easily quantifiable phenotypes, with many directly related to fitness. In addition, most of the study organisms are model species that we already have substantial ecological, genetic, and genomic information. However, with increasing technical advances in both hardware and software, many other biological traits in non-model organisms will likely be examined through experimental evolution studies. Furthermore, with ever decreasing costs, sequencing experimental evolution populations will likely be common in the coming years, thus ushering in an exciting phase of experimental evolutionary studies.

Comparisons between genome sequence evolution and phenotypic changes will provide unprecedented information to address many fundamental and applied questions. For example, while phenotypic changes differ in both the patterns and the rates between mutation accumulation populations and adaptation/directionally selected populations, do evolutionary patterns observed in the genomes also differ between these two types of experimental evolution populations? How frequent are epistatic interactions between novel mutations? Are there differences in the patterns of epistatic interactions among beneficial mutations, among deleterious mutations, and between beneficial and deleterious mutations? Are epistatic interactions responsible for the episodic phenotypic changes in both mutation accumulation and adaptation experiments? What is the impact of sex and recombination in genome structural changes in evolving experimental populations? How do those changes influence phenotypes differently between sexual and asexual populations? And, how can we prevent the degeneration of genetic stocks used in agriculture, forestry, animal husbandry, and food manufacturing? Many of these stocks are maintained in laboratory settings. Understanding their rates of evolution in artificial settings will help us design optimal strategies to conserve and even increase their qualities for future applications.

Acknowledgments

Research in my laboratory on experimental evolution has been supported by Natural Sciences and Engineering Research Council (NSERC) of Canada.

References

Barrick, J.E., Yu, D.S., Yoon, S.H., Jeong, H.Y., Oh, T.K., Schneider, D., et al. (2009) Genome evolution and adaptation in a long-term experiment with Escherichia coli. *Nature* **461**: 1243–9.

Bennett, A.F. and Lenski, R.E. (1999) Experimental evolution and its role in evolutionary physiology. *Amer Zool* **39**: 346–62.

Blount, Z.D., Borland, C.Z., and Lenski, R.E. (2008) Historical contingency and the evolution of a key innovation in an experimental population of Escherichia coli. *Proc Natl Acad Sci USA* **105**: 7899–906.

Bull, J.J., Badgett, M.R., Wichman, H.A., Huelsenbeck, J.P., Hillis, D.M., Gulati, A., et al. (1997) Exceptional convergent evolution in a virus. *Genetics* **147**: 1497–507.

Chao, L. (1990) Fitness of RNA virus decreased by Muller's ratchet. *Nature* **348**: 454–5.

Chao, L., Trang, T., and Matthews, C. (1992) Muller's ratchet and the advantage of sex in the RNA virus f6. *Evolution* **46**: 289–99.

Chou, H.H., Chiu, H.C., Delaney, N.F., Segre, D., and Marx, C.J. (2011) Diminishing returns epistasis among beneficial mutations decelerates adaptation. *Science* **332**: 1190–2.

Dallinger, W.H. (1887) The president's address. *J Roy Microscop Soc* **April**: 185–99.

Darwin, C. (1859) *On the origin of species*. London: Murray.

Garland, T. Jr. and Rose, M.R. (2009) *Experimental Evolution: Concepts, Methods, and Applications of Selection Experiments*. Berkeley, CA: University of California Press.

Grimberg, B. and Zeyl, C. (2005) The effects of sex and mutation rate on adaptation in test tubes and to mouse hosts by Saccharomyces cerevisiae. *Evolution* **59**: 431–8.

Houle, D., Hoffmaster, D.K., Assmacopoulos, S., and Charlesworth, B. (1992) The genomic mutation rate for fitness in Drosophila. *Nature* **359**: 58–60.

Johanssen, W. (1903) *Über Erblichkeit in Populationen und in reinen Linien*. Jena: Gustav Fischer (Cited in Garland and Rose 2009).

Keightley, P.D. and Caballero, A. (1997) Genomic mutation rates for lifetime reproductive output and lifespan in Caenorhabditis elegans. *Proc Natl Acad Sci USA* **94**: 3823–7.

Keightley. P.D. (1998) Genetic basis of response to 50 generations of selection on body weight in inbred mice. *Genetics* **148**: 1931–9.

Khan, A.I., Dinh, D.M., Schneider, D., Lenski, R.E., and Cooper, T.F. (2011) Negative epistasis between beneficial mutations in an evolving bacterial population. *Science* **332**: 1193–6.

Kibota, T. and Lynch, M. (1996) Estimate of the genomic mutation rate deleterious to overall fitness in Escherichia coli. *Nature* **381**: 694–6.

Lenski, R.E., Rose, M.R., Simpson, S.C. and Tadler, S.C. (1991) Long-term experimental evolution in Escherichia coli. I. Adaptation and divergence during 2,000 generations. *Amer Nat* **138**: 1315–41.

Lenski, R.E. (2012) The *E. coli* long-term experimental evolution project site. http://myxo.css.msu.edu/ecoli

Levin, B.R. and Anderson, R.M. (1999) The population biology of anti-infective chemotherapy and the evolution of drug resistance: more questions than answers. In S.C. Stern (Ed.) *Evolution in Health and Disease*, pp. 125–37. Oxford: Oxford University Press.

Mackay, T.F.C., Lyman, R.F., Jackson, M.S., Terzian, C., and Hill, W.G. (1992) Polygenic mutation in Drosophila melanogaster: estimates from divergence among inbred strains. *Evolution* **46**: 300–16.

Mackay, T.F.C., Fry, J.D., Lyman, R.F., and Nuzhdin, S.V. (1994) Polygenic mutation in Drosophila melanogaster: estimates from response to selection of inbred strains. *Genetics* **136**: 937–51.

MacLean, R.C. and Bell, G. (2003) Divergent evolution during an experimental adaptive radiation. *Proc R Soc Lond B* **270**: 1645–50.

Mukai, T. (1964) The genetic structure of natural populations of Drosophila melanogaster. I. Spontaneous mutation rate of polygenes controlling viability. *Genetics* **50**: 1–19.

Rose, M.R., Graves, J.L., and Hutchison, E.W. (1990) The use of selection to probe patterns of pleiotropy in fitness characters. In F. Gilbert (Ed.) *Insect Life Cycles: Genetics, Evolution and Co-ordination*, pp. 29–40. New York: Springer.

Rowe-Magnus, D. and Mazel, D. (2006) The evolution of antibiotic resistance. In H.S. Seifert and V.J. DiRita (Eds) *Evolution of Microbial Pathogens*, pp. 221–42. Washington, DC: American Society for Microbiology Press.

Schultz, S.T., Lynch, M., and Willis, J.H. (1999) Spontaneous deleterious mutation in Arabidopsis thaliana. *Proc Natl Acad Sci USA* **96**: 11393–8.

Shearman, J.R. and Wilton, A.N. (2011) Origins of the domestic dog and the rich potential for gene mapping. *Genet Res Int* article ID 579308, doi:10.4061/2011/579308.

Swallow, J.G., Garland, T. Jr., Carter, P.A., Zhan, W.Z., and Sieck, G.C. (1998) Effects of voluntary activity and genetic selection on aerobic capacity in house mice (Mus domesticus). *J Appl Physiol* **84**: 69–76.

Weber, K.E. and Diggins, L.T. (1990) Increased selection response in larger populations. II. Selection for ethanol vapor resistance in Drosophila melanogaster at two population sizes. *Genetics* **125**: 585–97.

Weinreich, D.M., Watson, R.A. and Chao, L. (2005) Perspectives: Sign epistasis and constraint on evolutionary trajectories. *Evolution* **59**: 1165–74.

Xu, J. (2002) Estimating the spontaneous mutation rate of loss of sex in the human pathogenic fungus Cryptococcus neoformans. *Genetics* **162**: 1157–67.

Xu, J. (2004) Genotype-environment interactions of spontaneous mutations affecting vegetative fitness in the human pathogenic fungus Cryptococcus neoformans. *Genetics* **168**: 1177–88.

Xu, J., Onyewu, C., Yoell, H.J., Ali, R.Y., Vilgalys, R., and Mitchell, T.G. (2001) Dynamic and heterogeneous mutations to fluconazole resistance in Cryptococcus neoformans. *Antimicrob Agents Chemother* **45**: 420–7.

Zeyl, C. and De Visser, J.A.G.M. (2001) Estimates of the rate and distribution of fitness effects of spontaneous mutation in Saccharomyces cerevisiae. *Genetics* **157**: 53–61.

PART II

Rapidly Evolving Genetic Elements

CHAPTER 6

Rapid evolution of low complexity sequences and single amino acid repeats across eukaryotes

Wilfried Haerty and G. Brian Golding

6.1 Introduction

According to the lock and key paradigm, the functionality of a protein region is defined by its propensity to fold into a stable three-dimensional structure. Unexpectedly, the most commonly shared polypeptide sequence among eukaryotes proteomes is composed of the repetition of few or a single amino acid that cannot fold into an identifiable stable three-dimensional structure with the consequence of an under-representation of these simple sequences in the protein databases (Huntley and Golding 2002). These simple sequences can be grouped into low complexity regions (LCRs), single amino acid repeats (homopolymers), as well as intrinsically disordered regions. The terms 'low complexity sequences' and 'intrinsically disordered regions' are often used interchangeably although they specify regions with different properties. 'Low complexity sequences' are sequences of low information content, whereas 'intrinsically disordered regions' specify sequences that are unfolded in their native state but fold into an ordered structure upon binding to another protein (Haerty and Golding 2010a).

The proportion of proteins including at least one low complexity sequence varies greatly among eukaryotes, from about 13% in *Caenorhabditis elegans* up to 34% in *Dictyostellium discoideum* (Haerty and Golding 2010a). Depending upon the parameters used to detect low complexity sequences up to 95% of the proteins in *Plasmodium falciparum* contain at least one simple sequence. These regions of low information content are known to diverge quickly both within and between species. Furthermore, because of the lack of detectable structure and the low information content of these regions, such sequences have been deemed non-functional and to evolve neutrally. However, the size variation of numerous LCRs has direct effects on the physiology and behavior of several species and in humans, it is also known to be associated with genetic disorders. Because of the rapid evolution of these sequences and their potential phenotypic effects as well as their influence upon their surrounding sequences, some authors proposed that these simple sequences act as 'evolution knobs', allowing in some cases rapid adaptation (Kashi and King 2006).

6.2 Rapid evolution of low complexity sequences

Simple sequences, including low complexity sequences, trinucleotide repeats, and intrinsically disordered regions are known to diverge rapidly between closely related species (Ellegren 2004). Likewise, variation in homopolymer content is found between the 12 fully sequenced *Drosophila* species. Depending upon the species analyzed, between 12–30% of the proteins have at least one homopolymer (Huntley and Clark 2007). These results found at the interspecific level also extend to within species; the analysis of 14 *Plasmodium falciparum* isolates shows that more than half of the 7711 aligned LCRs are variable in size in at least one isolate with a maximum size variation of 120 amino acids (Haerty and Golding 2011).

Rapidly Evolving Genes and Genetic Systems. First Edition. Edited by Rama S. Singh, Jianping Xu, and Rob J. Kulathinal.

6.2.1 Mutational processes

The estimated mutation rate of repeated sequences is 10–100,000 times greater than for point mutations (Gemayel et al. 2010). Because of the repetitive nature of low complexity sequences, the mutational processes involved are the same as those previously proposed to explain the evolution of microsatellite loci with the added selective constraints associated to coding sequences.

The main mutational process leading to the contraction or expansion of low complexity sequences is slipped strand mispairing. During replication, after the dissociation of the two strands, as a consequence of the repetitive nature of low complexity sequences, single-stranded DNA can fold and form stable hairpin loops which may cause the misalignment of the strands during the re-association. The size of the repeat will either contract if a hairpin is formed within the template strand or expand if the hairpin is located on the nascent strand (Ellegren 2004). This process is mostly modeled with the addition/subtraction of a single repeated unit in most of the mutational events. The probability that replication slippage occurs depends upon both the homogeneity and the size of the repeated sequence. Therefore, the occurrence of mutations within the repeat disrupting the motif leads to an increased stability by decreasing the opportunity for replication slippage to occur. Both theoretical and empirical observations confirmed this assumption (Kruglyak et al. 1998; Hancock and Simon 2005). Studies of repeated sequence instability within multiple taxa reported a significant correlation between repeat instability and the repeat homogeneity and length (Ellegren 2004).

Perfect repeats are proposed to evolve from a proto-repeat composed of the repetition of few motifs formed by single point mutation or insertions/deletion (indel) slippage following a double strand break (Buschiazzo and Gemmell 2006; Leclercq et al. 2010). This proto-repeat will act as a seed for the future repeat that will expand or contract due to replication slippage if a critical number of repeated units is reached. Through time the size of the repeat will vary due to slippage. As for microsatellites, there is an upper size boundary above which a mutational bias toward sequence contraction is expected and could partly be the consequence of the deleterious effects of the LCR (Hancock and Simon 2005; Usdin 2008). In addition the composition of the sequence will also change with the accumulation of point mutations which, in turn, will decrease the opportunity for replication slippage to occur. This latter process is expected to lead to the degeneracy of the repeat. However, it is still possible that enough repeated motifs remain to allow replication slippage to generate a new repeat (Buschiazzo and Gemmell 2006). This model of the formation of repeated sequences relies partly on a threshold size above which replication slippage will occur. Although this model has been widely accepted, a recent study showed that slippage can occur for repeats smaller than the expected threshold size (Leclercq et al. 2010).

Although a relationship between recombination rate and non-coding repeat sequences has been reported in multiple organisms, the causality of such a relationship is still unclear. The presence of repeats in recombination hotspots can be the consequence of the mutagenic effect of recombination (Ellegren 2004), or in contrast, the presence of long tandem repeats leading to the formation of Z-DNA can be the cause of double-strand breaks and recombination. However the role of recombination in low complexity sequence evolution has been increasingly suggested to explain large size variations between populations of the same species. Evidence for unequal crossing-over impacting the stability of low complexity sequences was first discovered for the poly-alanine repeat within the *HOXD13* gene whose size variation may cause synpolydactyly in human. In its normal state this repeat is known to be stable across generations and has a heterogeneous codon composition $(GCN)_n$, potentially reducing the probability for replication slippage to occur (Kruglyak et al. 1998).

Other factors either intrinsic or extrinsic to the repeat can also be directly involved in low complexity sequence instability. For instance, DNA demethylation has been found to directly affect the stability of trinucleotide repeats (Cleary et al. 2002). Furthermore, the distance to the origin of replication also affects the repeat stability: the closer the repeats are to the front of the origin of replication, the more unstable the repeats are (Cleary et al. 2002). Similarly, the orientation of the repeated sequence with respect to the origin of replication is also a factor influencing the repeat variation. For example, $(CTG)_n$ and $(CAG)_n$ did not show the same stability (Cleary et al. 2002). Transcription can also affect low complexity sequence stability. The formation of R-loops (RNA: DNA hybrid) at G+C rich regions during transcription leads to trinucleotide repeats instability as shown for both $(CAG)_n$ and $(CGG)_n$ repeats in human cells (Lin et al. 2010).

Size variation between repeats with the same codon composition can be influenced by the organism, the locus, as well as tissue specific factors. Differences in instability for the same repeat at the same locus have been found between cell lines depending upon their tissue of origin (Gomes-Pereira et al. 2001). Experiments have shown that repeats inserted in mouse are more stable than in human and longer repeated sequences are needed to observe repeat instability. Although most of the single unit mutational events will be targeted by the mismatch repair system (Harr et al. 2002), the repair mechanism has been proven ineffective for larger insertions/deletions in human (Panigrahi et al. 2010).

6.3 Rapid divergence of LCRs and their impact on surrounding sequences

One of the main characteristics of LCRs is their rapid divergence in size and amino acid composition both within and between species (Fig. 6.1). The repeat instability of the low complexity sequence/trinucleotide repeats is directly related to its length, composition, homogeneity, as well as the selective pressures acting on the genomic regions in which the LCR can be found. It is believed that the LCR will generally expand until it reaches a size above which it is deleterious (Hancock and Simon 2005). Because of the assumption of LCRs being non-functional and their rapid evolution, simple sequences are often considered to evolve neutrally.

Since the late 1990s, numerous studies have analyzed how simple sequences (trinucleotide repeats, sequences of low information content, intrinsically disordered regions) vary both within and between species. With the increasing number of fully sequenced genomes of closely related species, many reports have now validated previous observations made on a handful of loci. We have gained a better understanding of the forces shaping the divergence of these sequences between species. Furthermore, we now realize that the variation of low complexity sequences and trinucleotide repeats and more generally any repeats can have significant effects on their surrounding regions and in some cases on the overall genome structure.

6.3.1 LCRs as indicators of regions of lowered purifying selective pressures

Early in the study of microsatellites, observations were made of an increased polymorphism in the flanking sequences of the repeats, characterized by an increased density of polymorphic sites near the repeat boundaries (Brohede and Ellegren 1999). These observations have now been confirmed in multiple organisms for both coding and non-coding repeated sequences (Haerty and Golding 2011; Siddle et al. 2011).

Owing to their rapid divergence both in size and amino acid composition between species, low complexity sequences have long been considered to evolve neutrally or under relaxed constraints. These conclusions stem from the analyses of their composition, as well as the divergence between species of both low complexity sequences and their flanking regions.

Distribution biases of low complexity sequences between and within proteins can be interpreted as the consequence of lowered selective constraints. In most of the eukaryotes studied thus far low complexity sequences are mainly composed of hydrophilic amino acids (Huntley and Clark 2007), and are most often located on the surface of the proteins in contact with the solvent, away from the core of the protein in which buried residues are known to be under strong purifying selection. In addition to the preferential distribution of LCRs on the periphery of the three-dimensional structure of the protein, LCRs are enriched within alternatively spliced exons in nematode, fruitfly, zebrafish, mouse, and human relative to constitutively spliced exons (Haerty and Golding 2010b). Because alternatively spliced exons are found only

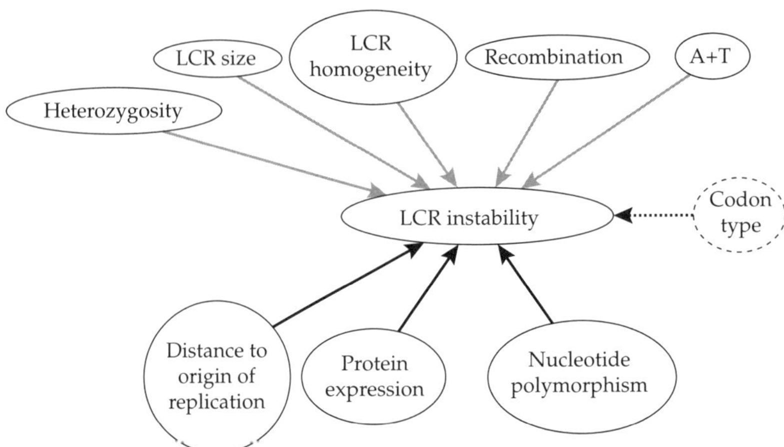

Figure 6.1 Relationship between low complexity sequence variation and different factors. Gray arrows indicate positive correlations, black arrows represent negative relationships, and dashed lines represent factors for which both positive and negative relationship with LCR instability are known.

in a fraction of the mature mRNA transcripts in comparison to constitutively spliced exons, lower selective constraints are expected to act on alternatively spliced exons (Haerty and Golding 2009). As expected within alternatively spliced exons, the codon composition of low complexity sequences is more homogeneous than that within constitutively spliced exons and LCRs are also more variable in size between species than those found within constitutively spliced exons (Haerty and Golding 2010a).

Among the many factors proposed to affect coding sequence evolution, gene expression has been found to be the major explanatory variable of sequence divergence between species. All observations thus far found highly expressed genes to be under stronger selective constraints than lowly expressed genes. In *P. falciparum* we found a negative correlation between the protein expression level and repeat size variation between isolates (Haerty and Golding 2011), as low complexity sequences within highly expressed genes are less variable than similar simple sequences within genes with lower expression levels. This observation, once again, agrees with the hypothesis of lowered selective constraints acting on low complexity sequences.

Likewise, the flanking sequences of low complexity sequences also show evidence for relaxation of selection. Using a small data set composed of human and mouse orthologs, Hancock et al. (2001) showed an increased substitution rate in the flanking sequences of homologous low complexity sequences. These results have since then been confirmed based on data from genome-wide analyses (Faux et al. 2007). A recent analysis of multiple *P. falciparum* genomes revealed an increasing density of single nucleotide polymorphisms towards the boundaries of low complexity sequences (Haerty and Golding 2011). Using a sample set of 31 proteins previously tested for evidence of selection (Huntley and Golding 2006), Huntley and Clark also showed that the flanking sequences of repeats affected by replication slippage were more divergent between species than sequences flanking repeats under stronger selective pressures. More generally, proteins hosting low complexity sequences are found to diverge faster between species than proteins without LCRs, even after removal of the simple sequences and controlling for functional annotation effects (Huntley and Clark 2007; Haerty and Golding 2011).

6.3.2 Mutagenic effect of LCRs

As mentioned earlier, it is likely that relaxation of selection may play an important role in the increased polymorphism within species and divergence between species of LCRs flanking sequences.

However, we cannot rule out a potential mutagenic effect of repeated sequences.

The increased polymorphism near low complexity sequence boundaries can be detected up to 150 nucleotides away from the repeats themselves in *P. falciparum*. Tian et al. (2008) reported similar ranges of increased polymorphism with respect to indels in multiple species including yeast, rice, *Drosophila*, rodents, and primates. The authors also found an increased SNP density with increasing size and abundance of indels. Using yeast genomes Tian et al. (2008) found an almost 35-fold increased mutation rate at heterozygous indels. The authors, after rejecting the potential mutagenic effects of recombination, suggest that heterozygosity at an indel is directly responsible for the increased polymorphism found in the surrounding DNA. According to this hypothesis, the presence of heterozygous indels locally affects chromosomal pairing leading to replication errors in the flanking sequences of the indel. This hypothesis has since been confirmed from the analysis of polymorphism relative to indels in plants with different rates of selfing. According to this model, the density of polymorphic sites close to indels should decrease as the rate of self-fertilization increases because the time an allele spends in the heterzygous state is strongly reduced in comparison to dioecious species (Hollister et al. 2010). Accordingly, Hollister et al. (2010) found a decrease of single nucleotide polymorphism (SNP) density close to indels as the selfing rate increases. The same hypothesis of a mutagenic effect of indels has also been invoked to explain similar observations within the flanking sequences of microsatellites (Amos 2010) and it is likely that a similar mechanism is responsible for the increased mutation rate observed in the flanking sequences of low complexity sequences in *P. falciparum*. Recently, using a population genomic approach in *Escherichia coli*, *Saccharomyces cerevisiae*, *Drosophila melanogaster*, and humans, McDonald et al. (2011) reported that the increased nucleotide substitution rate near insertion/deletions were not the consequence of indels per se but instead were directly associated to repeated sequences. The authors proposed the 'Repeat-sequence-induced Recurrent Repair' hypothesis to explain the surge of mutations near repeats. Because of the potential of repeated sequences to stall the replication fork and to cause double strand breaks, every time such events occur the surrounding DNA will be resynthesized by the error-prone DNA repair mechanism.

LCR instability can spread to other repeats within the genome. Blackwood et al. (2010) found instability at a $(CAG)_n$ repeat promoted variation at a tandem repeat located about 6.3 kb away in the *Escherichia coli* genome. Interestingly, Huntley and Clark (2007) found an increased density of sites with evidence of positive selection in the vicinity of LCRs and homopolymers in *Drosophila* species. This suggests that mutations induced by low complexity sequences can be the target of selection.

6.4 Low complexity sequences under selection

6.4.1 Deleterious effects of LCR size variation

Low complexity sequences and, more specifically, trinucleotide repeats are best known for the association between their size variation and several genetic disorders in humans. At least 26 diseases and genetic syndromes are associated with trinucleotide repeats, of which 17 are found within coding sequences and they mainly code for polyglutamine and polyalanine repeats (Mirkin 2007). For each of these syndromes, the size of the trinucleotide repeats directly influences the severity and the age of onset of the disease and the deleterious effect of the repeats can be both at the RNA level for simple sequences found within the intron, UTRs, and promoter regions and at the protein level (Mirkin 2007). In this latter case, the presence of an expanded repeat can result in protein toxicity associated with misfolding and increased stability of the protein (Usdin 2008). In addition proteins with poly-Q or poly-A can aggregate with other glutamine or alanine rich proteins to inactivate these proteins and result in the formation of cytotoxic protein inclusions and cell death.

6.4.2 DNA composition

Other evidence for the action of selection acting on low complexity sequences is the enrichment of coding regions in trinucleotide and hexanucleotide repeats in comparison to non-coding sequences, possibly as a consequence of purifying selection for frame preservation. In addition, there is a non-random distribution between coding and non-coding sequences of trinucleotide repeats depending upon their sequences. $(CAG)_n$ and $(CCG)_n$ repeats are significantly enriched within coding sequences in comparison to non-coding sequences in mammals and *Drosophila* (Huntley and Clark 2007; Bacolla et al. 2008). Huntley and Clark (2007) also noted that while $(CAG)_n$ repeats encoding poly-Q are over-represented within coding sequences, $(CAA)_n$ repeats also coding for poly-Q are more common within non-coding sequences. In a similar fashion the instability of a trinucleotide repeat also depends upon its composition. In *P. falciparum*, although poly-lysine repeats are abundant, they are among the least variable repeats. Because these repeats are mainly composed of $(AAA)_n$ repeats, replication slippage will likely cause frameshift mutations beyond the LCR repeat, hence the low variability of these repeats is probably the consequence of purifying selection (Haerty and Golding 2011).

6.4.3 LCR distribution

At the gene level, there is a non-random distribution of low complexity sequences among genes depending upon their function. Genes involved in development, transcription, DNA/RNA binding, reproduction, or immunity are enriched in low complexity sequences (Huntley and Clark 2007). In contrast, genes with enzymatic functions, or involved in protein synthesis have a lower proportion of low complexity sequences (Huntley and Clark 2007). The nature of the simple sequence was also found to differ depending upon both the exon splicing pattern (Haerty and Golding 2010b) and the gene function as poly-E, poly-L are enriched among genes involved DNA binding and transmembrane activity respectively, whereas poly-A, poly-G, poly-P, poly-S, and poly-Q are found mostly in genes with a transcription factor activity (Simon and Hancock 2009).

Some single amino acid repeats can be highly conserved between species although their surrounding sequences are rapidly diverging (Gojobori and Ueda 2011). Polylysine, polyarginine polythreonine, and polyleucine repeats are found to be conserved in length between mammals (Gojobori and Ueda 2011). The same result was observed by Simon and Hancock (2009) for the polyleucine repeats among eukaryotes. The authors associated the conservation of the repeat to a potential signal function. Polyhistidine has been found to be directly involved in protein localization in the nuclear speckles and is relatively conserved in primates (Salichs et al. 2009).

In *Drosophila*, the distribution of the low complexity sequences is non-random as the N- and C-terminal ends of the proteins are enriched in simple sequences (Huntley and Clark 2007). This non-random distribution can also be found in humans. Generally the first translated exon is enriched in poly-Q and poly-E (Bacolla et al. 2008). In *P. falciparum* low complexity sequences are found at a greater distance to the exon boundaries than expected by chance. Many studies have found exon boundaries to be under strong selective constraints due to the presence of splicing regulatory elements (Warnecke and Hurst 2007), and hence purifying selection may prevent low complexity regions colonizing exon boundaries.

In addition to the greater evolutionary age of constitutively spliced exons, purifying selection can also help explain the increased codon heterogeneity of low complexity sequences within constitutively spliced exons compared to alternatively spliced exons. If size variation of low complexity sequences is deleterious, point mutations that decrease the opportunity for replication slippage to occur are likely to be favored (Haerty and Golding 2010b).

6.4.4 Phenotypic effects of LCR size variation

The size variation of simple sequences within several developmental genes is directly associated with morphological variation and evolutionary novelty in eukaryotes. For instance, the size variation of two single amino acid repeats (poly-Q, poly-A) within the *Runx2* gene is tightly associated with the variation of skull morphology between dog breeds (Fondon and Garner 2004). In a similar fashion, an increased size of the polyalanine repeat within the *HOXD13* gene, which is directly responsible for the synpolydactyly in mouse and human, is likely responsible for the increased number of digits in the ancestor of the cetaceans (Wang et al. 2009).

Among the many genes with simple sequences in eukaryotes, the *Clock* genes that are responsible for the regulation of the physiological rhythms including circadian rhythm have been well studied. Across eukaryotes, the CLOCK protein has a polyglutamine repeat whose deletion will impede the transcription of downstream genes. This observation was found across many eukaryotes including human, mouse, and fruit fly. In addition the existence of an association between the length of the poly-Q repeat and circadian rhythm duration suggests their roles in environmental adaptation. A latitudinal gradient is observed in both the Chinook salmon and birds for the size of the repeat with longer alleles found in high frequency in northern latitudes whereas shortened alleles are found in high frequency in southern latitudes. This variation in the size of the poly-Q tract likely reflects adaptation to different daylight durations between habitats (O'Malley and Banks 2008).

Among the genes involved in biological rhythm regulation, the PERIOD protein in *Drosophila* harbors a proline–glycine repeat whose length variation not only affect the circadian rhythm duration but also multiple components of reproductive behavior (Kyriacou et al. 2008). Once again a latitudinal cline in repeat length is found in *Drosophila* populations with longer alleles found in northern latitudes and this likely reflects the effect of ecological adaptive selection (Kyriacou et al. 2008). A significant difference in circadian rhythm duration is found between alleles depending upon their length, longer alleles leading to shorter but more thermally stable periods whereas shorter alleles result in longer circadian periods that are, however, less stable at low temperature (Kyriacou et al. 2008).

6.4.5 Selection for low information content

Although the low complexity sequences within coding sequences may not be the direct target of selection, the function of the low complexity sequence can be under the scrutiny of selection. Low complexity sequences have been proposed to act as linkers between protein domains (Huntley and Golding 2000). Clarke et al. (2003) demonstrated that in *Plasmodium berghei* the low complexity protein segment found between the glucose-6-phosphate dehydrogenase (G6PD) and 6-phosphogluconolactonase domains is essential for the function of the G6PD domain. The deletion of the LCR resulted in the inactivation of the G6PD domain and the function can be rescued by the insertion of a LCR from another species even though there is low amino acid conservation.

6.5 Perspectives

Low complexity sequences have been suggested by some to act as 'evolutionary knobs' and their instability can generate genetic variation upon which selection may act (Kashi and King, 2006). The potential mutagenic effect of LCRs on their surrounding sequences (Haerty and Golding 2011) and the increased density of sites under positive selection in the vicinity of LCRs (Huntley and Clark, 2007) strengthen the hypothesis that LCRs may have a significant impact in evolution. However, this hypothesis needs to be properly addressed. Thus far, with the exception of a handful of studies, most of the analyses performed have used a single genome per species with the assumption that the variation within species is negligible with respect to the between-species divergence. Although this is true for single nucleotide variation, because of the high mutation rate of LCR, this assumption need not always hold (Haerty and Golding 2011). The increasing number of fully sequenced genomes of individuals of the same species as well as of closely related species will provide us unprecedented power to assess the evolution of low complexity sequences within a phylogenetic framework.

Acknowledgments

This work was supported by a Natural Sciences and Engineering Research Council of Canada and Canada Research Chair grant to G.B.G.

References

Amos, W. (2010) Heterozygosity and mutation rate: evidence for an interaction and its implications: the potential for meiotic gene conversions to influence both mutation rate and distribution. *Bioessays* **32**(1): 82–90.

Bacolla, A., Larson, J.E., Collins, J.R., Li, J., Milosavljevic, A., Stenson, P.D., et al. (2008) Abundance and length of simple repeats in vertebrate genomes are determined by their structural properties. *Genome Res* **18**(10): 1545–53.

Blackwood, J.K., Okely, E.A., Zahra, R., Eykelenboom, J.K., and Leach, D.R. (2010) DNA tandem repeat instability in the *Escherichia coli* chromosome is stimulated by mismatch repair at an adjacent CAG.CTG trinucleotide repeat. *Proc Natl Acad Sci USA* **107**(52): 22582–6.

Brohede, J. and Ellegren, H. (1999) Microsatellite evolution: polarity of substitutions within repeats and neutrality of flanking sequences. *Proc R Soc Lond B Biol Sci* **266**(1421): 825–33.

Buschiazzo, E. and Gemmell, N.J. (2006) The rise, fall and renaissance of microsatellites in eukaryotic genomes. *Bioessays* **28**(10): 1040–50.

Clarke, J.L., Sodeinde, O., and Mason, P.J. (2003) A unique insertion in *Plasmodium berghei* glucose-6-phosphate dehydrogenase-6-phosphogluconolactonase: evolutionary and functional studies. *Mol Biochem Parasitol* **127**(1): 1–8.

Cleary, J.D., Nichol, K., Wang, Y.H., and Pearson, C.E. (2002) Evidence of cis-acting factors in replication-mediated trinucleotide repeat instability in primate cells. *Nat Genet* **31**(1): 37–46.

Ellegren, H. (2004) Microsatellites: simple sequences with complex evolution. *Nat Rev Genet* **5**(6): 435–45.

Faux, N.G., Huttley, G.A., Mahmood, K., Webb, G.I., de la Banda, M.G., and Whisstock, J.C. (2007) RCPdb: An evolutionary classification and codon usage database for repeat-containing proteins. *Genome Res* **17**(7): 1118–27.

Fondon, 3rd J.W. and Garner, H.R. (2004) Molecular origins of rapid and continuous morphological evolution. *Proc Natl Acad Sci U S A* **101**(52): 18058–63.

Gemayel, R., Vinces, M.D., Legendre, M., and Verstrepen, K.J. (2010) Variable tandem repeats accelerate evolution of coding and regulatory sequences. *Annu Rev Genet* **44**: 445–77.

Gojobori, J. and Ueda, S. (2011) Elevated evolutionary rate in genes with homopolymeric amino acid repeats constituting nondisordered structure. *Mol Biol Evol* **28**(1): 543–50.

Gomes-Pereira, M., Fortune, M.T., and Monckton, D.G. (2001) Mouse tissue culture models of unstable triplet repeats: in vitro selection for larger alleles, mutational expansion bias and tissue specificity, but no association with cell division rates. *Hum Mol Genet* **10**(8): 845–54.

Haerty, W. and Golding, G.B. (2009) Similar selective factors affect both between-gene and between-exon divergence in Drosophila. *Mol Biol Evol* **26**(4): 859–66.

Haerty, W. and Golding, G.B. (2010a) Low-complexity sequences and single amino acid repeats: not just "junk" peptide sequences. *Genome* **53**(10): 753–62.

Haerty, W. and Golding, G.B. (2010b) Genome-wide evidence for selection acting on single amino acid repeats. *Genome Res* **20**(6): 755–60.

Haerty, W. and Golding, G.B. (2011) Increased polymorphism near low complexity sequences across the genomes of *Plasmodium falciparum* isolates. *Genome Biol Evol* 3: 539–50.

Hancock, J.M. and Simon, M. (2005) Simple sequence repeats in proteins and their significance for network evolution. *Gene* **345**(1): 113–18.

Hancock, J.M., Worthey, E.A., and Santibanez-Koref, M.F. (2001) A role for selection in regulating the evolutionary emergence of disease-causing and other coding CAG repeats in humans and mice. *Mol Biol Evol* **18**(6): 1014–23.

Hollister, J.D., Ross-Ibarra, J., and Gaut, B.S. (2010) Indel-associated mutation rate varies with mating system in flowering plants. *Mol Biol Evol* **27**(2): 409–16.

Huntley, M. and Golding, G.B. (2000) Evolution of simple sequence in proteins. *J Mol Evol* **51**(2): 131–40.

Huntley, M.A. and Clark, A.G. (2007) Evolutionary analysis of amino acid repeats across the genomes of 12 Drosophila species. *Mol Biol Evol* **24**(12): 2598–609.

Huntley, M.A. and Golding, G.B. (2002) Simple sequences are rare in the Protein Data Bank. *Proteins* **48**(1): 134–40.

Huntley, M.A. and Golding, G.B. (2006) Selection and slippage creating serine homopolymers. *Mol Biol Evol* **23**(11): 2017–25.

Kashi, Y. and King, D.G. (2006) Simple sequence repeats as advantageous mutators in evolution. *Trends Genet* **22**(5): 253–9.

Kruglyak, S., Durrett, R.T., Schug, M.D., and Aquadro, C.F. (1998) Equilibrium distributions of microsatellite repeat length resulting from a balance between slippage events and point mutations. *Proc Natl Acad Sci U S A* **95**(18): 10774–8.

Kyriacou, C.P., Peixoto, A.A., Sandrelli, F., Costa, R., and Tauber, E. (2008) Clines in clock genes: fine-tuning circadian rhythms to the environment. *Trends Genet* **24**(3): 124–32.

Leclercq, S., Rivals, E., and Jarne, P. (2010) DNA slippage occurs at microsatellite loci without minimal threshold length in humans: a comparative genomic approach. *Genome Biol Evol* **2**: 325–35.

Lin, Y., Dent, S.Y., Wilson, J.H., Wells, R.D., and Napierala, M. (2010) R loops stimulate genetic instability of CTG.CAG repeats. *Proc Natl Acad Sci U S A* **107**(2): 692–7.

McDonald, M.J., Wang, W.C., Huang, H.D., and Leu, J.Y. (2011) Clusters of nucleotide substitutions and insertion/deletion mutations are associated with repeat sequences. *PLoS Biol* **9**(6): e1000622.

Mirkin, S.M. (2007) Expandable DNA repeats and human disease. *Nature* **447**(7147): 932–40.

O'Malley, K.G. and Banks, M.A. (2008) A latitudinal cline in the Chinook salmon (*Oncorhynchus tshawytscha*) *Clock* gene: evidence for selection on PolyQ length variants. *Proc Biol Sci* **275**(1653): 2813–21.

Panigrahi, G.B., Slean, M.M., Simard, J.P., Gileadi, O., and Pearson, C.E. (2010) Isolated short CTG/CAG DNA slip-outs are repaired efficiently by hMutSbeta, but clustered slip-outs are poorly repaired. *Proc Natl Acad Sci U S A* **107**(28): 12593–8.

Salichs, E., Ledda, A., Mularoni, L., Alba, M.M., and de la Luna, S. (2009) Genome-wide analysis of histidine repeats reveals their role in the localization of human proteins to the nuclear speckles compartment. *PLoS Genet* **5**(3): e1000397.

Siddle, K.J., Goodship, J.A., Keavney, B., and Santibanez-Koref, M.F. (2011) Bases adjacent to mononucleotide repeats show an increased single nucleotide polymorphism frequency in the human genome. *Bioinformatics* **27**(7): 895–8.

Simon, M. and Hancock, J.M. (2009) Tandem and cryptic amino acid repeats accumulate in disordered regions of proteins. *Genome Biol* **10**(6): R59.

Tian, D., Wang, Q., Zhang, P., Araki, H., Yang, S., Kreitman, M., et al. (2008) Single-nucleotide mutation rate increases close to insertions/deletions in eukaryotes. *Nature* **455**(7209): 105–8.

Usdin, K. (2008) The biological effects of simple tandem repeats: lessons from the repeat expansion diseases. *Genome Res* **18**(7): 1011–19.

Wang, Z., Yuan, L., Rossiter, S.J., Zuo, X., Ru, B., Zhon H., et al. (2009) Adaptive evolution of 5′HoxD genes in the origin and diversification of the cetacean flipper. *Mol Biol Evol* **26**(3): 613–22.

Warnecke, T. and Hurst, L.D. (2007) Evidence for a trade-off between translational efficiency and splicing regulation in determining synonymous codon usage in *Drosophila melanogaster*. *Mol Biol Evol* **24**(12): 2755–62.

CHAPTER 7

Fast rates of evolution in bacteria due to horizontal gene transfer

Weilong Hao

7.1 Introduction

In the traditional teachings of evolution there are three major components: the generation of genetic variation, the action of population genetic processes on this variation, and the inheritance of the remaining genetic variation. In the generation of genetic variation, the major processes are mutations generating *de novo* genetic changes and processes such as recombination and segregation that generate new combinations of existing variation. Mutation is generally a rare process such that rates of mutation at a single base are usually 10^{-7}–10^{-8} per base pair per generation or less. It is only within RNA viruses that mutation rates tend to be larger than 10^{-5} (Drake 1993), but such a large mutation rate limits the size of the genome that can be maintained (Lynch 2010). But it is these rare mutations that the processes of selection and random drift act upon, and the resulting change in variation that is inherited by the next generation.

Bacteria have been known to have the ability to change rapidly in new environments. They have a large repertoire of biochemical abilities, but do not seem to have to wait to modify existing genes via slow mutations but instead adapt more rapidly to changing conditions. As just one example, antibiotics were first used during the Second World War and yet within a short period resistant bacteria were discovered. A short time later the mechanism for this resistance was discovered to be due to an 'R-factor' (Hotchkiss 1951). This resistance factor to penicillin turned out to be a plasmid that carried an antibiotic resistance gene and had been transferred into pneumococci. Later, it was discovered that large regions could be transferred between organisms, that these regions need not necessarily be carried on plasmids, and that collections of genes with related functions could be transferred as a whole. When involved with antibiotic resistance or adaptations that might lead to disease, these regions were termed pathogenicity islands (Blum et al. 1994). Now, it is widely accepted that antibiotic resistance genes are derived from a large and diverse gene pool presumably already present in environmental bacteria and that the spread of antibiotic resistance is strongly associated with horizontal gene transfer (D'Costa et al. 2007).

More recent efforts to document the importance of horizontal gene transfer in bacterial and archaeal evolution (Garcia-Vallvé et al. 2000; Ochman et al. 2000; Koonin et al. 2001; Gogarten et al. 2002) suggest that horizontal gene transfer plays a larger role in prokaryotic genome evolution than previously thought. For instance, it has been shown that as much as 81% of genes in prokaryotic genomes have been acquired by horizontal gene transfer (Dagan et al. 2008), and genes from all functional categories are subject to transfer (Zhaxybayeva et al. 2006).

A particularly noteworthy culmination of the massive effect of horizontal gene transfer was presented by Welch et al. (2002) in a comparison of three genomes of *Escherichia coli*. They compared the gene complement of uropathogenic strain CFT073, enterohemorrhagic strain EDL933 (O157:H7), and laboratory strain MG1655 (K-12). Despite the very close phylogenetic relationships of these organisms, this comparison found only 2996 genes were shared among a total gene content of 5016, 5063, and 4288, respectively, and these 2996 genes only constitute 39.2% of the combined (non-redundant) set of 7638 genes (Welch et al. 2002).

Rapidly Evolving Genes and Genetic Systems. First Edition. Edited by Rama S. Singh, Jianping Xu, and Rob J. Kulathinal.

The small proportion of shared genes among closely related bacteria suggests that horizontal gene transfer is not only a potent force leading to the acquisition of new genes and new functions but in addition, it must be a rapid process. How rapidly can this process take place in naturally occurring bacteria? To answer this question required the addition of more data from more genomes and a framework to put these data into perspective. Here we review some of the major breakthroughs in recent studies.

7.2 Quantifying horizontal gene transfer

A straightforward way of quantifying horizontal gene transfer is to count the identified foreign genes within each genome (Table 7.1). For instance, Koonin et al. (2001) found that the percentage of horizontally transferred genes varies from near 0% to 32.6% in the 22 complete bacterial genomes that they examined. The percentage of horizontally transferred genes is about 24% in the bacterium *Thermatoga maritima*, and most of the foreign genes are of archaebacterial origins (Nelson et al. 1999). Another way of quantifying horizontal gene transfer is to summarize the phylogenetic incongruence between each gene tree and a trusted reference tree or between independently inferred gene trees. Phylogenetic conflicts are very common in cyanobacteria and horizontal gene transfer has occurred for about 61% of genes from all functional categories (Zhaxybayeva et al. 2006; and Table 7.1). Both methods are powerful in addressing the significance of horizontal gene transfer in each genome or each gene family, but they become limited on the understanding of the big picture of the horizontal gene transfer process. Let's assume that a certain fraction (x) of genes in each of the y examined genomes are involved in horizontal transfer, the probability to have a gene tree identical to the species tree is $(1-x)^y$. The high frequency of phylogenetic conflicts, i.e. small $(1-x)^y$, could be due to either high percentage (x) of horizontally transferred genes or a large number (y) of examined genomes, or a combination of both. In fact, due to the accumulation of horizontally transferred genes over evolutionary timescales, minimum estimates of horizontally transferred genes across phylogenetic groups are generally higher than the fraction observed within an individual genome (Table 7.1). To obtain a clearer big picture of bacterial genome evolution, various methods have been developed to quantify horizontal gene transfer by mapping the changes onto a phylogeny (Snel et al. 2002; Kunin and Ouzounis, 2003; Mirkin et al. 2003).

Table 7.1 Quantifying horizontal gene transfer (HGT) within individual genomes and cross phylogenetic groups

Genome[a]/group[b]	Genes involved in HGT: in each genome	Reference
Treponema pallidum	32.6%	Koonin et al. 2001
Chlorobium tepidum	24.1%	Nakamura et al. 2004
Thermotoga maritima	24%	Nelson et al. 1999
E. coli K-12	18%	Lawrence and Ochman 1998
Bacillus subtilis	14.5%	Garcia-Vallvé et al. 2000
	across the group[c]	
Bacteria (181)	81%	Dagan et al. 2008
Proteobacteria (329)	75%	Kloesges et al. 2011
Prokaryotes (190)	75%	Dagan and Martin 2007
Prokaryotes (100)	62%	Puigbò et al. 2009
Cyanobacteria (11)	61%	Zhaxybayeva et al. 2006
Cyanobacteria (13)	53%	Shi and Falkowski 2008
Bacteria (98)	40%	Kunin et al. 2005
Lactobacillus (5)	40%	Nicolas et al. 2007
Prokaryotes (63)	34%	Cohen and Pupko 2010
Bacteria (144)	26%	Beiko et al. 2005

[a]The most HGT-rich genome identified in each reference
[b]With the number of studied taxa in each group shown in parentheses
[c]Minimum estimates across the group

Horizontal gene transfers could also be identified by detecting genes with an uneven presence or absence pattern, since gene transfer can result in the addition of novel genes in a particular genome. This simple criterion for identifying horizontally transferred genes has become increasingly popular and powerful due to the fast accumulation of multiple, closely related genome sequences. The number of gene gains/losses was inferred and mapped onto a phylogeny first using a parsimony approach (Snel et al. 2002) and the results uncovered a fast turnover of gene content during bacterial genome evolution. Taking similar approaches, Kunin and Ouzounis (2003) observed a pattern with substantially more lineage-specific gene gains than losses and demonstrated that, in order to keep genome size roughly constant, two or three gene losses per gene gain are needed to explain the observed pattern from the proteobacteria data. In a separate study, however, Mirkin et al. (2003) showed that the number of gene gains is approximately the same as the number of gene losses on a phylogeny that covers bacteria, archaea, and eukaryote. The notable discrepancy between these two studies turned out to be due to the fact that many recently acquired genes are rapidly lost during evolution (Hao and Golding 2004) and the significantly different depths in their phylogenies.

A better way to measure rates of gene gains and gene losses was developed based on the maximum likelihood paradigm (Hao and Golding 2006). The rates of gene gain and gene loss were assumed to be equal. The results suggest that the rates of horizontal transfer are comparable to the rate of nucleotide substitution. This suggests that gene gains and losses are common in bacterial genome evolution (Fig. 7.1a). Furthermore, when rates of gene gain and loss were separated on different parts of the phylogeny, much higher rates were observed on the most recent branches. The observed faster changes on recent branches are the consequence of high rates of gene turnover. Most acquired genes are rapidly lost and they do not persist for long. As a result, most anciently acquired genes are eventually deleted. When we examine extant genomes, we find, not surprisingly, only the genes that have been retained, which are predominantly the ones that have been recently acquired. This not only confirmed the conclusion of fast gene turnover in bacterial genome evolution drawn from previous parsimony studies but also put it on a firmer quantitative basis.

7.3 Understanding the variation of gene gain and loss

The fast accumulation of genome sequences and the successful development of advanced mathematical models open the door for a thorough investigation of gene gain and loss in bacterial genome evolution. First, the rates of gene gains and losses may not necessarily be equal and constant. Spencer and Sangaralingam (2009) demonstrated that the evolutionary pattern of gene gains and losses in some reductive genomes involves heterotachy (i.e. the change of evolutionary rate along the edges of a phylogenetic tree). A similar study on archaea has shown that most lineages are characterized by a net loss of gene families and major increases in gene repertoire have occurred only a few times (Csuros and Miklos 2009).

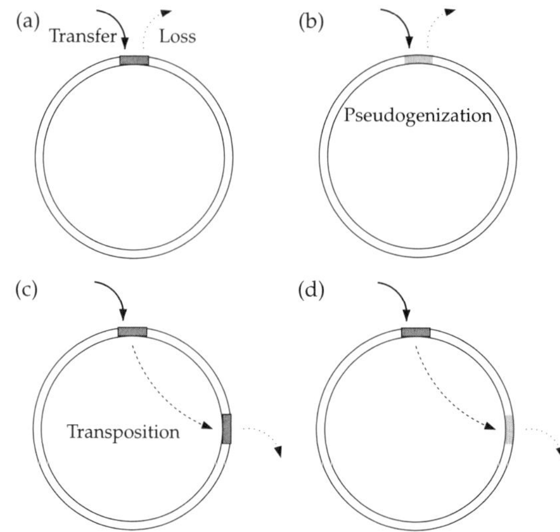

Figure 7.1 Cartoon model of gene movement during bacterial genome evolution. (a) Many recently acquired genes are deleted rapidly upon arrival. (b) Some horizontally transferred genes become pseudogenes and are deleted rapidly afterwards. (c) Some horizontally transferred genes are subject to transposition, and gene deletion consequently. (d) Transpositioned genes could be subject to pseudogenization and then gene deletion.

Second, the rates of gene gains and losses vary significantly across genes. For instance, many recently acquired genes are rapidly lost during evolution (Hao and Golding 2004), while some persist vertically for great lengths of time and even become characteristic of the group (Lerat et al. 2003). It has also been well documented that genes involved in metabolic pathways are more likely to be horizontally transferred than genes involved in information processing (Rivera et al. 1998). Incorporating variation in the rates of gene gains and gene losses, we have shown that while the difference in informational gene rates contributes to rate variation, it is only a small fraction of the variation present (Hao and Golding 2008b). We further demonstrated that a substantial amount of rate variation for gene gains and losses remains among both informational genes and among non-informational genes. Consistently, Cohen et al. (2011) also found a significant amount of rate variation for gene gains and losses across different gene functional categories.

Rate variation for gene gains and losses has also been accessed with respect to other gene propensities. For instance, in *E. coli/Shigella*, genes acquired from distantly related bacterial groups are less likely to persist than the gene with no existing homologs, known as ORFans (van Passel et al. 2008). The different persistence potentials in a genome suggest that many adaptive traits are conferred by completely novel genes that do not originate in distant bacterial genomes. Another factor concerns the number of interactions in which the encoded protein is involved. Recently acquired genes have fewer interaction partners compared to native genes in both regulatory and protein interaction networks, suggesting that transferred genes are gradually integrated into the regulatory network of their host over slow evolutionary time (Lercher and Pal 2008). In fact, connectivity has been shown to be an important and statistically significant factor in determining transferability (the ability for a gene to be horizontally transferred and maintained) (Cohen et al. 2011). Despite having fewer interaction partners, horizontally acquired proteins contain significantly more putative interaction sites than native proteins, so that the persisted ones would have a greater chance of forming new interactions in new species, thus integrating into existing networks (Gophna and Ofran 2011).

7.4 Horizontal gene transfer in duplicated genes

Gene duplication is believed to play a major role in genome evolution. In bacterial genome evolution, duplicated genes contribute to gene family expansion and hence genome size expansion (Ranea et al. 2004). However, duplicated genes were often ignored in most horizontal gene transfer studies mainly due to the high possibility of introducing misleading phylogenetic signals (Gogarten and Townsend 2005). In contrast to the rich literature on patterns of gene transfer, the association between gene transfer and duplication has not been extensively examined. In recent years, some evidence has emerged that there is a significant association between gene duplication and horizontal gene transfer, and this association is an important component of the rapid bacterial genome evolution.

Using both compositional and phylogenetic approaches, Hooper and Berg (2003) identified a number of proposed horizontally transferred genes and examined their frequency of duplication. Their results revealed that duplications are significantly over-represented among horizontally transferred genes compared to the native ones. The study suggested that horizontal transfer may accelerate the evolutionary process of duplication by bringing foreign genes that have mainly weak or no function into the genome (Hooper and Berg 2003). A similar pattern has been found in a recent study that a low duplicability of a gene is linked to a lower chance of being horizontally transferred (Wellner et al. 2007).

More recently, Treangen and Rocha (2011) moved one step further to distinguish the homologous members in protein families into xenologs (horizontally transferred genes) and paralogs (intrachromosomally duplicated genes). They have found that xenologs share fewer protein interactions, evolve faster, and persist longer in prokaryotic lineages possibly due to a higher/longer adaptive role; while duplicated genes are expressed more, and when persistent, they evolve slower. Furthermore, expansions are much more likely to arise by horizontal transfer than by intrachromosomal duplication (Treangen and Rocha 2011). Therefore, horizontal gene transfer could not only bring in a substantial number of completely novel genes, but also contribute significantly to sequence diversification in existing protein families.

7.5 Pseudogenization of horizontally transferred genes

Pseudogenes are commonly seen in various bacterial genomes, although the number of pseudogenes could vary significantly among different genomes.

By and large, pseudogenes are unique in each genome, suggesting that pseudogenes are formed and eliminated relatively rapidly from most bacterial genomes (Lerat and Ochman 2005). On the other hand, some highly degraded pseudogenes that have persisted correspond to genes with low expression levels and low connectivity in gene networks, supporting a view that pseudogenes need not be evolving in a strictly neutral manner (van Passel and Ochman 2007). A similar pattern was found in an extremely pseudogene-rich genome, *Mycobacterium leprae*, such that pseudogenes tend to be functionally less important and to be located in the 3′ half of the operon (Muro et al. 2011). Kuo and Ochman (2010) have proposed that selection acts to remove the highly expressed and connected pseudogenes to minimize the costs associated with transcribing and translating nonfunctional genes. To further understand the nature of selection on pseudogenes, research could be focused on examining the pseudogene retention time and the consequent substitution pattern after gene pseudogenization by comparing closely related pseudogene-rich genomes.

The presence of pseudogenes in bacterial genomes could potentially affect the rate estimation of gene gain and gene loss (Zhaxybayeva et al. 2007), since many pseudogenes in bacterial genomes are not properly recognized (e.g. some pseudogenes are annotated as whole ORFs, while some others are completely unannotated) (Lerat and Ochman 2004). In fact, the misrecognition of pseudogenes could result in complicated effects on estimates of gene gain and gene loss (Hao and Golding 2008a). Importantly, examination on the evolutionary histories of truncated protein sequences has revealed that gene truncation tends to take place preferentially in recently transferred genes (Hao and Golding 2008a; and as shown in Figure 1B). This suggests that gene truncation plays a role in facilitating the fast turnover of recently transferred genes. For these reasons, truncated genes were added as a new characteristic parameter in addition to gene presence and absence in a quantitative estimation and the results supported the notion that recently acquired genes are rapidly lost during bacterial genome evolution (Hao and Golding 2010). Furthermore, these results revealed that many recently truncated genes are in the process of being eliminated from the genome.

7.6 Mobile sequences and gene movement

Bacterial genomes in many lineages appear to undergo rapid change not only through gene gain and gene loss but also through gene transposition and large-scale genome rearrangement. Many of these genomic changes are associated with mobile sequences. There are three major types of mobile sequence: phages, insertion sequences (IS) elements (including transposases), and plasmids. All three types contribute significantly to genetic exchange between different genomes in various environmental conditions, but the contribution of each type to gene exchange might vary among organisms. In a systematic examination on gene families from 111 genomes, it has been shown that plasmids are overall the most dominant vectors of genetic exchange between bacterial chromosomes (Halary et al. 2010).

Many mobile sequences undergo fast rates of sequence turnover. For instance, when two closely related bacterial genomes have a large number of IS elements, they are most frequently of different types (Wagner 2006). Such an over-representation of recent IS elements results from their short persistence times in bacterial genomes, because gene transposition caused by IS elements tends to have deleterious consequences. The same trend seems to be true for phages and plasmids, since bacterial genomes from the same named species often have different phage sequences and variable plasmid genes when plasmids exist for these taxa.

Gene transposition has been found to be significantly associated with horizontally transferred genes (Fig. 7.1c and d), suggesting that gene transposition accelerates the evolution of horizontally transferred genes (Hao and Golding 2009). This observed association is partly because gene transposition is deleterious, many horizontally transferred genes are under relaxed selection, and hence transposition of horizontally transferred genes might be less deleterious compared to the situation with native genes. The association between gene transposition and horizontally transferred genes could also result from mechanisms other than selection. For instance, conjugative DNA transfer could induce the bacterial SOS response and then SOS induction facilitates gene transposition (Baharoglu et al. 2010).

Furthermore, IS elements play an important role in pseudogenization in genome evolution. For instance, large proportions of pseudogenes in *Shigella flexneri* 2a, are generated by IS element insertions. Yet these events seldom produce the pseudogenes present in the other genomes examined by Lerat and Ochman (2004). Interestingly, *Shigella* genomes generally have really high numbers of IS sequences (Touchon and Rocha 2007). Therefore, it would be of interest to systematically examine IS elements and pseudogenes in the IS element-rich genomes to quantitatively access the association between IS elements and pseudogenization in bacterial genome evolution.

7.7 Gene exchange goes fine-scale

Despite the wide use of the term 'horizontal gene transfer', horizontal transfer need not necessarily take place in the unit of gene. For instance, the penicillin-binding protein gene (*penA*) from penicillin-susceptible strains of *Neisseria meningitidis* and *N. gonorrhoeae* are very uniform, whereas those from penicillin-resistant strains consist of a mosaic of regions of different origins, suggesting that recombination can occur on a fine-scale (Spratt et al. 1992). Fine-scale recombination could also occur intricately to form a rather complex mosaic gene structure (Baldo et al. 2005). Recent systematic studies further demonstrated that the physical position for horizontal transfer or recombination to occur is not restricted by any functional units (e.g. motif, domain, or gene (Chan et al. 2009)).

Recombination rates vary substantially across genomes with respect to phylogeny and ecology (Vos and Didelot 2009). Many obligate intracellular parasites were found to exhibit low recombination rates, some well-studied species such as *Vibrio* and *Neisseria* have high recombination rates, and some lesser known species, *Flavobacterium* and *Pelagibacter*, were found to be even more recombinogenic. In the species *Vibrio vulnificus*, environmental isolates were found to have very high recombination rates, while disease-related lineages show lowered recombination rates (Bisharat et al. 2007). The latter is likely due to epidemic spread of a subset of virulent clones. In the genus *Neisseria*, commensals have an extensive repertoire of virulence alleles, a large fraction of which are exchanged on a fine scale with their pathogenic counterparts (Marri et al. 2010).

Fine-scale recombination plays a crucial role in introducing both sequence diversity and sequence divergence. In *E. coli*, bacterial diversity could be mostly due to recombination rather than mutation (Guttman and Dykhuizen, 1994). A recent study based on examining 240 *Streptococcus pneumoniae* genomes showed that 88% of all single-nucleotide polymorphisms (SNPs) were introduced by recombination rather than by mutation (Croucher et al. 2011). In a large-scale comparison of homologous recombination rates in bacteria and archaea, recombination was shown to contribute to more nucleotide changes than point mutation in over half (27 out of 48) of the examined species (Vos and Didelot 2009). Some of these changes introduced by recombination might lead to significant functional consequences.

7.8 Conclusions

During evolution, many genes are acquired into bacterial genomes via horizontal gene transfer. After transfer most genes are eventually deleted, during their transient stay, some might be subject to processes such as duplication, pseudogenization, and transposition. All integrated processes can lead to fast rates of bacterial genome evolution. From an evolutionary standpoint, the process of gene turnover, just like the process of nucleotide substitution, is the combination of the accumulation of changes over time and the outcome of ensuing selection.

Acknowledgments

This work was supported by startup funds from Wayne State University. I am grateful to Brian Golding for his extremely valuable input and comments, and I apologize to authors of work uncited due to space constraints.

References

Baharoglu, Z., Bikard, D., and Mazel, D. (2010) Conjugative DNA transfer induces the bacterial SOS response and promotes antibiotic resistance development through integron activation. *PLoS Genet* **6**: e1001165.

Baldo, L., Lo, N., and Werren, J.H. (2005) Mosaic nature of the Wolbachia surface protein. *J Bacteriol* **187**: 5406–18.

Beiko, R.G., Harlow, T.J., and Ragan, M.A. (2005) Highways of gene sharing in prokaryotes. *Proc Natl Acad Sci U S A* **102**: 14332–7.

Bisharat, N., Cohen, D.I., Maiden, M.C., Crook, D.W., Peto, T., and Harding, R.M. (2007) The evolution of genetic structure in the marine pathogen, Vibrio vulnificus. *Infect Genet Evol* **7**: 685 93.

Blum, G., Ott, M., Lischewski, A., Ritter, A., Imrich, H., Tschape, H., et al. (1994) Excision of large DNA regions termed pathogenicity islands from tRNA-specific loci in the chromosome of an Escherichia coli wild-type pathogen. *Infect Immun* **62**: 606–14.

Chan, C.X., Darling, A.E., Beiko, R.G., and Ragan, M.A. (2009) Are protein domains modules of lateral genetic transfer? *PLoS ONE* **4**: e4524.

Cohen, O. and Pupko, T. (2010) Inference and characterization of horizontally transferred gene families using stochastic mapping. *Mol Biol Evol* **27**: 703–13.

Cohen, O., Gophna, U., and Pupko, T. (2011) The complexity hypothesis revisited: connectivity rather than function constitutes a barrier to horizontal gene transfer. *Mol Biol Evol* **28**: 1481–9.

Croucher, N.J., Harris, S.R., Fraser, C., Quail, M.A., Burton, J., van der Linden, M., et al. (2011) Rapid pneumococcal evolution in response to clinical interventions. *Science* **331**: 430–4.

Csuros, M. and Miklos, I. (2009) Streamlining and large ancestral genomes in Archaea inferred with a phylogenetic birth-and-death model. *Mol Biol Evol* **26**: 2087–95.

Dagan, T. and Martin, W. (2007) Ancestral genome sizes specify the minimum rate of lateral gene transfer during prokaryote evolution. *Proc Natl Acad Sci U S A* **104**: 870–5.

Dagan, T., Artzy-Randrup, Y., and Martin, W. (2008) Modular networks and cumulative impact of lateral transfer in prokaryote genome evolution. *Proc Natl Acad Sci U S A* **105**: 10039–44.

D'Costa, V.M., Griffiths, E., and Wright, G.D. (2007) Expanding the soil antibiotic resistome: exploring environmental diversity. *Curr Opin Microbiol* **10**: 481–9.

Drake, J.W. (1993) Rates of spontaneous mutation among RNA viruses. *Proc Natl Acad Sci U S A* **90**: 4171–5.

Garcia-Vallvé, S., Romeu, A., and Palau, J. (2000) Horizontal gene transfer in bacterial and archaeal complete genomes. *Genome Res* **10**: 1719–25.

Gogarten, J.P., Doolittle, W.F., and Lawrence, J.G. (2002) Prokaryotic evolution in light of gene transfer. *Mol Biol Evol* **19**: 2226–38.

Gogarten, J.P. and Townsend, J.P. (2005) Horizontal gene transfer, genome innovation and evolution. *Nat Rev Microbiol* **3**:679–87.

Gophna, U. and Ofran, Y. (2011) Lateral acquisition of genes is affected by the friendliness of their products. *Proc Natl Acad Sci U S A* **108**: 343–8.

Guttman, D.S. and Dykhuizen, D.E. (1994) Clonal divergence in Escherichia coli as a result of recombination, not mutation. *Science* **266**: 1380–3.

Halary, S., Leigh, J.W., Cheaib, B., Lopez, P., and Bapteste, E. (2010) Network analyses structure genetic diversity in independent genetic worlds. *Proc Natl Acad Sci U S A* **107**: 127–32.

Hao, W. and Golding, G.B. (2004) Patterns of bacterial gene movement. *Mol Biol Evol* **21**: 1294–307.

Hao, W. and Golding, G.B. (2006) The fate of laterally transferred genes: Life in the fast lane to adaptation or death. *Genome Res* **16**: 636–43.

Hao, W. and Golding, G.B. (2008a) High rates of lateral gene transfer are not due to false diagnosis of gene absence. *Gene* **421**: 27–31.

Hao, W. and Golding, G.B. (2008b) Uncovering rate variation of lateral gene transfer during bacterial genome evolution. *BMC Genomics* **9**: 235.

Hao, W. and Golding, G.B. (2009) Does gene translocation accelerate the evolution of laterally transferred genes? *Genetics* **182**: 1365–75.

Hao, W. and Golding, G.B. (2010) Inferring bacterial genome flux while considering truncated genes. *Genetics* **186**: 411–26.

Hooper, S.D. and Berg, O.G. (2003) Duplication is more common among laterally transferred genes than among indigenous genes. *Genome Biol* **4**: R48.

Hotchkiss, R. (1951) Transfer of penicillin resistance in pneumococci by the desoxyribonucleate derived from resistant cultures. *Cold Spring Harb Symp Quant Biol* **16**: 457–61.

Kloesges, T., Popa, O., Martin, W., and Dagan, T. (2011) Networks of gene sharing among 329 proteobacterial genomes reveal differences in lateral gene transfer frequency at different phylogenetic depths. *Mol Biol Evol* **28**: 1057–74.

Koonin, E.V., Makarova, K.S., and Aravind, L. (2001) Horizontal gene transfer in prokaryotes: Quantification and classification. *Annu Rev Microbiol* **55**: 709–42.

Kunin, V. and Ouzounis, C.A. (2003) The balance of driving forces during genome evolution in prokaryotes. *Genome Res* **13**: 1589–94.

Kunin, V., Goldovsky, L., Darzentas, N., and Ouzounis, C.A. (2005) The net of life: reconstructing the microbial phylogenetic network. *Genome Res* **15**: 954–9.

Kuo, C.H. and Ochman, H. (2010) The extinction dynamics of bacterial pseudogenes. *PLoS Genet* 6(8): e1001050.

Lawrence, J.G., and Ochman, H. (1998) Molecular archaeology of the Escherichia coli genome. *Proc Natl Acad Sci U S A* **95**: 9413–17.

Lerat, E., Daubin, V., and Moran, N.A. (2003) From gene trees to organismal phylogeny in prokaryotes: The case of the gamma-Proteobacteria. *PLoS Biol* **1**: E19.

Lerat, E. and Ochman, H. (2004) $\Psi - \Phi$: exploring the outer limits of bacterial pseudogenes. *Genome Res* **14**: 2273–8.

Lerat, E. and Ochman, H. (2005) Recognizing the pseudogenes in bacterial genomes. *Nucleic Acids Res* **33**: 3125–32.

Lercher, M.J. and Pal, C. (2008) Integration of horizontally transferred genes into regulatory interaction networks takes many million years. *Mol Biol Evol* **25**: 559–67.

Lynch, M. (2010) Evolution of the mutation rate. *Trends Genet* **26**: 345–52.

Marri, P.R., Paniscus, M., Weyand, N.J., Rendon, M.A., Calton, C.M., Hernandez, D.R., et al. (2010) Genome sequencing reveals widespread virulence gene exchange among human *Neisseria* species. *PLoS One* **5**: e11835.

Mirkin, B.G., Fenner, T.I., Galperin, M.Y., and Koonin, E.V. (2003) Algorithms for computing parsimonious evolutionary scenarios for genome evolution, the last universal common ancestor and dominance of horizontal gene transfer in the evolution of prokaryotes. *BMC Evol Biol* **3**: 2.

Muro, E.M., Mah, N., Moreno-Hagelsieb, G., and Andrade-Navarro, M.A. (2011) The pseudogenes of Mycobacterium leprae reveal the functional relevance of gene order within operons. *Nucleic Acids Res* **39**: 1732–8.

Nakamura, Y., Itoh, T., Matsuda, H., and Gojobori, T. (2004) Biased biological functions of horizontally transferred genes in prokaryotic genomes. *Nat Genet* **36**: 760–6.

Nelson, K.E., Clayton, R.A., Gill, S.R., Gwinn, M.L., Dodson, R.J., Haft, D.H., et al. (1999) Evidence for lateral gene transfer between Archaea and bacteria from genome sequence of Thermotoga maritima. *Nature* **399**: 323–9.

Nicolas, P., Bessières, P., Ehrlich, S.D., Maguin, E., and van de Guchte, M. (2007) Extensive horizontal transfer of core genome genes between two Lactobacillus species found in the gastrointestinal tract. *BMC Evol Biol* **7**: 141.

Ochman, H., Lawrence, J.G., and Groisman, E.A. (2000) Lateral gene transfer and the nature of bacterial innovation. *Nature* **405**: 299–304.

Puigbò, P., Wolf, Y.I., and Koonin, E.V. (2009) Search for a 'Tree of Life' in the thicket of the phylogenetic forest. *J Biol* **8**: 59.

Ranea, J.A., Buchan, D.W., Thornton, J.M., and Orengo, C.A. (2004) Evolution of protein superfamilies and bacterial genome size. *J Mol Biol* **336**: 871–87.

Rivera, M.C., Jain, R., Moore, J.E., and Lake, J.A. (1998) Genomic evidence for two functionally distinct gene classes. *Proc Natl Acad Sci U S A* **95**: 6239–44.

Shi, T. and Falkowski, P.G. (2008) Genome evolution in cyanobacteria: the stable core and the variable shell. *Proc Natl Acad Sci U S A* **105**: 2510–15.

Snel, B., Bork, P., and Huynen, M.A. (2002) Genomes in flux: the evolution of archaeal and proteobacterial gene content. *Genome Res* **12**: 17–25.

Spencer, M and Sangaralingam, A. (2009) A phylogenetic mixture model for gene family loss in parasitic bacteria. *Mol Biol Evol* **26**(8): 1901–8.

Spratt, B.G., Bowler, L.D., Zhang, Q.Y., Zhou, J., and Smith, J.M. (1992) Role of interspecies transfer of chromosomal genes in the evolution of penicillin resistance in pathogenic and commensal *Neisseria* species. *J Mol Evol* **34**: 115–25.

Touchon, M. and Rocha, E.P. (2007) Causes of insertion sequences abundance in prokaryotic genomes. *Mol Biol Evol* **24**: 969–81.

Treangen, T.J. and Rocha, E.P. (2011) Horizontal transfer, not duplication, drives the expansion of protein families in prokaryotes. *PLoS Genet* **7**: e1001284.

van Passel, M.W., Marri, P.R., and Ochman, H. (2008) The emergence and fate of horizontally acquired genes in Escherichia coli. *PLoS Comput Biol* **4**: e1000059.

van Passel, M.W. and Ochman, H. (2007) Selection on the genic location of disruptive elements. *Trends Genet* **23**: 601–4.

Vos, M. and Didelot, X. (2009) A comparison of homologous recombination rates in bacteria and archaea. *ISME J* **3**: 199–208.

Wagner, A. (2006) Periodic extinctions of transposable elements in bacterial lineages: evidence from intragenomic variation in multiple genomes. *Mol Biol Evol* **23**: 723–33.

Welch, R.A., Burland, V., Plunkett, 3rd, G., Redford, P., Roesch, P., Rasko, D., et al. (2002) Extensive mosaic structure revealed by the complete genome sequence of uropathogenic Escherichia coli. *Proc Natl Acad Sci U S A* **99**: 17020–4.

Wellner, A., Lurie, M.N., and Gophna, U. (2007) Complexity, connectivity, and duplicability as barriers to lateral gene transfer. *Genome Biol* **8**: R156.

Zhaxybayeva, O., Gogarten, J.P., Charlebois, R.L., Doolittle, W.F., and Papke, R.T. (2006) Phylogenetic analyses of cyanobacterial genomes: Quantification of horizontal gene transfer events. *Genome Res* **16**: 1099–108.

Zhaxybayeva, O., Nesbo, C.L., and Doolittle, W.F. (2007) Systematic overestimation of gene gain through false diagnosis of gene absence. *Genome Biol* **8**: 402.

CHAPTER 8

Rapid evolution of animal mitochondrial DNA

Xuhua Xia

8.1 Introduction

Animal mitochondrial DNAs (mtDNAs) are interesting not only because they generally evolve much faster than nuclear genomes or plant mtDNAs (Gray et al. 1989; Lynch et al. 2006), but also because of the immense heterogeneity in their evolutionary rates. Even within mammalian species alone, the rate can vary by nearly two orders of magnitude (Nabholz et al. 2008). Such results serve to motivate evolutionary biologists to search for causes for the rapid and episodic evolution.

The rapid evolution in animal mtDNA relative to nuclear genomes used to be attributed to three factors, i.e. the high mutation rate associated with the production of free radicals in mitochondria (Balaban et al. 2005), the limited DNA repair function in mitochondria (Mason et al. 2003), and the multiple mitochondrial generations per cell generation coupled with a much higher replication error in mtDNA than in nuclear DNA (Johnson and Johnson 2001). However, it is not clear why these factors do not seem to apply to plant mtDNA (Lynch et al. 2006) which generally evolves much slower than plant nuclear genomes (Wolfe et al. 1987; Drouin et al. 2008).

Here I present genomic evidence to highlight three differences between animal and plant mtDNA, with the objective to further our understanding of not only the rapid evolution of animal mtDNA but also the rate heterogeneity among different animal mtDNA lineages. First, mtDNA replication in animals is not only error prone, but also leads to strong strand bias which, when coupled with strand switching of mitochondrial genes from one strand to the other, results in high rate of nucleotide substitution. I show that global strand bias is shared by nearly all animal groups except for poriferans and cnidarians. MtDNAs in these last two groups are similar to plant mtDNA in several ways, as will be shown later.

Second, in contrast to plant mtDNA with a single standard genetic code, animal mtDNAs feature a variety of different genetic codes. A change in genetic code represents a landmark change in evolution and is expected to contribute to accelerated and episodic evolution rates (Lynch et al. 2006). The slow-evolving mtDNAs in poriferans and cnidarians (Shearer et al. 2002) share a genetic code similar to the standard code, whereas fast-evolving mtDNAs in other metazoans have genetic codes which differ more from the standard code. I will show that a significant portion of amino acid and codon substitutions can be attributed to changes in genetic code.

Third, while plant mitochondria have efficient mechanisms for tRNA import (Salinas et al. 2006), most metazoan mitochondria typically depend entirely on tRNAs coded in mtDNA for translation, with cnidarian mitochondria being rare exceptions (Beagley et al. 1998). A gain/loss in tRNA genes in animal mtDNA often leads to significant changes at the coding sequences. I illustrate this point by contrasting mtDNAs with only $tRNA^{Met/CAU}$ (where CAU is the anticodon) and those with both $tRNA^{Met/CAU}$ and $tRNA^{Met/UAU}$ genes. The gain of $tRNA^{Met/UAU}$ leads to significantly increased AUA usage and AUG→AUA substitutions. The gain/loss of mitochondrial tRNA genes in plants does not yield detectable effect on coding sequences or codon usage in plant mtDNA.

Rapidly Evolving Genes and Genetic Systems. First Edition. Edited by Rama S. Singh, Jianping Xu, and Rob J. Kulathinal.

8.2 Mitochondrial replication, strand bias, and evolutionary rates

Most mutations occur during DNA replication, and different DNA replication mechanisms often leave distinct footprints in genomic strand asymmetric patterns (Xia 2012), which is typically measured by the GC skew (Lobry 1996; Marín and Xia 2008) defined as $(P_G-P_C)/(P_G+P_C)$. Bacterial species from *Bacillus subtilis* to *Escherichia coli* share the strand asymmetric pattern in Fig. 8.1a, which is a fingerprint of the single-origin bidirectional DNA replication shared by eubacterial species. The exceptionally slow-evolving sponge (Porifera) mtDNAs share the same strand asymmetric fingerprint as its eubacterial ancestor (Fig. 8.1b), so does the slow-evolving hydra (Cnidaria) mtDNA lineage. In contrast, the fast-evolving vertebrate mtDNAs share the strand asymmetric pattern (Fig. 8.1c and d) consistent with the strand-displacement model of DNA replication (Clayton 1982; Shadel and Clayton 1997; Clayton 2000; Bogenhagen and Clayton 2003; Brown et al. 2005) which, although challenged recently by a new proposal of a strand-coupled bidirectional replication (Yang et al. 2002; Yasukawa et al. 2005), is favored by current empirical evidence (Brown et al. 2005). The GC skew values for vertebrate mtDNA are all negative, implying global asymmetry in addition to the local asymmetric patterns.

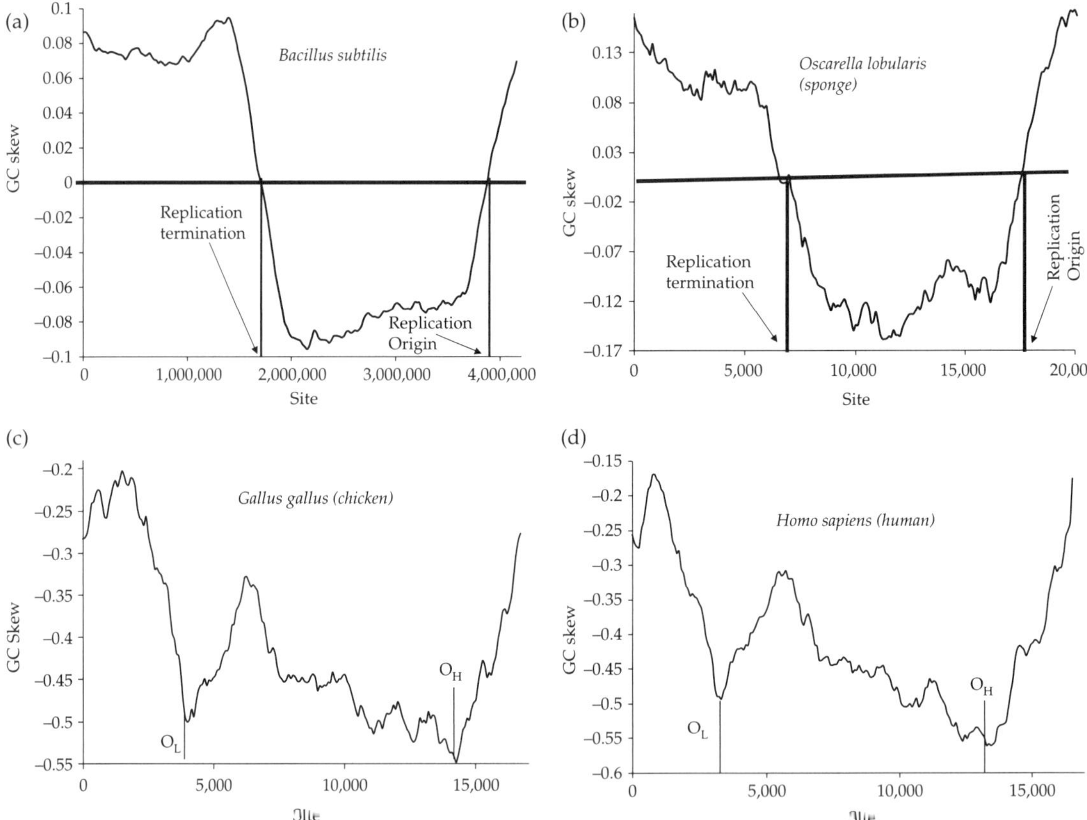

Figure 8.1 Genomic strand asymmetric patterns characterized by GC skew values along a sliding window, with inferred replication origins. The *Bacillus subtilis* pattern (a) is shared among all eubacterial species known to have single-origin bidirectional replication. The sponge mtDNA, which evolves slower than the nuclear DNA, has the strand asymmetric pattern similar to its eubacterial ancestor (b). Vertebrate mtDNAs are replicated by the highly derived, but error-prone, two-origin strand-displacement replication, and evolve much faster than the nuclear DNA. Modified from fig. 1, fig. 9a, fig. 9b and fig. 10c in Xia (2012), with permission from Bentham Science Publishers.

Mammalian mtDNA has two strands of different buoyant densities and consequently these are named the H-strand and the L-strand. The two strands have different nucleotide frequencies, with the H-strand being rich in G and T and the L-strand rich in A and C (Xia 2005). According to the strand-displacement model of DNA replication, the L-strand is first used as a template to replicate the daughter H-strand, while the parental H-strand was left single-stranded for an extended period because the DNA replication rate *in vivo* is only about 270 nucleotides/min (Clayton 1982; Shadel and Clayton 1997; Clayton 2000) which is about 200 times slower than that in *Escherichia coli*). Limited evidence suggests that selection for faster mtDNA replication is weak. For example, mutant human mitochondrial single-stranded DNA-binding protein (mtSSB) with two highly variable (presumably non-functional) regions deleted, which increases the mtSSB isoelectric point (pI) and presumably would also improve its electrostatic interaction with the negatively charged DNA, leads to more efficient DNA synthesis than the wild type mtSSB (Oliveira and Kaguni 2010). Thus, either the deletion mutant has never arisen naturally or it did but has not been favored by natural selection.

Spontaneous deamination of both A and C (Sancar and Sancar 1988; Lindahl 1993) occurs frequently in human mitochondrial DNA (Tanaka and Ozawa 1994), generating A→G and C→U mutations. Among these two types of spontaneous deamination, the C→U mutation occurs more frequently than the A→G mutation (Lindahl 1993). In particular, the C→U mutation mediated by the spontaneous deamination occurs in single-stranded DNA more than 100 times as frequently as in double-stranded DNA (Frederico et al. 1990). These C→U mutants will immediately be used as a template to replicate the daughter L-strand, leading to a G→A mutation in the L-strand after one round of DNA duplication. Therefore, the H-strand, left single-stranded for an extended period during DNA replication, tend to accumulate A→G and C→U mutations and become rich in G and T while the L-strand will become rich in A and C. This results in strong strand bias, i.e. the violation of Chargaff's parity rule 2.

Single-stranded DNA binding proteins (SSB) protects single-stranded DNA from nucleolytic degradations. In *E. coli*, this works best with the presence of Rec-A. SSB from *E. coli* also reduces the C–U deamination rate in single-stranded DNA by four- to fivefold (Lough et al. 2001). However, it is not known if mtSSB also has the equivalent Rec-A partner or if it also protects single-stranded DNA from deamination in mitochondria.

The vertebrate L-strand and H-strand, being AC-rich and GT-rich, respectively, would have a negative and a positive GC skew, respectively, but with the same absolute value. Therefore, we may use the absolute value of the global GC skew (with P_C and P_G values from the entire genome rather than from a sliding window) to characterize global strand asymmetry (GSA, Table 8.1). Plant mtDNAs exhibit little global strand asymmetry, with their GSA values close to zero. In contrast, animal mtDNAs typically have high GSA values (Table 8.1). The sponge (*Oscarella lobularis*) and the hydra (*Hydra oligactis*), representing Porifera and Cnidaria, respectively, are exceptional among animals in that their mtDNAs have GSA values similar to those in plant mtDNA. These two animal groups are also similar to plants in having slower evolutionary rates than their nuclear genomes (Shearer et al. 2002). The difference in GSA between plant mtDNA and animal mtDNA (excluding the sponge and hydra mtDNA) is highly significant (t-test, DF = 16, $p < 0.0001$).

The strong strand bias observed in animal mtDNA (Table 8.1, except for the sponge and the hydra mtDNA) suggest that a gene relocated from one strand to the other will experience a different mutation spectrum and consequently would evolve rapidly. In contrast, a strand switching is expected to have less effect on evolutionary rate in plant mtDNA.

Four tRNA genes ($tRNA^{Asp}$, $tRNA^{Glu}$, $tRNA^{Lys}$, and $tRNA^{Val}$) have switched strands in the eight decapod species (Fig. 8.2), whereas all other tRNA genes, as well as all protein-coding genes and rRNA genes, have not. I used this set of mtDNA to test the prediction that the four tRNA genes that have switched strands should evolve faster than homologous tRNA genes that have not. Because tRNA genes are too short for reliable phylogenetic reconstruction, I used the *COX1* sequences from the eight

Table 8.1 Nucleotide frequencies (P_A, P_C, P_G, and P_T) and global strand asymmetry (GSA) for representative metazoans and plants

Species	Accession	Length	P_A	P_C	P_G	P_T	GSA
Oscarella lobularis	NC_014863	20260	0.333	0.176	0.173	0.318	0.006
Hydra oligactis	NC_010214	16314	0.348	0.114	0.124	0.414	0.039
Caenorhabditis elegans	NC_001328	13794	0.314	0.089	0.149	0.448	0.253
Schistosoma japonicum	NC_002544	14085	0.249	0.084	0.206	0.462	0.422
Drosophila melanogaster	NC_001709	19517	0.418	0.103	0.076	0.404	0.150
Ciona intestinalis	NC_004447	14790	0.342	0.095	0.119	0.444	0.116
Branchiostoma lanceolatum	NC_001912	15076	0.269	0.159	0.214	0.358	0.148
Eptatretus burgeri	NC_002807	17168	0.328	0.229	0.106	0.337	0.366
Mitsukurina owstoni	NC_011825	17743	0.323	0.254	0.134	0.290	0.309
Danio rerio	NC_002333	16596	0.319	0.239	0.160	0.281	0.198
Xenopus laevis	NC_001573	17553	0.331	0.235	0.135	0.300	0.270
Alligator mississippiensis	NC_001922	16646	0.312	0.295	0.135	0.257	0.371
Gallus gallus	NC_001323	16775	0.303	0.325	0.135	0.238	0.412
Mus musculus	NC_005089	16299	0.345	0.244	0.124	0.287	0.328
Marchantia polymorpha	NC_001660	186609	0.285	0.210	0.214	0.291	0.009
Cycas taitungensis	NC_010303	414903	0.264	0.235	0.235	0.266	0.000
Arabidopsis thaliana	NC_001284	366924	0.279	0.225	0.222	0.273	0.006
Oryza sativa indica	NC_007886	491515	0.279	0.219	0.220	0.283	0.002
Sorghum bicolor	NC_008360	468628	0.281	0.220	0.217	0.282	0.008
Triticum aestivum	NC_007579	452528	0.279	0.221	0.222	0.278	0.002

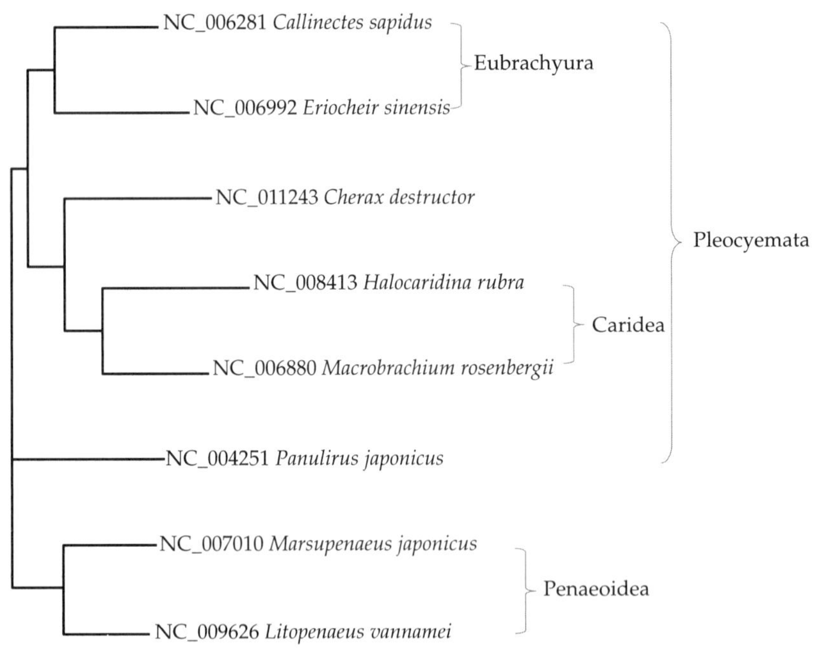

Figure 8.2 Phylogenetic tree of eight decapod species, built from the mitochondrial *COX1* gene. The leaves are labeled with both GenBank accession numbers and the species name. There is no strand switching event involving *COX1*.

mtDNA to construct a tree (Fig. 8.2) and used the *COX1* tree to constrain the topology of trees built with tRNA sequences, i.e. the tRNA sequences are used only to evaluate branch lengths. The prediction is that a tRNA gene that has switched strand in a lineage would evolve more rapidly than that in its sister lineage.

The four trees built with the four tRNA genes (Fig. 8.3) show a consistent pattern, i.e. the tRNA gene that switched strand evolved dramatically faster than its homologues that have not switched strands. Three of the strand-switching events involving tRNA genes (Fig. 8.3a–c) occurred in the lineage leading to *Eriocheir sinensis*. Other mitochondrial genes in *E. sinensis* that have not switched strand do not evolve faster than those in its sister taxon, suggesting that the faster evolutionary rate observed in Fig. 8.3a–c is not due to a generally increased evolutionary rate in the *E. sinensis* mtDNA. tRNA^{Val} switched strand in the lineage leading to *Cherax destructor*, and also exhibited a correspondingly accelerated evolutionary rate (Fig. 8.3d). The GSA values for *E. sinensis* and *C. destructor* are 0.248 and 0.280, respectively.

8.3 The change in genetic code and evolutionary rate

In contrast to plant mtDNA with the same standard genetic code, animal mtDNAs feature a variety of genetic codes (with translation tables 2, 4, 5, 9, 13, 14, 21, and 24) of which a sample is shown in Fig. 8.4. A change in translation table could have a profound effect on the evolution of amino acid

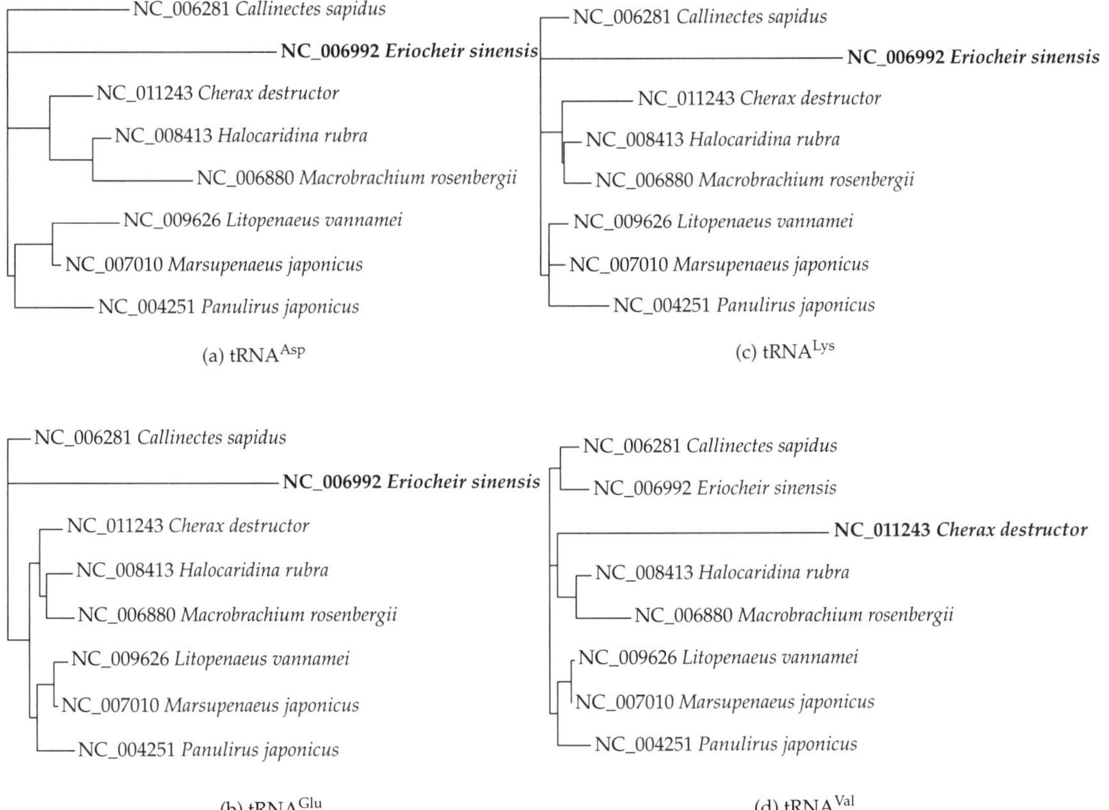

Figure 8.3 Trees constructed from tRNA genes, with the topology constrained by the *COX1* tree in Fig. 8.2. Each tree has one lineage (**in bold**) where the tRNA gene has switched strand, which is associated with a dramatically increased evolutionary rate.

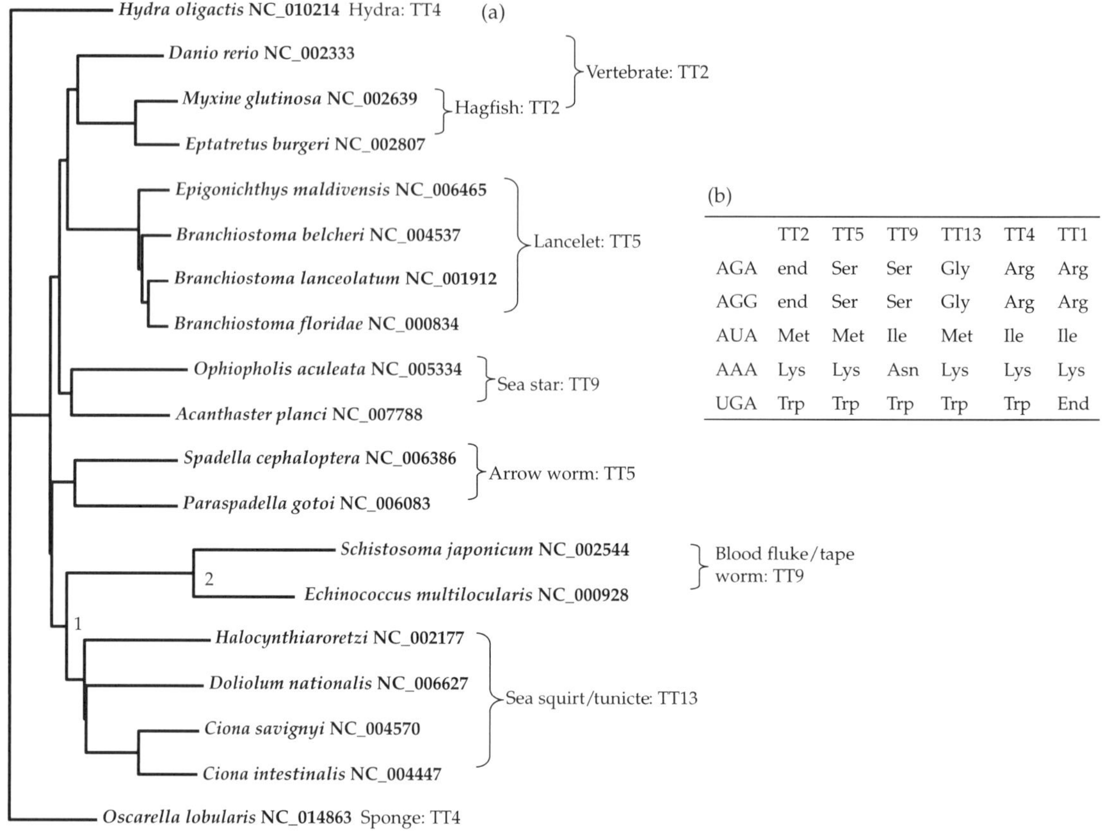

	TT2	TT5	TT9	TT13	TT4	TT1
AGA	end	Ser	Ser	Gly	Arg	Arg
AGG	end	Ser	Ser	Gly	Arg	Arg
AUA	Met	Met	Ile	Met	Ile	Ile
AAA	Lys	Lys	Asn	Lys	Lys	Lys
UGA	Trp	Trp	Trp	Trp	Trp	End

Figure 8.4 Different mitochondrial translation tables (TT) in representative animal species. The tree (a) is constructed from the amino acid sequence of the mitochondrial *COX1* gene and conforms largely to the common consensus except for the position of tunicates. However, the grouping of tunicates with blood flukes/tape worms is consistent with trees built from other mitochondrial protein-coding genes (e.g. Cyt-b, ND4, ND5) and ribosomal genes. Codons with different meanings in different animal mitochondrial genetic codes are tabled in (b). Also note that Porifera and Cnidaria represented by the sponge (*Oscarella lobularis*) and the hydra (*Hydra oligactis*), respectively, are known to have the slowest evolutionary rates among metazoans. They also have a genetic code (TT4) most similar to the standard code.

sequences and coding sequences leading to potentially many fortuitous mutation events (Lynch et al. 2006). This effect can manifest at two stages, illustrated as follows. First, the coding sequence '... AGA AGG AUA ...' coding for '... Arg Arg Ile ...' in genetic code 4 would code for '... Gly Gly Met ...' when the genetic code is changed to genetic code 13. Thus, we have three amino acid replacements without any change to the coding sequence. Second, if the original amino acid sequence '... Arg Arg Ile...' is functionally optimal, then there would be little purifying selection maintaining the new '... Gly Gly Met ...' sequence. This would increase the propensity of the new sequence to accumulate substitutions at these suboptimal Gly and Met sites. Also, if the original '... Arg Arg Ile...' is optimal, then there is positive selection favoring nonsynonymous substitutions to revert the new '... Gly Gly Met ...' to the original.

In short, the change in genetic code in the first stage leads to amino acid replacements without changing the coding sequence, whereas the second stage mediated by the weakened purifying selection or the positive selection for the new amino acid sequence to revert to the original would result in nonsynonymous substitutions in the coding sequences. In either case, an increased nonsynonymous substitution rate is expected. This suggests that the amino acids coded by codons whose meaning is changed by a change in genetic code would be frequently involved in amino acid substitutions.

Changes in genetic codes in animal mtDNA involve eight amino acids, Arg, Asn, Gly, Ile, Lys, Met, Ser, and Trp (Fig. 8.4b). They together represent 39.7% of all amino acids in the mtDNA in the 19 species included in Fig. 8.4. However, they participated in 72.7% of all amino acid replacements along the tree (which can be obtained by reconstructing the ancestral sequences for each node and making pairwise comparisons along the tree, e.g. between node 1 and node 2, between node 2 and *Schistosoma japonicum*, and between node 2 and *Echinococcus multilocularis* in Fig. 8.4). These amino acids are involved in a smaller percentage of amino acid replacements if comparisons are limited within clades sharing the same translation table, with or without the time of divergence controlled for.

The pattern of amino acid replacements can also help us infer the ancestral genetic code. For example, if the ancestral genetic code at node 1 (Fig. 8.4) is TT13, then all AGR (where R stands for A or G) codons will change meaning from Gly to Ser along the lineages leading to *S. japonicum* and *E. multilocularis* whose mtDNAs follow TT9. These non-adapted AGR codons will then be free to be replaced by alternative codons. Indeed, most of the amino acid replacements along the branches between nodes 1 and 2, between node 2 and the two leaves (*S. japonicum* and *E. multilocularis*) are Ser→Other replacements (where 'Other' stands for amino acids other than Ser). Of 85 nonsynonymous codon replacements inferred along the branch between nodes 1 and 2 (Fig. 8.4), 34 are Ser→Other replacements, and only five are Other→Ser replacements. Similarly, 11 of the 30 amino acid replacements between node 2 and *E. multilocularis* are Ser→Other replacements and only two are Other→Ser replacements. Of the 53 amino acid replacements inferred along the branch between node 2 and *S. japonicum*, 19 are Ser→Other replacements and only two are Other→Ser replacements.

Of the Ser→Other substitutions, the most frequent substitutions are Ser→Gly. This makes sense if the ancestral genetic code is genetic code 13. That is, the AGR codons, originally coding Gly in genetic code 13, changed meaning from Gly to Ser when the genetic code changed from 13 to 9. If the original Gly functions well, then there is positive selection to revert the new Ser back to Gly by an AGR→GGR mutation. This would explain why Ser→Gly replacements are the most frequent among Ser→Other replacements along the branches between nodes 1 and 2 and between node 2 and the two leaves (*S. japonicum* and *E. multilocularis*). However, to drive this point home, one would need to take into consideration of several other factors, such as Ser and Gly codon frequencies, and Ser and Gly similarities relative to the similarity between Ser and other amino acids.

While this result does not prove the effect of the genetic code change, it is surely consistent with the inferred effect of genetic code change. This indicates that a significant proportion of substitutions at the amino acid sequence or coding sequences can be attributed to changes in genetic code. As changes in genetic code are hardly gradual, codon substitutions mediated by such genetic code changes are expected to result in episodic changes at the coding sequence.

The diversification of genetic codes in animal mtDNAs is not surprising. Animal mtDNAs encode few genes and some codons are rare or even absent in mtDNA. A change in genetic code involving such codons will typically have little functional consequence. In addition, because of the existence of multiple mitochondrial genomes in a mitochondrion, nature is relatively free to experiment with different genetic codes through tRNA reassignment. A parallel diversification in genetic codon is observed in fungal mtDNA lineages with associated diversification in codon usage (Carullo and Xia 2008; Xia 2008). In contrast, a change in genetic code in the nuclear genome will affect many codons and many genes and is expected to have a major effect.

8.4 The change in tRNA genes and evolutionary rate

Plant mitochondria have complicated and efficient mechanisms for importing nucleus-encoded tRNAs into mitochondria (Salinas et al. 2006), leading to limited autonomy of the tRNA pool in plant mitochondria. In contrast, nucleus-encoded tRNA import into animal mitochondria is rare, except for cnidarians whose mtDNAs code only one or two tRNA genes (i.e. nuclear import is a necessity). A change in mitochondrial tRNA genes can often result in significant changes in coding sequences and coding strategy in animal mtDNA. I illustrate this with the Met codon family as follows.

MtDNA in most animal species code Met by AUA and AUG codons. In some animal species, e.g. vertebrates, these two codons are translated by a single $tRNA^{Met/CAU}$ species with a modified C (i.e. f^5C) at the first anticodon position (Grosjean et al. 2010). In other animal species, e.g. tunicates, an additional $tRNA^{Met/UAU}$ gene is present in the mtDNA. One would expect that, when $tRNA^{Met/UAU}$ is absent, Met should be preferably coded by AUG with reduced AUA usage. The gain of $tRNA^{Met/UAU}$ would favor more Met to be coded by AUA. Such a prediction can be readily tested by existing mtDNA data (Xia et al. 2007).

MtDNA in bivalve species have two $tRNA^{Met}$ genes. In some bivalve species (e.g. *Acanthocardia tuberculata*, *Crassostrea gigas*, *C. Virginica*, *Hiatella arctica*, *Placopecten magellanicus*, and *Venerupis philippinarum*), both $tRNA^{Met}$ genes have a CAU anticodon forming Watson–Crick base pairs with codon AUG. In some other bivalve species (e.g. *Mytilus edulis*, *M. galloprovincialis*, and *M. trossulus*), one $tRNA^{Met}$ has a CAU anticodon and the other has a UAU anticodon forming Watson–Crick base pairs with the AUA codon. One would predict that the latter should be more likely to code Met by AUA than the former, i.e. the proportion of AUA codon within the AUR codon family, designated P_{AUA}, should be greater in the latter with both a $tRNA^{Met/CAU}$ and a $tRNA^{Met/UAU}$ gene than in the former with a single $tRNA^{Met/CAU}$ gene in the mtDNA (Xia et al. 2007).

To test the prediction, I will use P_{UUA} (the proportion of UUA codon in the UUR codon family) as a reference control. Note that, at the same P_{UUA} level, P_{AUA} in the three *Mytilus* mtDNA with both a $tRNA^{Met/CAU}$ and a $tRNA^{Met/UAU}$ gene is significantly higher than that in the six bivalve species without a $tRNA^{Met/UAU}$ gene (Fig. 8.5a, ANCOVA

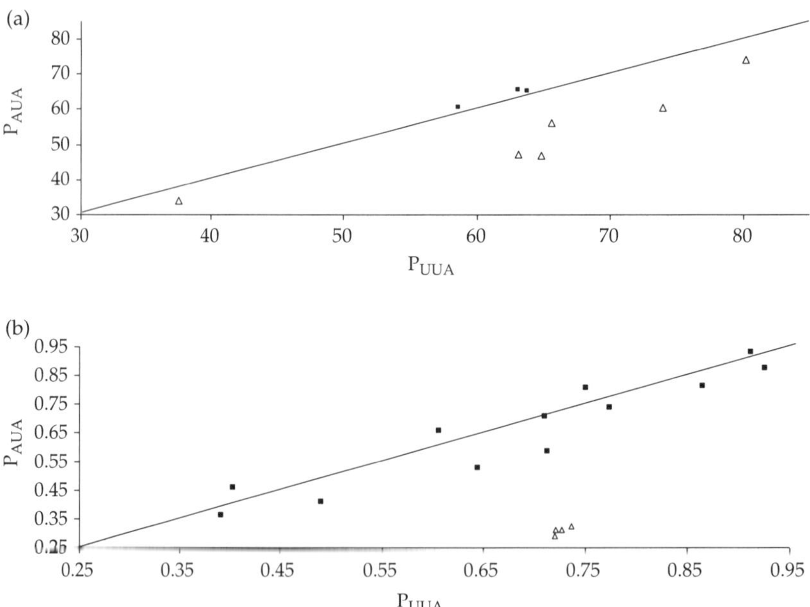

Figure 8.5 Relationship between P_{AUA} and P_{UUA}, highlighting the observation that P_{AUA} is greater when both a $tRNA^{Met/CAU}$ and a $tRNA^{Met/UAU}$ are present than when only $tRNA^{Met/CAU}$ is present in the mtDNA, for bivalve species (a) and chordate species (b). The filled squares are for mtDNA containing both $tRNA^{Met/CAU}$ and $tRNA^{Met/UAU}$ genes, and the open triangles are for mtDNA without a $tRNA^{Met/UAU}$ gene.

test, p = 0.0111). Thus, the presence of $tRNA^{Met/UAU}$ increases AUA usage significantly.

A similar comparison can be performed between the urochordates (tunicates, with both $tRNA^{Met/CAU}$ and $tRNA^{Met/UAU}$ genes in their mtDNA) and cephalochordates (lancelets, with only a $tRNA^{Met/CAU}$ gene in their mtDNA). Fig. 8.5b shows that P_{AUA} is much smaller in lancelets than in tunicates at the same P_{UUA} level. Thus, AUA usage is consistently increased by the gain of a $tRNA^{Met/UAU}$ gene (or consistently decreased by the loss of a $tRNA^{Met/UAU}$ gene) in animal mtDNA.

A gain of a $tRNA^{Met/UAU}$ gene is also associated with a surplus of AUG→AUA substitutions in animal mitochondrial coding sequences (results not shown). Similar associations can also be observed with other gain/loss of tRNA genes in animal mitochondrial. In contrast, a gain/loss of tRNA genes in plant mtDNA appears to have little effect on nucleotide substitutions or codon usage, presumably because such gain/loss events do not significantly alter the tRNA pool with a great deal of nuclear tRNA import into plant mitochondria.

8.5 Conclusions

Three factors may account for the rapid evolution, as well as the rate heterogeneity, among animal mtDNA lineages. First, animal mtDNAs, except for those in Porifera and Cnidaria, exhibit strong local and global strand bias and may share the error-prone strand-displacement replication documented in mammals. The strand bias, associated with genes switching from one strand to the other, contributes significantly to increased evolution rates. Poriferan and cnidarian mtDNAs, similar to plant mtDNA, do not exhibit global strand bias, have local strand asymmetric patterns similar to that of eubacterial species with single-origin replication, and also have extremely slow rates of evolution comparable to those in plant mtDNA and the nuclear genome. Second, in contrast to plant mtDNA with a single standard genetic code, animal mtDNAs feature a variety of different genetic codes and much of coding sequence evolution may be attributed to changes in genetic codes. Third, changes in tRNA pool in animal mitochondria, mediated by the gain/loss of tRNA genes in mtDNA, can contribute significantly to codon replacements in mitochondrial genes. All these factors are expected to result in accelerated and episodic evolution. Recent progresses in mtDNA research suggest that, while laboratory experiments remain important, many questions concerning mtDNA evolution can be addressed with the availability of genomic data and a comparative genomic approach.

Acknowledgments

This study is supported by the Discovery Grant from Natural Science and Engineering Research Council of Canada.

References

Balaban, R.S., Nemoto, S., and Finkel, T. (2005) Mitochondria, oxidants, and aging. *Cell* **120**: 483–95.

Beagley, C.T., Okimoto, R., and Wolstenholme, D.R. (1998) The mitochondrial genome of the sea anemone Metridium senile (Cnidaria): introns, a paucity of tRNA genes, and a near-standard genetic code. *Genetics* **148**: 1091–108.

Bogenhagen, D.F. and Clayton, D.A. (2003) The mitochondrial DNA replication bubble has not burst. *Trends Biochem Sci* **28**: 357–60.

Brown, T.A., Cecconi, C., Tkachuk, A.N., Bustamante, C., and Clayton, D.A. (2005) Replication of mitochondrial DNA occurs by strand displacement with alternative light-strand origins, not via a strand-coupled mechanism. *Genes Dev* **19**: 2466–76.

Carullo, M. and Xia, X. (2008) An extensive study of mutation and selection on the wobble nucleotide in tRNA anticodons in fungal mitochondrial genomes. *J Mol Evol* **66**: 484–93.

Clayton, D.A. (1982) Replication of animal mitochondrial DNA. *Cell* **28**: 693–705.

Clayton, D.A. (2000) Transcription and replication of mitochondrial DNA. *Hum Reprod* **15**: 11–17.

Drouin, G., Daoud H., and Xia, J. (2008) Relative rates of synonymous substitutions in the mitochondrial, chloroplast and nuclear genomes of seed plants. *Mol Phylogenet Evol* **49**: 827–31.

Frederico, L.A., Kunkel, T.A., and Shaw, B.R. (1990) A sensitive genetic assay for the detection of cytosine deamination: determination of rate constants and the activation energy. *Biochemistry (Mosc)* **29**: 2532–7.

Gray, M.W., Cedergren, R., Abel, Y., and Sankoff, D. (1989) On the evolutionary origin of the plant mitochondrion and its genome. *Proc Natl Acad Sci U S A* **86**: 2267–71.

Grosjean, H., de Crecy-Lagard, V., and Marck, C. (2010) Deciphering synonymous codons in the three domains of life: co-evolution with specific tRNA modification enzymes. *FEBS Lett* **584**: 252–64.

Johnson, A.A. and Johnson, K.A. (2001) Exonuclease proofreading by human mitochondrial DNA polymerase. *J Biol Chem* **276**: 38097–107.

Lindahl, T. (1993) Instability and decay of the primary structure of DNA. *Nature* **362**: 709–15.

Lobry, J.R. (1996) Asymmetric substitution patterns in the two DNA strands of bacteria. *Mol Biol Evol* **13**: 660–5.

Lough, J., Jackson, M., Morris, R., and Moyer, R. (2001) Bisulfite-induced cytosine deamination rates in E. coli SSB:DNA complexes. *Mutat Res* **478**: 191–7.

Lynch, M., Koskella, B., and Schaack, S. (2006) Mutation pressure and the evolution of organelle genomic architecture. *Science* **311**: 1727–30.

Marín, A. and Xia, X. (2008) GC skew in protein-coding genes between the leading and lagging strands in bacterial genomes: New substitution models incorporating strand bias. *J Theor Biol* **253**: 508–13.

Mason, P.A., Matheson, E.C., Hall, A.G., and Lightowlers, R.N. (2003) Mismatch repair activity in mammalian mitochondria. *Nucleic Acids Res* **31**: 1052–8.

Nabholz, B., Glémin, S., and Galtier, N. (2008) Strong variations of mitochondrial mutation rate across mammals—the longevity hypothesis. *Mol Biol Evol* **25**: 120–30.

Oliveira, M.T. and Kaguni, L.S. (2010) Functional roles of the N- and C-terminal regions of the human mitochondrial single-stranded DNA-binding protein. *PLoS One* **5**: e15379.

Salinas, T., Duchêne, A.-M., Delage, L., Nilsson, S., Glaser E., Zaepfel, M., et al. (2006) The voltage-dependent anion channel, a major component of the tRNA import machinery in plant mitochondria. *Proc Natl Acad Sci U S A* **103**: 18362–7.

Sancar, A. and Sancar, G.B. (1988) DNA repair enzymes. *Annu Rev Biochem* **57**: 29–67.

Shadel, G.S. and Clayton, D.A. (1997) Mitochondrial DNA maintenance in vertebrates. *Annu Rev Biochem* **66**: 409–35.

Shearer, T.L., Van Oppen, M.J., Romano, S.L., and Worheide, G. (2002) Slow mitochondrial DNA sequence evolution in the Anthozoa (Cnidaria). *Mol Ecol* **11**: 2475–87.

Tanaka, M. and Ozawa, T. (1994) Strand asymmetry in human mitochondrial DNA mutations. *Genomics* **22**: 327–35.

Wolfe, K.H., Li, W.H., and Sharp, P.M. (1987) Rates of nucleotide substitution vary greatly among plant mitochondrial, chloroplast and nuclear DNAs. *Proc Natl Acad Sci U S A* **84**: 9054–8.

Xia, X. (2005) Mutation and selection on the anticodon of tRNA genes in vertebrate mitochondrial genomes. *Gene* **345**: 1320.

Xia, X. (2008) The cost of wobble translation in fungal mitochondrial genomes: integration of two traditional hypotheses. *BMC Evol Biol* **8**: 211.

Xia, X. (2012) DNA replication and strand asymmetry in prokaryotic and mitochondrial genomes. *Curr Genomics* **13**: 16–27.

Xia, X., Huang, H., Carullo, M., Betran, E., and Moriyama, E.N. (2007) Conflict between translation initiation and elongation in vertebrate mitochondrial genomes. *PLoS ONE* **2**: e227.

Yang, M.Y., Bowmaker, M., Reyes, A., Vergani, L., Angeli, P., Gringeri, E., et al. (2002) Biased incorporation of ribonucleotides on the mitochondrial L-strand accounts for apparent strand-asymmetric DNA replication. *Cell* **111**: 495–505.

Yasukawa, T., Yang, M.-Y., Jacobs, H.T., and Holt, I.J. (2005) A bidirectional origin of replication maps to the major noncoding region of human mitochondrial DNA. *Mol Cell* **18**: 651–62.

CHAPTER 9

Rapid evolution of centromeres and centromeric/kinetochore proteins

Kevin C. Roach, Benjamin D. Ross, and Harmit S. Malik

9.1 Centromeres in 'the fast lane'

Accurate chromosome segregation is essential at each eukaryotic mitotic and meiotic cell division. The basic process of segregating chromosomes during cell division has remained virtually unchanged across millions of years of eukaryotic evolution (Malik and Henikoff 2009). This vital function is mediated by the kinetochore complex of proteins, which binds to the centromere and provides the attachment site for spindle microtubules. These microtubules pull apart sister chromatids, ensuring proper chromosome segregation. Incorrectly attached or unattached microtubules can trigger a cascade of signals that halts cell division. Given the high degree of functional constraint and the broad similarities across eukaryotes, the apparatus for chromosome segregation is expected to reflect purifying selection, wherein natural selection acts to conserve the sequence and function of most genes by removing deleterious mutations. Contrary to this expectation, centromeric DNA and a few key proteins required for chromosome segregation evolve rapidly across broad lineages of plants and animals.

A pattern of rapid evolution is often seen in other genes involved in recurrent adaptation or those that participate in genetic conflict. In the latter scenario, classically described as a 'Red Queen interaction', competing entities constantly vie for evolutionary dominance (van Valen 1973). Such 'Red Queen' scenarios explain rapid evolution in a wide range of biological phenomena including host–pathogen and sperm–egg interactions (discussed in Chapters 20 and 13 respectively). Red Queen interactions may also provide an explanation for the surprising finding that centromeric DNA and essential genes encoding components of the chromosome segregation apparatus evolve rapidly. Here, we highlight these observations and review a model that posits that competition between homologous chromosomes during female meiosis is the driving force behind this rapid evolution (Henikoff et al. 2001). This model provides broad taxonomic predictions for the evolution of centromeric DNA and proteins and introduces a general explanation for how rapid evolution at the protein–DNA interface might drive incompatibilities and reproductive isolation in animal species.

9.2 Rapidly evolving centromeric histones

Eukaryotic genomes are wrapped by nucleosomes that allow higher-order packaging of DNA into chromosomes (Malik and Henikoff 2003). Most nucleosomes are comprised of an octamer of histone proteins, two copies each of four canonical histones: H2A, H2B, H3, and H4. Because this packaging of DNA plays an important role in all aspects of chromosome structure, replication, and expression, it is unsurprising that canonical histones represent some of the most highly conserved proteins across eukaryotes. In contrast to the bulk of genomic DNA, centromeric DNA is wrapped into specialized nucleosomes. Multiple studies have shown that canonical histone H3 proteins are replaced by a centromere-specific variant, hereafter referred to as CenH3. Although CenH3s were discovered first by biochemical studies in mammals (where they are referred to as CENP-A)

Rapidly Evolving Genes and Genetic Systems. First Edition. Edited by Rama S. Singh, Jianping Xu, and Rob J. Kulathinal.

(Earnshaw and Rothfield 1985; Palmer et al. 1991), they are now believed to be present at the centromere in all eukaryotes. CenH3-containing nucleosomes, interspersed with canonical nucleosomes, are the fundamental units of centromeric chromatin (Blower et al. 2002; Panchenko and Black 2009). In situations where one of two potential centromeres on a chromosome is active (dicentric), or where a previously non-centromeric DNA acquires the ability to serve as a centromere (neocentromere), it is the presence of CenH3 proteins that serves as the diagnostic marker of active centromeric function. Functional CenH3 not only defines the basic unit of centromeric chromatin but also serves as part of a scaffold to recruit other kinetochore proteins and is thus essential for chromosome segregation (Panchenko and Black 2009).

CenH3s differ from canonical H3 proteins in three important ways (Malik and Henikoff 2003). First, canonical H3 proteins are highly conserved in both their histone fold domain and N-terminal tails, almost identical in sequence from *Entamoeba* to humans. In contrast, CenH3 N-terminal tails are variable in length and sequence to such a degree that they cannot even be considered homologous across different taxonomic groups. Second, the length of the loop 1 region of the histone fold domain, which contacts DNA, is longer in CenH3s than canonical H3 proteins. This might suggest a greater degree of sequence specificity in CenH3s when compared to canonical H3 proteins. Third, and perhaps most relevant from the perspective of this chapter, despite stringent functional constraint, CenH3s evolve rapidly across many lineages (Fig. 9.1).

Evolutionary studies of CenH3 genes in plants (*Arabidopsis*), insects (*Drosophila*), and mammals (primates) revealed strong evidence of adaptive evolution in the N-terminal tail and the loop 1 region of the histone tail fold domain (Malik and Henikoff 2001; Talbert et al. 2002; Schueler et al. 2010). Several lines of evidence imply that this rapid evolution is important for CenH3 function. For instance, chimeric swaps of the rapidly evolving Loop1 region of the histone fold domain of *Drosophila* reveal that Loop1 domains are required

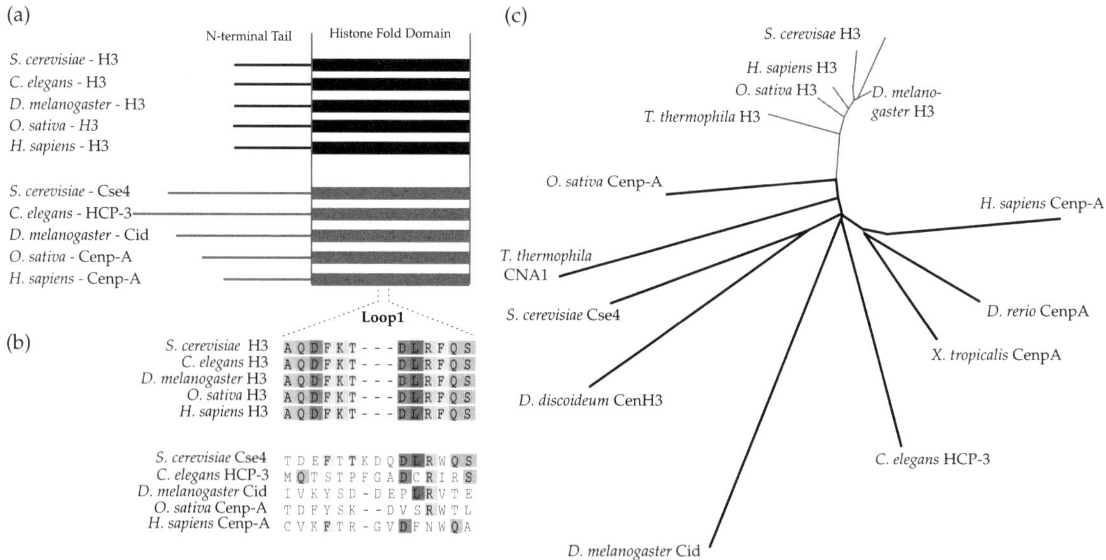

Figure 9.1 Centromeric histones diverge from canonical H3 proteins in sequence and function. (a) CenH3 proteins are distinguished from canonical histone H3s by several features. Canonical histones are highly conserved, while divergence of the N-terminal tail of CenH3 is so great that it is not possible to align these features across taxa. (b) Loop 1 domains within the histone hold domain also vary greatly in sequence and length in CenH3s but not canonical H3 proteins. (c) A neighbor-joining phylogeny of the histone-fold domains of select canonical and centromere proteins (CenH3s in bold lines) show the much faster evolution of CenH3s even in their histone fold domain.

for the correct localization of CenH3s to centromeres (Vermaak et al. 2002). Replacement of the Loop1 domain in *D. melanogaster* CenH3 with the orthologous domain from *D. bipectinata* abrogated correct centromeric localization. This implies that positive selection has actively shaped the centromeric protein–DNA interface. Similarly, CenH3 alleles from closely related plant species are not sufficient to confer full CenH3 function in *Arabidopsis* species (Ravi et al. 2010). Because CenH3s perform a conserved and essential function in eukaryotes, these findings of rapid evolution and functional divergence in CenH3s were quite surprising. Furthermore, rapid evolution in the Loop1 centromeric localization domain raised intriguing questions about the evolutionary forces that shape the interaction of CenH3s with centromeric DNA, which is also rapidly evolving in most species.

9.3 Bewildering centromeric DNA complexity and evolution

Despite carrying out identical function across eukaryotes, centromeric DNA varies widely among species (Malik and Henikoff 2009). The first centromeres characterized at the sequence level were the 'point' centromeres of *Saccharomyces cerevisiae.* One hundred and twenty-five base pairs (bp) of centromeric DNA is necessary and sufficient to recruit and assemble the protein components of the budding yeast kinetochore complex (Fig. 9.2). Thus, budding yeast centromeres are genetically defined. However, the simple 'point' centromeres of budding yeasts are exceptions to the rule even amongst fungi, many of which possess larger centromeres. Moreover, in most fungi (e.g., *S. pombe*), centromere identity is not dictated by the sequence of centromeric DNA but by the binding of centromeric proteins like CenH3s.

Centromeres in most multicellular organisms are even more complex, composed of large AT-rich repetitive sequences. These repetitive sequences, also termed 'satellite' DNA, were identified through early cloning and sequencing studies. The repetitive nature of centromeric DNA is challenging for modern sequencing technology and assembly. Current knowledge of metazoan centromeric DNA sequences is therefore based mostly on a few detailed studies of the centromeres of primates, *Drosophila*, and rice that required painstaking assembly and characterization over many years of effort.

Primate centromeres are composed of megabases of an AT-rich DNA sequence known as alpha-satellite. Alpha-satellite is a 171-bp monomeric repetitive sequence that was first identified as human DNA that disrupted chromosome segregation upon introduction into the chromosomes of African green monkey cultured cells (Haaf et al. 1992). Subsequent analysis of human centromeric sequences revealed that alpha-satellites in humans and primate relatives are arranged in higher-order arrays, where the array size varies from single alpha-satellites in most species (Cellamare et al. 2009) to a higher-order array in human centromeres, consisting of multiple tandemly-arranged monomers in repeat units. Conservation between monomers of the same array can be as low as 70–80% identity. In contrast, conservation between multimeric repeats is much higher (Rudd et al. 2006). The higher-order array structure appears to be evolutionarily young, found in only in some great apes. Moreover, some satellite-arrays are both evolutionarily young and chromosome-specific in human centromeres. For instance, the human X-chromosomal centromeric alpha-satellite array is a 2-kilobase repeat unit that is composed of 12 monomers of the DXZ1 alpha-satellite, an arrangement that is only found in the closest relatives of humans.

The evolution of centromeric and pericentric DNA sequences is sculpted by recombination (unequal crossing over and gene conversion), which acts to homogenize sequences in the center of centromeric arrays, whereas flanking pericentric sequences accumulate mutations and transpositions (Malik and Henikoff 2002). Repetitive monomers of alpha-satellite sequences are therefore not exclusive to primate centromeres; they are also found immediately adjacent to centromeres in pericentric heterochromatin. These pericentric sequences do not recruit centromeric proteins but still function to ensure proper chromosome segregation by recruiting cohesion proteins. Less pairwise sequence identity is observed among pericentric alpha-satellite monomers than between those

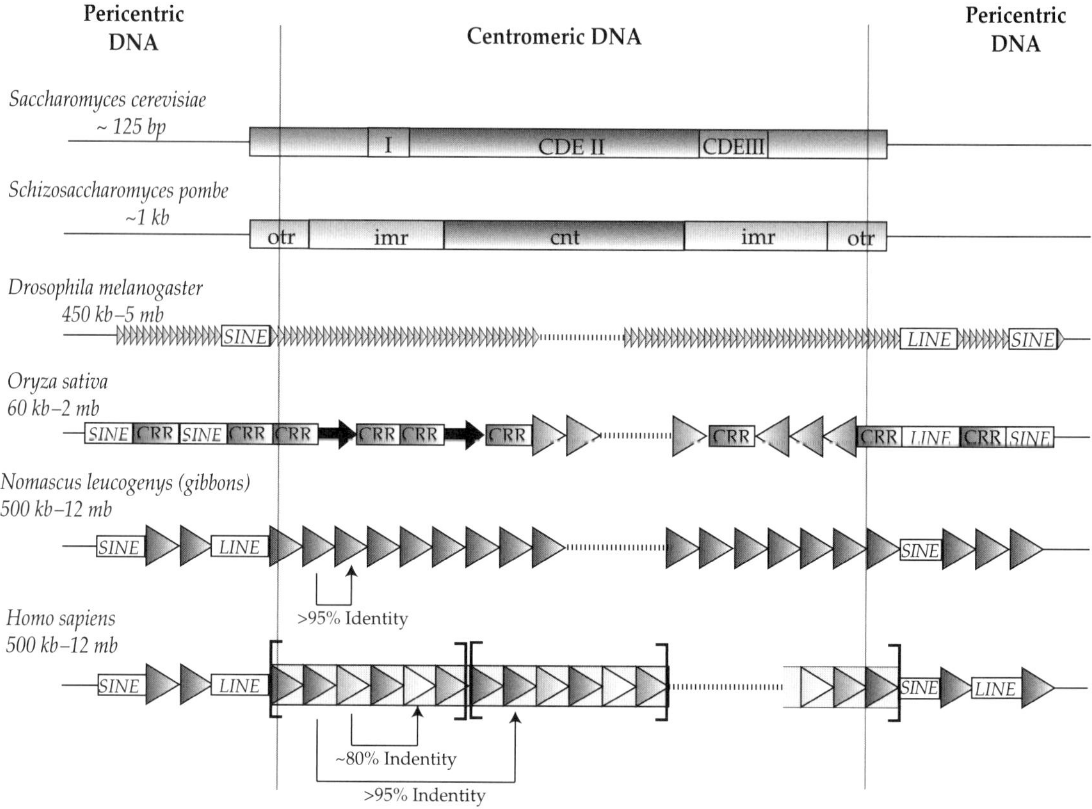

Figure 9.2 Dramatic variability in centromere size and sequence across eukaryotes. We present a few representative centromeres to highlight differences in their composition and size, ranging from 125 bp in *S. cerevisiae* to at least hundreds of kilobases in primates. Fungalcs centromeres are generally small. These may be specified genetically, as with *Saccharomyces cerevisiae*, or epigenetically as CenH3- bound DNA (e.g. *cnt* region in *S. pombe*). *Drosophila melanogaster* relies on regions of homogenous tandemly repeating pentameric repeats for CenH3 deposition. *Oryza sativa*, has a complex centromere common to many grasses. A 155-bp satellite, CentO, is arranged in long tandem arrays. Interspersed among arrays are centromeric retrotranspons in rice (CRRs) and actively transcribed genes. Both CentO and CRRs are bound by CenH3 and proposed to be functional centromeric sequences. Primate centromeres illustrate two types of arrays of alpha-satellites. *Nomascus leucogenys* centromeres consist of tandemly repeated monomers of 171-bp alpha-satellites, each with very high identity with its neighbors. In contrast, *Homo sapiens* alpha-satellites are arranged in arrays or repeat units (indicated by parentheses) spanning many alpha-satellites. The repeat units are highly identical within an array, although the neighboring alpha satellites within a repeat unit are quite dissimilar to each other.

found in centromeric arrays, which may reflect relaxed constraint or less efficient homogenization. It has been suggested that pericentric alpha-satellite monomers represent older centromeric satellites that were replaced by newly arisen variants in the middle of the centromere. In the process, the alpha-satellite monomers were gradually displaced to the edges of the centromeric array. Thus, these pericentric sequences serve as fossil records of ancestral centromeric sequences. The best-studied example of this phenomenon is found in the pericentric region of the human X-chromosome, where the oldest alpha-satellite domains are the furthest from the current centromere (Rudd et al. 2006). Homogenization of alpha-satellites is not always limited to a single chromosomal array. Indeed, a higher-order array can arise at the centromere of one chromosome during recent primate evolution, spread to other chromosomes by transposition, and become fixed (Schueler and Sullivan

2006). Surprisingly, centromeric satellite sequences are more divergent between species than are pericentric satellites (Rudd et al. 2006). The functional centromeric sequences are thus the most rapidly evolving between species, despite being most functionally constrained by their role in chromosome segregation.

In *D. melanogaster*, centromeric DNA from a minichromosome was found to be primarily composed of repetitive pentameric sequences interspersed with transposable elements. Eighty-five percent of the centromeric sequence was found to be AATAT and AAGAG satellites, with very low sequence variation (Sun et al. 1997). While the sequence composition of centromeric satellites seems to be invariant within species, the size of satellite arrays can vary dramatically within members of the same *Drosophila* species (Bachmann and Sperlich 1993). However, centromeric satellites differ even more dramatically between species. For example, there is a hundred-fold difference in abundance for the AAGAG satellite between *D. melanogaster* and *D. erecta*, which shared a common ancestor only 5–10 million years ago (Lohe and Brutlag 1987). Furthermore, some satellites present in the *D. melanogaster* genome are completely absent in the genome of *D. simulans*, suggesting complete turnover of centromeric sequences in less than 2.5 million years (Sawamura et al. 1995).

Rapid evolution of centromeric DNA has also been observed in plants. In *Oryza sativa*, centromeric regions are largely composed of two components that are interspersed with each other: a 155-bp centromeric CentO satellite, and a centromeric specific CRR (centromeric retrotransposon in rice) retrotransposon. Using chromatin immunoprecipitation experiments, investigators pulled down DNA associated with rice CenH3 and found a high level of sequence divergence at the centromeres of closely related species of wild rice. Some species of rice completely lack centromeric CentO. Comparative genomics revealed that the CentO satellites represent evolutionarily young inventions that supplanted ancestral centromeric satellites during recent evolution in rice species (Lee et al. 2005). This implies that plants, like primates and *Drosophila*, have experienced dynamic evolution of their centromeric DNA.

9.4 The 'centromere paradox': conflict, not coevolution

Observations from primates, *Drosophila*, and plants reveal a dynamic picture of centromeric DNA evolution and rapid evolution of centromeric proteins, despite an essential conserved role in chromosome segregation. How do we reconcile this rapid evolution in the face of extreme functional constraint? We consider three scenarios that may explain this 'centromere paradox'.

In the first scenario, higher mutation rates may introduce rapid changes in centromeric DNA, which then have to be accommodated by changes in centromeric proteins. It is conceivable that centromeric satellite repeats might be subject to a higher mutation rate, perhaps a result of the unique chromatin environment they are in or as a result of their unique AT-rich nucleotide composition. Indeed, recent reports from budding yeast suggest that centromeric DNAs might be subject to elevated mutation rates (Bensasson et al. 2008). Nonetheless, for newly arisen mutant centromeric sequences to survive under this coevolutionary scenario, they would have to encounter rare compensatory mutations in a centromere binding protein to avoid being eliminated by purifying selection. Thus, even if centromeric mutation rates were higher, the frequency of compensatory mutations in centromeric proteins would be rare, so this coevolutionary scenario appears unlikely to account for the centromere paradox.

In a second scenario, biased gene conversion may rapidly alter centromeric DNA satellites, followed by coevolutionary accommodation by centromeric proteins (Dover 2002). A new centromeric allele could arise which was favored by recombination (or by biased gene conversion). This new centromeric variant could then spread throughout the satellite arrays of homologous centromeres in the species. If such biased gene conversion events result in the fixation of a new centromeric DNA array that compromises accurate chromosome segregation, strong selection will act on centromere proteins to restore function. Since centromeric DNA is strongly impacted by recombination, this scenario has some explanatory power. However, the coevolutionary process should stop once a satellite variant has driven to fixation and centromeric proteins have coevolved. For it to start anew, new recombinational bias would have to be invoked in which the previously successful centromeres were replaced by newer versions. This process therefore seems unlikely to produce the recurrent patterns of rapid evolution observed across taxa. Furthermore, under this scenario, the new centromere is fixed not because of increased centromeric ability but would have to be fixed in spite of decreased centromeric ability to account for the positive selection of centromeric proteins.

We favor a third possibility in which increased centromeric ability translates directly to *increased transmission* during chromosome segregation (Henikoff et al. 2001; Henikoff and Malik 2002; Malik and Henikoff 2009). This opportunity for increased transmission arises from the unique nature of female meiosis in both plants and animals. Unlike mitosis or male meiosis, female meiosis in plants and animals is an asymmetric cell division. While mitosis produces two identical daughter cells and male meiosis results in four viable gametes, only one of the four products of female meiosis can be passed on to the next generation in the oocyte. There is, therefore, an opportunity for competition among loci on homologous chromosomes to compete for positioning and inclusion in the oocyte. Centromeres are ideally positioned to compete during female meiosis. Under this model, centromeres competitively orient towards the 'preferred pole' during meiosis I (Fig. 9.3), perhaps by recruiting more microtubules and biasing the acentrosomal spindle in female meiosis, resulting in a transmission advantage. It bears mentioning that unlike postmeiotic dysfunction following male meiosis, centromere drive incurs no fertility cost to females (Malik 2005). We posit that this competition, or centromere-drive, is the underlying genetic conflict that explains the rapid evolution of centromeric DNA. If this transmission advantage in female meiosis has a subsequent negative consequence in male meiosis, it would require centromeric proteins to adapt to restore male fertility. The key advantages of the centromere-drive model (over the biased gene conversion

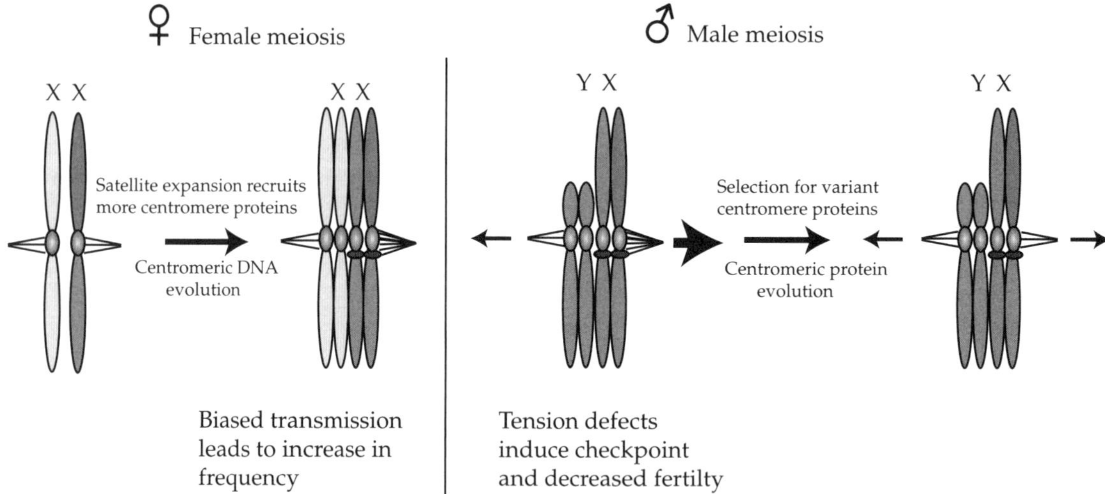

Figure 9.3 Centromere drive and suppression. In this illustration, a centromeric satellite expansion on a chromosome, for instance an X chromosome, allows for increased recruitment of centromeric proteins and microtubule attachment (a). This now 'stronger' centromere can alter the orientation of the meiotic spindle and increase its transmission during the asymmetric female meiotic divisions relative to its non-driving homolog. Such a variant would quickly increase in frequency in a population. However, the consequences of centromere drive are revealed during the symmetrical male meiotic divisions (b), here illustrated with paired X and Y chromosomes. The 'stronger' X centromere could induce tension defects that lead to reduced male fertility through the activation of a checkpoint or increased non-disjunction. Thus, strong selective pressure will favor alleles of centromere proteins that can restore tension equality and alleviate centromere drive. Thus, rapid evolution of centromeric DNA and proteins can be explained in two subsequent steps of a classic 'Red Queen' conflict.

scenario) are twofold. First, the new centromeres have increased rather than compromised centromere function. Second, it has the dynamics of a classical meiotic drive-suppression system, in which the cheating centromere 'wins' initially, but is 'suppressed' in subsequent steps via adaptation. A key discriminator between these models is that 'centromere drive' should be highly subject to taxonomic differences in meiotic programs, whereas biased gene conversion should not. As we highlight here, the genetic and taxonomic patterns of centromeric protein evolution support this model overwhelmingly.

9.5 Support for the centromere drive model

The first cytological evidence that chromosomes can exploit asymmetries in female meiosis came from studies of B chromosomes in grasshoppers, which exploit this asymmetry to enhance their own transmission (Hewitt 1973). There is now increasing evidence that many animals and plant chromosomes are shaped by similar biases in female meiosis. One of the more dramatic examples is from Robertsonian chromosomes in humans. Robertsonians are formed by the fusion of two acrocentric chromosomes into a single metacentric chromosome. Offspring of heterozygous individuals (carriers) receive either the Robertsonian fusion or the two wild type acrocentric chromosomes. In humans, male Robertsonian carriers transmit the wild-type chromosomes and the Robertsonian fusion to their offspring at equal rates, whereas the Robertsonian fusion chromosome is preferentially transmitted to 58% of the offspring from a female carrier (Pardo-Manuel de Villena and Sapienza 2001).

Further evidence for centromere drive comes from genetic studies in *Mimulus* (monkeyflower) species. In intraspecies crosses of *M. guttatus*, there is a transmission bias associated with the *D* locus, which is thought to be an expansion/duplication of a centromeric region. Chromosomes bearing the *D* locus were preferentially transmitted at 58% through female meiosis, whereas no distortion was seen in male meiosis (Fishman and Saunders 2008). In earlier crosses involving interspecies F1 hybrids between *M. guttatus* and *M. nasutus*, the *D* locus exhibited a 98% transmission bias in female meiosis (Fishman and Willis 2005). In the absence of viability differences, such strong transmission bias could only result from distortion during meiosis I, leaving the centromere as the most probable candidate for the *D* locus (Fishman and Willis 2005; Malik 2005).

Both of these examples highlight a key point about consequences of transmission distortion in female meiosis i.e. an accompanying defect in male meiosis. For instance, human male carriers of Robertsonian fusions suffer a high rate of fertility defects (Daniel 2002). Similarly, *M. guttatus* males that are homozygous for the *D* locus suffer 20% lower pollen counts (Fishman and Saunders 2008). This suggests that either heterozygosity or homozygosity of a driving centromere can be deleterious to male meiosis. One likely explanation is that unequal centromere strengths might result in tension inequity and increased non-disjunction in male meiosis (Henikoff et al. 2001; Henikoff and Malik 2002). While this could conceivably occur between any pair of chromosomes, the sex chromosomes might be especially susceptible to the deleterious effects of centromere drive. For instance, in XY systems such as mammals, the Y chromosome does not undergo female meiosis and is thus not subjected to centromere drive. Repeated rounds of competition and drive on X chromosomes could result in a 'super-X' centromere competing against a much weaker Y centromere. This inequity in centromere affinity can lead to greater rates of non-disjunction in XY male meiosis and potentially result in male sterility. Thus, selection will favor alleles of centromere binding proteins that alleviate drive and restore meiotic parity by adaptively altering their DNA-binding specificity.

9.6 Taxonomic differences in susceptibility to centromere drive

The concept of intragenomic conflict between selfish centromeric DNA sequences and centromere binding proteins is a general principle that can be extrapolated across taxa that undergo asymmetrical meiotic divisions. While patterns of evolution at the centromere play out differently in different lineages, the underlying selective forces may be universal across such taxa. Centromere drive therefore provides specific predictions about where we might expect to encounter rapid evolution of centromeres and their proteins, based on taxonomic differences in meiotic programs.

For instance, plants and animals undergo both female and male meiosis, rendering them susceptible to both centromere drive and its suppression. This is consistent with findings that both centromeric DNA and proteins evolve rapidly in plants and animals (Malik and Henikoff 2001; Talbert et al. 2002; Schueler et al. 2010). In contrast, fungi only undergo symmetric 'male' meiosis. As a result, they are not expected to undergo centromere-drive and therefore should show no evidence of centromere-drive suppression. Patterns of evolution appear to largely bear out this prediction; in particular, fungal centromeric proteins do not appear to evolve rapidly (Talbert et al. 2004). Other eukaryotic taxa experience only asymmetric 'female' meiosis. For instance, ciliated protozoans like *Tetrahymena thermophila* undergo mating between two partners, each of which undergoes asymmetric 'female' meiosis, before exchanging gametes. Intriguingly, we find no evidence of positive selection on *Tetrahymena* centromeric histones (Elde et al. 2011). This might indicate that suppressing deleterious effects in male meiosis might be the primary driver of centromeric protein evolution in taxa like plants and animals. More taxonomic sampling in lineages that have either male or female meiosis, or neither, will shed further light into the universality of the centromere-drive model.

9.7 Rapid evolution of other centromeric proteins

Based on the centromere drive model, centromeric histones need not be the only proteins that evolve to suppress the driving centromeres. Other proteins of the kinetochore, especially those that bind or are more closely associated with the centromeric DNA are also candidates to suppress centromere drive.

CENP-C is another well-studied centromeric protein that, like CenH3, is essential and found in all eukaryotes that have been investigated thus far (Fukagawa et al. 1999). CENP-C binds both centromeric DNA and CenH3s in humans, in addition to acting as a scaffold to link the outer kinetochore proteins to the centromere. Evidence from mammals showed that while CENP-C recognizes and binds centromeric DNA directly, it is still dependent on CenH3 for proper localization (Politi et al. 2002; Trazzi et al. 2009). However, in *Drosophila*, CenH3 and CENP-C have an interdependent relationship. For instance, depletion of CENP-C by RNA interference affects the deposition of new CenH3 at *Drosophila* centromeres (Orr and Sunkel 2011). Consistent with the predictions of the centromere-drive model, CENP-C evolves rapidly in both plants and animals. CENP-C evolution is often much more rapid than that of CenH3, likely reflecting the higher evolutionary constraints imposed on CenH3s (Talbert et al. 2004; Schueler et al. 2010).

The kinetochore is a trilaminar structure, consisting of an inner and outer kinetochore with a fibrous corona forming the connection to the microtubules (Cheeseman and Desai 2008). All three kinetochore layers contain proteins that are essential for its function and for chromosome segregation. However, the centromere-drive model predicts that only *inner* kinetochore proteins that directly interact with centromeric DNA will have an opportunity to suppress driving centromeres, be subject to selection, and rapidly evolve. The constitutive centromere-associated network (CCAN) is the principal group of proteins in the inner kinetochore (Hori et al. 2008). As the name suggests, the 16 proteins that constitute the CCAN are found bound to the centromere during the entire cell cycle. Based on their proximity to centromeric DNA, the CCAN proteins are ideally positioned to act as suppressors of centromere drive. Thus, systematic evolutionary analysis of the CCAN proteins, together with those found in the outer kinetochore and corona might lend further support to the predictions of the centromere drive model.

Heterochromatin proteins that participate in satellite-DNA binding proximal to centromeric regions could also be candidates for novel suppressors of centromere drive. Like CenH3s, heterochromatic satellite-binding proteins are often essential yet evolve rapidly. It is conceivable that similar genetic conflicts may shape the evolution of both centromeric and heterochromatic proteins. By adapting to bind newly arisen centromeric satellites, these heterochromatin proteins could prevent recruitment of centromeric proteins and thereby suppress centromere drive.

9.8 Centromere drive and postzygotic isolation between species

It is remarkable that a process as essential as chromosome segregation could nonetheless be shaped by lineage-specific genetic conflict and rapid evolution. This opens up the possibility that protein–DNA interactions crucial for meiosis and mitosis may not function identically across even closely related species. In hybrids, this could create incompatibilities resulting from a gain or loss of protein–DNA interactions. In crosses between *Drosophila melanogaster* and *D. simulans* (two species that have diverged for about 2.5 million years), hybrid inviability is caused by allelic mismatches of a heterochromatin protein *Lethal hybrid rescue* (*Lhr*) gene in one direction (Brideau et al. 2006), and caused by a species-specific satellite repeat called *Zygotic hybrid rescue* (*Zhr*) in the other direction (Sawamura et al. 1995; Ferree and Barbash 2009). Altered heterochromatic DNA-binding specificity also appears to play a key role in the male hybrid sterility phenotypes associated with the *OdsH* gene in hybrids between *D. simulans* and *D. mauritiana* (two species that diverged less than 0.5 million years ago) (Bayes and Malik 2009). Although such cases represent only a few examples, association with centromeric/heterochromatic function is highly overrepresented in the dataset of genes that have been found to be implicated in hybrid inviability and sterility, suggestive of a broader role for centromere drive in postzygotic isolation of incipient species (Malik and Henikoff 2009). It is indeed an intriguing possibility that genetic conflicts that drive the rapid evolution of centromeric DNA and proteins might underlie the hybrid sterility and inviability on which species concepts are most commonly based.

9.9 Future directions

It is clear that centromeric DNA and centromere-binding proteins evolve rapidly in many taxa. While the centromere-drive model has high explanatory power, several key questions remain unanswered, leaving this field ripe for further investigation. For instance, while CenH3 (and to some extent CENP-C) evolution has been investigated, most other centromere and kinetochore proteins have not been similarly examined. Evolutionary characterization of the entire complement of proteins that function in chromosome segregation could provide support for the idea that conflict with centromeric DNA sequences drives the evolution of genes encoding centromere-binding proteins. Such an evolutionary systems analysis could not only confirm this prediction of the centromere-drive model, but also identify novel DNA-binding components of the kinetochore. Another prediction of the centromere-drive model is that centromere binding proteins like CenH3 will only evolve rapidly in taxa with both asymmetric and symmetric meiotic programs. Evolutionary investigation of taxa that lack meiosis or with unusual meiosis programs could provide further support or modifications to the model. Finally, to understand the functional consequences of the rapid evolution of centromeric proteins and DNA, it will be very informative to genetically test whether adaptive changes in centromere proteins like CenH3 and Cenp-C affect chromosome segregation during meiosis or mitosis using model systems such as *Drosophila* or *Arabidopsis*.

Acknowledgments

We thank Nitin Phadnis for his comments on the manuscript. This work was supported by predoctoral fellowships from the National Science Foundation (to KCR and to BDR), grant R01-GM74108 from the National Institutes of Health (to HSM) and a grant from the Mathers Foundation (to HSM). HSM is an Early Career Scientist of the Howard Hughes Medical Institute.

References

Bachmann, L. and Sperlich, D. (1993) Gradual evolution of a specific satellite DNA family in *Drosophila* ambigua, D. tristis, and D. obscura. *Mol Biol Evol* **10**(3): 647–59.

Bayes, J.J. and Malik, H.S. (2009) Altered heterochromatin binding by a hybrid sterility protein in Drosophila sibling species. *Science* **326**(5959): 1538–41.

Bensasson, D., Zarowiecki, M., Burt, A., and Koufopanou, V. (2008) Rapid evolution of yeast centromeres in the absence of drive. *Genetics* **178**(4): 2161–7.

Blower, M.D., Sullivan, B.A., and Karpen, G.H. (2002) Conserved organization of centromeric chromatin in flies and humans. *Dev Cell* **2**(3): 319–30.

Brideau, N.J., Flores, H.A., Wang, J., et al. (2006) Two Dobzhansky-Muller genes interact to cause hybrid lethality in Drosophila. *Science* **314**(5803): 1292–5.

Cellamare, A., Catacchio, C.R., Alkan, C., et al. (2009) New insights into centromere organization and evolution from the white-cheeked gibbon and marmoset. *Mol Biol Evol* **26**(8): 1889–900.

Cheeseman, I.M. and Desai, A. (2008) Molecular architecture of the kinetochore-microtubule interface. *Nat Rev Mol Cell Biol* **9**(1): 33–46.

Daniel, A. (2002) Distortion of female meiotic segregation and reduced male fertility in human Robertsonian translocations: consistent with the centromere model of co-evolving centromere DNA/centromeric histone (CENP-A). *Am J Med Genet* **111**(4): 450–2.

Dover, G. (2002) Molecular drive. *Trends in Genetics* **18**(11): 587–9.

Earnshaw, W.C. and Rothfield, N. (1985) Identification of a family of human centromere proteins using autoimmune sera from patients with scleroderma. *Chromosoma* **91**(3): 313–21.

Elde, N., Roach, K., Yao, M.C., and Malik, H.S. (2011) Absence of positive selection on centromeric histones in Tetrahymena suggests unsuppressed centromere-drive in lineages lacking male meiosis. *J Mol Evol* **72**: 510–20.

Ferree, P.M. and Barbash, D.A. (2009) Species-specific heterochromatin prevents mitotic chromosome segregation to cause hybrid lethality in Drosophila. *PLoS Biol* **7**(10): e1000234.

Fishman, L. and Saunders, A. (2008) Centromere-associated female meiotic drive entails male fitness costs in monkeyflowers. *Science* **322**(5907): 1559–62.

Fishman, L. and Willis, J.H. (2005) A novel meiotic drive locus almost completely distorts segregation in mimulus (monkeyflower) hybrids. *Genetics* **169**(1): 347–53.

Fukagawa, T., C. Pendon, Morris, J., and Brown, W. (1999) CENP-C is necessary but not sufficient to induce formation of a functional centromere. *EMBO J* **18**(15): 4196–209.

Haaf, T., Warburton, P.E., and Willard, H.F. (1992) Integration of human [alpha]-satellite DNA into simian chromosomes: Centromere protein binding and disruption of normal chromosome segregation. *Cell* **70**(4): 681–96.

Henikoff, S., Ahmad, K., and Malik, H.S. (2001) The centromere paradox: stable inheritance with rapidly evolving DNA. *Science* **293**(5532): 1098–102.

Henikoff, S. and Malik, H.S. (2002) Centromeres: selfish drivers. *Nature* **417**(6886): 227.

Hewitt, G. (1973) Variable transmission rates of a B-chromosome in Myrmeleotettix maculatus (Thumb.) Acrididae: Orthoptera). *Chromosoma* **40**(1): 83–106.

Hori, T., Amano, M., Suzuki, A., et al. (2008) CCAN makes multiple contacts with centromeric DNA to provide distinct pathways to the outer kinetochore. *Cell* **135**(6): 1039–52.

Lee, H.-R., Zhang, W., Lnagdon, T., et al. (2005) Chromatin immunoprecipitation cloning reveals rapid evolutionary patterns of centromeric DNA in Oryza species. *Proc Natl Acad Sci U S A* **102**(33): 11793–8.

Lohe, A.R. and Brutlag, D.L. (1987) Identical satellite DNA sequences in sibling species of Drosophila. *J Mol Biol* **194**(2): 161–70.

Malik, H.S. (2005) Mimulus finds centromeres in the driver's seat. *Trends Ecol Evol* **20**(4): 151–4.

Malik, H.S. and Henikoff, S. (2001) Adaptive evolution of Cid, a centromere-specific histone in Drosophila. *Genetics* **157**(3): 1293–8.

Malik, H.S. and Henikoff, S. (2002) Conflict begets complexity: the evolution of centromeres. *Curr Opin Genet Dev* **12**(6): 711–18.

Malik, H.S. and Henikoff, S. (2003) Phylogenomics of the nucleosome. *Nat Struct Biol* **10**(11): 882–91.

Malik, H.S. and Henikoff, S. (2009) Major evolutionary transitions in centromere complexity. *Cell* **138**(6): 1067–82.

Orr, B. and Sunkel, C.E. (2011) Drosophila CENP-C is essential for centromere identity. *Chromosoma* **120**(1): 83–96.

Palmer, D.K., O'Day, K., Trong, H.L., Charbonneau, H., and Margolis, R.L. (1991) Purification of the centromere-specific protein CENP-A and demonstration that it is a distinctive histone. *Proc Natl Acad Sci U S A* **88**(9): 3734–8.

Panchenko, T. and Black, B.E. (2009). The epigenetic basis for centromere identity. *Prog Mol Subcell Biol* **48**: 1–32.

Pardo-Manuel de Villena, F. and Sapienza, C. (2001) Transmission ratio distortion in offspring of heterozygous female carriers of Robertsonian translocations. *Hum Genet* **108**(1): 31–6.

Politi, V., Perini, G., Trazzi, S., et al. (2002) CENP-C binds the alpha-satellite DNA in vivo at specific centromere domains. *Journal of Cell Science* **115**(Pt 11): 2317–27.

Ravi, M., Kwong, P.N., Menorca, R.M., et al. (2010) The rapidly evolving centromere-specific histone has stringent functional requirements in Arabidopsis thaliana. *Genetics* **186**(2): 461–71.

Rudd, M.K., Wray, G.A., and Willard, H.F. (2006) The evolutionary dynamics of alpha-satellite. *Genome Res* **16**(1): 88–96.

Sawamura, K., Fujita, A., Yokoyama, R., et al. (1995) Molecular and genetic dissection of a reproductive isolation gene, zygotic hybrid rescue, of Drosophila melanogaster. *Jpn J Genet* **70**(2): 223–32.

Schueler, M.G. and Sullivan, B.A. (2006) Structural and functional dynamics of human centromeric chromatin. *Annu Rev Genomics Human Genet* **7**: 301–13.

Schueler, M.G., Swanson, W., Thomas, P.J.; NISC Comparative Sequencing Program, Green, E.D. (2010) Adaptive evolution of foundation kinetochore proteins in primates. *Mol Biol Evol* **27**(7): 1585–97.

Sun, X., Wahlstrom, J., and Karpen, G. (1997) Molecular structure of a functional Drosophila centromere. *Cell* **91**(7): 1007–19.

Talbert, P.B., Bryson, T.D., and Henikoff, S. (2004) Adaptive evolution of centromere proteins in plants and animals. *J Biol* **3**(4): 18.

Talbert, P.B., Masuelli, R., Tyagi, A.P., Comai, L., and Henikoff, S. (2002) Centromeric localization and adaptive evolution of an Arabidopsis histone H3 variant. *Plant Cell* **14**(5): 1053–66.

Trazzi, S., Perini, G., Bernardoni, R., et al. (2009) The C-terminal domain of CENP-C displays multiple and critical functions for mammalian centromere formation. *PLoS ONE* **4**(6): e5832.

van Valen, L. (1973) A new evolutionary law. *Evolutionary Theory* **1**: 1–30.

Vermaak, D., Hayden, H.S., and Henikoff, S. (2002) Centromere targeting element within the histone fold domain of Cid. *Mol Cell Biol* **22**(21): 7553–61.

CHAPTER 10

Rapid evolution via chimeric genes

Rebekah L. Rogers and Daniel L. Hartl

10.1 Introduction

Adaptation depends ultimately upon genetic variation. While single amino acid substitutions are extremely common, their ability to explore all adaptive possibilities is limited (Carneiro and Hartl 2010). Similarly, duplicate genes, although key players in adaptive evolution and the origins of developmental complexity, do not provide a substantial source of functional novelty in the near term (Rogers and Hartl 2011). Duplicate genes create redundancy, which allows genes to diverge via neofunctionalization and subfunctionalization (Lynch and Conery 2000). However, each of these fates requires long periods of time in which genes accumulate point mutations, thus providing a very limited source of novel functions in the face of sudden or rapidly changing selective pressures.

More unusual mutations that combine different coding sequences, while somewhat rare, are able to explore greater distances in protein folding space, rendering accessible a greater number of adaptive peaks (Giver and Arnold 1998; Cui et al. 2002). Chimeric genes, which are created when portions of two distinct gene sequences unite to form a novel open reading frame, offer a means whereby single mutations can produce substantial flexibility that can allow organisms to explore a vast range of mutational space.

In this chapter, we describe how chimeric genes serve as a substantial source of new genes, which can effect rapid and drastic genetic changes in the face of selective pressures. We also explore the forces that result in the formation and maintenance of chimeric genes within the genome as well as how these forces differ among organisms.

10.2 Mechanisms of formation

The mutational mechanisms that form chimeric genes can have distinct impacts on the structure, function, and regulation of chimeric genes as well as the rates at which they are formed. In *D. melanogaster*, the youngest chimeric genes have formed through tandem duplications that have not respected gene boundaries. There are two molecular mechanisms that explain the placement and structure of the youngest chimeric genes. One involves the large-loop mismatch repair system, whereas the other results from a process similar to replication slippage (Rogers et al. 2009). There are no known chimeric genes in *D. melanogaster* that appear to have formed from retrogenes, a pattern in stark contrast to the *alcohol dehydrogenase (Adh)* chimeras that are found in other species of *Drosophila* (Wang et al. 2000; Jones and Begun 2005; Jones et al. 2005) but highly consistent with low levels of transposable element (TE) activity in the *D. melanogaster* lineage (Zhou et al. 2008; Rogers et al. 2009).

Similarly at odds with *D. melanogaster*, taxa outside the genus *Drosophila* show a marked association between chimeras and TE or retro-element activity, with a number of chimeric genes forming through retrotransposition. In rice, many newly formed retrogenes recruit new exons, adding material to the 3′ end of the gene (Wang et al. 2006). In primates, a number of chimeric retrogenes exist (Vinckenbosc et al. 2006; Virgen et al. 2008; Zhang et al. 2009), although segmental duplication can produce chimeric genes as well (Marques-Bonet et al. 2009). In these species, higher numbers of TEs and other repetitive elements can also facilitate recombination, allowing for the formation

Rapidly Evolving Genes and Genetic Systems. First Edition. Edited by Rama S. Singh, Jianping Xu, and Rob J. Kulathinal.

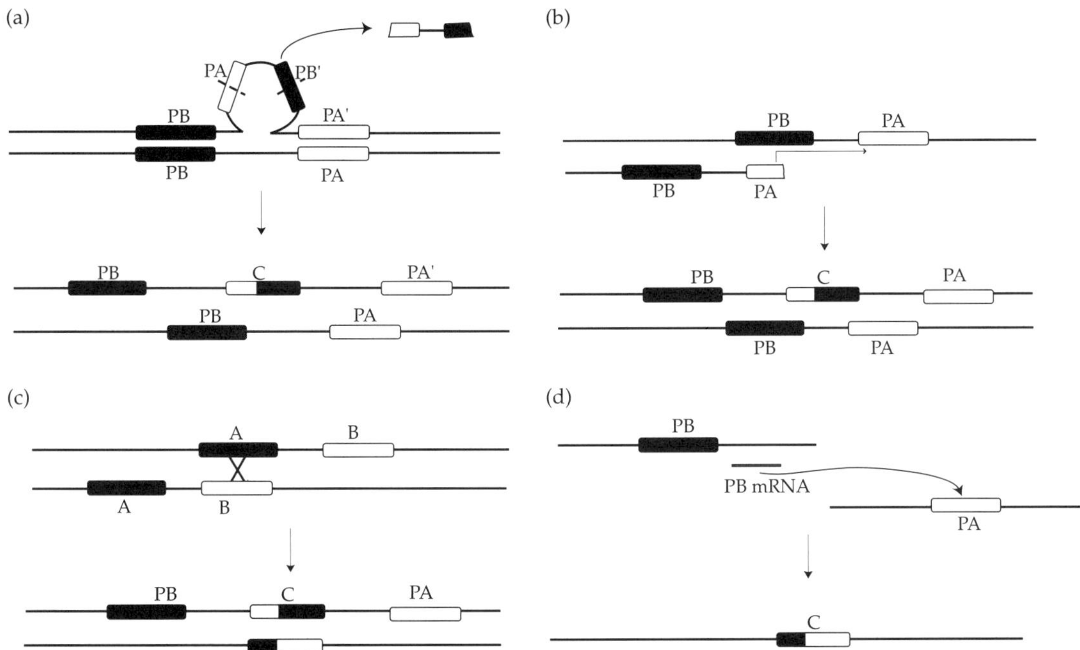

Figure 10.1 Four putative mechanisms for the formation of chimeric genes. A. Large loop mismatch repair. During either meiosis or mitosis a chromatid containing a duplication pairs with an unduplicated chromatid. The extruding loop of unpaired DNA invokes the action of the large loop mismatch repair system, which excises the segment imprecisely, producing a chimeric gene. B. Replication slippage. During replication the polymerase stalls and dissociates after partially replicating parental gene A. When replication continues, parental genes A and B misalign, producing a chimeric gene. C. Ectopic recombination. Similar nucleotide sequences in parental genes A and B facilitate pairing and recombination, resulting in a chimeric gene. D. Retrotransposition. The mRNA of one parental gene is reverse transcribed by a retroelement. The new retrogene lands within or adjacent to a second gene sequence, recruiting new exons to produce a chimeric gene.

of new genes, whereas the compact genome of *D. melanogaster* is governed more by local DNA mutations. Furthermore, retrotransposition will be strongly biased towards genes that are highly expressed, as supported by the large number of *Adh* chimeras in the *Drosophila*. Hence, retrogenes may influence genome content in different ways from errors in DNA replication and repair.

In mammals, where genomes contain long introns and large amounts of repetitive DNA, a small number of chimeric genes have been identified that have formed through ectopic recombination or ectopic gene conversion (Sedman et al. 2008; Opazo et al. 2009). These observations afford a compelling case that even small changes in cellular machinery and TE profiles can have profound influences on the genetic makeup of various organisms and their abilities to adapt to environmental changes.

Classical views on exon-shuffling predict that introns could provide reasonable breakpoints so that different domains can combine without disrupting protein folding patterns (Gilbert 1978; Patthy 2003). Yet, in *D. melanogaster* we have found that breakpoints within exons are common, that the boundaries of chimera formation do not often respect protein domains (Rogers et al. 2009; Rogers and Hartl 2011) and that many of these mid-domain breaks are selectively favored (Rogers and Hartl 2011). Furthermore, the chimeric gene *Qtzl* carries a segment inherited out of frame with respect to the parental genes, yet it has demonstrable phenotypic effects (Rogers et al. 2010). Hence, the limits of protein space may be more flexible than widely assumed, and the limits of protein modularity may fall at a level below conserved functional domains (Rogers and Hartl 2011).

However, the ability of chimeric genes to produce variation goes beyond obvious protein rearrangements. In *D. melanogaster*, chimeric genes appear to be commonly associated with changes in RNA transcription or stability, as well as changes in cellular targeting (Rogers and Hartl 2011). By shuffling different upstream and downstream sequences, both tissue-specific and timepoint-specific expression can change independently (Rogers and Hartl 2011). This ability to effect changes in gene regulation as well as peptide sequence can have distinct evolutionary consequences.

Sequences that are expressed in multiple tissues or in both sexes may be highly constrained by pleiotropic effects (Van Dyken and Wade 2010; Yampolsky and Bouzinier 2010). New changes that can be advantageous in one tissue or life stage can cause detrimental effects in another. Chimeric genes offer a means whereby single mutations can produce substantial flexibility that will allow organisms to explore a wide range of mutational space. The ability to change regulatory patterns along different axes immediately and independently can free chimeric genes from a large number of pleiotropic constraints and allow rapid evolution. Thus, chimeric genes can allow not only for immediate changes, but they can also potentially allow sequences to explore a greater range of mutational space and adaptive possibilities than other types of mutations (Rogers and Hartl 2011).

10.3 Selection

Evolutionary theory predicts that initial adaptive steps come from mutations of large effect, which are then followed by mutations of smaller effect that offer minor functional adjustments (Orr 2005). Chimeric genes provide for large-scale genetic change, which in some cases can translate to extreme selective effects. Estimates of selection coefficients for chimeric genes can be as high as 1% (Rogers et al. 2010), a massive selective impact for *D. melanogaster* relative to its effective population size ($N_e \approx 10^6$). Roughly 20% of newly-formed chimeric genes in *D. melanogaster* appear to be adaptive, a fivefold enrichment for selective sweeps compared to newly formed duplicate genes (Rogers and Hartl 2011). The appearance of these genes is often followed by a series of adaptive amino acid replacements that can modify the function of a chimeric gene (Jones and Begun 2005; Jones et al. 2005). Furthermore, chimeric genes form at high enough rates to provide a steady stream of adaptive changes (Rogers et al. 2009). Thus, chimeric genes, while seemingly unconventional, are important factors in adaptive evolution as well as serious contributors to genomic content and organization in *D. melanogaster*.

Although the role of chimeric genes in adaptation has not been systematically surveyed in any organism outside *Drosophila*, there are a few well-defined cases of adaptive chimeric genes in other organisms. In humans the chimeric gene *PIPSL* formed from retrotransposition to produce a new coding sequence that is expressed in the testes, and has experienced adaptive bursts similar to those seen in several *Adh*-derived chimeras (Zhang et al. 2009). In contrast, many chimeric retrogenes in rice show d_N/d_S values that signal strong selective constraint (Wang et al. 2006). The frequency with which chimeric genes can foster adaptive changes in organisms outside *Drosophila* remains to be seen. Differences in environmental stability, developmental plasticity, and cellular tolerance for abnormal genetic changes could easily result in different genetic profiles across phyla, and may therefore influence organisms' ability to adapt to sudden environmental change.

10.4 Genomic stability

Intimately coupled with selective pressures and the cellular mechanisms that produce chimeric genes are the rates at which they form and are preserved, or stably incorporated into the genome. Estimated rates of formation and preservation are at this point available only for *D. melanogaster*, but they indicate that chimeric genes from at relatively high rates of about 76 per million years (Rogers et al. 2009). In contrast, systematic chimeric gene searches in *S. cerevisiae* and *C. albicans* have netted no chimeric genes (Rogers, Bedford, and Hartl, unpublished). More permissive chimeric gene searches in *C. elegans* and rice have netted large numbers of chimeric genes, possibly indicating even higher rates of formation than in *D. melanogaster* (Katju, and Lynch

2006; Wang et al. 2006). The extent to which these differences are due to annotation methods for existing gene models or indicate genuine absence is not entirely clear. Still, there could be stark differences in the rates at which these genes form among organisms, which could in turn influence the organisms' potential to respond to selective changes.

Estimates of chimeric gene preservation in *D. melanogaster* indicate that 1.4% of chimeric genes that form become stably incorporated into the genome, a rate of preservation that is roughly equivalent to the rate of preservation for duplicate genes (Rogers et al. 2009) but far lower than the rate at which chimeric genes fix due to selection (Rogers and Hartl 2011). These preserved chimeric genes all appear to have formed from related peptides that are currently differentially regulated (Rogers and Hartl 2011). Shuffling portions of distantly related proteins has been shown to produce novel phenotypic effects in yeast (Mody et al. 2005), and it is possible that while these parental peptides appear similar, their chimeric rearrangements could produce fully distinct functions. Still, our results also suggest that regulatory differences may often be essential for chimera preservation.

These preserved genes that are maintained in the genome due to selective forces are also in stark contrast to *jingwei* and the other young chimeras that have been associated with recent selective sweeps, which all form from drastically unrelated peptides (Wang et al. 2000; Rogers and Hartl 2011). Such a disparity implies that in *D. melanogaster* the forces that shape genome content over long periods of time may differ from the forces that are active in short-term adaptation to newly arising selective pressures. While many new chimeric genes form from unrelated parental genes, virtually all of these types of chimeras seem to disappear over time, leaving only those that have formed from distantly related paralogs (Rogers and Hartl 2011).

This disparity between old and young chimeras implies that advantageous genes conferring novel functions may readily fix, but often will not be maintained when selective pressures shift. In the absence of selection to maintain newly fixed genes, a deletion-biased genome like that of *D. melanogaster* (Petrov et al. 1996) is likely to lose genetic factors that were once advantageous but are not currently favored. Moreover, fixation alone is not sufficient to result in the preservation of genes over long time periods. Rather, preservation occurs when constant selective pressures prevent the removal of the new sequences. Such selective constraints can result either from partitioning ancestral gene functions or through the development of new functions (Lynch and Conery 2000). Therefore, the distinction between genes that are maintained over time and those that are removed from the genome hinges on the persistence of selective pressures.

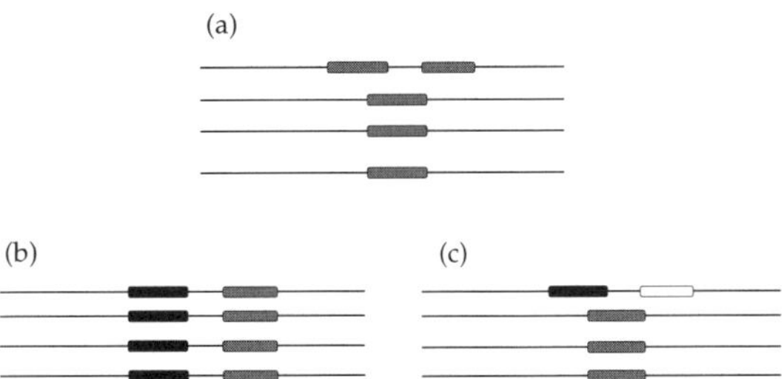

Figure 10.2 Putative fates of a duplicated gene in a population of four haploid individuals. A. Formation. A duplication forms on a single chromosome in the population. This duplication may then be lost or preserved, depending on functional impacts, selective impacts, and population dynamics. B. Fixation. One duplicate copy neofunctionalizes and becomes fixed in a population due to selection. If selective pressures remain constant, this gene will be preserved. Alternatively, if selective pressures change, the gene may be removed from the genome through nonfunctionalizing mutations. In *D. melanogaster* 3.8% of new duplicate genes fix due to selective sweeps compared with 19.3% of newly formed chimeras. C. Preservation. Duplicate copies subfunctionalize, such that each copy performs a subset ancestral gene functions. Neither copy can be removed from the genome without incurring a selective disadvantage, and hence the gene will be preserved in the genome regardless of selective impacts or frequency in the population. Estimates from *D. melanogaster* suggest that 1.4% of chimeric genes are eventually preserved compared with 4.1% of duplicates.

10.5 Function

One of the least understood aspects of chimeric genes concerns the ways in which different protein segments contribute to novel and adaptive functions. One of the first chimeric genes discovered in *Drosophila*, *jingwei*, is derived from a retrotransposed copy of the *Adh* locus, with a new 5′ end derived from *yande* (Wang et al. 2000). This novel peptide performs dehydrogenase activity like

Adh, but the new residues confer a different specificity towards long-chain primary alchols to produce a novel phenotype (Zhang et al. 2004). Hence, the adaptive value of *jingwei* derives not from an entirely novel chemical activity, but rather from a change in the substrate upon which the enzyme acts.

While it is clear that many of these chimeric peptides are beneficial, it is difficult to partition their molecular impacts with respect to selection. We do not fully understand how often interactions between domains produce novel functions or whether chimeras are favored primarily due to the regulatory changes they can effect upon existing domains. In the formation of *jingwei*, changes in specificity due to new amino acid residues have resulted in a new phenotype, but it is still unclear how easily such changes can take place with other peptides or how they might be coordinated with regulatory changes.

For many of the older chimeric genes it is equally difficult to infer the original selective impacts of a gene from its current functions. The chimeric gene *pannier* has been implicated in the differentiation of cardiac mesoderm, while its parental genes *GATAe* and *grn* are known to be involved in the development of the endoderm and ectoderm respectively (Rogers et al. 2009). However, each of these genes has additional functions in subsequent developmental stages, with *pnr* involved in eye development (Oros et al. 2010). A hypothesis of initial benefits through basic developmental pathways followed by resurrection for alternative functions in later stages may seem compelling. However, the full dimensions of this particular story will not be elucidated without determining the functions of the parental genes in distant relatives, which are not currently tractable models.

In the case of the *D. melanogaster* chimera *Qtzl* we have limited functional data. This gene has appeared only recently and swept to fixation around 15,000 years ago. Hence, it is unlikely to have diverged from its original adaptive functions. Overexpression of *Qtzl* is sufficient to partially suppress defects in neurodevelopment (Laviolette et al. 2005; Rogers et al. 2010), a factor that is suggestive of a role for *Qtzl* in neurons under normal conditions. More importantly, however, it demonstrates that even seemingly unlikely candidates such as this frame-shifted chimera can produce phenotypic effects. The two other genes in our study that show signals of selection, *CG18853* and *CG18217*, have *P*-element insertion lines that are viable and fertile, but which have not been assessed for other phenotypic effects (Rogers and Hartl 2011).

For the remaining chimeras identified in *D. melanogaster*, progress may require a substantial amount of effort and creativity, as many of these are not readily amenable to standard approaches of RNAi or targeted knockouts. However, *CG11961* is an older chimeric gene now expressed in the testes, where both of its parental genes are silenced. This gene is also highly expressed in late larvae, as well as whole adult females in which, again, both parental genes are silenced. *CG11961* is separated from its parental genes by several mutations, and may therefore be an amenable target for RNAi. As this gene has a regulatory profile that is differentiated from both of its parental genes, it is a strong candidate for neofunctionalization.

10.6 Non-coding DNA

Sdic is a chimeric gene functioning in the testes that has experienced a selective sweep in *D. melanogaster* (Nurminsky et al. 1998). However, *Sdic* appears to be unusual in that several exons are derived wholly or partly from previously non-coding sequence. A large number of similar constructs have been identified in *C. elegans*, suggesting that such chimeras may not be unusual. Similarly, stop and start codon shifting is common in *D. melanogaster*, showing that the addition of previously non-coding DNA to an open reading frame is not necessarily detrimental. Yet, it is uncertain how many of these have beneficial effects like *Sdic*. Even *Qtzl*, although created from two coding sequences, includes a large segment that is inherited in a different reading frame (Rogers et al. 2010). While selective benefits from previously non-coding material may be rare, they are not as highly unlikely as was once thought.

Many chimeric genes, including those that were involved in selective sweeps, display mid-domain breaks in putative protein folding domains, even when the chimera joins drastically different peptides (Rogers and Hartl 2011). These mid-domain breaks are often thought to be detrimental, resulting in dysfunctional peptides that can harm the cell. Yet, the genes encoding some of these peptides are found at or near the center of a strong selective sweep, making it unlikely that the chimeric protein causes massive cellular problems (Rogers and Hartl 2011). While whole-domain shuffling can clearly produce new, functional peptides, these results suggest that mid-domain breaks could be equally important for development of novel functions both in an evolutionary context and in protein engineering.

10.7 Future directions

We have firmly established that chimeric genes can indeed contribute to adaptation and that they have the potential to serve as an exceptionally rich source of functional novelty in *Drosophila*. We also know that chimeras are reasonably common in most higher eukaryotes, although cursory searches suggest that the number and type of chimeric genes vary greatly across taxa. Still, the ways in which this standing variation in chimeric genes influences evolutionary outcomes has not yet been fully elucidated. Furthermore, if different cellular mechanisms can produce a bias in the types of chimeric genes produced, it is not yet understood how these differences in genome content can provide for different adaptive solutions in the face of environmental change. Systematic surveys of chimeric genes and their contributions to adaptation in a variety of taxa is essential to understand the broad impacts complex mutations have during evolution.

A second major open question concerns the role of chimeric genes and other complex mutations in the evolution of genome content and organismic complexity. The observed decoupling of selective sweeps and adaptation from the forces that define genome content over long periods of time would suggest that the factors which influence genomic complexity are largely undefined. Understanding the functional consequences of chimeric gene formation as well as their propensity to provide for neofunctionalization or adaptive subfunctionalization should help determine why some genes are preserved. More importantly, discovering which types of genetic and functional changes are likely to persist for millions of years rather than providing for transient selective sweeps will advance our understanding of the evolutionary forces that shape organisms and their genomes.

References

Carneiro, M. and Hartl, D. L. (2010) Colloquium papers: Adaptive landscapes and protein evolution. *Proc Natl Acad Sci, USA* **107** Suppl 1: 1747–51.

Cui, Y., Wong, W. H., Bornberg-Bauer, E., and Chan, H. S. (2002) Recombinatoric exploration of novel folded structures: a heteropolymer-based model of protein evolutionary landscapes. *Proc Natl Acad Sci USA* **99**: 809–14.

Gilbert, W. (1978) Why genes in pieces? *Nature* **271**: 501.

Giver, L. and Arnold, F. H. (1998) Combinatorial protein design by in vitro recombination. *Curr Opin Chem Biol* **2**: 335–8.

Jones, C. D. and Begun, D. J. (2005) Parallel evolution of chimeric fusion genes. *Proc Natl Acad Sci, USA* **102**: 11373–8.

Jones, C. D., Custer, A. W., and Begun, D. J. (2005) Origin and evolution of a chimeric fusion gene in Drosophila subobscura, D. madeirensis and D. guanche. *Genetics* **170**: 207–19.

Katju, V. and Lynch, M. (2006) On the formation of novel genes by duplication in the Caenorhabditis elegans genome. **23**: 1056–67.

Laviolette, M. J., Nunes, P., Peyre, J. B., Aigaki, T., and Stewart, B. A. (2005) A genetic screen for suppressors of Drosophila NSF2 neuromuscular junction overgrowth. *Genetics* **170**: 779–92.

Lynch, M. and Conery, J. S. (2000) The evolutionary fate and consequences of duplicate genes. *Science* **290**: 1151–55.

Marques-Bonet, T., Girirajan, S., and Eichler, E. E. (2009) The origins and impact of primate segmental duplications. *Trends genet* **25**: 443–54.

Mody, A., Weiner, J., and Ramanathan, S. (2009) Modularity of MAP kinases allows deformation of their signalling pathways. *Nat Cell Biol* **11**: 484–91.

Nurminsky, D. I., Nurminskaya, M. V., Aguiar, D. D., and Hartl, D. L. (1998) Selective sweep of a newly evolved sperm-specific gene in Drosophila. *Nature* **396**: 572–575.

Opazo, J. C., Sloan, A. M., Campbell, K. L., and Storz, J. F. (2009) Origin and ascendancy of a chimeric fusion gene: the beta/delta-globin gene of paenungulate mammals. *Mol Biol Evol* **26**: 1469–78.

Oros, S. M., Tare, M., Kango-Singh, M., and Singh, A. (2010) Dorsal eye selector pannier (pnr) suppresses the eye fate to define dorsal margin of the Drosophila eye. *Developmental Biol* **346**: 258–71.

Orr, H. A. (2005) The genetic theory of adaptation: a brief history. *Nat Rev Genet* **6**: 119–127.

Patthy, L. (2003) Modular assembly of genes and the evolution of new functions. *Genetica* **118**: 217–231.

Petrov, D. A., Lozovskaya, E. R., and Hartl, D. L. (1996) High intrinsic rate of DNA loss in Drosophila. *Nature* **384**: 346–9.

Rogers, R. L., Bedford, T., and Hartl, D. L. (2009) Formation and longevity of chimeric and duplicate genes in Drosophila melanogaster. *Genetics* **181**: 313–22.

Rogers, R. L., Bedford, T., Lyons, A. M., and Hartl, D. L. (2010) Adaptive impact of the chimeric gene Quetzalcoatl in Drosophila melanogaster. *Proc Natl Acad Sci USA* **107**: 10943–8.

Rogers, R. L. and Hartl, D. L. (2011) Chimeric genes as a source of rapid evolution in Drosophila melanogaster. *Mol Biol Evol*. **29**: 517–29.

Sedman, L., Padhukasahasram, B., Kelgo, P., and Laan, M. (2008) Complex signatures of locus-specific selective pressures and gene conversion on human growth hormone/chorionic somatomammotropin genes. *Hum Mut* **29**: 1181–93.

Van Dyken, J. D. and Wade, M. J. (2010) The genetic signature of conditional expression. *Genetics* **184**: 557–70.

Vinckenbosch, N., Dupanloup, I., and Kaessmann, H. (2006) Evolutionary fate of retroposed gene copies in the human genome. *Proc Natl Acad Sci USA* **103**: 3220–5.

Virgen, C. A., Kratovac, Z., Bieniasz, P. D., and Hatziioannou, T. (2008) Independent genesis of chimeric TRIM5-cyclophilin proteins in two primate species. *Proc Natl Acad Sci USA* **105**: 3563–8.

Wang, W., Zhang, J., Alvarez, C., Llopart, A., and Long, M. (2000) The origin of the Jingwei gene and the complex modular structure of its parental gene, yellow emperor, in Drosophila melanogaster. *Mol Biol Evol* **17**: 1294–301.

Wang, W., Zheng, H., Fan, C., Li, J., Shi, J., Cai, Z., et al. (2006) High rate of chimeric gene origination by retroposition in plant genomes. *Plant Cell* **18**: 1791–802.

Yampolsky, L. Y. and Bouzinier, M. A. (2010) Evolutionary patterns of amino acid substitutions in 12 Drosophila genomes. *BMC Genomics* **11** Suppl 4, S10.

Zhang, J., Dean, A. M., Brunet, F., and Long, M. (2004) Evolving protein functional diversity in new genes of Drosophila. *Proc Natl Acad Sci USA* **101**: 16246–50.

Zhang, Y., Lu, S., Zhao, S., Zheng, X., Long, M., and Wei, L. (2009) Positive selection for the male functionality of a co-retroposed gene in the hominoids. *BMC Evol Bic* **9**: 252.

Zhou, Q., Zhang, G., Zhang, Y., Xu, S., Zhao, R., Zhan, Z., et al. (2008) On the origin of new genes in Drosophila. *Genome Res* **18**: 1446–55.

CHAPTER 11

Evolutionary interactions between sex chromosomes and autosomes

Manyuan Long, Maria D. Vibranovski, and Yong E. Zhang

11.1 Introduction

The sex chromosomes offer a genetic apparatus involved in the sex determination in many dioecious organisms. There can be heterogametically defined males and homogametically defined females (the X–Y systems, e.g. humans and *Drosophila*) or vice versa as heterogametically females and homogametically males (the Z–W systems, e.g. chickens and silkworms). The origin and evolution of sex chromosomes has been a classic topic in evolutionary genetics that has led to many interesting observations and various theories with predicting powers. From a retrospective view, three stages of pursuit with respect to the evolution of sex chromosomes have provided much progress in understanding of the process, patterns, and evolutionary forces involved. In the first stage, attention was paid more often to the member of the sex chromosomal pair with genetically suppressed recombination, Y and W. It was proposed that these highly diverged, often degenerate, chromosomes originated from autosomes (Muller 1932; Ohno 1967; Charlesworth 1978, 1991; Lucchesi 1994), with mounting evidence recently from various genetic and genomic comparisons (Charlesworth and Charlesworth 2000). In the second stage, an active exploration examined the evolutionary changes that occurred on the X chromosome. For example, the rapid-X hypothesis, with its evolutionary dynamics that interpret the rapid change of X-linked genes (Charlesworth et al. 1987), explained whether or not sexual antagonistic mutations prefer an X-linked environment (Rice 1984). These two stages of exploration gave insight into the process and mechanisms of chromosomal evolution. Since several reviews (e.g. Vicoso and Charlesworth 2006; Ellegren and Parsch 2007; Ellegren 2011) provide clear overviews of these major lines of research, we will not simply repeat what these reviews have already summarized but will focus on the discussion of a new picture that is recently emerging in a third stage of sex chromosome evolution research: the interaction between sex chromosomes and autosomes.

Whereas investigations based on the specific biology of sex chromosomes gave exciting insight and generated valuable data about the evolution of sex chromosomes, this third stage of research—exploring the interaction between coevolving sex chromosomes and autosomes—started a decade ago when a directional copying process through retroposition was observed (Betran et al. 2002). The central question raised was no longer how the two members of the sex chromosome pair and the genes encoded in them evolve by themselves, or how the sex chromosomes affect each other during evolution. The new question is whether or not the sex chromosomes and autosomes directly affect each other over evolutionary timescales. In other words, how is the evolution of the entire genome determined by evolutionary interactions between the sex chromosomes and autosomes? Three questions were derived from this newly defined problem: (1) what is the global genome-wide pattern associated with this interaction process? (2) How does the evolution of sex chromosomes globally change the gene content across the whole genome? (3) What is the evolutionary mechanism that underlies sex chromosome–autosome interactions? The pursuit of these questions presented a new angle to view, together, sex chromosome and genome

Rapidly Evolving Genes and Genetic Systems. First Edition. Edited by Rama S. Singh, Jianping Xu, and Rob J. Kulathinal.

evolution. Evidence revealed that the evolution of sex chromosomes was no longer solely a consequence of the unique genetics of the sex chromosomes themselves, but a result of global interactions between sex chromosomes and autosomes. In this chapter, we will provide an overview of this emerging area of genome evolution, often rapid in nature, which has been driven by evolutionary interactions between the sex chromosomes and the autosomes.

11.2 Gene traffic between sex chromosome and autosomes

Male-biased genes are a class of rapidly evolving elements with high rates of origination (e.g. Swanson et al. 2001, 2004; Ellegren and Parsch 2006; Vicoso and Charlesworth, 2006). Early theoretical works described predictions of the chromosome locations of mutations from various genetic models. Notably, Rice (1984) discussed the genetic conditions in which the mutation for sexual antagonism with advantageous male but disadvantageous female effects would more likely be X-linked if it was recessive. Charlesworth et al. (1987) compared the fixation probabilities between sex chromosomal and autosomal mutations in various genetic models, showing that for recessive mutations, X-linked loci possess a fixation probability higher than autosomal loci. These theoretical results led to a conventional belief in the early 2000s that most genes for male functions might be on the X-chromosome with very limited data of genomic locations (e.g. Wang et al. 2001; Bainbridge 2003). However, such predictions on the chromosomal locations of male-biased genes were soon put into question with the analyses of genome sequence data from *Drosophila melanogaster* and humans. These genomic sequence data and analyses revealed unexpected interaction between the X chromosome and autosomes, which also impacted the genomic locations of sex-biased genes and the evolution of sex chromosomes.

11.2.1 Gene traffic in *Drosophila*

Soon after *D. melanogaster*, the first multicellular organism, was sequenced in 2000 (Adams et al. 2000), a computational approach to identify new genes was developed (Betran et al. 2002). Because there was only one fruit fly assembly available, the team decided to focus on the paralogous comparisons for identifying new genes created from RNA-based duplication, e.g. retroposition (Brosius 1993), in which a parental gene transcribes, processes out introns, and then adds a poly-A tail to the 3′ end of the retrogene and a pair of short duplicate sequences flanking the retrosequence. The ancestral relationship between parental and new copies can be easily discriminated by looking at the exon–intron structures: the parental copy contains introns whereas the new copy is intronless and may carry a poly-A track. Through a pairwise comparison of all annotated genes, recent gene duplicates with protein sequence identities higher than 70% were identified in the *D. melanogaster* genome. Further inspection of intron presence or absence in these duplicates identified 24 pairs of new retrogenes and parental gene pairs (Betran et al. 2002). A reduction to 50% protein identity uncovered 81 interchromosomal retroposition events (Dai et al. 2007).

This dataset of interchromosomal retropositions revealed two unexpected patterns (Table 11.1). First, there were 53% retrogenes that originated from the X-linked (X) parental copies. The proportion of X→A retropositions was remarkably higher than the proportion of genes on the X (17% in the fly genome). An expectation of neutrality that assumed the random generation and insertion of retrosequences predicted that the frequency of interchromosomal retroposition should be proportional to the numbers of genes and the lengths of chromosomes (the formula for calculating expectation was developed in Betran et al. (2002)). Thus, the observed and expected rates of retroposition across chromosomes differed significantly. Second, it was observed that 90% of new genes from the X→A retroposition evolved testis expression. This suggests that the origin of retrogenes would be related to the evolution of *de novo* testis function. By using genomic sequences of multiple *Drosophila* species, Bai et al. (2007) estimated the rate of retroposition throughout different evolutionary periods in ancestral genomes of *Drosophila* and detected no origination bursts,

Table 11.1 Retrogenes prefer autosomal locations in *Drosophila* (Dai et al. 2007)

	X→A	A→X	Ai → Aj	Total
Observation	43	10	28	81
Expectation	18	16	47	
Excess (%)	132	−37	−40	
	$\chi^2 = 39.13$, df = 2, $P = 2 \times 10^{-8}$			

indicating that the process of retrosposition is a stable process with a constant rate within *Drosophila* lineages.

Retroposition also occurs within chromosomes, i.e. both retrogene and parent copies are located on the same chromosome. Dai et al. (2007) showed that retroposition events within autosomes 2 and 3 in *D. melanogaster* were actually more frequent than the retroposition between autosomes (46:28). However, contrary to the autosomes, the parental genes that are located on the X appeared to avoid inserting its retrogenes onto the X chromosome. Among 44 X-derived retrogenes, only one was re-inserted onto the X whereas the other 43 moved to autosomes 2 and 3. Thus, these within-chromosomal data further supported the interchromosomal analysis: the X chromosome tended to be avoided as an insertion site of retrogenes while a large excess of its genes fathered the retrogenes.

These described studies were primarily conducted on the retrogenes found from a single species, *D. melanogaster*, so the system was underutilized: the *Drosophila* genus consists of more than 2000 species (Powell 1997). Is the X→A retrogene traffic a general phenomenon in the entire genus? After the genomes of 12 *Drosophila* species, representative of the species in the two subgenera of *Drosophila*, were sequenced (Clark et al. 2007), Vibranovski et al. (2009a) and Meisel et al. (2009) independently investigated this problem. The former study took advantage of a gene relocation database including RNA-based duplicates independently identified by Bhutkar et al. (2007) in the 12 species and the latter created their own retrogene database via a comparison of the 12 genome sequences. Both studies revealed significant X→A retrogene traffics in non-*D. melanogaster* lineages, suggesting that this is a general phenomenon in the genus.

11.2.2 Gene traffic in mammals

Soon after the initial observations of X→A traffic in *Drosophila*, attempts were made to investigate whether or not a similar process of retrogene origination also existed in the genomes of humans and other mammalian species. However, two issues from previous analyses of the genomes of *Drosophila* and humans had to be considered. First, Venter et al. (2001) failed to find a pattern in their genomic analysis of retroposition between the X chromosome and autosomes, because no attempt was made to construct a theoretical expectation as a baseline for comparison with the observation. Second, in the derivation of the expected chromosomal distribution of the *Drosophila* retrogenes, the expectation that retroposition number was proportional to both the gene number and the length of donor and recipient chromosomes assumed random mutation. It was unlikely to directly test this hypothesis in *Drosophila* because of a lack of functionless retrogenes, i.e. the processed pseudogenes (Harrison et al. 2003), although it seems to be so in an indirect inference (Betran et al. 2004) by examining the distribution of the LINE-like retrotransposons (Kaminker et al. 2002).

Stimulated by these considerations, Emerson et al. (2004) investigated the chromosomal distribution of retroposition mutations by surveying the distribution of the retropseudogenes and their parents in humans. Because retropseudogenes are not functional, their fixation probabilities should follow the prediction of the neutral theory of molecular evolution (Kimura 1983), that mutation rate is equal to the rate of neutral substitution. An examination of 1859 retropseudogenes and their parents in the human genome revealed a highly significant linear regression with the number of genes per chromosome as donors and chromosome length as recipients. This finding strongly suggested that retroposition in mammalian genomes is a random process with respect to their chromosomal distribution. Comparing 94 and 105 functional retrogenes in, respectively, human and mouse, created by interchromosomal retroposition with expected random frequencies, Emerson et al. (2004) revealed patterns unexpected from the previous analysis with *Drosophila*. Similar to *Drosophila*, there is an excess of X-linked parental genes that were copied as a retrogene onto autosomes. Different from *Drosophila*, there is an excess of retrogenes on the X-chromosome, in sharp contrast to a low rate of retroposition between autosomes (Table 11.2). Thus, the gene traffic in mammals are two-way processes between the X chromosome and autosomes. However, looking at the expression of these genes unveiled interesting patterns: the vast majority of the autosomal retrogenes which originated from the X-linked parental genes were found to be expressed in testis while an excess of the retrogenes on the X were non-sexually expressed or female-expressed (Potrzebowski et al. 2008). These bidirectional movements of retrogenes revealed the mutual impact of the X chromosome and autosomes in the fixation of new retrogenes and reorganizing the landscape of sex genes and non-sexual genes in the mammalian genome, as also seen in the mouse genome (Emerson et al. 2004).

When did the retrogene traffic start to emerge? Comparative genomic analysis of multiple mammalian species mapped the retroposition events on various branches of the mammalian phylogenetic tree (Potrzebowski et al. 2008). A high rate of retrogene origination (also see Vinckenbosch et al. 2006) was observed close to the eutherian–marsupial split 180 million years ago (mya), coincidently when the nascent sex chromosomes were formed. The expression analysis of these retrogenes out of the X chromosome was found to compensate for the silencing of their X-linked parental genes during male meiotic sex chromosome inactivation (MSCI), indicating that the MSCI is a main selective target to drive the retrogenes into the autosomes.

11.2.3 The cause and consequence of gene traffic

The non-random distribution of retrogenes and their parental genes discussed in the earlier sections in mammals and flies indicated that the mutation distribution was not the cause. Further functional analyses based on tissue expression revealed a potential target for natural selection: compensation for MSCI on the X chromosome that often silences the expression of X-linked parental genes is likely a selective advantage that can directionally fix the retrogenes on the autosomes. A recent population genomic analyses using the McDonald–Kreitman test (McDonald and Kreitman 1991) on retroposed loci in *D. melanogaster* detected positive selection responsible for the significant excess of fixed X-origination events of retroposition (Schrider et al. 2011).

Retrosposition represents a copying mechanism that can transfer genes between ectopic chromosomal locations, e.g. between the X chromosome and autosomes. There are other copying mechanisms that can also facilitate gene movement to autosomes, including DNA-based duplication (e.g. Vibranovski et al. 2009a; Zhang et al. 2010a) and

Table 11.2 Retroposition between the X chromosome and autosomes in humans (Emerson et al. 2004)

Retroposition	Expected	Observed	Excess (%)	*P-value*
Parental copies from chromosomes:				
X→	3.76	15	299	0.00012
A→	90.24	79	−12	
Retrogene insertion into chromosomes:				
→X	3.61	13	260	0.00244
→A	90.39	81	−10	

selective gene extinction on the X chromosome (Sturgill et al. 2007). These copying mechanisms, if under positive selection over a long evolutionary timescale, predict an enrichment of male expression genes on autosomes. This predicted consequence has been detected in the genomes of mammals (Khil et al. 2004) and *Drosophila* (Parisi et al. 2003; Ranz et al. 2003) in which an under-representation of male genes was observed on the X chromosome, resulting in dominant male genes on the autosomes. Recently, it was also observed that excess female genes moved to autosomes in birds (Ellegren 2011), which can be interpreted as the earlier observed depletion of female-biased genes on the Z (Kaiser and Ellegren 2006; Storchova and Divina 2006; Mank and Ellegren 2008). The bidirectional retrogene movement between the Z chromosome and autosomes was recently found to be associated with an excess of female retrogenes from the Z chromosome and an excess of male retrogenes onto the Z chromosome (Wang et al. 2011), which confirms a previously observed over-representation of testis-specific genes on the Z chromosome in these organisms (Arunkumar et al. 2009). Thus, the evolution of sex chromosomes clearly impacted the numbers and functional properties of genes in autosomes; these two chromosomes extensively interacted in the past.

11.3 The generality of gene traffic out of the X in the genus *Drosophila*

Gene traffic associated with testis expression raised the possibility that natural selection may have played an essential role in the distribution of sex-biased genes, suggesting that the 'out of the X' movement pattern should not be limited on the particular lineage toward *D. melanogaster* or on the particular molecular mechanism to generate new gene duplicates. Similar gene traffic should also be observed in non-*D. melanoagster* species and non-RNA-based duplication such as DNA-based duplication. Testing the generality of gene traffic requires a multiple-species genomic comparison in order to assess the ancestral and derived states of gene duplicates, which fortunately is supported by the availability of the 12 sequenced *Drosophila* species (Clark et al. 2007).

11.3.1 Gene traffic in Drosophilidae and RNA-based and DNA-based duplication

Vibranovski et al. (2009b) analyzed the duplicate gene database from an independent group (Bhutkar et al. 2007) who identified all duplicate events by comparing the genome sequences of the 12 *Drosophila* species (Clark et al. 2007). They differentiated between newly created copies derived from RNA-based and DNA-based duplications and mapped the traffic patterns between the X chromosome and autosomes onto the phylogenetic tree of this genus (Fig. 11.1). The distributions of the RNA-based and DNA-based duplication events in the phylogenetic tree compared to neutral expectations (Fig. 11.1) revealed that: (1) RNA-based duplication events in the non-*D. melanogaster* lineages showed significant X→A movements, as was previously found in the paralogous analysis in the *D. melanogaster* lineage (Betran et al. 2002). This analysis suggests that the gene traffic generated by RNA-based duplication is not a specific property of the *D. melanogaster* genome, but a general phenomenon in the *Drosophila* genus as represented by the sequenced twelve species. (2) Surprisingly, DNA-based duplication events identified from the 12 species also showed significant out-of-X moment. By pooling all 203 events, 85 moved from the X chromosome to autosomes, significantly more frequently than expected at a 61.7% excess (Table 11.3).

11.3.2 Independent tests of gene traffic

Meisel et al. (2009) also generated a gene duplicate database using the 12 *Drosophila* species' genomes. In this valuable effort, they provided an independent test of similar issues. First, they confirmed the X–A patterns in the RNA-based duplication in these species. Second, they reported no significant X→A excess in interchromosomal distribution of DNA-based duplication events except for the excess of DNA-based movement out of the neo-X chromosome in *D. pseudoobscura*. Thus, while most observations in the Vibranovski et al. (2009b) were confirmed, there was a difference regarding most lineages in the DNA-based duplicates in this study. This difference was embedded in the different tests

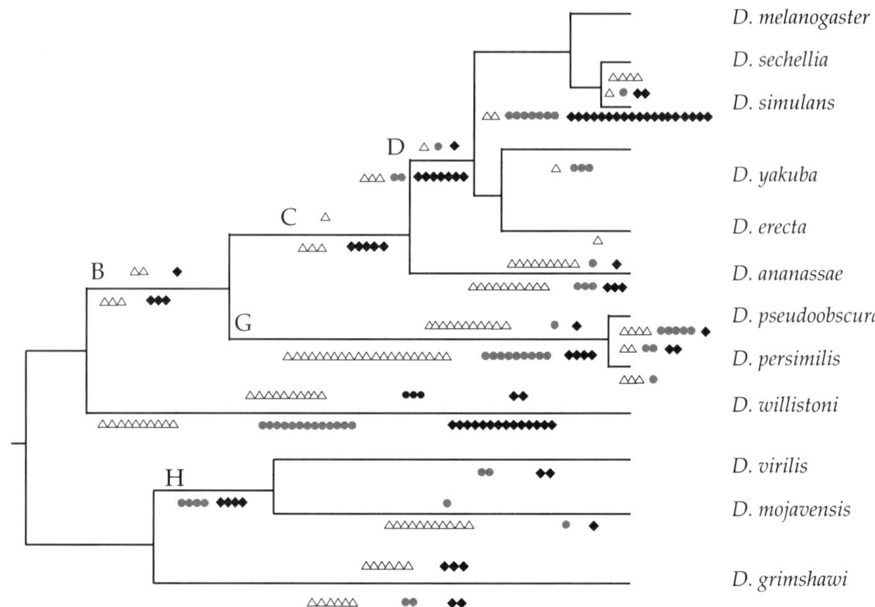

Figure 11.1 The phylogenetic distribution of new genes created by RNA-based and DNA-based duplication in the *Drosophila* genus (Vibranovski et al. 2009). Relocations based on RNA and DNA are located above and below the branch lines, respectively. Movements between chromosomes are presented as follows: (△) X→ A; (•) A→X; (♦) Ai→Aj. The average expected proportions of these relocations are 21:23:56, respectively. For species bearing neo-X chromosome the average expected proportions are 35:34:31.

Table 11.3 The analysis of the new genes which originated through RNA-based duplication and DNA-based duplication within the *Drosophila* genus

	Observation	RNA-based duplication Expectation	Excess	Observation	DNA-based duplication Expectation	Excess
X→A	39	18	121%	85	53	62%
A→X	9	16	−43%	52	62	−16%
Ai→Aj	11	26	−57%	66	89	−26%
		$x^2 = 36.29$, df = 2, $P = 1.32 \times 10^{-8}$			$x^2 = 27.28$, df = 2, $P = 1.19 \times 10^{-6}$	

Note: all the branches toward *D. melanogaster* were excluded for the RNA-based duplication.

that dealt with sample size and lineage distribution. In the Meisel et al. (2009) study, individual lineages were tested individually, many of which had very small numbers of duplication events rendering the tests with very low statistical power. For example, in *D. melanogaster*, only 9–12 events were detected and used in a statistical analysis while in *D. grimshawi*, a sample size as small as four events was used. Even so, when checked, the excess in the analysis of 15 individual cases for DNA-based duplication, most of the cases (11) demonstrated a positive excess for X→A, implicating a pattern in support of the conclusion drawn by Vibranovski et al. (2009a). It appears to be safe to conclude that the two different databases independently created by Bhutkar et al. (2007) and Meisel et al. (2009) support the same conclusion: X→A gene traffic is a general property in the *Drosophila* genus, independent of duplication mechanisms (i.e. RNA-based vs. DNA-based duplication).

11.4 Mechanisms underlying gene traffic out of the X: the detection of meiotic sex chromosome inactivation

A few evolutionary genetic models have been proposed to discuss the roles of related evolutionary mechanisms that drive the accumulation of male-biased genes on the autosomes and can be used to interpret the gene traffic between sex chromosomes and autosomes. These include models built at the population genetic level and the molecular mechanistic level.

11.4.1 Evolutionary genetic models

The evolutionary models commonly discussed include sexual antagonism, faster-X evolution, and the meiotic drive model. All these models, under the assumptions of certain genetic conditions, can provide interpretations for the observed interactions between the sex chromosome and autosomes. However, no statistical tests were developed for the quantitative analyses of gene movement. In sexual antagonism, the original version, as proposed in Rice (1984), predicted the X-enrichment of antagonistic alleles that favor males and were undesirable for females if such alleles were recessive. This was not the case for the distribution of male-biased genes (Parisi et al. 2003; Ranz et al. 2003), which might represent the resolution of the conflict (Innocenti and Morrow 2010). But this, similar to the faster-X evolution for recessive advantageous alleles (Charlesworth et al. 1987), is consistent with the initial stage of the traffic, a temporal excess of young male genes, as demonstrated by Zhang et al. (2010a). Assuming the dominance of antagonistic alleles, a prediction is a higher fixation probability in autosomes, which provides an explanation of the excess X→A traffic. Recently, duplication was proposed as a mechanism to resolve the sexual antagonism in which different copies can evolve male- and female-specific functions (Ellegren and Parsch 2007; Gallach et al. 2010; Gallach and Betran 2011). An analysis of the duplication model of sexual antagonism revealed that dominance was not needed to interpret these patterns of gene movement (Connallon and Clark 2011). The meiotic-drive alternative proposed by Tao (2007a, b) predicted that autosomal retrogenes might serve as an autosomal repressor to suppress the X-linked distorter in order to ensure a normal sex ratio. In this model, the excess of autosomal retrogenes can be a result of selection against meiotic-drive.

11.4.2 Molecular mechanistic models

Currently, there are two mechanistic processes which may serve as target selection to avoid: MSCI which was used in Betran et al. (2002) and dosage compensation (DC) recently proposed by Vicoso and Charlesworth (2009) and Bachtrog et al. (2010). Both hypotheses are based on the idea that if some functional process is occurring on the X that prevent or reduce the expression of male-biased genes, then natural selection will favor those mutations which relocate these genes onto autosomes. The DC and MSCI hypotheses complement each other by restricting the localization of male-biased genes on the X because both are not complete processes, as shown in the observation that those X-linked regions expressing MSCI were in the regions less compensated between the sites initiating DC in *D. melanogaster* (Bachtrog et al. 2010).

The phenomenology of MSCI has been well established in mammals (Richler et al. 1992; Ayoub et al. 1997) and observed in nematodes (Kelly et al. 2002; Reinke et al. 2004) and birds (Shoenmakers et al. 2009). The recent origination of MSCI in therian was found to correlate with the starting stage of gene movement out of the X in the similar period (Potrzebowski et al. 2008) (Fig. 11.1). It should be noted that the inactivation is by no means complete, showing various degrees of reduction in the expression level in different chromosomal regions and different organisms. However, the same phenomenon and its evolutionary role were not so straightforward in the exploration.

The possibility that MSCI may exist in *Drosophila* can be traced back to the early 1970s when Lifschytz and Lindsley (1972) analyzed the relationship between sterility and chromosomal translocations in *Drosophila*. While MSCI has been identified in mammals and nematodes (Richler et al. 1992; Kelly et al. 2002), it was not until recently that supporting evidence for MSCI in *Drosophila* and chicken has been demonstrated (Hense et al. 2007; Vibranovski et al. 2009a; Schoenmakers et al. 2009). In *Drosophila*, two studies used different approaches to measure gene activity in spermatogenesis to show the downregulation of X-linked genes (Hense et al. 2007; Vibranovski et al. 2009a). In the first study, a testis-specific reporter gene construct was inserted into different positions of the genome (Hense et al. 2007). Revealed by β- galactosidase enzymatic assays and RT-PCR (reverse transcription polymerase chain reaction) in whole *Drosophila* testis, the X-linked insertions showed significantly lower expression than those of the autosomal-linked ones, thus supporting the MSCI in *Drosophila* (Hense et al. 2007). The insertion positions were later expanded to construct a fine-scale map of the X-chromosome demonstrating that the inactivation phenomenon is spread along the entire chromosome (Kemkemer et al. 2011). In the second study, a global gene expression profile from mitotic and meiotic and postmeiotic cells from male germline was characterized via microarrays (Vibranovski et al. 2009b). Although the cells from the two first stages of spermatogenesis were not completely separated, the Bayesian analysis provided a more powerful approach than regular means-based comparisons (e.g. Sturgill et al. 2007; Meiklejoin et al. 2011) detecting a significant downregulation of X-linked genes in meiosis (Fig. 11.2). Furthermore, this analysis also revealed the compensatory expression between parental genes and retrogenes in mitotic and meiotic stages. These data, based on expression differences between mitotic and meiotic stages, provided the first opportunity to place the MSCI phenomenon in a specific meiotic phase of spermatogenesis (Vibranovski et al. 2009b). Recently, the reanalysis of *Drosophila* testis transcriptional profiles in a mutant that terminated the development of spermatogenesis in early stages of mitosis revealed a significant reduction in the expression of the X-linked genes compared to autosomes in the wild-type males, suggesting both dosage compensation in mitosis and X-inactivation in meiosis (Deng et al. 2011).

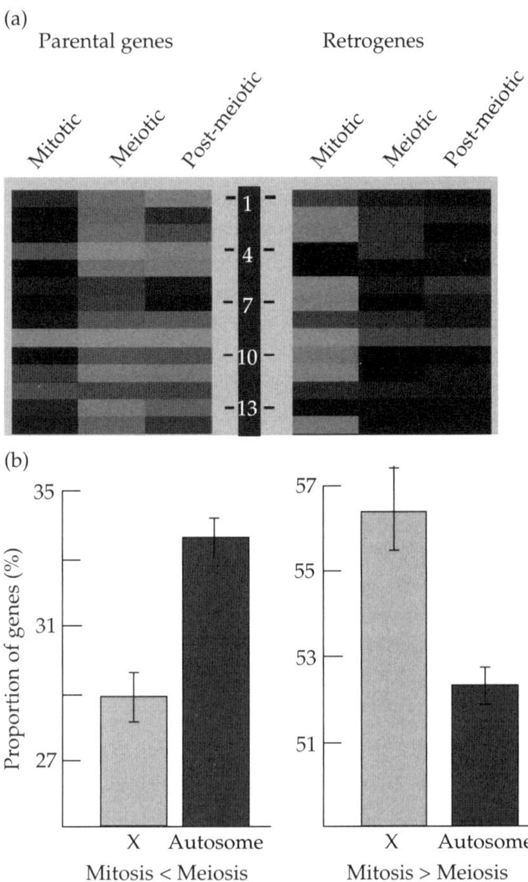

Figure 11.2 The gene locations and MSCI. (a) Mouse parental and retrogenes (from Potrzebowski et al. 2008). The transcription profile for the 14 genes in spermatogenesis which show that the retrogenes, which are copied onto autosomes, express in the MSCI stages of meiosis and postmeiosis while the parental genes on the X-chromosome express only in the mitosis stage before MSCI. (b) *Drosophila* genes that are expressed in spermatogenesis (from Vibranovski et al. 2009b). Bayesian comparison of the genes that show higher expression in meiosis compared to mitosis revealed a significant enrichment on autosomes (the left panel: meiosis > mitosis). In contrast, a comparison of the genes that show a higher expression in mitosis than meiosis revealed that most genes are X-linked (the right panel: meiosis < mitosis). Significantly more genes show the complementary expression patterns between the mitosis and meiosis stages, as shown in mouse (a).

11.5 The X-recruitment of young male-biased genes and gene traffic out of the X chromosome

The previous analyses revealed that male-biased genes are under-represented on the X chromosome of *D. melanogaster* (Parisi et al. 2003; Ranz et al. 2003). In mammals, Khil et al. (2004) found that the genes expressed during the meiotic stage in the male germline were also under-represented on the X-chromosome. This genome-wide pattern was further confirmed in multiple *Drosophila* species (Sturgill et al. 2007). Consistently, gene traffic studies showed an out-of-X gene traffic pattern where both DNA- and RNA-level autosomal duplicates tend to be male-biased if they have X-linked parental genes (Betran et al. 2002; Betran et al. 2004; Vibranovski et al. 2009a). These results may be interpreted in the sexual antagonistic model of dominant male-beneficial and female-undesirable alleles (Rice 1984) or the MSCI (Vibranovski et al. 2009b). However, a number of studies identified X-linked young testis-specific genes including *Sdic* and *Hun* originated by DNA-level duplication (Nurminsky et al. 1998; Arguello et al. 2006) and *Hydra* and four other *de novo* genes (Levine et al. 2006; Chen et al. 2007). Are these observations contradictory with the observed out-of-X gene traffic?

11.5.1 Age-dependence in *Drosophila*

This line of evidence suggests that the X chromosome is actively recruiting new male-biased genes regardless of its overall paucity of male-biased genes and out-of-X male-biased gene traffic. In order to test this hypothesis, we developed a genome-wide dating strategy to infer gene ages based on syntenic genomic alignments (Zhang et al. 2010a). We classified 12,856 protein-coding genes into seven different age groups and 947 (7%) young genes originated after *Sophophora* and *Drosophila* subgenus split.

We next profiled the transcriptional bias of new genes based on FlyAtlas microarray data (Chintapalli et al. 2007). After removing probes mapping to both parental gene and daughter genes (Dai et al. 2005) and identifying genes differentially expressed between testis and ovary (Gentleman et al. 2004; Smyth 2004), a stage-specific distribution of new genes with distinctive expression pattern was observed. As shown in Fig. 11.3, X-linked young genes are significantly more male-biased compared to autosomal young genes. Interestingly, a majority (70%) of recently evolved X-linked genes postdating the *D. melanogaster* and *D. yakuba* split are male-biased. With the elapse of evolutionary time, this proportion steadily declined. In contrast, autosomal young genes show a relatively stable proportion of male-biased genes. We also performed a genome-wide analysis without partitioning genes into different age groups and confirmed the overall demasculization of the X-chromosome where only 19% of X-linked genes are male-biased in contrast to 26% of autosomal genes which were

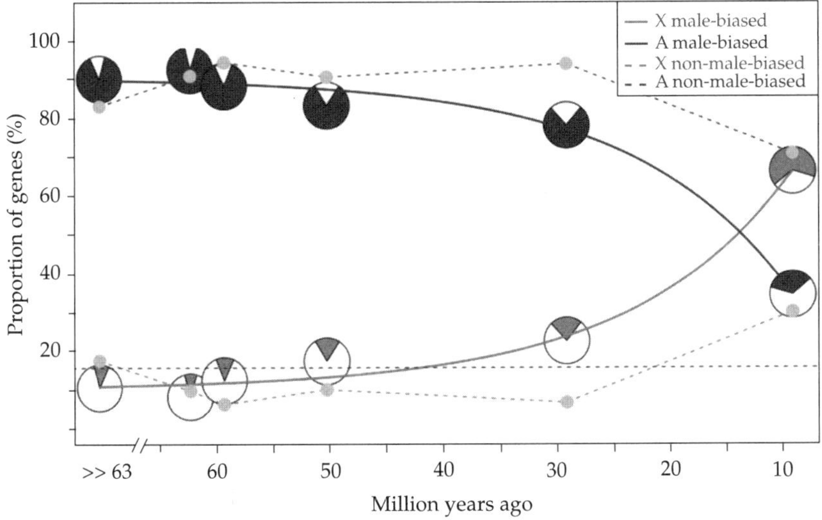

Figure 11.3 The shift of gene expression between the X-chromosome and autosomes over evolutionary time in the *Drosophila* genus, as shown by the proportions of male-biased and non-male-biased (female-biased and unbiased combined) genes originating in different evolutionary periods (Zhang et al. 2010a). For male-biased genes, we calculated the male-biased proportion as the number of male-biased genes in a given chromosome out of the whole genome. Analogously, we calculated non-male-biased gene proportions. Pie plots mark chromosomal proportions of male-biased genes. The proportions were calculated across six different evolutionary timeframes (0–6) from ancestral lineages towards present-day *D. melanogaster*.

male-biased. We, thus, detected the early stages in which X-linked male-biased genes are dominant: X-linked dominance then decreases over time until autosomal male-biased genes establish their dominance.

11.5.2 Age-dependence in mammals

Since mammals and flies possess a similar XY system that may be subject to similar evolutionary processes such as sexual antagonism, faster-X, and MSCI, we expect a similar, if not identical, pattern of new gene origination with their sex-biased expression as observed in evolution of *Drosophila* new genes. Indeed, both human and mouse data show that young X-linked genes are enriched with male-biased genes and the trend becomes reversed as gene age increases until evolutionary old male-biased genes become dominant on the autosomes (Fig. 11.3; Zhang et al. 2010b). However, because the number of young genes (hominoid-specific or primate-specific) is only 10% the fraction of the older male-biased genes, the general pattern seen from whole testis transcriptomes in humans is that autosomal male-biased genes are in significant excess over X-linked male-biased genes. Consistent with the previous genome-wide analysis (Namekawa et al. 2006), the majority of X-linked genes (i.e. 489 old X-linked genes) are subject to MSCI whereas significantly more autosomal genes are transcribed in spermatocytes. This trend extends later to spermatids. In contrast, 35 young genes are not subject to MSCI where a similar proportion of young genes are expressed in spermatocyte (29% vs. 23%) and a significantly higher proportion of X-linked young genes are expressed in spermatids (71% vs. 29%). In rodent genomes, 55% of the testis genes which showed the X-linkage originated in the 10 million years (my) after mouse diverged from rat and the rest, 25%, are rodent-specific although they originated after mouse–rat divergence (Mueller et al. 2008; Zhang et al. 2010b).

11.5.3 The slow enrichment of X-linked female genes

In both flies and mammals, the enrichment of female-biased genes on the X-chromosome was observed in the older gene group. In *Drosophila*, for all the genes older than 63 my (predating the divergence of the two subgenera, *Sophophora* and *Drosophila*), the proportion of female genes on the X chromosome is 11% higher than autosomal females genes. On the other hand, for new genes that have originated within 63 my, only 9% of X-linked new genes evolved female-biased expression and only a few new autosomal genes were detected to have female-biased expression (Fig. 11.4). These data revealed a slow pace of female gene evolution and their preferential fixation onto the X-chromosome. A similar evolutionary process of ovary-biased genes was also

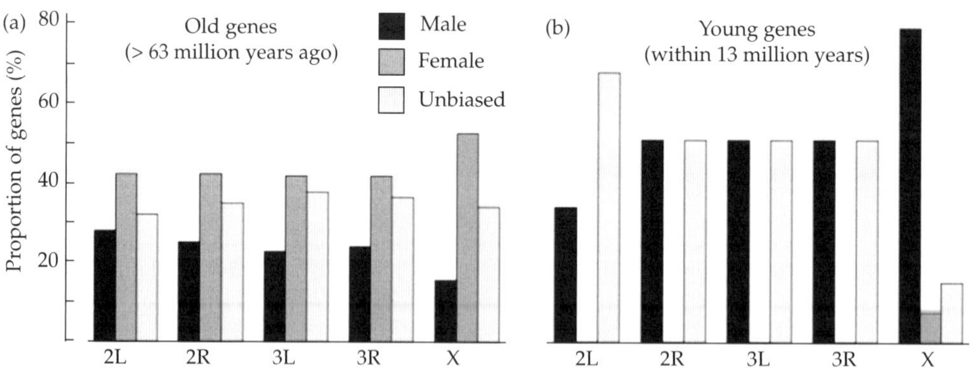

Figure 11.4 The female enrichment of genes in *Drosophila*, as shown in the chromosomal distributions of male-biased, female-biased, and unbiased *D. melanogaster* genes. (a) Evolutionary older genes that originated 63 mya before the *Sophophora–Drosophila* subgenus split. (b) Evolutionary recent genes that originated within recent 13 my.

observed in mammalian genomes, leading to significantly higher proportion of ovary-biased genes fixed in the X than autosomes, ~180 mys (before the placental–marsupial split) (Zhang et al. 2010). However, for the genes which originated within primates, more ovary-biased genes are fixed in autosomes than the X-chromosome. In genes originating between the therian and primate divergence, there are no significant differences in female-biased gene fixation between the X and autosomes. This pattern is consistent with the detected low rate of copying autosomal parental genes onto the X by the mechanism of retroposition (Emerson et al. 2004), revealing a long evolutionary time before the X-chromosome establishes a significant enrichment of female-biased genes. Interestingly, in the chicken genome, it was observed that testis-biased genes originating during avian evolution appeared to have moved to the Z-chromosomes, thus leading to over-representation on the Z chromosome (Ellegren, 2011), consistent with the female-biased gene fixation patterns in XY sex chromosomal systems. However, it was observed that the excess of old genes which are expressed in somatic ovarian cells (Granulosa) were enriched on the Z chromosome of chicken too (Morkovsky et al. 2010). These data suggest that X/Z chromosomes have been experiencing a similar functional reorganization towards an enrichment of heterogametic sex functions, since their origination from ancestral autosomes.

11.6 Concluding remarks

In this chapter, we summarize over a decade of major findings from the cross-chromosomal gene traffic literature, after the initial findings that an excess of retrogenes was found copied onto autosomes from X-linked parental genes in *Drosophila*. Through these observations and analyses, extended from *Drosophila* to mammals, birds, silkworms, and nematodes, a new concept is emerging: the interaction between the sex chromosomes and autosomes has impacted the evolution of genes and genomes, continuously changing the structure of genomes in terms of gene content and their reproductive functions in the sex chromosomes and autosomes. Over a longer evolutionary timescale, genes with heterogametic sex-biased expression will establish a dominant presence in autosomes and a lower but significant excess of genes with homogametic sex-biased expression in the sex chromosomes, X and Z. In *Drosophila* and mammalian genomes with independent origins of sex chromosomes, processes of gene evolution share an evident pattern: both started from the X-linkage of dominant young male-biased genes before the trend shifted towards an autosomal dominance of male-biased genes. Furthermore, the diverse genomes of XY and ZW genetic systems evolved via symmetrical patterns of gene movements through long evolutionary processes even though their heterogameties (or homogameties) define opposite sexes. However, although it may be safe to conclude that the underlying evolutionary force to drive the interaction is positive selection, the evolutionary genetic mechanisms and selective targets are far from clear. The current data reveal that there are likely multiple factors responsible, including population genetic processes and molecular mechanisms. Mechanistic processes such as meiotic sex chromosomal inactivation and dosage compensation are better understood than population genetic processes in which no explicit statistical tests have been developed. These leave new and challenging questions to pursue the understanding of evolutionary interactions between the sex chromosome and autosomes and their roles in driving the evolution of genes, genomes, and genetic systems such as sex and reproduction.

References

Adams, M.D., Celniker, S.E., Holt, R.A., Evans, C.A., Gocayne, J.D., Amanatides, P.G., et al. (2000) The genome sequence of *Drosophila melanogaster*. *Science* **287**: 2185–95.

Arguello, J.R., Chen, Y., Yang, S., Wang, W., and Long, M. (2006) Origination of an X-linked testes chimeric gene by illegitimate recombination in Drosophila. *PLoS Genet* **2**(5): e77.

Arunkumar, K.P., Mita, K., and Nagaraju, J. (2009) The silkworm Z chromosome is enriched in testis351 specific genes. *Genetics* **182**: 493–501.

Ayoub, N., Richler, C., and Wahrman, J. (1997) Xist RNA is associated with the transcriptionally inactive XY body in mammalian male meiosis. *Chromosoma* **106**: 1–10.

Bai, Y.S., Casola, C., Feschotte, C., and Betrán, E. (2007) Comparative genomics reveals a constant rate of origination and convergent acquisition of functional retrogenes in Drosophila. *Genome Biology* **8**: R11.

Bachtrog, D., Toda, N.R., and Lockton, S. (2010) Dosage compensation and demasculinization of X chromosomes in Drosophila. *Curr Biol* **20**(16): 1476–81.

Bainbridge, D. (2003) *The X in Sex – How the X chromosome Controls our Lives*. Cambridge, MA: Harvard University Press.

Betrán, E., Thornton, K., and Long, M. (2002) Retroposed new genes out of the X in Drosophila. *Genome Res* **12**, 1854–9.

Betrán, E., Emerson, J.J., Kaessmann, H., and Long, M. (2004) Sex chromosomes and male functions: Where do new genes go? *Cell Cycle* **3**: 873–5.

Bhutkar, A., Russo, S.M., Smith, T.F., and Gelbart, W.M. (2007) Genome-scale analysis of positionally relocated genes. *Genome Res* **17**: 1880–7.

Brosius, J. (1991) Retroposons – seeds of evolution. *Science* **251**: 753.

Charlesworth, B. (1978) A model for the evolution of Y chromosomes and dosage compensation. *Proc Natl Acad Sci U S A* **75**: 5618–22.

Charlesworth, B. (1991) The evolution of sex chromosomes. *Science* **251**: 1030–3.

Charlesworth, B. and Charlesworth, D. (2000) The degeneration of Y chromosomes. *Philos Trans R Soc Lond B Biol Sci* **355**(1403): 1563–72.

Charlesworth, B., Coyne, J.A., and Barton, N.H. (1987) The relative rates of evolution of sex chromosomes and autosomes. *Am Nat* **130**(1): 113–46.

Chen ST, Cheng HC, Barbash DA, and Yang HP. (2007) Evolution of hydra, a recently evolved testis-expressed gene with nine alternative first exons in Drosophila melanogaster. *PLoS Genet* **3**(7): e107.

Chintapalli, V.R., Wang, J., and Dow, J.A.T. (2007) Using FlyAtlas to identify better Drosophila melanogaster models of human disease. *Nat Genet* **39**(6): 715–20

Clark, A.G., Eisen, M.B., Smith, D.R., Bergman, C.M., Oliver, B., Markow, T.A., et al. (2007) Evolution of genes and genomes on the Drosophila phylogeny. *Nature* **450**: 203–18.

Connallon, T. and Clark, A.G. (2011) The resolution of sexual antagonism by gene duplication. *Genetics* **187**: 919–937.

Dai, M., Wang, P., Boyd, A.D., Kostov, G., Athey, B., Jones, E.G., et al. (2005) Evolving gene/transcript definitions significantly alter the interpretation of GeneChip data. *Nucleic Acids Res* **33**(20): e175.

Deng, X.X., Hiatt, J.B., Nguyen, D.K., Ercan, S., Sturgill, D., Hillier, L.W., et al. (2011) Evidence for compensatory upregulation of expressed X-linked genes in mammals, Caenorhabditis elegans and Drosophila melanogaster. *Nature Genet* **43**(12): 1179–85.

Ellegren, H. (2011a) Sex chromosome evolution: recent progress and the influence of male and female heterogamety. *Nat Rev Genet* **12**:157–66.

Ellegren, L. (2011b) Emergence of male-biased genes on the chicken Z-chromosome: Contrasts between male and female heterogametic systems. *Genome Res* **21**(12): 2082–6.

Ellegren, H. and Parsch, J. (2007) The evolution of sex-biased genes and sex-biased gene expression. *Nat Rev Genet* **8**: 689–98.

Emerson, J.J., Kaessmann, H., Betran, E., and Long, M. (2004) Extensive gene traffic on the mammalian X chromosome. *Science* **303**: 537–40.

Gallach, M. and Betrán, E. (2011) Gene duplication might resolve intralocus sexual conflict. *Trends Ecol Evol* **26**: 558–9.

Gallach, M., Chandrasekaran, C., and Betrán, E. (2010) Analyses of nuclearly encoded mitochondrial genes suggest gene duplication as a mechanism for resolving intralocus sexually antagonistic conflict in Drosophila. *Genome Biol Evol* **2**: 835–50.

Gentleman, R.C., Carey, V.J., Bates, D.M., Bolstad, B., Dettling, M., Dudoit, S., et al. (2004) Bioconductor: open software development for computational biology and bioinformatics. *Genome Biol* **5**(10): R80.

Harrison, P.M., Milburn, D., Zhang, Z.L., Bertone, P., and Gerstein, M. (2003) Identification of pseudogenes in the Drosophila melanogaster genome. *Nucleic Acids Res* **31**: 1033–7.

Hense, W., Baines, J.F., and Parsch, J. (2007) X chromosome inactivation during Drosophila spermatogenesis. *PLoS Biol* **5**: e273.

Innocenti, P. and Morrow, E.H. (2010) The sexually antagonistic genes of Drosophila melanogaster. *PLoS Biol* **8**(3): e1000335.

Kaiser, V.B., and Ellegren, H. (2006) Nonrandom distribution of genes with sex-biased expression in the chicken genome. *Evolution* **60**: 1945–51.

Kaminker, J.S., Bergman, C.M., Kronmiller, B., et al. (2002) The transposable elements of the Drosophila melanogaster euchromatin: A genomics perspective. *Genome Biol* **3**: research0084.1–84.2.

Kelly, W.G., Schaner, C.E., Dernburg, A.F., Lee, M.H., Kim, S.K., Villeneuve, A.M., et al. (2002) X-chromosome silencing in the germline of C. elegans. *Development* **129**: 479–92.

Kemkemer, C., Hense, W., and Parsch, J. (2011) Fine-scale analysis of X chromosome inactivation in the male germline of Drosophila melanogaster. *Mol Biol Evol* **28**(5): 1561–63.

Khil, P.P., Smirnova, N.A., Romanienko, P.J., and Camerini-Otero, R.D. (2004) The mouse X chromosome is enriched for sex-biased genes not subject to selection by meiotic sex chromosome inactivation. *Nat Genet* **36**: 642–6.

Kimura, M. (1983) *The Neutral Theory of Molecular Evolution*. Cambridge: Cambridge University Press.

Levine, M.T., Jones, C.D., Kern, A.D., Lindfors, H.A., and Begun, D.J. (2006) Novel genes derived from noncoding DNA in Drosophila melanogaster are frequently X-linked and exhibit testis-biased expression. *Proc Natl Acad Sci U S A* **103**(26): 9935–9.

Lifschytz, E. and Lindsley, D.L. (1972) The role of X-chromosome inactivation during spermatogenesis. *Proc Natl Acad Sci U S A* **69**: 182–6.

Lucchesi, J.C. (1994) The evolution of heteromorphic sex chromosomes. *BioEssays* **16**: 81–3.

Mank, J.E. and Ellegren, H. (2009) Sex-linkage of sexually antagonistic genes is predicted by female, but not male, effects in birds. *Evolution* **63**: 1464–72.

Meiklejohn, C.D., Landeen, E.L., Cook, J.M., Kingan, S.B., and Presgraves, D.C. (2011) Sex chromosome-specific regulation in the Drosophila male germline but little evidence for chromosomal dosage compensation or meiotic inactivation. *PLoS Biol* **9**(8): e1001126.

Meisel, R.P., Han, M.V., and Hahn, M.W. (2009) A complex suite of forces drives gene traffic from Drosophila X chromosomes. *Genome Biol Evol* **1**: 176–88.

Mořkovský, L., Storchová, R., Plachý, J., Ivánek, R., Divina, P., and Hejnar, J. (2010) The chicken Z chromosome is enriched for genes with preferential expression in ovarian somatic cells. *J Mol Evol* **70**(2): 129–36.

Mueller, J., Mahadevaiah, S., Park, P., Warburton, P.E., Page, D.C., and Turner, J.M. (2008) The mouse X chromosome is enriched for multicopy testis genes showing postmeiotic expression. *Nat Genet* **40**: 794–9.

Muller, H.J. (1932) Some genetic aspects of sex. *Am Nat* **66**: 118–38.

Namekawa, S.H., Park, P.J., Zhang, L.F., Shima, J.E., McCarrey, J.R., Griswold, M.D., et al. (2006) Postmeiotic sex chromatin in the male germline of mice. *Curr Biol* **16**(7): 660–7.

Namekawa, S.H. and Lee, J.T. (2009) XY and ZW: Is meiotic sex chromosome inactivation the rule in evolution? *PLoS Genet* **5**: e1000493

Nurminsky, D.I., Nurminskaya, M.V., De Aguiar, D., and Hart, D.L. (1998) Selective sweep of a newly evolved sperm-specific gene in Drosophila. *Nature* **396**(6711): 572–5.

Ohno, S. (1967) *Sex Chromosomes and Sex-Linked Genes*. Berlin: Springer.

Parisi, M., Nuttall, R., Naiman, D., Bouffard, G., Malley, J., Andrews, J., et al. (2003) Paucity of genes on the Drosophila X chromosome showing male-biased expression. *Science* **299**: 697–700.

Potrzebowski, L., Vinckenbosch, N., Marques, A.C., Chalmel, F., Jegou, B., and Kaessmann, H. (2008) Chromosomal gene movements reflect the recent origin and biology of therian sex chromosomes. *PLoS Biol* **6**: e80.

Powell, J.P. (1997) *Progress and Prospects in Evolutionary Biology—The Drosophila Model*. New York: Oxford University Press.

Ranz, J.M., Castillo-Davis, C.I., Meiklejohn, C.D., and Hartl, D.L. (2003) Sex-dependent gene expression and evolution of the Drosophila transcriptome. *Science* **300**: 1742–5.

Reinke, V., Gil, I.S., Ward, S., and Kazmer, K. (2004) Genome-wide germline enriched and sex-biased expression profiles in Caenorhabditis elegans. *Development* **131**: 311–23.

Rice, W.R. (1984) Sex chromosomes and the evolution of sexual dimorphism. *Evolution* **38**: 735–42.

Richler, C., Soreq, H., and Wahrman, J. (1992) X inactivation in mammalian testis is correlated with inactive X-specific transcription. *Nature Genet* **2**: 192–5.

Schoenmakers, S., Wassenaar, E., Hoogerbrugge, J.W., Laven, J.S., Grootegoed, J.A., and Baarends, W.M. (2009) Female meiotic sex chromosome inactivation in chicken. *PLoS Genet* **5**: e1000466.

Smyth, G.K. (2004) Linear models and empirical Bayes methods for assessing differential expression in microarray experiments. *Stat Appl Genet Mol Biol* **3**: Article3.

Storchova, R. and Divina, P. (2006) Nonrandom representation of sexbiased genes on chicken Z chromosome.*J Mol Evol* **63**: 676–81.

Sturgill, D., Zhang, Y., Parisi, M., and Oliver, B. (2007) Demasculinization of X chromosomes in the Drosophila genus. *Nature* **450**: 238–41.

Swanson, W.J., Clark, A.G., Waldrip-Dail, H.M., Wolfner, M.F., and Aquadro, C.F. (2001) Evolutionary EST analysis identifies rapidly evolving male reproductive proteins in Drosophila. *Proc Natl Acad Sci U S A* **98**: 7375–9.

Swanson, W.J., Wong, A., Wolfner, M.F., and Aquadro, C.F. (2004) Evolutionary expressed sequence tag analysis of Drosophila female reproductive tracts identifies genes subjected to positive selection. *Genetics* **168**: 1457–65.

Tao, Y., Masly, J.P., Araripe, L., Ke, Y., and Hartl, D.L. (2007a) A new sex-ratio meiotic drive system in

Drosophila simulans. I. Characterization of an autosomal suppressor. *PLoS Biology* **5**(11): e292.

Tao, Y., Araripe, L., Kingan, S.B., Ke, Y., Xiao, H.L., and Hartl, D.L. (2007b) *A sex-ratio* meiotic drive system in Drosophila simulans II: An X-linked disorder. *PLoS Biology* 5(11): e293.

Venter, J.C., Adams, M.D., Myers, E.W., Li, P.W., Mural, R.J., Sutton, G.G., et al. (2001) The sequence of the human genome. *Science* **291**: 1304–51.

Vibranovski, M.D., Zhang, Y., and Long, M. (2009a) General gene movement off the X chromosome in the Drosophila genus. *Genome Res* **19**(5): 897–903.

Vibranovski, M.D., Lopes, H.F., Karr, T.L., and Long, M. (2009b) Stage-specific expression profiling of Drosophila spermatogenesis suggests that meiotic sex chromosome inactivation drives genomic relocation of testis-expressed genes. *PLoS Genet* **5**(11): e1000731.

Vibranovski, M.D., Chalopin, D.S., Lopes, H.F., Long, M., and Karr, T.L. (2010) Direct evidence for postmeiotic transcription during Drosophila melanogaster spermatogenesis. *Genetics* **186**(1): 431–3.

Vicoso, B. and Charlesworth, B. (2006) Evolution on the X chromosome: unusual patterns and processes. *Nat Rev Genet* **7**: 645–53.

Vicoso, B. and Charlesworth, B. (2009) The deficit of male-biased genes on the D. melanogaster X chromosome is expression-dependent: A consequence of dosage compensation? *J Mol Evol* **68**: 576–83.

Wang, J., Vibranovski, M., and Long, M. (2012) The gene traffic out of Z in silkworm. *J Mol Evol* (in press).

Wang, P.J., McCarrey, J.R., Yang F., and Page, D.C. (2001) An abundance of X-linked genes expressed in spermatogonia. *Nat Genet* **27**: 422–6.

CHAPTER 12

Evolutionary signatures in non-coding DNA

Dara G. Torgerson and Ryan D. Hernandez

12.1 Introduction

Approximately 97–98% of the human genome is estimated to be non-coding in nature, yet the majority of evolutionary studies continue to focus on protein-coding regions to search for signatures of adaptive evolution. In fact, it was originally thought that sequencing the vast amounts of non-coding DNA was a waste of resources, as the majority was likely to be functionally devoid. Results from high-throughput genomic analyses have changed this opinion, although many hurdles remain.

Part of the challenge is that non-coding DNA is involved in a diverse array of biological functions that often remain elusive. Although it's commonly referred to as 'non-coding' DNA, our understanding to date is that it codes for a variety of functional elements predominantly related to gene regulation and genome architecture, including transcription factor binding sites (TFBSs), non-coding RNAs, microRNA binding sites, splice sites, and histone/nucleosome binding sites. The occurrence of non-random patterns in non-coding DNA with respect to nucleotide composition, methylation patterns, recombination rates, divergence, and genetic diversity give a general indication of its functional relevance. Even the once-termed 'junk DNA' may also function as a genetic 'pool' for evolutionary novelty, including ancestral repeats and pseudogenes. As we work towards a more detailed annotation of the human genome, we expect that an increasing proportion of the non-coding genome will be identified as being functional.

This leads to the question of why it is essential to study evolutionary patterns in non-coding DNA, despite the majority of it awaiting annotation. One could argue that, first and foremost, identifying deviations from neutral evolution can lead the way in identifying novel functional regions. Natural selection must act on an expressed phenotype, which is why identifying non-coding regions subject to natural selection has frequently paved the way for functional studies to prioritize regions that are more likely to be functionally relevant today. Identifying functional non-coding DNA can, in turn, advance our understanding of basic nuclear structure and processes, including the regulation of gene expression as it extends from studies of general organismal biology to understanding the mechanisms of human disease.

One could equally argue that until we fully integrate evolutionary studies of coding and non-coding regions of the genome, we will never fully understand the mechanisms of evolution itself, how speciation occurs, or the genetic structure and history of populations. In fact, despite the initial focus of studying evolutionary patterns in protein-coding DNA, it is possible that the majority of adaptive evolutionary change has occurred in non-coding DNA. It has long been proposed that phenotypic differences between human and chimp are more likely a result of differences in gene regulation as compared to differences in the actual protein sequence itself (King and Wilson, 1975), yet we are only beginning to test that hypothesis. There are several examples whereby adaptation to novel environments has involved changes in gene regulation, and there is evidence to suggest that a greater extent of adaptive evolution has taken place in non-coding as compared to protein-coding DNA.

In this chapter we will discuss the challenges and opportunities of studying the evolution of

Rapidly Evolving Genes and Genetic Systems. First Edition. Edited by Rama S. Singh, Jianping Xu, and Rob J. Kulathinal.

non-coding DNA, and review the emerging patterns of evolutionary signatures in different classes of non-coding DNA in the human genome. We will conclude with a brief discussion of future scientific prospects as we undergo an exponential growth in the amount of genomic data available for evolutionary analyses in non-coding DNA.

12.2 Challenges to studying the evolution of non-coding DNA

One of the fundamental questions in evolutionary genetics is to what degree natural selection has acted on non-coding DNA. Because natural selection must act on an expressed phenotype, it is important to distinguish between functional and non-functional non-coding DNA—analogous to distinguishing between synonymous and nonsynonymous sites in studies of the evolution of protein-coding DNA. Equally non-trivial is deciding on whether existing evolutionary models are suitably transferable to studies of non-coding DNA, despite the potential for increased heterogeneity in the genomic environment (such as differential GC content, mutation and recombination rates) and the presence of novel or potentially unique modes of evolution (such as increased turnover of functional sites). Lastly, many of the existing methods for detecting non-neutral evolution have been geared towards protein-coding regions, which often rely on using non-coding DNA as a neutral standard. In this section we provide a brief overview of what we perceive as the three major challenges for identifying signatures of non-neutral evolution in non-coding DNA.

12.2.1 Identifying functional non-coding DNA

Methods for detecting selection rely on identifying patterns of diversity and/or divergence in functionally relevant sequences that deviate from a neutral model. In protein-coding DNA the functional unit of study is typically defined as nonsynonymous sites within a single gene (but occasionally a single exon or even codon), or by grouping these sites across all genes or those belonging to a similar pathway or functional group. While it is intuitive to assume that mutations that alter the amino acid sequence may have a functional consequence, defining a unit of study in non-coding DNA is logistically more difficult. Progress has been made by grouping non-coding sites into phylogenetically conserved and non-conserved classes of sites, and by identifying those more likely to have a gene regulatory function using either experimental or computational approaches.

High-throughput experimental approaches have greatly accelerated the functional annotation of non-coding DNA over the past few years, including chromatin immunoprecipitation followed by microarray analysis (ChIP-chip) or by sequencing (ChIP-seq). By cross-linking proteins that are bound to the DNA sequence and determining their location, specific regions in the genome have been identified where proteins interact with non-coding DNA to regulate transcription. As a result of this technology, the general location of numerous histone modification and TFBSs have been identified, however, with the caveat that differences in temporal and experimental conditions can affect the location and types of regulatory proteins that are bound to the DNA sequence. For example, differences in the tissue or cell type examined (including *in vitro*- vs. *in vivo*-based experiments), developmental stage, time of day, diet, and numerous other environmental factors can lead to differential patterns of gene expression. Regardless, the application of high-throughput experimental approaches to genome annotation has notably enhanced the annotation of non-coding DNA in the human genome, notably by the efforts of the National Human Genome Research Institute (NHGRI) Encyclopedia of DNA Elements (ENCODE) Project (Birney et al. 2007).

Various computational methods have been developed to facilitate the identification of functional non-coding DNA. For example, identifying signals of phylogenetic conservation and non-neutral evolution based on the observation that functionally relevant sites are more likely to be conserved throughout evolution (Birney et al. 2007). Of course, not all functional regions are expected to be conserved, and therefore the interpretation of the results of evolutionary studies must consider the means by which putatively functional non-coding DNA was identified. Another common approach makes use of a sample of experimentally determined TFBSs to predict specific sequence motifs where transcription factors are more likely to form direct interactions. The inferred motifs can then be used to search for novel occurrences of predicted TFBSs throughout the genome. However, because TFBSs are typically short, degenerate sequences, this method is subject to a high false positive rate. Ways of getting around this have been explored, for example by conditioning on the clustering of multiple TFBSs to identify a *cis*-regulatory module (Blanchette et al. 2006).

These are only a few of the methods used to distinguish between functional and putatively non-functional, non-coding DNA, and each have their own limitations for evolutionary studies.

12.2.2 Estimating the neutral evolutionary rate

Characterizing the effects of natural selection throughout the genome requires an accurate depiction of expected patterns of polymorphism and divergence under neutrality. The neutral evolutionary rate is generally estimated from regions of the genome that are putatively non-functional, but several challenges exist. For example, it is recognized that most genomes are composed of heterogeneous patterns of GC content, which is correlated with patterns of recombination, which in turn are both correlated with levels of diversity and substitution rates (Duret and Arndt 2008). Moreover, it has been hypothesized that one of the major drivers of evolutionary rates is GC-biased gene conversion—a byproduct of the double-stranded break repair mechanism required during meiosis.

In addition to neutral processes confounding estimates of overall evolutionary rates, unidentified targets of natural selection can also cause misleading results. Certain targets of natural selection can also have an effect on evolutionary rates at linked neutral sites, particularly when comparing closely related species (e.g. background selection (Charlesworth 1994)). In the end, characterizing the neutral evolutionary rate that should be used as a baseline for inferring natural selection is a complicated mixture of partially correlated effects and confounding factors. Overcoming these difficulties is a central challenge for inferring the effects of non-neutral evolution in non-coding DNA. While in coding regions one can default to using synonymous sites as a neutral proxy, no such category of sites exists outside of genes.

12.2.3 Limitations of identifying rapid evolution in non-coding DNA

Comparative and population genomics have set the stage for inferring the effects of natural selection across the human genome, yet the majority of methods are either more appropriate, or have increased power for the analysis of protein-coding DNA. By statistically comparing the genomes of a wide range of species, it is possible to identify regions that are highly conserved across species and those that are rapidly evolving. However, it is often more difficult to distinguish rapidly evolving sequences from paralogous sequences in non-coding as compared to protein-coding sequences, due to a greater heterogeneity in evolutionary patterns in the former. Furthermore, it is reasonable to assume that a greater proportion of rapidly evolving non-coding sequences have no alignable orthologous sequences, resulting in much of the analysis of non-coding DNA being limited to conserved non-coding sequences (CNCs). While this surely has an effect on the analysis of both coding and non-coding sequences, alignments of rapidly evolving protein-coding DNA can be augmented through an alignment of the translated amino acid sequence itself whereas non-coding sequences cannot. Methods based on comparing patterns of polymorphism to divergence are subject to the same limitations, with the added caveat that historically there was little data available on human non-coding variation that was uniformly ascertained.

12.3 Patterns of evolution in non-coding DNA

Recent advances in high-throughput sequencing have lead to an accelerating amount of publicly available, mammalian full-genome sequences. We have also begun quantifying human variation at the genome-wide level in a less-biased fashion, enabling new opportunities for evolutionary analysis on non-coding DNA. Together these advances have lead to an explosion of studies with the common goal of identifying signatures of natural selection in the non-coding region of the human genome. As with all evolutionary genetics studies, approaches taken in the analysis of non-coding DNA are comparative in nature, with the primary differences between studies residing in the types of sites being studied, and the types of comparisons being made.

12.3.1 Selection in conserved non-coding sequences?

Phylogenetic conservation is a good predictor of biological function that is useful for identifying non-coding sequences that are involved in gene regulation (Birney et al. 2007). However, there are various definitions and methods for identifying CNCs, ranging from a simple estimate of pairwise divergence between two species at orthologous sites to employing a model-based estimate of conservation using multispecies alignments (Siepel et al. 2005). One surprising finding is that the majority of conserved regions lie in non-coding regions of the genome (Waterston et al. 2002), raising the possibility that a greater degree of functionally relevant DNA is non-coding in nature. Moreover, the existence of 'ultraconserved' non-coding elements (Bejerano et al. 2004) that have been subject to purifying selection three times as strong as nonsynonymous sites (Katzman et al. 2007), and 'human accelerated regions' (HARs) that are predominantly non-coding in nature and otherwise conserved in mammals (Pollard et al. 2006). Over the past few years, studies aimed at identifying signatures of natural selection in conserved non-coding sequences have begun to reveal some interesting patterns in the evolution of non-coding DNA.

In 2005 it was being debated as to whether non-coding DNA in humans was subject to natural selection at all. Comparisons of human and chimp divergence in non-coding DNA in the upstream regions of genes by Keightley and others (2005b) revealed less conservation in CNCs compared to rodents, suggesting a general relaxation of natural selection on gene regulatory regions in primates. Patterns of human polymorphism suggested that the reduced conservation between human/chimp could not be attributed to increased adaptive evolution in primates, but rather to differences in the effective population size of primates and rodents. However, similar comparisons by Bush and Lahn (2005) suggested that purifying selection has remained an active force on conserved non-coding DNA throughout primate evolution. Three independent studies examining patterns of human genetic variation in the Perlegen (Keightley et al. 2005a), HapMap (Drake et al. 2006), and a more recent resequencing study (Torgerson et al. 2009) came to the same conclusion. All observed an excess of low frequency derived mutations within CNCs compared to a neutral standard (e.g. Fig. 12.1), suggesting that purifying selection has acted to maintain weakly deleterious mutations at low frequencies. But the question remained: was there any evidence for widespread adaptive evolution in CNCs in humans?

While it may seem counterintuitive to look for signatures of rapid, adaptive evolution in conserved sequences (as we assume they are conserved due to increased selective constraint on functionally relevant regions), a valid strategy is to define conserved regions in a manner that is nested, and/or independent of the test being used to identify rapid evolution. In many studies this has involved searching for accelerated evolution along the branch leading to modern humans as compared to the rest of the phylogenetic tree. Alternatively, several have taken the approach of identifying CNCs using more distantly related species (interspecific comparisons), but comparing patterns of genetic variation (intraspecific comparisons) and divergence to a more closely related species.

Beginning in 2006, genomic analyses began to identify signatures of adaptive evolution in CNCs in mammals, and to find non-random patterns with respect to the types of genes that reside next to rapidly evolving CNCs. Several studies compared evolutionary rates along different branches of the mammalian tree in order to test for human-specific acceleration. Prabhakar and others (2006) found an excess of human-specific substitutions within CNCs than expected under a model of selective constraint, suggesting that CNCs are not all evolutionarily constrained. Interestingly, they found that rapidly evolving CNCs along both the human and chimp

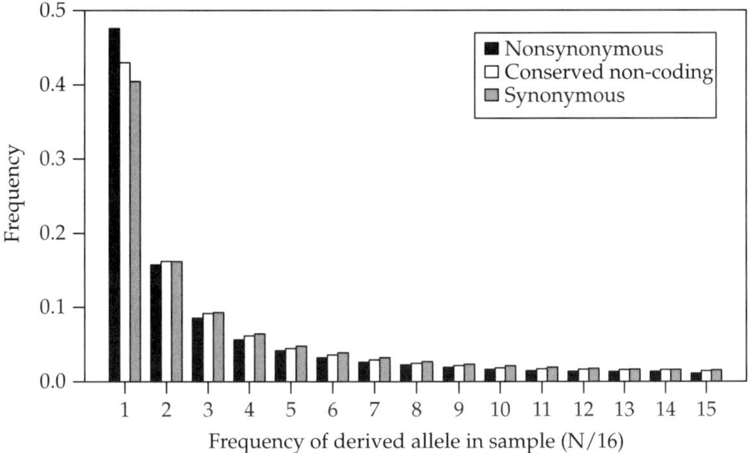

Figure 12.1 Plot of the site frequency spectrum in coding and conserved non-coding sites from the resequencing of 15 African Americans by Celera Genomics (data from Torgerson et al. 2009). The proportion of low-frequency-derived alleles (frequency of 1/16) is highest in nonsynonymous sites, followed by conserved non-coding and then synonymous sites. An excess of low-frequency-derived alleles suggests the presence of selective constraint on functionally relevant sites.

branches were more likely to be found near genes involved in neuronal cell adhesion, despite limited overlap in the location of accelerated CNCs in either species. Their results suggested that gene regulation associated with brain development and function is subject to adaptive evolution in both the human and chimp, but with a different set of genes.

The existence of rapid evolution in CNCs via positive selection was corroborated by two independent studies using different datasets and methods of analysis. Bird and others (2007) applied a relative rate test and concluded that 15% of CNCs showed evidence of accelerated evolution, and that accelerated CNCs were enriched in regions that had undergone segmental duplications. In 2007, Kim and Pritchard developed a shared rates test to identify lineage specific rate variation across the mammalian tree, and estimated that 32% of CNCs deviated from neutrality by either showing faster or slower rates of evolution (or both) (Kim and Pritchard 2007). Interestingly they observed a slight excess of 'speed ups' as compared to 'slow downs' (55% vs. 45%), hinting that adaptive evolution may, in fact, be more common than selective constraint in CNCs. Moreover, their data suggested that rapid evolution in CNCs was more likely to occur in short bursts of adaptation as compared to a consistent rate along a phylogenetic branch. Kim and Pritchard (2007) excluded regions of segmental duplications, however, their results and those from Bird and others (2007) both revealed that numerous CNCs have been subject to positive selection throughout human evolution.

Patterns of human variation are also consistent with the existence of positive selection driving the rapid evolution of CNCs. Bird and others (2007) observed an excess of high-frequency derived alleles in accelerated CNCs using an expanded set of HapMap snps (phase I and II) than examined previously (Drake et al. 2006), consistent with the actions of positive selection driving advantageous alleles to high frequency. Furthermore, SNPs within accelerated CNCs showed patterns of higher F_{ST} (fixation index) between human populations as compared to SNPs in non-accelerated CNCs, suggesting that rapid evolution of CNCs has likely occurred more recently in human evolution leading to population-specific differences in allele frequencies. Lastly, they identified a significant excess of expression quantitative trait loci (eQTLs) in accelerated CNCs, suggesting that positive selection was in fact driven by changes in gene regulation.

One of the challenges in examining patterns of human variation using data from the HapMap project is that of ascertainment bias (Clark et al. 2005), however, resequencing studies have come to similar conclusions. In the early 2000s, Celera Genomics resequenced CNCs in the flanking regions of over 15,000 genes in 39 humans (Torgerson et al. 2009). Patterns in the site frequency spectrum suggest that CNCs are subject to stronger selective constraints as compared to synonymous sites (Fig. 12.1). Gene-specific analyses revealed that CNCs near genes expressed in the fetal brain had significantly higher probabilities of positive selection as compared to CNCs near genes expressed in other tissues. Together with the results of Prabhakar and others (2006), non-random patterns observed in human variation and lineage-specific substitution rates in CNCs suggest that gene regulation associated with brain development and function is more often subject to adaptive evolution.

Not surprisingly, evolution occurs by processes other than single nucleotide substitutions, including those only recently recognized that remain to be integrated into a unified model of evolutionary change in non-coding DNA. For example, in 2011 McLean and others (McLean et al. 2011) identified over 500 highly conserved sequences at least 1 kb in size that underwent human-specific deletion. Almost all were non-coding in nature, and many were found near genes involved in steroid hormone signaling and neural function. Two of these regions involved the deletion of tissue-specific enhancers, including one that lead to a loss of androgen-dependent sensory vibrissae and penile spines in humans, and another that was correlated to the expansion of specific brain regions in humans. Clearly we have only begun to understand the dynamics and extent of natural selection acting on conserved non-coding regions, and as we narrow in on the specific functions of these regions we are likely to discover the true extent of rapid and adaptive evolution in non-coding DNA.

However, the patterns and process of human evolution via natural selection remains controversial. Hernandez and others (2011) discovered that substitutions in CNCs (and protein-coding sequences) along the human lineage are largely devoid of any convincing evidence for positive selection using pilot data from the 1000 Genomes Project. They concluded that much of human adaptation has occurred through modes of selection that are inconsistent with the classic models of positive selection, or has occurred in non-coding regions of the genome that are not evolutionarily conserved.

12.3.2 Detecting selection in promoters and TFBSs

Although conserved non-coding DNA is often functional, not all functional non-coding DNA is conserved (Birney et al. 2007). Therefore the search for non-neutral evolution in non-coding DNA is incomplete without looking at functional elements identified through means other than conservation. Several studies have approached this by examining specific classes of non-coding sequences, including the entire 5′ upstream region of genes regardless of conservation, core promoter regions, experimentally and computationally predicted TFBSs, microRNA binding sites, and the microRNAs themselves.

A typical model of gene regulation involves the binding of specific combinations of transcription factors to short, degenerate TFBSs that are often just upstream of the transcription start site. While patterns of human variation suggest that non-coding sequences in the 5′ upstream (promoter) region of genes have experienced more purifying selection compared to other non-coding sequences (Torgerson et al. 2009), these findings are based on conserved non-coding sequences and may not be representative of upstream regions as a whole. A study by Haygood and others (2007) compared rates of human evolution within the entire promoter region of genes as defined by 5 kb upstream of the transcription start site (TSS), without prior conditioning on phylogenetic conservation (that is, apart from requiring a whole-genome alignment of orthologous human–chimp–macaque sequences). They identified at least 250 promoter regions with evidence of positive selection enriched for genes involved in neural development and function, consistent with trends observed in CNCs (Prabhakar et al. 2006; Torgerson et al. 2009). Therefore, the hypothesis that more adaptive evolution has occurred in the regulation of genes involved in human cognition is supported regardless of conservation.

Haygood and others (2007) also made the observation that the promoter regions of genes involved in nutrition (notably glucose metabolism) show a excess of positive selection along the human lineage, suggesting that dietary shifts in humans may have driven the rapid evolution of non-coding DNA. Three additional studies lend support to this hypothesis. First, Sethupathy and others (2008) identified signatures of adaptive evolution within the upstream/promoter regions of genes by examining patterns of human polymorphism, also without conditioning on conservation. However, they restricted their analysis to include only those non-coding sites within computationally predicted TFBSs (as compared to the entire 5′ promoter region as in Haygood et al. (2007)), and identified an enrichment of positive selection in the promoters of genes involved in protein metabolism. Liang and others (2008) identified signatures of positive selection in the core promoters of 24 genes enriched for biosynthetic and metabolic processes by comparing substitution patterns in core promoters to that of proximal promoters. Planas and Serrat (2010) applied the identical method to Haygood et al. (2007) but restricted their analysis to non-coding sites within the proximal promoter region (1 kb vs. 5 kb upstream of the TSS). Overall they found that 21% of genes with positively selected proximal promoters were involved in protein metabolism and metabolism in general. Therefore, multiple studies support the hypothesis that rapid, adaptive evolution in non-coding DNA is driven by dietary shifts in humans.

12.3.3 Emerging trends in microRNA binding sites

Non-coding RNAs are also heavily involved in the regulation of gene expression, with microRNAs receiving the majority of attention in studies of human evolution. MicroRNAs are a class of non-coding RNAs involved in post-translational gene regulation through repression of translation and mRNA degradation. While TFBSs are typically found within the upstream regions of genes and thought to initiate gene transcription, microRNA binding sites are typically located in the 3′ UTRs of mRNAs and act to suppress gene translation, adding another layer of complexity to gene regulation. A typical microRNA binding site is similarly quite short (~22 nucleotides in length), although it is often less degenerate than a TFBS. While not all human genes appear to be regulated by microRNAs, the ones that are tend to evolve under strong selective constraint (Nielsen et al. 2009).

As with TFBS, microRNA binding sites (Chen and Rajewsky 2006; Mu et al. 2011) and the microRNAs themselves (Quach et al. 2009) show evidence for selective constraint based on patterns of human variation, however there is also evidence to suggest these non-coding sites may also evolve under rapid, adaptive evolution. In fact, Chen and Rajewsky (2006) identified a SNP in a predicted microRNA binding site of the *Map1lc3b* gene that showed a high degree of differentiation between human populations (F_{ST} in the 99.8 percentile of HapMap and Perlegen SNPs present in 3′ UTRs with the same heterozygosity), suggesting recent adaptive evolution in a population-specific manner. Notably, post-transcriptional misregulation of *Map1lc3b* has been implicated in giant axonal neuropathy and fragile X syndrome, and is important for neurogenesis. Quach and others (2009) noted the existence of a non-coding RNA-rich island on chromosome 14 that contained SNPs with independent evidence for population-specific positive selection in Europeans and East Asians. Moreover, data from the 1000 Genomes Pilot Project suggests that microRNAs with more predicted targets are more often subject to positive selection based on having reduced polymorphism, higher divergence, and an excess of high-frequency derived alleles (Mu et al. 2011). Therefore, rapid evolution through positive selection is likely occurring in non-coding DNA involved in microRNA gene regulation as well.

12.3.4 Coding versus non-coding

Genomic studies are beginning to reveal differences in evolutionary patterns in coding versus non-coding regions. Initial genomic comparisons suggested that non-coding DNA contains a greater *absolute number* of constrained sites compared to protein-coding sites. Gaffney and Keightley (2006) identified three times as many constrained sites in non-coding compared to coding DNA by comparisons of human/mouse substitution rates, and that the majority of sites were located in intergenic regions over 5 kb away from any known genes. They also found that genes involved in developmental and neuronal processes tend to have a greater number of constrained non-coding sites in their proximity as compared to electron transport and a variety of metabolic processes. While this appears to contradict the findings of more adaptive evolution in non-coding DNA near genes involved in brain development and function (Prabhakar et al. 2006; Haygood et al. 2007; Torgerson et al. 2009), the presence of selective constraint as estimated from human/mouse comparisons does not preclude the possibility for lineage-specific positive selection, nor the existence of positive selection at nearby sites.

However, the *proportion* of protein-coding sites under constraint appears to be much higher than the proportion of non-coding sites, both in conserved non-coding sites (Torgerson et al. 2009) and TFBSs (Mu et al. 2011). Comparisons of human polymorphism and divergence suggest that nonsynonymous sites are under stronger selective constraint as compared to CNCs and synonymous sites, however, CNCs notably in the 5′ upstream region of genes are more constrained than synonymous sites (Torgerson et al. 2009). Selective constraint also appears to be higher in TFBSs as compared to synonymous sites, yet not as high as synonymous sites (Mu et al. 2011). However, comparisons of the extent of non-neutral evolution in coding vs. non-coding sequences have yet to be made on equal grounds, as the delimitation of functional non-coding DNA has remained much more elusive as compared to protein-coding DNA.

Comparisons of the extent of adaptive evolution in coding versus the flanking non-coding regions have also been attempted, but are subject to the same limitation. For example, Torgerson et al. (2009) identified higher probabilities of positive selection on CNCs as compared to nonsynonymous sites through an analysis of human polymorphism. However, comparisons with synonymous sites suggested the difference may be driven by more neutral rather than more adaptive evolution in CNCs. This could in part be due to differences in power to detect natural selection, as CNCs likely contain a large proportion of non-functional sites. Using a phylogenetic-based comparison, Planas and Serrat (2010) found no significant difference in the absolute percentage of genes showing signatures of positive selection in both promoter and coding regions. However, this result is surprising as the likely inclusion of a greater proportion of neutral, non-functional sites in the analysis of non-coding regions is predicted to underestimate the extent of positive selection.

Evidence for a decoupling of selection on coding and non-coding DNA stems from comparisons of the strength and direction of selection, and from differences in the kinds of genes that are enriched for signatures of natural selection. Although there is a slight correlation between estimates of the probability of natural selection in coding and flanking CNCs (which may be driven by linkage disequilibrium), there is little predictive power to infer selection on non-coding sites based on patterns observed in protein-coding sites and vice versa (Torgerson et al. 2009). Furthermore, the correlation between selective constraint and breadth of gene expression (Gaffney and Keightley 2006), and the proportion of positively selected genes in the center versus the periphery of protein networks (Planas and Serrat 2010) appear to be in the opposite direction for coding versus non-coding sites. However, Torgerson and others (2009) report a higher probability of positive selection in CNCs near genes expressed in the fetal brain, but find no such enrichment at corresponding nonsynonymous sites. Planas and Serrat (2010) found only a small number of positively selected genes within enriched functional classes, however, they found a non-overlapping set of positively selected genes between coding and proximal promoter regions. Therefore, while direct comparisons of the degree to which human adaptation has been driven by changes in coding versus non-coding DNA are being refined, an emerging trend is that there are differential effects of natural selection on coding and regulatory regions of genes.

12.4 Future prospects

As we gain a more thorough understanding of the function of non-coding DNA, we will undoubtedly make additional insights into human evolutionary history. Facilitating this process are the rapid technological advances in data collection that has made full genome sequencing more commonplace. Our understanding of evolutionary patterns is being accelerated through sequencing the genomes of a thousand species (e.g. Genome 10K Community of Scientists 2009), and of thousands of individuals within a single species (e.g. the 1000 Genomes Project (Durbin et al. 2010)). However, many challenges exist in developing and refining experimental, statistical, and computational techniques for studying patterns of evolution in non-coding DNA. For example, a more complete integration of comparative and population genomics approaches is required to combine long-term evidence of rapid evolution with very recent effects of natural selection. Moreover, novel strategies will be necessary to better compare evolutionary patterns in coding and non-coding DNA. Despite these challenges, there are many lines of evidence to already suggest that non-coding DNA has in many instances undergone rapid, adaptive evolution in humans. Future scientific advances in data collection and analytical methods are likely to reveal this is only the beginning.

References

Alexander, R.P., Fang, G., Rozowsky, J., Snyder, M., and Gerstein, M.B. (2010) Annotating non-coding regions of the genome. *Nat Rev Genet* **11**: 559–71.

Bejerano, G., Pheasant, M., Makunin, I., Stephen, S., Kent, W. J., Mattick, J.S., et al. (2004) Ultraconserved elements in the human genome. *Science* **304**: 1321–5.

Bird, C.P., Stranger, B.E., Liu, M., Thomas, D.J., Ingle, C.E., Beazley, C., et al. (2007) Fast-evolving non-coding sequences in the human genome. *Genome Biol* **8**: R118.

Birney, E., Stamatoyannopoulos, J.A., Dutta, A., Guigo, R., Gingeras, T.R., Margulies, E.H., et al. (2007) Identification and analysis of functional elements in 1% of the human genome by the ENCODE pilot project. *Nature* **447**: 799–816.

Blanchette, M., Bataille, A.R., Chen, X., Poitras, C., Laganiere, J., Lefebvre, C., et al. (2006) Genome-wide computational prediction of transcriptional regulatory modules reveals new insights into human gene expression. *Genome Res* **16**: 656–68.

Bush, E.C., and Lahn, B.T. (2005) Selective constraint on noncoding regions of hominid genomes. *PLoS Comput Biol* **1**: e73.

Charlesworth, B. (1994) The effect of background selection against deleterious mutations on weakly selected, linked variants. *Genet Res* **63**: 213–27.

Chen, K. and Rajewsky, N. (2006) Natural selection on human microRNA binding sites inferred from SNP data. *Nat Genet* **38**: 1452–6.

Clark, A.G., Hubisz, M.J., Bustamante, C.D., Williamson, S.H., and Nielsen, R. (2005) Ascertainment bias in studies of human genome-wide polymorphism. *Genome Res* **15**: 1496–502.

Drake, J.A., Bird, C., Nemesh, J., Thomas, D.J., Newton-Cheh, C., Reymond, A., et al. (2006) Conserved non-coding sequences are selectively constrained and not mutation cold spots. *Nat Genet* **38**: 223–7.

Durbin, R.M., Abecasis, G.R., Altshuler, D.L., Auton, A., Brooks, L.D., Durbin, R.M., et al. (2010) A map of human genome variation from population-scale sequencing. *Nature* **467**: 1061–73.

Duret, L. and Arndt, P.F. (2008) The impact of recombination on nucleotide substitutions in the human genome. *PLoS Genet* **4**: e1000071.

Gaffney, D.J., Blekhman, R., and Majewski, J. (2008) Selective constraints in experimentally defined primate regulatory regions. *PLoS Genet* **4**: e1000157.

Gaffney, D.J., and Keightley, P.D. (2006) Genomic selective constraints in murid noncoding DNA. *PLoS Genet* **2**: e204.

Genome 10K Community of Scientists. (2009) Genome 10K: a proposal to obtain whole-genome sequence for 10,000 vertebrate species. *J Hered* **100**: 659–74.

Haygood, R., Fedrigo, O., Hanson, B., Yokoyama, K.D., and Wray, G.A. (2007) Promoter regions of many neural- and nutrition-related genes have experienced positive selection during human evolution. *Nat Genet* **39**: 1140–4.

Hernandez, R.D., Kelley, J.L., Elyashiv, E., Melton, S.C., Auton, A., McVean, G., et al. (2011) Classic selective sweeps were rare in recent human evolution. *Science* **331**: 920–4.

Katzman, S., Kern, A.D., Bejerano, G., Fewell, G., Fulton, L., Wilson, R.K., et al. (2007) Human genome ultraconserved elements are ultraselected. *Science* **317**: 915.

Keightley, P.D., Kryukov, G.V., Sunyaev, S., Halligan, D.L., and Gaffney, D.J. (2005a) Evolutionary constraints in conserved nongenic sequences of mammals. *Genome Res* **15**: 1373–8.

Keightley, P.D., Lercher, M.J., and Eyre-Walker, A. (2005b) Evidence for widespread degradation of gene control regions in hominid genomes. *PLoS Biol* **3**: e42.

Kim, S.Y., and Pritchard, J.K. (2007) Adaptive evolution of conserved noncoding elements in mammals. *PLoS Genet* **3**: e147.

King, M.C., and Wilson, A.C. (1975) Evolution at two levels in humans and chimpanzees. *Science* **188**: 107–16.

Liang, H., Lin, Y.S., and Li, W.H. (2008) Fast evolution of core promoters in primate genomes. *Mol Biol Evol* **25**: 1239–44.

McLean, C.Y., Reno, P.L., Pollen, A.A., Bassan, A.I., Capellini, T.D., Guenther, C., et al. (2011) Human-specific loss of regulatory DNA and the evolution of human-specific traits. *Nature* **471**: 216–19.

Mu, X.J., Lu, Z.J., Kong, Y., Lam, H.Y., and Gerstein, M.B. (2011) Analysis of genomic variation in non-coding elements using population-scale sequencing data from the 1000 Genomes Project. *Nucleic Acids Res* **39**(16): 7058–76.

Nielsen, R., Hubisz, M.J., Hellmann, I., Torgerson, D., Andres, A.M., Albrechtsen, A., et al. (2009) Darwinian and demographic forces affecting human protein coding genes. *Genome Res* **19**: 838–49.

Planas, J., and Serrat, J.M. (2010) Gene promoter evolution targets the center of the human protein interaction network. *PLoS One* **5**: e11476.

Pollard, K.S., Salama, S.R., King, B., Kern, A.D., Dreszer, T., Katzman, S., et al. (2006) Forces shaping the fastest evolving regions in the human genome. *PLoS Genet* **2**: e168.

Prabhakar, S., Noonan, J.P., Paabo, S., and Rubin, E.M. (2006) Accelerated evolution of conserved noncoding sequences in humans. *Science* **314**: 786.

Quach, H., Barreiro, L.B., Laval, G., Zidane, N., Patin, E., Kidd, K.K., et al. (2009) Signatures of purifying and local positive selection in human miRNAs. *Am J Hum Genet* **84**: 316–27.

Sethupathy, P., Giang, H., Plotkin, J.B., and Hannenhalli, S. (2008) Genome-wide analysis of natural selection on human cis-elements. *PLoS ONE* **3**: e3137.

Siepel, A., Bejerano, G., Pedersen, J.S., Hinrichs, A.S., Hou, M., Rosenbloom, K., et al. (2005) Evolutionarily conserved elements in vertebrate, insect, worm, and yeast genomes. *Genome Res* **15**: 1034–50.

Torgerson, D.G., Boyko, A.R., Hernandez, R.D., Indap, A., Hu, X., White, T.J., et al. (2009) Evolutionary processes acting on candidate cis-regulatory regions in humans inferred from patterns of polymorphism and divergence. *PLoS Genet* **5**: e1000592.

Waterston, R.H., Lindblad-Toh, K., Birney, E., Rogers, J., Abril, J.F., Agarwal, P., et al. (2002) Initial sequencing and comparative analysis of the mouse genome. *Nature* **420**: 520–62.

PART III

Sex- and Reproduction-Related Genetic Systems

CHAPTER 13

Evolution of sperm–egg interaction

Melody R. Palmer and Willie J. Swanson

13.1 Introduction

The success of fertilization and reproduction in sexual species is dependent on the interaction between sperm and egg. Some sperm–egg interactions succeed between distant species, creating hybrids, but others fail between closely related species at one or more stages. Given the importance of fertilization in the maintenance of species integrity and boundaries, it is surprising that reproductive proteins are often evolving more rapidly than the genome-wide average (Swanson and Vacquier 2002). In this review we will discuss the evolution of sperm–egg interaction in terms of the mechanisms, proteins, and selective pressures involved. Experimental approaches are rapidly changing with new genomic and proteomic technologies, and there is a correspondingly rapid increase in the amount of new data on fertilization-associated proteins. This data will improve our understanding of the evolution of sperm–egg interacting proteins, for which variation or conservation may determine the degree of reproductive isolation between populations or species. There are exciting opportunities to identify the molecular players in the process of speciation: the more species studied in greater detail, the more proteins could be identified.

13.2 Evolution at each step of sperm–egg interaction

Sperm and eggs in the organisms we will discuss are similar in their basic structures, but diverse in their molecules and mechanisms of recognition, binding, entry, and fusion. At each step, we see variation in the degree of species specificity and rate of protein evolution. Is this a result of variation in selective pressure at the different stages of

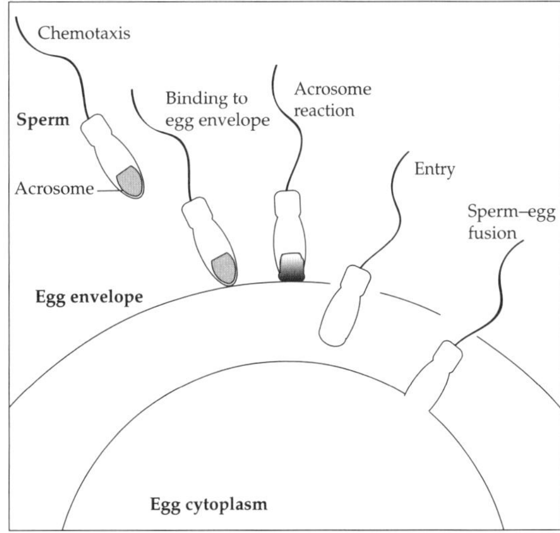

Figure 13.1 Simplified depiction of the basic steps of mammalian and invertebrate fertilization.

fertilization? Looking at the specific steps (Fig. 13.1) will give insights into the stages of fertilization that are under the strongest selection in rapidly evolving systems. For example, the marine mollusk abalone sperm proteins lysin and sp18 are both evolving rapidly, but sp18, which acts closer to the end of fertilization, has higher divergence between species than lysin (Metz et al. 1998). Non-sperm ejaculated proteins, such as seminal fluid proteins (SFPs) in insects (*Drosophila*; Findlay et al. 2009) and mammals (rodents; Ramm et al. 2009) also evolve rapidly, and their function is the subject of active research.

In order to physically interact, the sperm must first find the egg. An important factor in this step is chemoattraction—where the egg releases an attractant that stimulates sperm to swim in the direction of the egg. Chemotaxis of sperm to eggs has

Rapidly Evolving Genes and Genetic Systems. First Edition. Edited by Rama S. Singh, Jianping Xu, and Rob J. Kulathinal.

been observed *in vitro* in organisms from free-spawning marine invertebrates to mammals (Eisenbach 1999). A diverse range of *in vitro* attractants has been identified, including small molecules, peptides, low molecular mass proteins, and a lipid molecule, but few have been associated with eggs outside of marine invertebrates (Ward et al. 1985; Miller and Vogt 1996; Spehr et al. 2003; Kaupp et al. 2008). The sperm's environment and path to the egg differ between internal and external fertilizers, which may impact the evolution of the attractant molecules and receptor proteins. In internal fertilizers such as mammals, *in vivo* chemoattraction is not well understood. Researchers have identified a few candidate receptors in mammalian sperm, but none are confirmed or well characterized. One group of candidates is testis-expressed olfactory receptors. A small group of these are implicated in sperm signal response, such as hOR17-4, but no natural ligands are known (Spehr et al. 2003). In mammals, testis-expressed ORs are highly conserved, particularly in the ligand binding domain (Branscomb et al. 2000). Consistent with this, Sun et al. (2003) showed that chemotaxis toward follicular factors is not species specific in rabbit and human sperm.

Chemoattraction is particularly important in free-spawning marine organisms because of the heightened possibility of encountering heterospecific gametes and the dilution of sperm and eggs in the ocean. For example, corals undergo mass spawning events where gametes from multiple species are present in the water at once. One study showed that three species of *Acropora* have species-specific sperm motility initiation by eggs, but no molecular players are known (Morita et al. 2006). In abalone, chemotaxis is also species specific (Riffell et al. 2004). *Haliotis rufescens* eggs release L-tryptophan, which acts as a chemoattractant. Sperm from *H. fulgens* do not respond to egg factors from *H. rufescens*, and the same for the reciprocal. This could be a result of selection against hybridization in species with overlapping habitats and spawning seasons. In sea urchin species, the egg peptide speract varies in sequence and sperm response (Guerrero et al. 2010). Conversely, in frogs and mice, allurin, a member of the Crisp protein family, appears to have conserved chemotactic induction (Burnett et al. 2008). More candidate receptor proteins will be discovered with new proteomic techniques. More sequence information may reveal interesting patterns of selection at the stage of chemotaxis.

The next step is binding and passage through the egg envelope, one of the best-studied molecular mechanisms of sperm–egg interaction. The known molecules that mediate this interaction are diverse. Egg envelopes are composed of large glycoproteinaceous matrices. They share structural properties across taxa from marine invertebrates to mammals, such as the ZP domain (Monné et al. 2006), but are divergent in sequence. A recent analysis showed similarity in the functionally important residues of structurally homologous regions of ZP domains in human, mollusk, and yeast. This suggests that the overall mechanisms may be similar. In contrast, for sperm egg-binding and entry proteins there is generally no homology between distantly related species. The striking diversity of sperm proteins involved in egg-coat entry reflects the variation of selective pressures and speciation events over a long evolutionary history.

Before the sperm penetrates the egg envelope it must undergo the acrosome reaction (AR), the release of the contents of a vesicle in the sperm head. In marine invertebrates there is variation in the importance of this step in the species-specificity of fertilization. The sea urchin sperm receptor for egg jelly (suREJ1) and egg fucose sulfate polymers interact to induce the AR (Vacquier and Moy 1997). The fucan polymers vary between urchin species and are species-specific inducers of the AR (Vilela-Silva et al. 2008). However, their species-specificity and importance as a barrier to hybridization differ across species pairs. Sea stars also have an egg jelly polysaccharide recognition and AR induction system, but it is so far only found to be subfamily specific (Nakachi et al. 2006). A later step, mediated by bindin in sea urchins, may be more important for species-specific fertilization. Receptor suREJ1 has a mammalian homolog, Pkdrej (Hughes et al. 1999), which localizes to the sperm head and is evolving under positive selection in primates (Hamm et al. 2007). However, it appears to be involved in the timing of the AR in mammals rather than having a direct interaction with the egg (Sutton et al. 2008).

Sperm–egg envelope interactions are best characterized in urchin, abalone, and mouse. The egg envelope is a major hybridization barrier in mammal *in vitro* fertilization; its removal allows greater heterospecific membrane fusion (Yanagimachi 1994). It appears that a complex interaction of the ZP proteins is important for species-specific recognition in mammalian eggs (Yauger et al. 2011). Abalone lysin and vitelline envelope receptor for lysin (VERL) was the first pair of interacting fertilization proteins to be identified (Swanson and Vacquier 1997). Lysin creates a hole in the egg vitelline envelope through non-enzymatic interactions with VERL (Lewis et al. 1982). Sperm proteins lysin and sea urchin bindin are both rapidly evolving under positive selection (Lee and Vacquier 1992; Lee et al. 1995; Metz and Palumbi 1996). Egg receptor proteins have more recently been sequenced and found to also be under positive selection. For example, abalone VERL and mammalian ZP2 and ZP3 show positive selection (Galindo et al. 2003; Swanson et al. 2001; Turner and Hoekstra 2006), and urchin EBR1 activity is species-specific (Kamei and Glabe 2003). Other abalone ZP domain proteins of unknown function are also under positive selection (Aagaard et al. 2006). Rapidly evolving residues can often be correlated to ligand-receptor binding domains, implicating selection on the interaction between sperm and egg. Sea urchin bindin has been used extensively to study the evolutionary dynamics of sperm-egg interaction. For example, Palumbi et al. (1999) found that the success of a sperm depends on its and the female's bindin genotype (Palumbi 1999). Levitan and Stapper found that this effect is dependent on population densities (Levitan and Stapper 2010). Lysin and VERL, along with mammalian sperm sp56 and egg envelope protein ZP3, show evolutionary patterns that indicate coevolution (Clark et al. 2009; Rohlfs et al. 2010).

Sperm–egg membrane fusion mechanisms are less well understood. Known and putative gamete fusagens vary in their sequence conservation. HAP2-GCS1 is a gamete fusagen in *Plasmodium*, *Chlamydomonas*, and *Arabidopsis*. It is hypothesized to be the ancestral gene for gamete fusion, due to its presence in many animal genomes, and its expression in the testis of *Hydra*, a basal cnidarian (Wong and Johnson 2010). However, it appears to have been lost in many lineages, suggesting redundant molecular mechanisms for gamete fusion, maybe as a result of positive selection. Positive selection is seen in fusagenic sperm proteins in mammals and marine invertebrates. Sea urchin bindin and abalone sp18 are both acrosome proteins with *in vitro* fusagenic properties, and both evolve rapidly and localize to the acrosome (Ulrich et al. 1998; Swanson and Vacquier 1995). In mouse, 23 membrane proteins from the sperm surface and acrosomal membrane show evidence of positive selection (Dorus et al. 2010), and in mammals five sperm surface ADAM proteins show positive selection in their adhesion domains (Finn and Civetta 2010). Phenotypes in mouse knockout studies frequently contradict *in vitro* results. One current hypothesis is that egg CD9 plays a role in fusion in mice, and Izumo is a candidate interacting protein (for a recent review of mammalian sperm–egg fusion, please see Rubinstein et al. (2006)). Like other gamete interaction proteins, CD9 has sites under positive selection (Swanson et al. 2003). In *Drosophila* the sperm enters the egg and must go through plasma membrane breakdown (PMBD). Two proteins, Snky and Mfr, are required for efficient PMBD, but their specific functions are still unclear (Wilson et al. 2006; Smith and Wakimoto 2007). Both appear to have homologs in distant species, but have not been found in studies of mosquito testis transcriptomes and sperm proteomes (Krzywinska and Krzywinski 2009; Sirot et al. 2011). The prevalence of positive selection on fusagenic proteins may vary based on the fertilization system. If there is selection for a barrier to fertilization and rapid evolution at earlier steps results in successful adaptations, there may be relaxed selection on the later steps.

Another class of proteins exhibiting rapid evolution in internally fertilizing species is SFPs, which are ejaculated along with sperm. Ramm et al. (2009) found interspecific diversity in SFP composition and sequence within muroid rodents. Seminal fluid and sperm protein evolution can also be compared by dissection and partitioning of male reproductive tissues for comparative proteomics and EST (expressed sequence tag) sequencing. In mice, seminal vesicle proteins evolve rapidly on average, but proteins from the other male reproductive tract tissues are under evolutionary constraint when compared to the whole genome (Dean et al. 2009). In insects, cross-species studies show that accessory gland proteins have increased evolutionary rates (Wagstaff and Begun 2005; Andres et al. 2006). In field crickets, positive selection is seen in SFPs between closely related species (Andres et al. 2006; Marshall et al. 2011), and there appears to be little conservation in overall insect SFP composition (Walters and Harrison 2010). Two *Gryllus* species of crickets that have rapidly diverging seminal fluid proteomes, but are otherwise closely related, are an example of how we can use these systems to study the evolution of reproductive isolation in hybrid zones (Andres et al. 2008). The specific functions and interactions of seminal fluid proteins are still largely uncharacterized. Accessory gland proteins in *Drosophila* modulate female postmating behavior, such as causing an increase in egg laying (Wolfner 2009). Work in *Drosophila* on SFPs is addressed in Chapter 15.

13.3 Causes of rapid evolution

We have long observed the pattern of rapidly evolving gamete interaction proteins in a wide range of taxa (Swanson and Vacquier 2002). Identification of putative sperm–egg interaction proteins under positive selection is becoming easier as new technologies allow faster and cheaper DNA sequencing and protein identification by mass spectrometry. However, testing hypotheses about why these proteins evolve rapidly remains a challenge. What selective pressures are acting in each situation? It is likely that an interplay of forces act in a variable way in each group of species. Here we will discuss the various hypotheses to explain rapid gamete interaction protein evolution, and the data that support or refute each hypothesis.

Some proposed mechanisms for rapid evolution of reproductive proteins deal with the within-species effects such as sexual conflict and sperm competition (Fig. 13.2). High population density can drive sexual conflict over the mating rate, which may result in several possible patterns of evolution. Mathematical models demonstrate that runaway coevolution can result in the evolution of reproductive barriers within species (Gavrilets 2000). In one model, females diversify into two distinct groups due to sexual conflict. In response, males may either be stuck in an intermediate fitness state or diversify to match each genotype, resulting in sympatric speciation (Gavrilets and Waxman 2002). Experimental studies of the relationship of sperm densities, the genotype of a protein involved in sperm–egg interaction, and reproductive success are limited to sea urchin bindin. Reproductive success in *Echinometra* sea urchins is dependent on male bindin genotype, and varies according to female bindin genotype (Palumbi 1999). Levitan and collegues have conducted many studies on the effect of male and female density on reproductive success. In *Strongylocentrotus franciscanus*, there is an optimum range of female reproductive success between sperm limitation and polyspermy, and a relationship between sperm bindin genotype success and density (Levitan and Ferrell 2006). Consistent with predictions of sexual conflict, rare male alleles are more successful in high-density situations. In a high-density living *Strongylocentrotus* species, common bindin alleles were generally more advantageous, but selection maintains some less common, highly successful, variants (Levitan and Stapper 2010). These results would be even more interesting if they were able to also include the EBR1 sequence of the females, and attempt to better understand the evolution of the interaction itself.

We are also beginning to see these patterns in mammals. Sperm competition can affect the development of fertilization barriers between closely related rodents (Martín-Coello et al. 2009). As female proteins that prevent polyspermy are selected for, diversification arises, and a byproduct of that is differentiation between species. Sperm competition may also affect primate seminal fluid genes. Positive selection in a main component of the semen coagulum in primates correlates with higher female promiscuity (Dorus et al. 2004).

Alternatively, the reinforcement hypothesis states that diversifying selection for fertilization proteins to prevent hybridization when spawning or copulation between species overlaps causes rapid evolution (for review see Noor 1999). This hypothesis can apply to behavioral and ecological factors as well as the proteins involved in direct gamete interaction. The latter is best studied in free-spawning

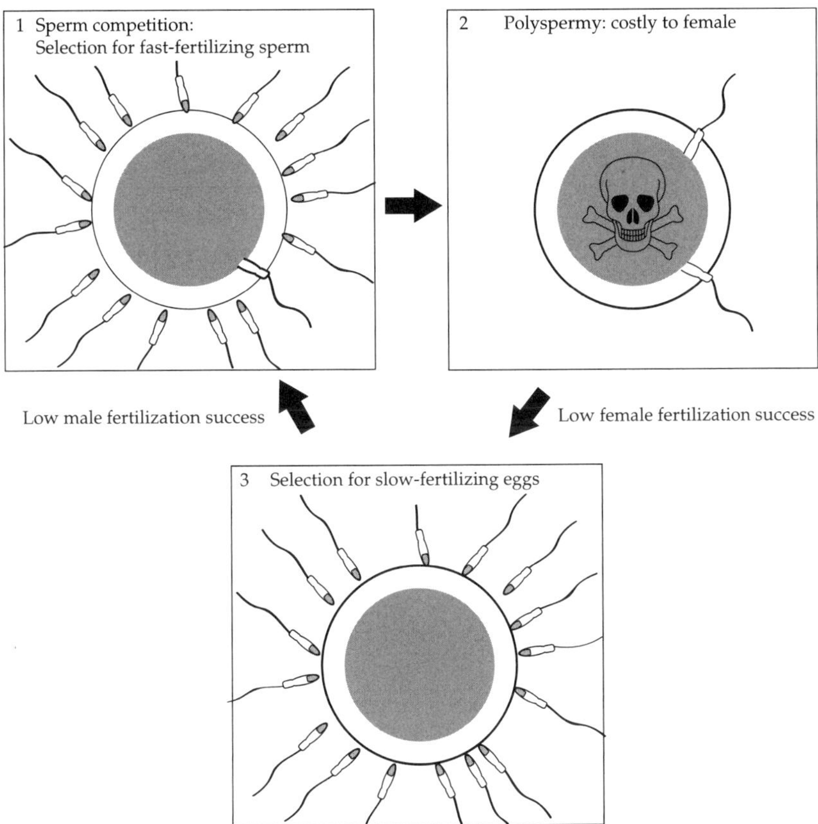

Figure 13.2 A cycle of sexual conflict can drive rapid coevolution between sperm and egg proteins in sperm-dense situations. Sperm competition results in selection for the sperm that can most rapidly fertilize the egg. That may lead to an increase in polyspermy rates, which results in selection for eggs that have stronger blocks to fertilization.

marine organisms. Reinforcement in species such as mussels, abalone, and sea urchin can be tested by comparing the rates of evolution of a gamete interaction protein between sympatric and allopatric populations. The data produced to date do not consistently support this hypothesis. Single studies have found no evidence for reinforcement in abalone lysin (Clark et al. 2007) or ascidian gamete recognition proteins (Nydam and Harrison 2011). In *Mytilus* species of mussels, Springer and Crespi (2007) found lysin-M7 divergence patterns supporting the reinforcement hypothesis, but in another study Slaughter et al. (2008) tested gamete compatibility and found greater compatibility in sympatry, suggesting reinforcement is not the dominant selective force. Riginos et al. (2006) also did not detect reproductive character displacement between sympatric and allopatric species of *Mytilus*. The most data in this field is from studies of sea urchin and its sperm protein bindin. Bindin evolution studies vary in their implications. Two species of South Pacific *Echinometra* have divergence and positive selection in sympatric populations, but some share alleles in allopatric populations, supporting reinforcement (Geyer and Palumbi 2003). However, further studies of *Echinometra* species found bindin positive selection and heterospecific incompatibility in allopatric populations (Metz et al. 1994; McCartney and Lessios 2004; Geyer and Lessios 2009). These results suggest that some interspecific selective pressures such as sexual conflict or sperm competition may be driving the rapid evolution of bindin. In addition, an understanding of the evolution of the female receptor protein for bindin in each case would also be useful because reinforcement suggests selective pressure should be stronger on the egg, because it is most negatively affected by non-specific fertilization.

13.4 Methods to identify interacting proteins

Sexual conflict and sperm competition drive rapid divergence between sperm–egg interacting proteins. These are potential mechanisms of sympatric speciation (Gavrilets and Waxman 2002). We must identify the proteins responsible for species isolation in order to study past and current speciation events. Identification methods include biochemical, immunochemical, and genetic approaches. Gamete interaction proteins were first identified with biochemical purification, starting with sea urchin bindin (Vacquier and Moy 1977). This is why abalone and sea urchin, in which large amounts of gametes can be easily obtained, were the earliest models of fertilization. Mammalian interactions have been tested with *in vitro* antibody blocking studies, but these may be non-specific in their results. Genetic knockouts in mouse are useful because you can remove a single protein and test the specific result. However, sometimes they give results that conflict with *in vitro* protein function predictions (e.g. Baba et al. 1994). One reason could be that the interaction relies on a combination of proteins that remain functional with the loss of one member, or some other form of redundancy. In knockout studies, it is important to also characterize the ability of mutant sperm to compete with wild type. This was key in revealing a phenotype for mutant Pkdrej mice (Sutton et al. 2008). A benefit of using free-spawning invertebrates as model systems is that you can more easily observe fertilization under natural conditions than in those with internal fertilizations such as mammals. Recently, mass spectrometry has allowed the characterization of numerous sperm and seminal fluid proteomes (mouse: Stein et al. 2006; human: Baker 2007; fruit fly: Dorus et al. 2006). More should be done to describe the proteomes of the egg coat and egg plasma membrane (like Aagaard et al. 2006, 2010; Yamada et al. 2009).

Detecting putative interacting protein pairs may be possible without first finding clear biochemical evidence. Experimental data have shown that interacting protein pairs often contain signatures of coevolution. Using that signal, Clark et al. (2009) and Rohlfs et al. (2010) presented sequence analysis methods to predict protein–protein interaction in reproductive proteins that are physically un-linked. Clark et al. used the known interacting pair lysin and VERL in eight species of abalone. They showed a correlation of dN/dS values along the branches that were not seen in non-interacting proteins, and determined statistical significance with likelihood ratio tests. Rohlfs et al. detected selection for allele matching using human population genotype data and composite linkage disequilibrium. They found that the human putative interacting proteins egg ZP3 and sperm ZP3R showed allelic associations at more SNPs than those at background genomic levels and random gene pairs. Their result is consistent with the prediction that interacting proteins will fix compensatory mutations as each protein evolves in order to maintain allele matching.

13.5 Conclusions

We have described the current state of knowledge of gamete interaction proteins, their evolution, and methods of analysis and identification. Many more candidate sperm–egg interaction proteins will be identified as DNA sequencing and mass spectrometry become more accessible. Thorough phylogenetic sampling will allow us to assess the diversity and evolution of these candidates. However, it is important to understand their function in gamete interaction with genetic and biochemical studies as well. In addition, we should work to characterize the evolution of more egg proteins, to better understand selection acting on proteins from both sides. Knowing the sequences and binding domains of interacting protein pairs will be particularly useful for understanding the evolution, divergence, and mechanisms that lead to reproductive isolation.

References

Aagaard, J.E., Yi, X., MacCoss, M.J., and Swanson, W.J. (2006) Rapidly evolving zona pellucida domain proteins are a major component of the vitelline envelope of abalone eggs. *Proc Natl Acad Sci U S A* **103**(46): 17302–7.

Aagaard, J.E., Vacquier, V.D., MacCoss, M.J., and Swanson, W.J. (2010) ZP domain proteins in the abalone egg coat include a paralog of VERL under positive selection that binds lysin and 18-kDa sperm proteins. *Mol Biol Evol* **27**(1): 193–203.

Andres, J.A., Maroja, L.S., Bogdanowicz, S.M., Swanson, W.J., and Harrison, R.G. (2006) Molecular evolution of seminal proteins in field crickets. *Mol Biol Evol* **23**(8): 1574–84.

Andres, J.A., Maroja, L.S., and Harrison, R.G. (2008) Searching for candidate speciation genes using a proteomic approach: seminal proteins in field crickets. *Proc Roy Soc B Biol Sci* **275**(1646): 1975–83.

Baba, T., Azuma, S., and Kashiwabara, S. (1994) Sperm from mice carrying a targeted mutation of the acrosin gene can penetrate the oocyte zona pellucida and effect fertilization. *J Biol Chem* **269**(50): 31845–9.

Baker, M.A., Reeves, G., Hetherington, L., Müller, J., Baur, I., and Aitken, R.J. (2007) Identification of gene products present in Triton X-100 soluble and insoluble fractions of human spermatozoa lysates using LC-MS/MS analysis. *Proteomics Clin Appl* **1**(5): 524–32.

Branscomb, A., Seger, J., and White, R.L. (2000) Evolution of odorant receptors expressed in mammalian testes. *Genetics* **156**(2): 785–97.

Burnett, L.A., Xiang, X., Bieber, A.L., and Chandler, D.E. (2008) Crisp proteins and sperm chemotaxis: discovery in amphibians and explorations in mammals. *Int J Dev Biol* **52**(5–6): 489–501.

Clark, N.L., Gasper, J., Sekino, M., Springer, S.A., Aquadro, C.F., and Swanson, W.J. (2009) Coevolution of interacting fertilization proteins. *PLoS Genet* **5**(7), e1000570.

Clark, N.L., Findlay, G.D., Yi, X., MacCoss, M.J., and Swanson, W.J. (2007) Duplication and selection on abalone sperm lysin in an allopatric population. *Mol Biol Evol* **24**(9): 2081–90.

Dean, M.D., Clark, N.L., Findlay, G.D., Karn, R.C., Yi, X., Swanson, W.J., et al. (2009) Proteomics and comparative genomic investigations reveal heterogeneity in evolutionary rate of male reproductive proteins in mice (Mus domesticus). *Mol Biol Evol* **26**(8): 1733–43.

Dorus, S., Busby, S.A., Gerike, U., Shabanowitz, J., Hunt, D.F., and Karr, T.L. (2006) Genomic and functional evolution of the Drosophila melanogaster sperm proteome. *Nat Genet* **38**(12): 1440–5.

Dorus, S., Evans, P.D., Wyckoff, G.J., Choi, S.S., and Lahn, B.T. (2004) Rate of molecular evolution of the seminal protein gene SEMG2 correlates with levels of female promiscuity. *Nat Genet* **36**(12): 1326–9.

Dorus, S., Wasbrough, E.R., Busby, J., Wilkin, E.C., and Karr, T.L. (2010) Sperm proteomics reveals intensified selection on mouse sperm membrane and acrosome genes. *Mol Biol Evol* **27**(6): 1235–46.

Eisenbach, M. (1999) Sperm chemotaxis. *Rev Reprod* **4**(1): 56–66.

Findlay, G.D., MacCoss, M.J., and Swanson, W.J. (2009) Proteomic discovery of previously unannotated, rapidly evolving seminal fluid genes in Drosophila. *Genome Res* **19**(5): 886–96.

Finn, S. and Civetta, A. (2010) Sexual selection and the molecular evolution of ADAM proteins. *J Mol Evol* **71**(3): 231–40.

Galindo, B.E., Vacquier, V.D., and Swanson, W.J. (2003) Positive selection in the egg receptor for abalone sperm lysin. *Proc Natl Acad Sci U S A* **100**(8): 4639–43.

Gavrilets, S. (2000) Rapid evolution of reproductive barriers driven by sexual conflict. *Nature*, **403**(6772): 886–9.

Gavrilets, S. and Waxman, D. (2002) Sympatric speciation by sexual conflict. *Proc Natl Acad Sci U S A* **99**(16): 10533–8.

Geyer, L. and Palumbi, S. (2003) Reproductive character displacement and the genetics of gamete recognition in tropical sea urchins. *Evolution* **57**(5): 1049–60.

Geyer, L.B. and Lessios, H. (2009) Lack of character displacement in the male recognition molecule, bindin, in Altantic sea urchins of the genus Echinometra. *Mol Biol Evol* **26**(9): 2135–46.

Guerrero, A., Nishigaki, T., Carneiro, J., Yoshiro, T., Wood, C.D., and Darszon, A. (2010) Tuning sperm chemotaxis by calcium burst timing. *Dev Biol* **344**(1): 52–65.

Hamm, D., Mautz, B.S., Wolfner, M.F., Aquadro, C.F., and Swanson, W.J. (2007) Evidence of amino acid diversity-enhancing selection within humans and among primates at the candidate sperm-receptor gene PKDREJ. *Am J Hum Genet* **81**(1): 44–52.

Hughes, J., Ward, C.J., Aspinwall, R., Butler, R., and Harris, P.C. (1999) Identification of a human homologue of the sea urchin receptor for egg jelly: a polycystic kidney disease-like protein. *Hum Mol Genet* **8**(3): 543–9.

Kamei, N. and Glabe, C.G. (2003) The species-specific egg receptor for sea urchin sperm adhesion is EBR1,a novel ADAMTS protein. *Genes Dev* **17**(20): 2502–7.

Kaupp, U.B., Kashikar, N.D., and Weyand, I. (2008) Mechanisms of sperm chemotaxis. *Annu Rev Physiol* **70**: 93–117.

Krzywinska, E. and Krzywinski, J. (2009) Analysis of expression in the Anopheles gambiae developing testes reveals rapidly evolving lineage-specific genes in mosquitoes. *BMC Genomics* **10**: 300.

Lee, Y. and Vacquier, V. (1992) The divergence of species-specific abalone sperm lysins is promoted by positive Darwinian selection. *Biol Bull* **182**(1): 97–104.

Lee, Y., Ota, T., and Vacquier, V. (1995) Positive selection is a general phenomenon in the evolution of abalone sperm lysin. *Mol Biol Evol* **12**(2): 231–8.

Levitan, D.R. and Ferrell, D.L. (2006) Selection on gamete recognition proteins depends on sex, density, and genotype frequency. *Science* **312**(5771): 267–9.

Levitan, D.R. and Stapper, A.P. (2010) Simultaneous positive and negative frequency-dependent selection on sperm bindin, a gamete recognition protein in the sea urchin Strongylocentrotus purpuratus. *Evolution* **64**(3): 785–97.

Lewis, C.A., Talbot, C.F., and Vacquier, V.D. (1982) A protein from abalone sperm dissolves the egg vitelline layer by a nonenzymatic mechanism. *Dev Biol* **92**(1): 227–39.

Marshall, J.L., Huestis, D.L., Garcia, C., Hiromasa, Y., Wheeler, S., Noh, S., et al. (2011) Comparative proteomics uncovers the signature of natural selection acting on the ejaculate proteomes of two cricket species isolated by postmating, prezygotic phenotypes. *Mol Biol Evol* **28**(1): 423–35.

Martín-Coello, J., Benavent-Corai, J., Roldan, E.R., and Gomendio, M. (2009) Sperm competition promotes asymmetries in reproductive barriers between closely related species. *Evolution* **63**(3): 613–23.

McCartney, M.A. and Lessios, H.A. (2004) Adaptive evolution of sperm bindin tracks egg incompatibility in neotropical sea urchins of the genus Echinometra. *Mol Biol Evol* **21**(4): 732–45.

Metz, E. and Palumbi, S. (1996) Positive selection and sequence rearrangements generate extensive polymorphism in the gamete recognition protein bindin. *Mol Biol Evol* **13**(2): 397–406.

Metz, E., Kane, R.E., Yanagimachi, H., and Palumbi, S.R. (1994) Fertilization between closely-related sea-urchins is blocked by incompatibilities during sperm-egg attachment and early stages of fusion. *Biol Bull* **187**(1): 23–34.

Metz, E.C., Robles-Sikisaka, R., and Vacquier, V.D. (1998) Nonsynonymous substitution in abalone sperm fertilization genes exceeds substitution in introns and mitochondrial DNA. *Proc Natl Acad Sci U S A* **95**(18): 10676–81.

Miller, R.L. and Vogt, R. (1996) An N-terminal partial sequence of the 13 kDa Pycnopodia helianthoides sperm chemoattractant "startrak" possesses sperm-attracting activity. *J Exp Biol* **199**(Pt 2): 311–18.

Monné, M., Han, L., and Jovine, L. (2006) Tracking down the ZP domain: From the mammalian zona pellucida to the molluscan vitelline envelope. *Semin Reprod Med* **24**(4): 204–16.

Morita, M., Nishikawa, A., Nakajima, A., Iguchi, A., Sakai, K., and Takemura, A. (2006) Eggs regulate sperm flagellar motility initiation, chemotaxis and inhibition in the coral Acropora digitifera, A. gemmifera and A. tenuis. *J Exp Biol* **209**(Pt 22): 4574–9.

Nakachi, M., Moriyama, H., Hoshi, M., and Matsumoto, M. (2006) Acrosome reaction is subfamily specific in sea star fertilization. *Dev Biol* **298**(2): 597–604.

Noor, M.A. (1999) Reinforcement and other consequences of sympatry. *Heredity* **83**(Pt 5): 503–8.

Nydam, M.L. and Harrison, R.G. (2011) Reproductive protein evolution in two cryptic species of marine chordate. *BMC Evol Biol* **11**: 18.

Palumbi, S. (1999) All males are not created equal: Fertility differences depend on gamete recognition polymorphisms in sea urchins. *Proc Natl Acad Sci U S A* **96**(22): 12632–7.

Ramm, S.A., McDonald, L., Hurst, J.L., Beynon, R.J., and Stockley, P. (2009) Comparative proteomics reveals evidence for evolutionary diversification of rodent seminal fluid and its functional significance in sperm competition. *Mol Biol Evol* **26**(1): 189–98.

Riffell, J.A., Krug, P.J., and Zimmer, R.K. (2004) The ecological and evolutionary consequences of sperm chemoattraction. *Proc Natl Acad Sci U S A* **101**(13): 4501–6.

Riginos, C., Wang, D., and Abrams, A.J. (2006) Geographic variation and positive selection on M7 lysin, an acrosomal sperm protein in mussels (Mytilus spp.). *Mol Biol Evol* **23**(10): 1952–65.

Rohlfs, R.V., Swanson, W.J., and Weir, B.S. (2010) Detecting coevolution through allelic association between physically unlinked loci. *Am J Hum Genet* **86**(5): 674–85.

Rubinstein, E., Ziyyat, A., Wolf, J.P., Le Naour, F., and Boucheix, C. (2006) The molecular players of sperm-egg fusion in mammals. *Semin Cell Dev Biol* **17**(2): 254–63.

Sirot, L.K., Hardstone, M.C., Helinski, M.E., Ribeiro, J.M., Kimura, M., Deewatthanawong, P., et al. (2011) Towards a semen proteome of the dengue vector mosquito: protein identification and potential functions. *PLoS Negl Trop Dis* **5**(3), e989.

Slaughter, C., McCartney, M.A., and Yund, P.O. (2008) Comparison of gamete compatibility between two blue mussel species in sympatry and in allopatry. *Biol Bull* **214**(1): 57–66.

Spehr, M., Gisselmann, G., Poplawski, A., Riffell, J.A., Wetzel, C.H., Zimmer, R.K., et al. (2003) Identification of a testicular odorant receptor mediating human sperm chemotaxis. *Science* **299**(5615): 2054–8.

Springer, S.A. and Crespi, B.J. (2007) Adaptive gamete-recognition divergence in a hybridizing Mytilus population. *Evolution* **61**(4): 772–83.

Stein, K., Go, J.C., Lane, W.S., Primakoff, P., and Myles, D.G. (2006) Proteomic analysis of sperm regions that mediate sperm-egg interactions. *Proteomics* **6**(12): 3533–43.

Sun, F., Giojalas, L.C., Rovasio, R.A., Tur-Kaspa, I., Sanchez, R., and Eisenbach, M. (2003) Lack of species-specificity in mammalian sperm chemotaxis. *Dev Biol* **255**(2): 423–7.

Sutton, K.A., Jungnickel, M.K., and Florman, H.M. (2008) A polycystin-1 controls postcopulatory reproductive selection in mice. *Proc Natl Acad Sci U S A* **105**(25): 8661–6.

Swanson, W. and Vacquier, V. (1995) Extraordinary divergence and positive Darwinian selection in a fusagenic protein coating the acrosomal process of abalone spermatozoa. *Proc Natl Acad Sci U S A* **92**(11): 4957–61.

Swanson, W. and Vacquier, V. (1997) The abalone egg vitelline envelope receptor for sperm lysin is a giant multivalent molecule. *Proc Natl Acad Sci U S A* **94**(13): 6724–9.

Swanson, W. and Vacquier, V. (2002) Reproductive protein evolution. *Annu Rev Ecol Systemat* **33**: 161–79.

Swanson, W.J., Yang, Z., Wolfner, M.F., and Aquadro, C.F. (2001) Positive Darwinian selection drives the evolution of several female reproductive proteins in mammals. *Proc Natl Acad Sci U S A* **98**(5): 2509–14.

Swanson, W.J., Nielsen, R., and Yang, Q. (2003) Pervasive adaptive evolution in mammalian fertilization proteins. *Mol Biol Evol* **20**(1): 18–20.

Turner, L.M. and Hoekstra, H.E. (2006) Adaptive evolution of fertilization proteins within a genus: variation in ZP2 and ZP3 in deer mice (Peromyscus). *Mol Biol Evol* **23**(9): 1656–69.

Ulrich, A.S., Otter, M., Glabe, C.G., and Hoekstra, D. (1998) Membrane fusion is induced by a distinct peptide sequence of the sea urchin fertilization protein bindin. *J Biol Chem* **273**(27): 16748–55.

Vacquier, V.D. and Moy, G.W. (1977) Isolation of bindin: the protein responsible for adhesion of sperm to sea urchin eggs. *Proc Natl Acad Sci U S A* **74**(6): 2456–60.

Vacquier, V.D. and Moy, G.W. (1997) The fucose sulfate polymer of egg jelly binds to sperm REJ and is the inducer of the sea urchin sperm acrosome reaction. *Dev Biol* **192**(1): 125–35.

Vilela-Silva, A.-C.E.S., Hirohashi, N., and Mourao, P.A.S. (2008) The structure of sulfated polysaccharides ensures a carbohydrate-based mechanism for species recognition during sea urchin fertilization. *Int J Dev Biol* **52**(5–6): 551–9.

Wagstaff, B.J. and Begun, D.J. (2005) Molecular population genetics of accessory gland protein genes and testis-expressed genes in Drosophila mojavensis and D. arizonae. *Genetics* **171**(3): 1083–101.

Walters, J.R. and Harrison, R.G. (2010) Combined EST and proteomic analysis identifies rapidly evolving seminal fluid proteins in Heliconius butterflies. *Mol Biol Evol* **27**(9): 2000–13.

Ward, G.E., Brokaw, C.J., Garbers, D.L., and Vacquier, V.D. (1985) Chemotaxis of Arbacia punctulata spermatozoa to resact, a peptide from the egg jelly layer. *J Cell Biol* **101**(6): 2324–9.

Wolfner, M.F. (2009) Battle and ballet: molecular interactions between the sexes in Drosophila. *J Hered* **100**(4): 399–410.

Wong, J.L. and Johnson, M.A. (2010) Is HAP2-GCS1 an ancestral gamete fusogen? *Trends Cell Biol* **20**(3): 134–41.

Yamada, L., Saito, T., Taniguchi, H., Sawada, H., and Harada, Y. (2009) Comprehensive egg coat proteome of the ascidian Ciona intestinalis reveals gamete recognition molecules involved in self-sterility. *J Biol Chem*, **284**(14): 9402–10.

Yauger, B., Boggs, N.A., and Dean, J. (2011) Human ZP4 is not sufficient for taxon-specific sperm recognition of the zona pellucida in transgenic mice. *Reproduction*, **141**(3): 313–19.

Yanagimachi R. (1994) Mammalian fertilization. In E. Knobil and J.D. Neill (Eds) *The Physiology of Reproduction, vol. 1*, pp. 189–317. New York: Raven Press.

CHAPTER 14

Rates of sea urchin bindin evolution

H. A. Lessios and Kirk S. Zigler

14.1 Introduction

Reproduction at the level of gametic interactions involves activation and attraction of the sperm by egg compounds, induction of the acrosome reaction by the egg jelly, adhesion of the sperm to the egg, and fusion of the two membranes in order to permit the transmission of genetic material. All of these interactions are mediated by molecules. Some of these molecules, such as sea urchin speract, carry out their functions indiscriminately, even if sperm and egg belong to distantly related taxa (Vieira and Miller 2006). Others function in a species-specific or even genotype-specific manner. Selectivity between sperm gamete recognition molecules and their egg receptors is particularly important in organisms with external fertilization, because in the absence of copulation, there are few other opportunities for exercising mate choice. Consequently, such molecules are exposed to the action of selection more directly than molecules with the same function in organisms with internal fertilization. The DNA that codes for gamete recognition molecules often, but not always, evolves rapidly, displaying ratios of amino acid replacement to synonymous substitutions larger than unity, a signature of positive (diversifying) selection (Swanson and Vacquier 2002a, b; Vacquier and Swanson 2011). As a rule, such positive selection is targeted at certain regions of each molecule, presumably involved in gamete selectivity, whereas the rest of the sequence may evolve conservatively under purifying selection, because it performs basic functions essential for fertilization.

The first gamete recognition protein to be characterized was sea urchin bindin (Vacquier and Moy 1977). Bindin DNA was subsequently amplified and sequenced in *Strongylocentrotus purpuratus* by Gao et al. (1986), and then studied with regards to its intra- and interspecific polymorphism with special attention given to detecting positive selection in its exons. These topics have been extensively reviewed (Vacquier et al. 1995; Swanson and Vacquier 2002a, b; Lessios 2007, 2011; Zigler 2008; Palumbi 2009; Vacquier and Swanson 2011). In this chapter, we explore what bindin sequences from various sea urchin species reveal about the rate of evolution of this molecule. Does bindin really evolve in the fast lane?

14.2 Function and structure of bindin

Sea urchin bindin is a protein that coats the acrosome process of sperm after the acrosomal reaction occurs. It interacts with the egg bindin receptor, EBR1, a glycoprotein (Kamei and Glabe 2003), to attach the sperm to the egg's vitelline layer and to fuse membranes of the gametes. The full-length precursor of bindin is cleaved after translation to form the mature molecule. Among the sea urchin species that have been studied to date, the length of mature bindin ranges from 193–418 amino acids (Zigler and Lessios 2003a). The single sea star in which bindin has been characterized was found to contain 793 amino acids (Patino et al. 2009). In both sea urchins and sea stars, there is a single intron separating two exons. Bindins of 11 species of sea urchins from six orders contain a conserved region in the second exon that codes for approximately 55 amino acids. Eighteen amino acids in this conserved region, thought to be involved in membrane fusion (Rocha et al. 2008), have not changed since the extant orders of Echinoidea split from each other, 250 million years ago (mya). Only one amino acid in this region has changed between sea stars and sea urchins in the 500 million years (my)

Rapidly Evolving Genes and Genetic Systems. First Edition. Edited by Rama S. Singh, Jianping Xu, and Rob J. Kulathinal.

that the two echinoderm classes have been evolving independently (Patino et al. 2009; Vacquier and Swanson 2011). The reputation of bindin as a fast-evolving protein is owed to two regions flanking the conserved core, which in some genera have accumulated many point mutations and insertions–deletions. These are the regions that most likely confer fertilization species-specificity (Lopez et al. 1993). The protein moiety of EBR1, which contains 3713–4595 amino acids, has only been sequenced in two species of *Strongylocentrotus* (Kamei and Glabe 2003).

14.3 Rate of bindin evolution

Bindin has been sequenced in 11 genera of sea urchins, but intrageneric variation, which permits insights in the evolution of the molecule, has been studied in only seven: *Echinometra* (Metz and Palumbi 1996; McCartney and Lessios 2004), *Strongylocentrotus* (Biermann 1998), *Arbacia* (Metz et al. 1998a), *Tripneustes* (Zigler and Lessios 2003b), *Heliocidaris* (Zigler et al. 2003), *Lytechinus* (Zigler and Lessios 2004), and *Paracentrotus* (Calderon et al. 2009, 2010). Selection on bindin in all of these genera has been studied as the ratio of amino acid replacement to silent substitutions ($\omega = d_N/d_S$). By this criterion, there is evidence of positive selection ($\omega > 1$) in *Echinometra*, *Strongylocentrotus*, *Heliocidaris*, and *Paracentrotus*, but not in *Arbacia*, *Tripneustes*, and *Lytechinus*. In addition to being an indication of selection at the nucleotide level, the ω ratio would also be a good measure of relative rates of adaptive evolution if silent sites evolved at the same rate in all genera. This, however, is not the case in bindin. Bindins with higher rates of nonsynonymous substitution also have higher rates of synonymous substitution (Zigler and Lessios 2003b). This correlation has also been observed in other molecules such as alcohol dehydrogenase, ATP synthetase, cyclophilin 1, or enolase (e.g. Dunn et al. 2001), and there are a number of hypotheses as to its cause. While it is typically thought to arise from some form of codon bias, codon usage in sea urchin bindin is very equitable (Zigler and Lessios 2003a). Thus, due to different codon biases, comparing ω ratios between bindins of different genera may lead to erroneous conclusions regarding evolutionary rates. To compare the absolute rate of evolution between genera we need to determine the number of nonsynonymous substitutions per nonsynonymous site that accumulate per unit time. Such a calculation requires evidence of dates of divergence. In this chapter, we will use the interspecific divergence of cytochrome oxidase I (COI) as a proxy for the time since speciation. Calibrated by the rise of the Isthmus of Panama, approximately 3 mya, COI of sea urchins diverges at an average rate of 3.6 % per my (Lessios 2008).

Gauged by divergence in COI, average rates of adaptive divergence of bindin within a genus vary between 2.80×10^{-3} nonsynonymous substitutions per nonsynonymous site per my ($d_N my^{-1}$) in *Arbacia* and 22.4×10^{-3} $d_N my^{-1}$ in *Strongylocentrotus* (Table 14.1). As one might expect, genera in which bindin evolves under positive selection, show amino acid divergence rates almost four times higher than genera in which bindin appears to be under purifying selection: the average substitution rate in *Strongylocentrotus*, *Echinometra*, and *Heliocidaris* is 20.4×10^{-3} $d_N my^{-1}$ whereas in *Arbacia*, *Tripneustes*, *Lytechinus*, *Pseudoboletia*, and *Diadema*, it is 5.96×10^{-3} $d_N my^{-1}$. The question we would like to answer is how these rates of adaptive evolution compare with those of other proteins, both of those that have been deemed to evolve rapidly in other taxa, and those that carry out other functions in sea urchins.

Fig. 14. 1 presents a comparison of the rates of adaptive evolution of bindin to seven other classes of reproductive proteins from five groups of organisms. These are all proteins that are generally considered as fast-evolving. Because COI in different taxa evolves at different rates, it is necessary to apply taxon-specific calibrations to calculate divergence rates. To estimate absolute rates of protein evolution, we have assumed that COI evolves at an average rate of 3.6% per my in sea urchins (Lessios 2008), 2.7% per MY in gastropods (Lessios 2008), 2.3% per my in insects (Papadopoulou et al. 2010), and 1.6% per my in hominids (Kumar et al. 2005). Estimated in this manner, the evolutionary rates of bindins in different genera of sea urchins, even those found to be under selection, are slower than that of reproductive proteins of gastropods or insects. They are more comparable to those of

Table 14.1 Pairwise divergence in bindin and in cytochrome oxidase I (COI) of selected species of sea urchin genera in which bindin variation has been studied. **K2P**: Kimura two-parameter distance; $\mathbf{d_N}$: amino acid substitutions per non-synonymous site; $\mathbf{d_S}$: synonymous substitutions per synonymous site; **MY**: million years. Estimated rates of divergence of bindin are based on the assumption that COI in sea urchins diverges at a rate of 0.036 per site per my.

Genus	Species	Species	Bindin d_N	COI d_S	Bindin d_N/ K2P	Bindin d_S/ COI K2P	Bindin d_N/ COI K2P	MY	Reference
Arbacia	*lixula*	*punctulata*	0.007	0.069	0.090	0.072	0.764	0.0026	Metz et al. 1998a
Arbacia	*lixula*	*stellata=incisa*	0.007	0.096	0.134	0.053	0.716	0.0019	
Arbacia	*lixula*	*dufresnei*	0.016	0.071	0.124	0.129	0.570	0.0046	
Arbacia	*punctulata*	*stellata=incisa*	0.003	0.088	0.139	0.022	0.635	0.0008	
Arbacia	*punctulata*	*dufresnei*	0.011	0.059	0.124	0.085	0.477	0.0031	
Arbacia	*stellata=incisa*	*dufresnei*	0.013	0.071	0.119	0.105	0.597	0.0038	
Heliocidaris	*erythrogramma*	*tuberculata*	0.069	0.149	0.147	0.469	1.014	0.0169	Zigler et al. 2003
Tripneustes	*ventricosus*	*gratilla+depressus*	0.016	0.026	0.087	0.187	0.293	0.0067	Zigler and Lessios 2003
Echinometra	*oblonga*	*mathaei*	0.021	0.054	0.023	0.905	2.328	0.0326	Metz and Palumbi 1996
Echinometra	*oblonga*	*type A*	0.024	0.076	0.032	0.757	2.371	0.0273	
Echinometra	*mathaei*	*type A*	0.028	0.051	0.024	1.169	2.107	0.0421	
Echinometra	*lucunter*	*viridis*	0.022	0.047	0.050	0.440	0.940	0.0158	McCartney and Lessios 2004
Echinometra	*lucunter*	*vanbrunti*	0.026	0.046	0.102	0.255	0.451	0.0092	
Echinometra	*viridis*	*vanbrunti*	0.014	0.083	0.126	0.111	0.659	0.0040	
Lytechinus	*pictus*	*variegatus*	0.013	0.105	0.135	0.096	0.778	0.0035	Zigler and Lessios 2004
Lytechinus	*variegatus*	*williamsi*	0.006	0.022	0.017	0.353	1.294	0.0127	
Lytechinus	*semituberculatus*	*pictus*	0.025	0.073	0.114	0.219	0.640	0.0079	
Lytechinus	*euerces*	*Sphaerechinus granularis*	0.019	0.100	0.089	0.213	1.124	0.0077	
Pseudoboletia	*indiana*	*maculata*	0.006	0.024	0.073	0.082	0.329	0.0030	Zigler et al. (in press)
Strongylocentrotus	*purpuratus*	*pallidus*	0.021	0.062	0.072	0.287	0.863	0.0103	Biermann 1998
Strongylocentrotus purpuratus	*pallidus*	*droebachiensis*	0.031	0.086	0.075	0.418	1.148	0.0150	
Strongylocentrotus	*purpuratus*	*H. pulcherrimus*	0.073	0.158	0.104	0.704	1.514	0.0253	
Strongylocentrotus	*pallidus*	*droebachiensis*	0.025	0.036	0.035	0.715	1.011	0.0257	
Strongylocentrotus	*pallidus*	*H. pulcherrimus*	0.066	0.119	0.070	0.941	1.696	0.0339	
Strongylocentrotus	*droebachiensis*	*H. pulcherrimus*	0.063	0.139	0.094	0.672	1.481	0.0242	

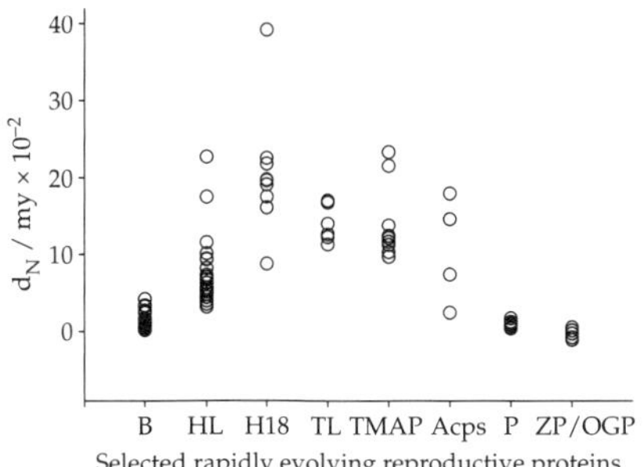

Figure 14.1 Bindin evolution relative to known fast-evolving reproductive proteins from other taxa. Non-synonymous substitutions per non-synonymous site (d_N) per million years, between congeneric species (except in hominids, in which they are within the same family) in sea urchin bindin (B) (data from references in Table 14.1), abalone lysin (HL) and 18 kD protein (H18) (data from Metz et al. 1998b), *Tegula* lysin (TL), and the mature region of TMAP protein (TMAP) (data from Hellberg and Vacquier 1999; Hellberg et al. 2000), *Drosophila* Acp26Aa and Acp36DE (Acps) (data from Tsaur and Wu 1997), hominid protamine 1 and 2 (P), ZP2, ZP3 and oviductal glycoprotein (ZP/OGP) (data from Wyckoff et al. 2000).

protamines, zona pellucida proteins, and oviductal glycoprotein in hominids. Adjustments to the assumed rate of COI evolution, or even an assumption of a universal COI clock, would not change this conclusion. Thus, by the standard of other fast-evolving reproductive proteins from other invertebrates, bindin evolves only at moderate rates.

How do rates of bindin evolution compare to rates of evolution among other sea urchin proteins? To answer this question, we compared all protein coding DNA sequences of *Lytechinus variegatus* in GenBank to their closest matches in the *Strongylocentrotus purpuratus* complete genome. With the exception of *S. purpuratus*, more genes have been sequenced from *Lytechinus variegatus* than any other species of sea urchin. *Lytechinus* and *Strongylocentrotus* diverged approximately 60 mya. Sequences were available for 90 *L. variegatus* genes. The protein sequence of each gene was compared between the two species via protein-protein BLAST to GenBank's 'non-redundant (nr) protein sequences' database. The closest match to a *S. purpuratus* protein was noted, and the two protein sequences were aligned using Clustal in MEGA (v. 4.0). We then used MEGA to calculate the p-distance between the aligned protein sequences. We identified matches for 85 of the 90 *Lytechinus* genes. The five genes that did not have a match may be: (1) missing from the annotated *Strongylocentrotus* genome; (2) lost in the *Strongylocentrotus* lineage; or (3) mis-annotated in their original *Lytechinus* entry. The set of genes that we compared contained proteins with various functions, including many involved in reproduction, and also in development, cytoskeleton formation, cell attachment, and stress responses. After ranking the divergences of the 85 proteins, that of bindin was the sixth largest, with a p-distance of 0.326 for the full-length molecule and 0.314 for the mature portion. Of the five proteins with divergence values higher than bindin, vitellogenin and SFE-1 also carry out functions related to reproduction, whereas the other three were involved in development. Considering the inevitable bias of proteins available for comparison, the conclusion from this comparison is that bindin evolves at moderately fast rates in relation to other sea urchin proteins.

14.4 Possible reasons for different evolutionary rates in bindin

Why does bindin in four sea urchin genera evolve more rapidly under strong positive selection, than in three other genera in which it is subject to purifying selection? In the absence of data regarding variation in its egg receptor, the answer can only be speculative. Possible reasons for this lack of pattern have been thoroughly reviewed (Lessios 2007, 2011; Zigler 2008; Palumbi 2009). Here we present a summary of the hypotheses that have been proposed so far.

One possibility is that positive selection of bindin arises from the need for species recognition when two closely related species are in danger of hybridizing with each other. We will call this the 'reinforcement hypothesis.' This name does not imply that speciation by reinforcement has actually taken place, but rather that bindin alleles resembling those of a sympatric species—and thus allowing gamete wastage in inferior hybrids—have been selected against. A broad-brush picture of comparisons between genera is consistent with this hypothesis. When bindin rates of divergence of species that are entirely allopatric with respect to congeners are compared to those of species that may have a higher probability of hybridization, those of the former are clustered around lower values than those of the latter (Fig. 14.2). Genera with many sympatric species, such as *Strongylocentrotus*, and *Echinometra* tend to have the highest rates of interspecific bindin divergence. Not all the data, however, are consistent with the reinforcement hypothesis. Contrary to what is expected from selection for species recognition, bindin is polymorphic and shows the signature of positive selection not just between species but also between alleles of the same species (Metz and Palumbi 1996; Lessios 2007, 2011). A pattern of character displacement is present in one species of Pacific *Echinometra* (Geyer and Palumbi 2003) in partial geographic overlap with its sister species but not in an Atlantic species of the same genus that also needs to contend with the challenge of a sister species existing over part of its range (Geyer and Lessios 2009). Given the present evidence, the hypothesis that reinforcement in sympatry accelerates bindin divergence is as likely as the hypothesis that divergence in bindin, due to other causes, allows for sympatric coexistence.

Another possibility for the differences in rates of bindin evolution could be that they are correlated to the relative age of species in different sea urchin genera. If, as Civetta and Singh (1998) have suggested, episodes of divergence in reproductive molecules are concentrated at the time of speciation, and if selection on these molecules is subsequently relaxed, younger species would show higher rates of bindin differentiation than older ones. This hypothesis is not supported by the data. Sea urchins tend to conform to 'Jordan's rule' (Jordan 1905). Young sister species tend to be distributed on either side of a geographic barrier, and only older species become sympatric with the passage of time (Lessios 2010). Thus, allopatric species are, in general, younger than sympatric ones, and if bindin divergence were accelerated during speciation then slowed down, they should show more differences in this molecule per unit time than sympatric ones. The opposite is true (Fig. 14.2).

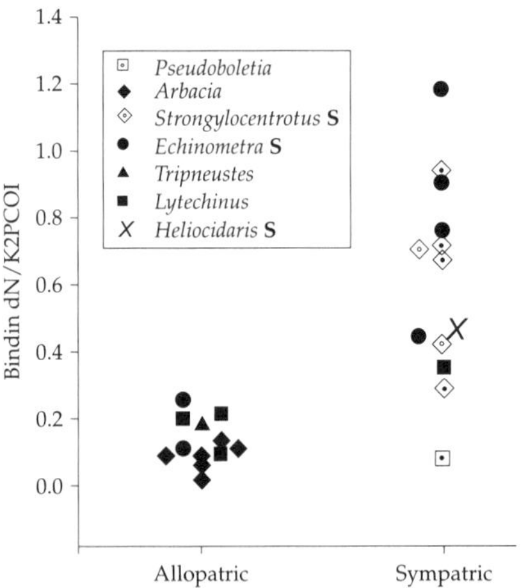

Figure 14.2 Comparison of interspecific rates of bindin divergence between genera. Amino acid replacement substitutions (d_N) per replacement site in bindin divided by Kimura-two-parameter distance in cytochrome oxidase I (COI K2P) in allopatric and sympatric species of eight genera of sea urchins. A species is considered as 'allopatric' if its range does not overlap with that of another member of the same genus. Genera in which bindin has been shown to be under selection are marked in the legend with **S**.

The most credible hypothesis to date for differences in the rates of bindin evolution is that they are caused by differences in the intensity of sexual selection and sexual conflict. Using variation in bindin genotypes of females as a proxy for variation in the bindin receptor (with which bindin is expected to show linkage disequilibrium), Palumbi (1999) has found that sexual selection exists in *Echinometra mathaei*. Eggs are fertilized at higher rates by sperm carrying the same bindin allele. Using the same proxy, Levitan and Farrell (2006) and Levitan and Stapper (2010) showed in *Strongylocentrotus franciscanus* and *S. purpuratus* that sperm density and the danger of polyspermy establish different selective regimes for various bindin alleles. At low sperm densities, most offspring are produced by the union of sperm and egg possessing bindin alleles that are most common in the population. At high sperm densities, rare alleles leave behind the most offspring, because common alleles, causing fast fertilization, result in polyspermic zygotes, which fail to develop. Thus, there is always selection on males to effect fast fertilization, but females in high sperm densities benefit from having alleles that retard fertilization: a typical sexual conflict situation. Depending on ecological conditions, sexual conflict can occur in some populations but not others, thus resulting in different rates of bindin evolution.

14.5 Conclusions and future prospects

In comparison to other invertebrate reproductive proteins, bindin evolves moderately rapidly in some genera and slowly in others. Selective reasons for the differences that cause these dissimilarities in rates are still the subject of speculation, but they may well be related to fertilization environments and intraspecific processes. Interspecific processes, such as reinforcement, can also not be ruled out. There may well be no universal explanation for the presence or absence of positive selection in different sea urchin taxa. Gametic proteins are often brought up as examples of rapid evolution. Fast evolution is certainly true for each of these proteins in the particular genus in which they have been studied. However, in a great many of the documented cases of fast molecular evolution, the evidence comes only from a small fraction of taxa. Data on sea urchin bindin, though far from covering the entire echinoid class, derive from multiple genera. This broader taxonomic coverage alone may explain why more diversity in the mode of evolution of this molecule has been documented than has been found in other invertebrate reproductive proteins.

Future laboratory studies linking the structure of different bindin alleles with the specificity of fertilization would be of great benefit in understanding the evolution of this molecule. We already know which amino acids evolve under selection, but we will need to determine the functional reasons for such selection. Additional understanding of the sources of natural selection on this molecule and the rate of its evolution would come from comparative studies that link fertilization ecology in nature with the success of particular bindin alleles. Simply characterizing species as sympatric or allopatric on the basis of their geographic distribution is not adequate for determining the role of reinforcement or other interspecific processes in bindin evolution. Ultimately, interest in the evolution of bindin and similar molecules stems from our desire to understand the process of speciation and the role of sexual selection in the evolution of reproductive isolation. In that respect, assessing the importance of bindin as a reproductive isolation barrier between species relies on studies that are not aimed directly at this molecule alone. Whether bindin is involved in speciation depends not just on the species-specificity of its interactions with its receptor but on the probability that gametes of two closely related sea urchin species will encounter each other in nature. Even when gametic interactions are, in fact, species-specific, it is still necessary to determine whether bindin or some other molecule, acting earlier in the sequence of fertilization, is responsible. Thus, information on habitat separation, reproductive timing, and pre-spawning chemical communication as well as on the role of other reproductive molecules is important in understanding whether intra- or interspecific interactions mold the evolution of the bindin. Most of all, we will need to link variation of bindin to variation in its egg receptor. The study of EBR1 has been retarded by its enormous size. Recent advances in techniques for massive DNA sequencing have made it practical to gather data on individual variation in large stretches of genetic material, and will no doubt soon be applied to this problem.

Acknowledgments

We thank Laura Geyer and Santosh Jagadeeshan for comments on the manuscript.

References

Biermann, C.H. (1998) The molecular evolution of sperm bindin in six species of sea urchins (Echinoida: Strongylocentrotidae). *Mol Biol Evol* **15**: 1761–71.

Calderon, I., Turon, X., and Lessios, H.A. (2009) Characterization of the sperm molecule bindin in the sea urchin genus Paracentrotus. *J Mol Evol* **68**: 366–76.

Calderon, I., Ventura, C.R.R., Turon, X., and Lessios, H.A. (2010) Genetic divergence and assortative mating between colour morphs of the sea urchin Paracentrotus gaimardi. *Mol Ecol* **19**: 484–93.

Civetta, A. and Singh, R.S. (1998) Sex-related genes, directional sexual selection, and speciation. *Mol Biol Evol* **15**: 901–9.

Dunn, K.A., Bielawski, J.P., and Yang, Z. (2001) Substitution rates in Drosophila nuclear genes: implications for translational selection. *Genetics* **157**: 295–305.

Gao, B., Klein, L.E., Britten, R.J., and Davidson, E.H. (1986) Sequence of mRNA coding for bindin, a species-specific sea urchin sperm protein required for fertilization. *Proc Natl Acad Sci U S A* **83**: 8634–8.

Geyer, L.B. and Palumbi, S.R. (2003) Reproductive character displacement and the genetics of gamete recognition in tropical sea urchins. *Evolution* **57**, 1049–60.

Geyer, L.B. and Lessios, H.A. (2009) Lack of character displacement in the male recognition molecule, bindin, in Altantic sea urchins of the genus Echinometra. *Mol Biol Evol* **26**: 2135–46.

Hellberg, M.E. and Vacquier, V,D. (1999) Rapid evolution of fertilization selectivity and lysin cDNA sequences in teguline gastropods. *Mol Biol Evol* **16**: 839–48.

Hellberg, M.E., Moy, G.W., and Vacquier, V.D. (2000) Positive selection and propeptide repeats promote rapid interspecific divergence of a gastropod sperm protein. *Mol Biol Evol* **17**: 458–66.

Jordan, D. S. (1905) The origin of species through isolation. *Science* **22**: 545–62.

Kamei, N. and Glabe, C.G. (2003) The species-specific egg receptor for sea urchin sperm adhesion is EBR1, a novel ADAMTS protein. *Genes Dev* **17**: 2502–7.

Kumar, S., Filipski, A., Swarna, V., Walker, A., and Hedges, S.B. (2005) Placing confidence limits on the molecular age of the human-chimpanzee divergence. *Proc Natl Acad Sci U S A* **102**: 18842–7.

Lessios, H.A. (2007) Reproductive isolation between species of sea urchins. *Bull Mar Sci* **81**: 191–208.

Lessios, H.A. (2008) The Great American Schism: Divergence of marine organisms after the rise of the Central American Isthmus. *Annu Rev Ecol Evol Systema* **39**: 63–91.

Lessios, H.A. (2010) Speciation in sea urchins. In L.G. Harris, S.A. Böttger, C.W. Walker, and M.P. Lesser (Eds) *Echinoderms: Durham. Proceedings of the 12th Echinoderm Conference, Durham, New Hampshire*, pp. 91–101. London: CRC Press.

Lessios, H.A. (2011) Speciation genes in free-spawning marine invertebrates. *Integr Comp Biol* **51**(3): 456–65.

Levitan, D.R. and Ferrell, D.L. (2006) Selection on gamete recognition proteins depends on sex, density, and genotype frequency. *Science* **312**: 267–9.

Levitan, D.R. and Stapper, A.P. (2010) Simultaneous positive and negative frequency-dependent selection on sperm bindin, a gamete recognition protein in the sea urchin *Strongylocentrotus purpuratus*. *Evolution* **64**: 785–97.

Lopez, A., Miraglia, S.J., and Glabe, C.G. (1993) Structure/function analysis of the sea-urchin sperm adhesive protein bindin. *Dev Biol* **156**: 24–33.

McCartney, M.A. and Lessios, H.A. (2004) Adaptive evolution of sperm bindin tracks egg incompatibility in neotropical sea urchins of the genus Echinometra. *Mol Biol Evol* **21**: 732–45.

Metz, E.C. and Palumbi, S.R. (1996) Positive selection and sequence rearrangements generate extensive polymorphism in the gamete recognition protein bindin. *Mol Biol Evol* **13**: 397–406.

Metz, E.C., Gomez-Gutierrez, G., and Vacquier, V.D. (1998a) Mitochondrial DNA and bindin gene sequence evolution among allopatric species of the sea urchin genus *Arbacia*. *Mol Biol Evol* **15**: 185–95.

Metz, E.C., Robles-Sikisaka, R., and Vacquier, V.D. (1998b) Nonsynonymous substitution in abalone sperm fertilization genes exceeds substitution in introns and mitochondrial DNA. *Proc Natl Acad Sci U S A* **95**: 10676–81.

Palumbi, S.R. (1999) All males are not created equal: fertility differences depend on gamete recognition polymorphisms in sea urchins. *Proc Natl Acad Sci U S A* **96**: 12632–7.

Palumbi, S.R. (2009) Speciation and the evolution of gamete recognition genes: Pattern and process. *Heredity* **102**: 66–76.

Papadopoulou, A., Anastasiou, I., and Vogler, A.P. (2010) Revisiting the insect mitochondrial molecular clock: The mid-Aegean Trench calibration. *Mol Biol Evol* **27**: 1659–72.

Patino, S., Aagaard, J.E., MacCoss, M.J., Swanson, W.J., and Hart, M.W. (2009) Bindin from a sea star. *Evol Dev* **11**: 376–81.

Rocha, S., Lucio, M., Pereira, M.C., Reis, S., and Brezesinski, G. (2008) The conformation of fusogenic B18 peptide in surfactant solutions. *J Peptide Sci* **14**: 436–41.

Swanson, W.J. and Vacquier, V.D. (2002a) The rapid evolution of reproductive proteins. *Nat Rev Genet* **3**: 137–44.

Swanson, W.J. and Vacquier, V.D. (2002b) Reproductive protein evolution. *Annu Rev Ecol Systemat* **33**: 161–79.

Tsaur, S.-C. and Wu, C.-I. (1997) Positive selection and the molecular evolution of a gene of male reproduction, Acp26Aa of Drosophila. *Mol Biol Evol* **14**: 544–9.

Vacquier, V.D. and Moy, G.W. (1977) Isolation of bindin: The protein responsible for adhesion of sperm to sea urchin eggs. *Proc Natl Acad Sci U S A* **74**: 2456–60.

Vacquier, V.D. and Swanson, W.J. (2011) Selection in the rapid evolution of gamete recognition proteins in marine invertebrates. *Cold Spring Harb Perspect Biol* **3**: a002931.

Vacquier, V.D., Swanson, W.J., and Hellberg, M.E. (1995) What have we learned about sea urchin sperm bindin? *Dev Growth Differ* **37**: 1–10.

Vieira, A. and Miller, D.J. (2006) Gamete interaction: Is it species-specific? *Mol Reprod Dev* **73**: 1422–9.

Wyckoff, G.J., Wang, W., and Wu, C.I. (2000) Rapid evolution of male reproductive genes in the descent of man. *Nature* **403**: 304–8.

Zigler, K.S. (2008) The evolution of sea urchin sperm bindin. *Int J Dev Biol* **52**: 791–6.

Zigler, K.S. and Lessios, H.A. (2003a) 250 million years of bindin evolution. *Biol Bull* **205**: 8–15.

Zigler, K.S. and Lessios, H.A. (2003b) Evolution of bindin in the pantropical sea urchin *Tripneustes*: Comparisons to bindin of other genera. *Mol Biol Evol* **20**: 220–31.

Zigler, K.S. and Lessios, H.A. (2004) Speciation on the coasts of the new world: Phylogeography and the evolution of bindin in the sea urchin genus *Lytechinus*. *Evolution* **58**: 1225–41.

Zigler, K.S., Raff, E.C., Popodi, E., Raff, R.A., and Lessios, H.A. (2003) Adaptive evolution of bindin in the genus *Heliocidaris* is correlated with the shift to direct development. *Evolution* **57**: 2293–302.

Zigler, K.S., Byrne, M., Raff, E.C., Lessios, H.A., and Raff, R.A. (in press) Natural hybridization in the sea urchin genus Pseudoboletia between species without apparent barriers to gamete recognition. Evolution.

CHAPTER 15

Evolution of *Drosophila* seminal proteins and their networks

Alex Wong and Mariana F. Wolfner

15.1 Introduction

Since the early days of comparative genetics and biochemistry, researchers have noted striking variation in rates of molecular evolution among proteins (e.g. Zuckerkandl and Pauling 1965). At one end of the spectrum, the sequences of proteins such as histones evolve extremely slowly, with very few changes over long periods of time (e.g. DeLange et al. 1969), indicating strong selection against amino acid change. At the other extreme, the antigen-recognition domain of the mammalian class I major histocompatibility complex shows extensive diversity both within and between species, due to balancing selection (Hughes and Nei 1988). In recent years, general patterns concerning the functions of proteins on this rapidly evolving end of the spectrum have emerged from a growing wealth of comparative genomic data. For example, proteins involved in reproduction, as well as those involved in immunity, are consistently identified among the most rapidly evolving proteins encoded in the genomes of animals (e.g. in *Drosophila*; Civetta and Singh 1995; Clark et al. 2007).

In this chapter, we focus on members of one group of proteins that has been an example of rapid evolution—seminal fluid proteins (SFPs). Using *Drosophila* SFPs as a case study, we briefly review evidence for the rapid evolution of some of these reproductive proteins, and discuss potential causes of this rapid divergence. We also note that within the category of seminal proteins there are members that have evolved more slowly than others. We then propose that the evolutionary dynamics of some SFPs are more properly viewed in the context of the networks in which they participate. For example, a network can include members that are conserved and ones that are rapidly evolving; this can still lead to rapid evolution of the network's function, while preserving conserved biochemical activities.

15.2 *Drosophila* seminal fluid as a model system for rapidly evolving proteins

The SFPs of *Drosophila* have been studied in great detail, from both functional and evolutionary perspectives. SFPs are produced in several secretory organs in the male reproductive tract, including the ejaculatory bulb, the ejaculatory duct, and the paired accessory glands; products of the latter are referred to as Acps. SFPs are transferred to the female during mating along with sperm and non-protein components of the seminal fluid. In *Drosophila melanogaster*, SFPs are required for a wide range of behavioral and physiological postmating changes in females (reviewed in Chapman, 2008; Sirot et al. 2009; Avila et al. 2011), including increased egg-production and egg-laying, decreased receptivity to remating, decreased lifespan and sleep, changes in uterine conformation, and increased feeding. Moreover, some SFPs have demonstrated effects on sperm storage and sperm competition (see Avila et al. 2011 for a review).

To date, well over 150 SFPs have been identified in *D. melanogaster* (reviewed in Avila et al. 2011). Initially, SFPs were identified in two general ways: based on phenotypes produced after introduction into females, or based on the proteins' tissue- and sex-specific gene expression. For exam-

Rapidly Evolving Genes and Genetic Systems. First Edition. Edited by Rama S. Singh, Jianping Xu, and Rob J. Kulathinal.

ple, fractionated extracts of male accessory glands were tested for the ability to induce egg production or to decrease mating receptivity upon injection into females. This identified several proteins in *Drosophila*, such as the sex peptide (SP, Acp70A) in *D. melanogaster* (Chen et al. 1988). Gene expression studies included differential cDNA hybridization approaches, in which genes were identified on the basis of exclusive (or much higher) RNA abundance in male accessory glands, and further chosen on the basis of encoding a protein with a predicted signal sequence. Once '-omic' methods became available, EST (expressed sequence tag) studies, microarrays, and proteomics studies identified additional seminal proteins (reviewed in Avila et al. 2011). Noteworthy among the proteomics studies is that of Findlay et al. (2008), since its authors were able to directly identify proteins that had been transferred to females during mating: females were labeled with heavy nitrogen, rendering their proteins undetectable by standard mass spectrometry. The females were then mated to unlabeled males, and proteins from the mated females' reproductive tracts were subjected to mass spectrometry. Only the proteins they had received from males were detectable. This method has now been successful in identifying seminal proteins in other animals as well (mosquito: Sirot et al. 2011; mouse: Dean et al. 2011). Finally, recent guilt-by-association studies have identified several new SFPs as genes whose expression correlates with known SFPs across a series of wild-derived lines (Ayroles et al. 2011).

SFPs identified in *D. melanogaster* fall into a variety of biochemical classes, as inferred from primary sequence and from comparative modeling (e.g. Mueller et al. 2004 for Acps). Classes include a variety of proteases and protease inhibitor types, lectins, cysteine-rich secretory proteins (CRISPs), lipases, and prohormone-like proteins and peptides. Notably, the same classes of proteins are found in the ejaculates of all animals that have been tested thus far, indicating a broad conservation of biochemical functions associated with seminal fluid.

Although biochemical classes of SFPs are conserved between insects and mammals, population genetic and molecular evolutionary studies have documented rapid evolution and positive selection on a number of *Drosophila* SFPs. The *Drosophila* SFP 'ovulin' in particular, has been surveyed intensively, with a series of studies documenting high within- and between-species diversity, as well as evidence for pervasive directional selection on this protein (e.g. Aguadé et al. 1992; Tsaur et al. 1998; Findlay et al. 2008; Wong et al. 2012). Since the pioneering work of Aguadé and colleagues on ovulin's evolutionary history, evidence for positive selection has been reported for a number of other SFPs (Table 15.1). Rapidly evolving SFPs participate in a wide range of processes, including induction of ovulation (ovulin), sperm storage and/or sperm competition (Acp36DE, Acp29AB, CG9997), proteolysis (CG9997), and immunity (CG32382; Kambris et al. 2006) (Fig. 15.1), but functions of other rapidly evolving SFPs have yet to be elucidated. Consistent with findings of rapid evolution at many individual SFP genes, large-scale sequence comparisons in several species have shown that, on average, SFP loci evolve more rapidly and/or are more likely to experience positive selection than are genes not encoding SFPs (Civetta and Singh 1995; Haerty et al. 2007; Findlay and Swanson 2010). Formally-analogous rapid-evolutionary patterns have been demonstrated in rodents (Turner et al. 2008; Dean et al. 2009) and primates (Clark and Swanson 2005; Wong 2010).

Table 15.1 Positive selection on diverse seminal fluid proteins in *Drosophila melanogaster* and *D. simulans*; selected examples

SFP	Functions/phenotypes
Ovulin	Induction of ovulation 24 hours postmating
CG9997	Release of sperm from storage; long-term response; modulation of sex-peptide activity
Acp36DE	Sperm storage; sperm competition
Acp29AB	Sperm storage; sperm competition
Acp62F	Sperm competition; toxic; ovulin processing
CG32382 (*Sphinx2*)	Immunity
CG10363	Immunity
29 other SFPs	Unknown

In addition to the rapid evolution of SFP genes at the sequence level, it is becoming evident that the overall complement of SFPs changes rapidly: Studies in different species of *Drosophila* have found numerous species- or clade-specific SFPs (e.g. Wagstaff and Begun 2007; Findlay et al. 2008, 2009;

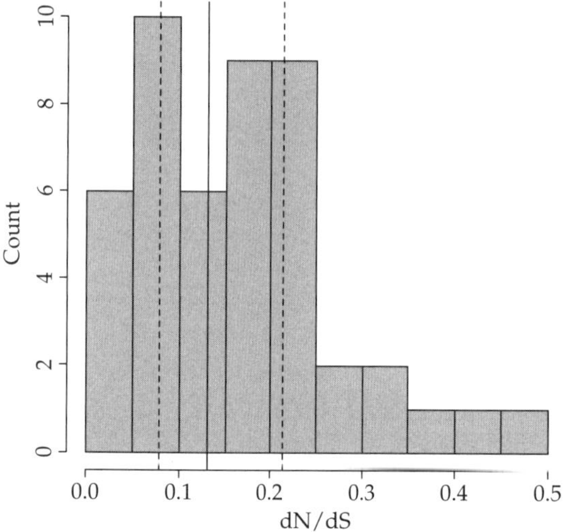

Figure 15.1 Distribution of dN/dS (omega) values for 47 SFP genes in the *Drosophila melanogaster* subgroup (estimated from six *Drosophila* species by Larracuente et al. 2008). The solid vertical line represents an average dN/dS for tissue-specific genes in the *Drosophila* genome (AW unpublished data), with the dotted lines marking the quartiles of the distribution of dN/dS for tissue-specific genes.

Almeida and Desalle 2009), and clear orthologs of many *D. melanogaster* SFPs are apparently absent from other species with fully sequenced genomes (Mueller et al. 2005; Wagstaff and Begun 2005; Haerty et al. 2007). Similarly, gene duplicates are common amongst SFPs, with several recent duplicates identified in *D. melanogaster* and other species (e.g. Wagstaff and Begun 2007). The apparently rapid turnover of SFP genes may be indicative of positive selection for novel SFPs. Positive selection for new SFP genes might occur under sexual conflict if, for example, novel proteins are able to manipulate previously unexploited pathways in females. Alternatively, rapid loss of some SFP genes could be due to a relatively minor fitness cost associated with the loss of individual SFP genes, perhaps due to redundancy between SFPs.

The rapid sequence evolution of individual SFPs (and possibly of the complement of SFPs), and that of reproductive proteins more widely, has primarily been ascribed to two processes in animals: postcopulatory sexual selection, and immune interactions (for reviews and perspectives, see Simmons 2005; Singh and Kulathinal 2005; Panhuis et al. 2006; Lawniczak et al. 2007; Chapman 2008). Postcopulatory sexual selection refers to any process occurring after the act of mating that affects gamete usage. For example, in species in which females mate with several males, postcopulatory sexual selection includes competition between ejaculates from different males (sperm competition), as well as biased patterns of sperm usage by females (sperm preference or cryptic female choice). In addition, since male and female reproductive interests may not always coincide—for example, it will typically be in a male's interest to prevent a female from remating, but it will sometimes be in a female's interest to remate—sexual conflict over a female's postmating behavior and physiology is another important variety of postcopulatory sexual selection. Given that a number of *Drosophila* SFPs are known to be important for sperm competition, sperm storage, and the control of postmating female behaviours (reviewed in Sirot et al. 2009; Avila et al. 2011), and that variation in SFP genes is associated with sperm competitive ability and with variation in postmating female behavior (e.g. Fiumera et al. 2005), it is likely that they are subject to strong postcopulatory sexual selection.

Host–pathogen interactions may also underlie the rapid evolution of some SFPs. The seminal fluids of *Drosophila* and of other insect species contain anti-bacterial proteins (reviewed in Avila et al. 2011), and SFPs alter the expression of immune genes in females after mating (e.g. Lawniczak and Begun 2004; McGraw et al. 2004; Innocenti and Morrow 2009; Peng et al. 2005, Mack et al. 2006, Kapelnikov et al. 2008). Moreover, at least two *Drosophila* SFPs have known roles in immune cascades (CG32382 and CG32383; Kambris et al. 2006). Interestingly, while the expression of immune genes is generally increased in females following mating, systemic immune function is actually reduced (Fedorka et al. 2007), suggesting that reproductive-tract specific assays of immune function will be required. Host–pathogen interactions are thought to drive the rapid evolution of immune proteins in a wide variety of species (e.g. Hughes and Nei 1988; Sackton et al. 2007), and so immune processes occurring in the female reproductive tract could result in positive selection on SFP proteins involved in pathogen response present (see Lawniczak et al. 2007 for a review).

15.3 Extensive variation in rates of SFP evolution

Most work on the evolution of SFPs (and other reproductive proteins) has focused on their rapid evolution, documenting high evolutionary rates and evidence for positive selection in many species (reviewed in Panhuis et al. 2006), and investigating the causes of rapid evolution (e.g. Clark et al. 2009; Finn and Civetta 2010; Civetta, Chapter 17, this volume). This focus is entirely appropriate given observations of high rates of molecular evolution and evidence for positive selection on many SFPs. Nonetheless, extensive *variation* in rates of SFP evolution has been overlooked as a consequence of this attention to rapid evolution: Not all SFPs evolve rapidly or show evidence of positive selection, and indeed some are highly conserved. Fig. 15.1 shows a histogram of dN/dS values for 47 Acp genes for which comparative sequence data were analyzed by (Clark et al. 2007; Larracuente et al. 2008). Also shown on Fig. 15.1 is the mean dN/dS value (solid vertical line) for tissue-specific genes for ~8000 *Drosophila* genes (Larracuente et al. 2008); tissue-specific genes like Acps tend to evolve more rapidly in general than do broadly-expressed genes, possibly due to lower levels of pleiotropy (e.g. Larracuente et al. 2008; Wong et al. unpublished data). In the present context, it is interesting to note that 21 of these Acps have a dN/dS value lower than the average for tissue-specific genes, indicating substantial sequence conservation. Thus, while a number of SFPs evolve rapidly, many do not (see also Findlay and Swanson 2010). Such SFPs may not be subject to sexual selection, sexual conflict, or immunity, and thus would not experience positive selection from these sources. Alternatively, even if these slowly evolving SFPs are involved in processes that promote rapid evolution, pleiotropy, and/or particularly rigid structural constraints may lead to an overall pattern of conservation.

Sex-peptide (SP, also known as Acp70A) represents a particularly interesting case study. SP, a 36-amino acid peptide, was first identified as a potent inducer of egg-laying and inhibition of receptivity upon injection into unmated female *Drosophila* (Chen et al. 1988). Subsequent studies have shown that SP is necessary for many other

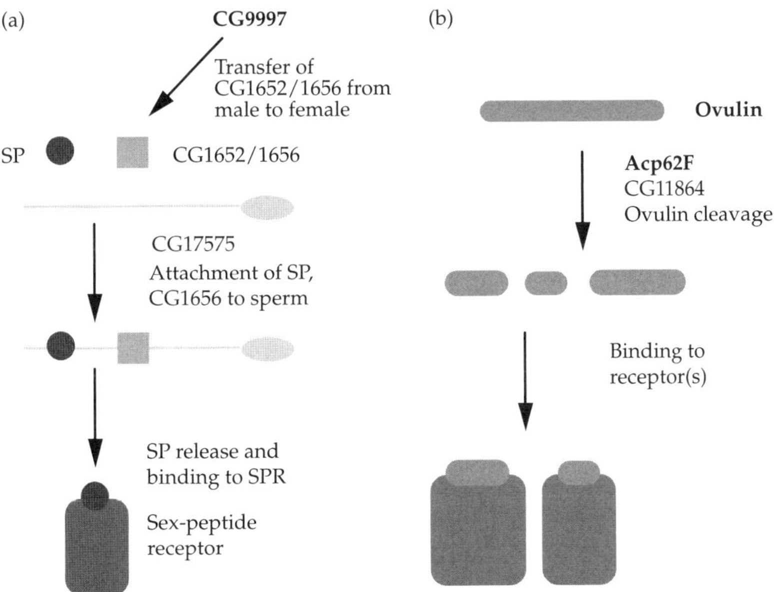

Figure 15.2 Interaction networks of the *Drosophila* SFPs sex-peptide (a) and ovulin (b). Proteins encoded by genes showing evidence of positive selection are highlighted in boldface type.

postmating effects in the female (reviewed in Avila et al. 2011), such as reduced lifespan, increases in feeding, activity, and juvenile hormone synthesis postmating, proper sperm release from storage, and that it may modulate the female immune system. Given the effects of SP on mated females, we might expect this protein to be subject to strong positive selection—each of these effects could be implicated in sexual conflict, sperm competition, and/or host–pathogen interactions.

Comparative genomic and population genetic data, however, give little evidence for positive selection on SP. dN/dS for the SP gene in the *melanogaster* subgroup is about 0.28 (AW, unpublished data), and this relatively high value of dN/dS falls in the top 12% of dN/dS values for tissue-specific genes (Wong unpublished data; Larracuente et al. 2008; note that many of the ~6000 genes not analyzed by Larracuente et al. are likely to be rapidly evolving, and so SP's dN/dS is probably less elevated than this 12% figure suggests). While SP's elevated dN/dS is intriguing, it is not itself indicative of positive selection (for which dN/dS > 1 would be required), and statistical tests of neutrality do not indicate a history of positive selection on SP. For example, Wong et al. (2012) surveyed molecular polymorphism data at the SP locus in African populations of *D. melanogaster* and its close relative *D. simulans*. Neutrality tests aimed at detecting selection on a variety of timescales, using different features of the data, failed to detect any signature of selection. Furthermore, in a polymorphism survey in a European population of *D. melanogaster*, Cirera and Aguadé (1997) did identify unusual haplotype structure upstream of the SP locus, with two distinct haplotypes distinguishable by 22 polymorphisms over an ~300-base pair region. However, given the recent colonization of Europe by *D. melanogaster* (Haddrill et al. 2005), demographic events, rather than selection, are a likely explanation for these patterns.

Several factors could contribute to the apparent lack of positive selection at the SP locus. One possibility is simply lack of statistical power: given that the SP locus is very short, some methods—particularly those that contrast nonsynonymous and synonymous substitutions within a gene—may lack power to detect selection. The power of other methods that detect selection from patterns of polymorphism should not be affected by the short length of the SP locus, however, since polymorphisms in flanking sequences are also used in the inference of selection.

More interestingly, substantial constraints may be imposed on SP evolution by interactions with its receptor, SPR (Yapici et al. 2008). This G-protein coupled receptor has been shown by genetic studies to act through neurons in the reproductive tract to mediate SP's action on egg production and receptivity (Häsemeyer et al. 2009; Yang et al. 2009). Interestingly, SPR has several ligands in addition to SP (Kim et al. 2010; Poels et al. 2010; Yamanaka et al. 2010). SPR is highly conserved, having been identified in such distantly related insects as moths (Hanin et al. 2011). It is possible that SPR's evolution is constrained by its interactions with several ligands. Given the conservation of SPR, features of SP itself that are involved in interactions with SPR may be under considerable constraint.

The apparent lack of positive selection on SP itself seemed unexpected given this molecule's involvement in processes relevant to sexual conflict, sperm competition, and immunity, which are traditionally thought to result in positive selection. SP does not act alone, however. Recent studies have identified four SFPs that are required to associate SP with sperm, an association that is necessary to retain SP in mated females (Ravi Ram and Wolfner 2007, 2009). Subsequent slow release of SP from the sperm allows it to continue to exert its effect in females (as long as they contain sperm and SP), a phenomenon referred to as the long-term response to mating (LTR) (Peng et al. 2005). The four SFPs needed to associate SP with sperm are the predicted lectins CG1652 and CG1656, the CRISP (cysteine-rich secretory protein) CG17575, and the predicted protease homolog CG9997. RNA interference (RNAi) knockdown of any of these genes eliminates the LTR to a similar degree as the knockout/knockdown of SP itself (Ravi Ram and Wolfner 2007). These four proteins, as well as SP, exhibit a complex set of interactions, leading Ravi Ram and Wolfner (2009) to propose that they constitute a LTR network, possibly in conjunction with other as yet unidentified partners.

15.4 Selection on a network?

The discovery of the LTR network raises the possibility that positive selection might act on some members of the network, but not others, e.g. not on SP itself. Different members of the network may be subject to different constraints, either for intrinsic, structural reasons, or owing to pleiotropic consequences of new mutations. Thus, even if few adaptive mutations are available for SP itself for the reasons suggested earlier, selection may act on other proteins involved with SP function. In this way, network function might evolve in response to sperm competition or sexual conflict, but with minimal negative outcomes for non-selected phenotypes. Consistent with this line of thought, the protease CG9997 has been subject to positive selection, with an excess of amino acid substitutions between *D. melanogaster* and *D. simulans* (Wong et al. 2012). While the functional consequences of this rapid evolution are not known, we propose that changes at CG9997 may modulate aspects of the LTR without perturbing the entire system (see also Findlay and Swanson (2010) for a discussion of the evolution of SP and the LTR network).

In addition to interactions between members of the LTR network, a number of interactions have been described amongst other SFPs. Ovulin, for example, is proteolytically cleaved following transfer to the female, and at least two other SFPs, the protease CG11864 and the protease inhibitor Acp62F, are required for its normal proteolysis (Ravi Ram et al. 2006; Mueller et al. 2008). Both ovulin and one molecule that may mediate its activity (Acp62F) are subject to positive selection (Table 15.1), while an additional potential modulator (CG11864) shows no evidence for positive selection (Wong et al. 2012). This example stands in contrast to the LTR network, where the active hormone (SP) does not show evidence for positive selection. Different interaction networks may therefore show markedly different patterns of selection.

Seminal protein activity is likely modulated in a variety of ways, extending beyond direct interactions (such as between SP and CG9997, or between CG11864 and ovulin). For example, Peng et al. (2005) proposed that between-species variation in the length of the sperm tail may have consequences for SP function: the long-term response results from a slow release of SP from sperm, so males producing longer sperm may induce a stronger or longer-lasting LTR if longer sperm 'store' more SP. In addition, Wigby et al. (2009), Fedorka et al. (2011), and Sirot et al. (2011) showed that males adjust the levels SFPs produced, or transferred to females during mating, in response to the presence of competitor males and female mating status. Thus, physiological and behavioral factors can also modulate seminal fluid function without changes in the protein sequence of seminal fluid proteins.

Consideration of selection on network function suggests that a more comprehensive approach will be important in understanding the evolution of reproductive proteins. Selection ultimately acts on reproductive phenotypes, such as female remating rate. While this outcome may be altered at the level of direct male–female interactions, for example, SFPs and their receptors in females, selection may act upon many other parts of the network. As the preceding examples suggest, network function might be altered by changing the availability of key SFPs, through changes in amounts transferred or stored, or through changes in protein stability. In addition, the activity of SFPs could be modulated through interactions with other binding partners, or through post-translational modifications such as proteolysis. For their part, females could control remating rate through sequence changes of the sex-peptide receptor that affect SP binding (direct interactions), or by modulating downstream events such as signal transduction and transcriptional activation. Alterations in the sequence or expression level of downstream effector genes could also play important roles. Neurological and endocrine pathways, as well as systems involved in resource partitioning, are likely to contribute.

It is unclear *a priori* whether the same pathways, and the same components thereof, will contribute to changes in reproductive phenotypes in different lineages. Indeed, the rapid turnover of SFP genes may indicate that different pathways or pathway components are the focus of postmating sexual selection in different species. Under a conflict scenario, for example, a male might manipulate his mates' responses, and a female may distinguish between different potential fathers, using a variety of mechanisms. If females evolve an effective mechanism to avoid male manipulation via one pathway, then male ejaculate proteins targeting that pathway may become ineffective. If a particular ejaculate protein loses effectiveness due to a female counter-adaptation, then selection on that protein will be relaxed, i.e. inactivating mutations will have no fitness effects. Consequently, male–female conflict may result in the pseudogenization of genes encoding proteins that are no longer effective in manipulating female postmating responses.. Novel ejaculate proteins targeting other pathways (or other components of the same pathway) might then become favored, leading to an accumulation of different ejaculate components in different lineages.

15.5 Conclusions

Reproductive proteins are consistently identified as some of the most rapidly evolving members of animal proteomes. In *Drosophila*, extensive genetic, biochemical, and evolutionary studies have generally supported the hypotheses that postcopulatory sexual selection and/or immune interactions are responsible for the rapid sequence evolution of SFPs. Nonetheless, not all SFPs evolve rapidly. Moreover, some SFPs that might be expected to be subject to sexual selection, such as SP, show no evidence for positive selection. We have suggested that consideration of the networks in which SFPs operate may help to explain the heterogeneous nature of selection on these molecules, as well as the high rate of loss and gain of SFP genes. Identification and characterization of SFP networks will thus be important for a full picture of the evolution of SFPs and reproductive proteins more widely.

Acknowledgments

We thank Rama Singh, Jianping Xu, and Rob Kulathinal for inviting our contribution to this book. Frank Avila, Geoff Findlay, and Jessica Sitnik provided valuable comments on the manuscript. We apologize to authors whose work could not be cited directly (but is rather cited through its inclusion in review articles) due to limitations on the number of references permitted for this chapter. We acknowledge funding support from the NIH (MFW: grant R01-HD038921) and the Banting Postdoctoral Fellowship program (AW).

References

Aguadé, M., Miyashita, N., and Langley, C.H. (1992) Polymorphism and divergence in the Mst26A male accessory gland gene region in Drosophila. *Genetics* **132**: 755–70.

Almeida, F.C. and Desalle, R. (2009) Orthology, function and evolution of accessory gland proteins in the Drosophila repleta group. *Genetics* **181**: 235–45.

Avila, F.W., Sirot, L.K., LaFlamme, B.A., Rubinstein, C.D., and Wolfner, M.F. (2011) Insect seminal fluid proteins: identification and function. *Annu Rev Entomol* **56**: 21–40.

Ayroles, J.F., LaFlamme, B.A., Stone, E.A., Wolfner, M.F., and Mackay, T.F.C. (2011) Functional genome annotation of Drosophila seminal fluid proteins using transcriptional genetic networks. *Genet Res (Camb)* **93**(6): 387–95.

Chapman, T. (2008) The soup in my fly: evolution, form and function of seminal fluid proteins. *PLoS Biol* **6**: e179.

Chen, P.S., Stumm-Zollinger, E., Aigaki, T., Balmer, J., Bienz, M., and Böhlen, P. (1988) A male accessory gland peptide that regulates reproductive behavior of female D. melanogaster. *Cell* **54**: 291–8.

Cirera, S. and Aguadé, M. (1997) Evolutionary history of the sex-peptide (Acp70A) gene region in Drosophila melanogaster. *Genetics* **147**: 189–97.

Civetta, A. and Singh, R. (1995) High divergence of reproductive tract proteins and their association with postzygotic reproductive isolation in Drosophila melanogaster and Drosophila virilis group species. *J Mol Evol* **41**: 1085–95.

Clark, A.G., Eisen, M.B., Smith, D.R., Bergman, C.M., Oliver, B., Markow, T.A., et al. (2007) Evolution of genes and genomes on the Drosophila phylogeny. *Nature* **450**: 203–18.

Clark, N.L. and Swanson, W.J. (2005) Pervasive adaptive evolution in primate seminal proteins. *PLoS Genetics* **1**: e35.

Clark, N.L., Gasper, J., Sekino, M., Springer, S.A., Aquadro, C.F., and Swanson, W.J. (2009) Coevolution of interacting fertilization proteins. *PLoS Genet* **5**: e1000570.

Dean, M.D., Clark, N.L., Findlay, G.D., Karn, R.C., Yi, X., Swanson, W.J., et al. (2009) Proteomics and comparative genomic investigations reveal heterogeneity in evolutionary rate of male reproductive proteins in mice (Mus domesticus). *Mol Biol Evol* **26**: 1733–43.

Dean, M.D., Findlay, G.D., Hoopmann, M.R., Wu, C.C., Maccoss, M.J., Swanson, W.J., et al. (2011) Identification of ejaculated proteins in the house mouse (Mus domesticus) via isotopic labeling. *BMC Genomics* **12**: 306.

DeLange, R.J., Fambrough, D.M., Smith, E.L., and Bonner, J. (1969) Calf and pea histone IV. 3. Complete amino acid sequence of pea seedling histone IV; comparison with the homologous calf thymus histone. *J Biol Chem* **244**: 5669–79.

Fedorka, K.M., Linder, J.E., Winterhalter, W., and Promislow, D. (2007) Post-mating disparity between potential and realized immune response in *Dro*sophila melanogaster. *Proc Roy Soc Lond B Biol Sci* **274**: 1211–17.

Fedorka, K.M., Winterhalter, W.E., and Ware, B. (2011) Perceived sperm competition intensity influences seminal fluid protein production prior to courtship and mating. *Evolution* **65**: 584–90.

Findlay, G.D. and Swanson, W.J. (2010) Proteomics enhances evolutionary and functional analysis of reproductive proteins. *Bioessays* **32**: 26–36.

Findlay, G.D., Yi, X., Maccoss, M.J., and Swanson, W.J. (2008) Proteomics reveals novel Drosophila seminal fluid proteins transferred at mating. *PLoS Biol* **6**: e178.

Findlay, G.D., MacCoss, M.J., and Swanson, W.J. (2009) Proteomic discovery of previously unannotated, rapidly evolving seminal fluid genes in Drosophila. *Genome Res* **19**: 886–96.

Finn, S. and Civetta, A. (2010) Sexual selection and the molecular evolution of ADAM proteins. *J Mol Evol* **71**: 231–40.

Fiumera, A.C., Dumont, B.L., and Clark, A.G. (2005) Sperm competitive ability in Drosophila melanogaster associated with variation in male reproductive proteins. *Genetics* **169**: 243–57.

Haddrill, P.R., Thornton, K.R., Charlesworth, B., and Andolfatto, P. (2005) Multilocus patterns of nucleotide variability and the demographic and selection history of Drosophila melanogaster populations. *Genome Res* **15**: 790–9.

Haerty, W., Jagadeeshan, S., Kulathinal, R.J., Wong, A., Ravi Ram, R., Sirot, L.K., et al. (2007) Evolution in the fast lane: rapidly evolving sex-related genes in Drosophila. *Genetics* **177**: 1321–35.

Hanin, O., Azrielli, A., Zakin, V., Applebaum, S., and Rafaeli, A. (2011) Identification and differential expression of a sex-peptide receptor in Helicoverpa armigera. *Insect Biochem Mol Biol* **41**: 537–44.

Hughes, A.L. and Nei, M. (1988) Pattern of nucleotide substitution at major histocompatibility complex class I loci reveals overdominant selection. *Nature* **335**: 167–70.

Häsemeyer, M., Yapici, N., Heberlein, U., and Dickson, B.J. (2009) Sensory neurons in the Drosophila genital tract regulate female reproductive behavior. *Neuron* **61**: 511–18.

Innocenti, P. and Morrow, E.H. (2009) Immunogenic males: a genome-wide analysis of reproduction and the cost of mating in Drosophila melanogaster females. *J Evol Biol* **22**: 964–73.

Kambris, Z., Brun, S., Jang, I.H., Nam, H.-J., Romeo, Y., Takahashi, K., et al. (2006) Drosophila immunity: a large-scale in vivo RNAi screen identifies five serine proteases required for Toll activation. *Curr Biol* **16**: 808–13.

Kapelnikov A., Zelinger E., Gottlieb Y., Rhrissorrakrai K., Gunsalus K.C., Heifetz Y. (2008) Mating induces an immune response and developmental switch in the Drosophila oviduct. *Proc Natl Acad Sci USA* **105**: 13912–7.

Kim, Y.-J., Bartalska, K., Audsley, N., Yamanaka, N., Yapici, N., Lee, J.-Y., et al. (2010) MIPs are ancestral ligands for the sex peptide receptor. *Proc Natl Acad Sci USA* **107**: 6520–5.

Larracuente, A.M., Sackton, T.B., Greenberg, A.J., Wong, A., Singh, N.D., Sturgill, D., et al. (2008) Evolution of protein-coding genes in Drosophila. *Trends Genet* **24**: 114–23.

Lawniczak, M.K.N. and Begun, D.J. (2004) A genome-wide analysis of courting and mating responses in *Drosophila melanogaster* females. *Genome* **47**: 900–10.

Lawniczak, M.K.N., Barnes, A.I., Linklater, J.R., Boone, J.M., Wigby, S., and Chapman, T. (2007) Mating and immunity in invertebrates. *Trends Ecol Evol* **22**: 48–55.

Mack P.D., Kapelnikov A., Heifetz Y., Bender M. (2006) Mating-responsive genes in reproductive tissues of female Drosophila melanogaster. *Proc Natl Acad Sci USA* **103**: 10358–63.

McGraw, L.A., Gibson, G., Clark, A.G., and Wolfner, M.F. (2004) Genes regulated by mating, sperm, or seminal proteins in mated female Drosophila melanogaster. *Curr Biol* **14**: 1509–14.

Mueller, J.L., Ripoll, D.R., Aquadro, C.F., and Wolfner, M.F. (2004) Comparative structural modeling and inference of conserved protein classes in Drosophila seminal fluid. *Proc Natl Acad Sci USA* **101**: 13542–7.

Mueller, J.L., Ravi Ram, S.W., McGraw, L.A., Bloch-Qazi, M.C., Siggia, E.D., Clark, A.G., et al. (2005) Cross-species comparison of Drosophila male accessory gland protein genes. *Genetics* **171**: 131–43.

Mueller, J.L., Linklater, J.R., Ravi Ram, K., Chapman, T., and Wolfner, M.F. (2008) Targeted gene deletion and phenotypic analysis of the Drosophila melanogaster seminal fluid protease inhibitor Acp62F. *Genetics* **178**: 1605–14.

Panhuis, T.M., Clark, N.L., and Swanson, W.J. (2006) Rapid evolution of reproductive proteins in abalone and Drosophila. *Phil Trans Roy Soc Lond B Biol Sci* **361**: 261–8.

Peng, J., Chen, S., Büsser, S., Liu, H., and Honegger, T. (2005) Gradual release of sperm bound sex-peptide controls female postmating behavior in Drosophila. *Curr Biol* **15**: 207–13.

Peng, J., Zipperlen, P., Kubli, E. (2005) Drosophila sex-peptide stimulates female innate immune system after mating via the Toll and Imd pathways. *Curr Biol* **15**: 1690–4.

Poels, J., van Loy, T., Vandersmissen, H.P., van Hiel, B., van Soest, S., Nachman, R.J., et al. (2010) Myoinhibiting peptides are the ancestral ligands of the promiscuous Drosophila sex peptide receptor. *Cell Mol Life Sci* **67**: 3511–22.

Ravi Ram, K. and Wolfner, M.F. (2007) Sustained post-mating response in Drosophila melanogaster requires multiple seminal fluid proteins. *PLoS Genetics* **3**: e238.

Ravi Ram, K. and Wolfner, M.F. (2009) A network of interactions among seminal proteins underlies the long-term postmating response in Drosophila. *Proc Natl Acad Sci USA* **106**: 15384–9.

Ravi Ram, K., Sirot, L.K., and Wolfner, M.F. (2006) Predicted seminal astacin-like protease is required for processing of reproductive proteins in Drosophila melanogaster. *Proc Natl Acad Sci USA* **103**: 18674–9.

Sackton, T.B., Lazzaro, B.P., Schlenke, T.A., Evans, J.D., Hultmark, D., and Clark, A.G. (2007) Dynamic evolution of the innate immune system in Drosophila. *Nat Genet* **39**: 1461–8.

Simmons, L.W. (2005) The evolution of polyandry: Sperm competition, sperm selection, and offspring viability. *Annu Rev Ecol Evol Syst* **36**: 125–46.

Singh, R.S. and Kulathinal, R.J. (2005) Male sex drive and the masculinization of the genome. *Bioessays* **27**: 518–25.

Sirot, L.K., Wolfner, M.F., and Wigby, S. (2011) Protein-specific manipulation of ejaculate composition in response to female mating status in Drosophila melanogaster. *Proc Natl Acad Sci U S A* **108**: 9922–6.

Sirot, L.K., LaFlamme, B.A., Sitnik, J.L., Rubinstein, C.D., Avila, F.W., Chow, C.Y., et al. (2009) Molecular social interactions: Drosophila melanogaster seminal fluid proteins as a case study. *Adv Genet* **68**(January): 23–56.

Sirot, L.K., Hardstone, M.C., Helinski, M.E., Ribeiro, J.M., Kimura, M., Deewatthanawong, P., et al. (2011) Towards a semen proteome of the dengue vector mosquito: protein identification and potential functions. *PLoS Negl Trop Dis* **5**: e989.

Tsaur, S.C., Ting, C.T., and Wu, C.I. (1998) Positive selection driving the evolution of a gene of male reproduction, Acp26Aa, of Drosophila: II. Divergence versus polymorphism. *Mol Biol Evol* **15**: 1040–6.

Turner, L.M., Chuong, E.B., and Hoekstra, H.E. (2008) Comparative analysis of testis protein evolution in rodents. *Genetics* **179**: 2075–89.

Wagstaff, B.J. and Begun, D.J. (2005) Comparative genomics of accessory gland protein genes in Drosophila melanogaster and D. pseudoobscura. *Mol Biol Evol* **22**: 818–32.

Wagstaff, B.J. and Begun, D.J. (2007) Adaptive evolution of recently duplicated accessory gland protein genes in desert Drosophila. *Genetics* **177**: 1023–30.

Wigby, S., Sirot, L.K., Linklater, J.R., Buehner, N., Calboli, F.C., Bretman, A., et al. (2009) Seminal fluid protein allocation and male reproductive success. *Curr Biol* **19**: 751–7.

Wong, A. (2010) Testing the effects of mating system variation on rates of molecular evolution in primates. *Evolution* **64**: 2779–85.

Wong, A., Turchin, M.C., Wolfner, M.F., and Aquadro, C.F. (2012) Temporally variable selection on proteolysis related reproductive tract proteins in Drosophila. *Mol Biol Evol* **29**(1): 229–38.

Yamanaka, N., Hua, Y.-J., Roller, L., Spalovská-Valachová, I., Mizoguchi, A., Kataoka, H., et al. (2010) Bombyx prothoracicostatic peptides activate the sex peptide receptor to regulate ecdysteroid biosynthesis. *Proc Natl Acad Sci U S A* **107**: 2060–5.

Yang, C.-H., Rumpf, S., Xiang, Y., Gordon, M.D., Song, W., Jan, L.Y., et al. (2009) Control of the postmating behavioral switch in Drosophila females by internal sensory neurons. *Neuron* **61**: 519–26.

Yapici, N., Kim, Y.-J., Ribeiro, C., and Dickson, B.J. (2008) A receptor that mediates the post-mating switch in Drosophila reproductive behaviour. *Nature* **451**: 33–7.

Zuckerkandl, E. and Pauling, L. (1965) Evolutionary divergence and convergence in proteins. In V. Bryson and H.J. Vogel (Eds) *Evolving genes and proteins*, pp. 97–166. New York: Academic Press.

CHAPTER 16

Evolutionary genomics of the sperm proteome

Timothy L. Karr and Steve Dorus

16.1 Introduction

The evolution of sexual reproduction and its associated systems has been a prominent topic of scientific research and debate since the inception of evolutionary theory. At the cellular level, the sperm and the egg are the *sine qua non* of sexual reproduction and, as the central players of animal fitness and species survival, these two cell types have understandably occupied human inquiry for millennia. Historically, a multitude of perspectives, hypotheses, and experimental approaches relating to the evolution of sex have converged upon spermatozoa. This 'sperm-centric' perspective arose in large part because sperm are physically responsible for the delivery of male contributions required for fertilization and are produced in vast numbers, thus making them easy to isolate and study. They are also the first cell type upon which an entire evolutionary theory, sperm competition, was based (Parker et al. 1972). Sperm competition has been extensively researched, both theoretically and empirically, and the evolutionary outcomes of different 'scenarios' related to sperm motility, storage, utilization, and inheritance have been explored in great detail (reviewed in Birkhead et al. 2008). Therefore, our understanding of the basic principles and likely selective forces associated with sperm competition and sperm evolution far exceed our understanding of other cellular systems. Despite this, the impact of selection associated with sperm competition upon the molecular building blocks of sperm and the genome as a whole have, until recently, remained enigmatic.

Although there has been widespread interest in the molecular evolution of reproductive genes (reviewed in Swanson and Vacquier 2002), analyses of sperm lagged significantly behind other reproductive genetic systems due largely to the fact that mature spermatozoa are transcriptionally silent (Hecht 1990). This characteristic of sperm rendered gene expression assays largely uninformative regarding the ultimate molecular composition of sperm. Evolutionary analyses were thus conducted upon testis expressed genes often without direct links to sperm form or function. This obstacle was ultimately overcome with the application of mass spectrometry (MS) to the study of the sperm proteome. These studies, which began in earnest a decade ago, were initially targeted studies aiming to identify sperm components which differed between fertile and infertile males (Ficarro et al. 2003) or proteins which undergo post-translational modification during sperm capacitation (Pixton et al. 2004) and have progressed, in step with advances in proteomic methodologies, to whole-cell proteome characterization in *Drosophila* (Dorus et al. 2006; Wasbrough et al. 2010) and a range of mammalian taxa (reviewed in Oliva et al. 2009).

In this review, we discuss the remarkable increase in our understanding of the molecular composition of sperm achieved primarily through mass spectrometry-based proteomics and the associated revelations concerning the evolutionary genomics of sperm. We begin with an overview of our characterization of the *Drosophila melanogaster* sperm proteome and the surprising insights it has provided about the selective forces acting upon sperm, as well as the dynamic role gene creation has played in sperm evolution. Turning our focus to positive selection acting upon mammalian sperm, we then discuss the accelerated evolution of sperm cell

Rapidly Evolving Genes and Genetic Systems. First Edition. Edited by Rama S. Singh, Jianping Xu, and Rob J. Kulathinal.

membrane and acrosome proteins revealed in an analysis of subcellular sperm protein localization. In conclusion, we summarize evidence supporting the theory of compartmentalized adaptation in response to sexual selection, a theory which has emerged from our evolutionary genomic analyses of sperm and from complementary studies of other male reproductive tissues.

16.2 Characterization of the *Drosophila* sperm proteome

Our initial characterization of the *Drosophila melanogaster* sperm proteome (Dorus et al. 2006; termed the DmSP-I) provided the first whole-cell catalog of integral proteins comprising insect sperm and the first evolutionary and functional genomic analysis of a whole eukaryotic cell type. Since then, dramatic advances in MS instrumentation and associated computational tools, has allowed us to reanalyze the original DmSP-I using improved methodologies that have significantly increased the depth of proteome coverage and reproducibility across biological replicates (Wasbrough et al. 2010). This reanalysis has resulted in an expanded sperm proteome (termed the DmSP-II) containing 1108 proteins and represents the most robust database of insect sperm proteins for evolutionary genomic analyses.

On a mass basis, consistent with our previous two-dimensional quantitative gel analysis, the DmSP-II is dominated by two categories of protein families, the tubulins and a diverse family of sperm leucyl aminopeptidases, the S-LAP protein family (Dorus et al. 2011). Additionally, the proteome is enriched for proteins functioning in diverse central metabolic pathways, including proteins involved in glycolysis and gluconeogenesis, the citric acid cycle, and those with oxidative reductase activity. Other highlights from the DmSP-II include three heat shock proteins (*Hsp26*, *Hsp68*, *Hsp83*). *Hsp83* has previously been shown to function in sperm axoneme assembly, an observation supporting its identification in the DmSP-II (Yue et al. 1999). Interestingly, 22 ribosomal proteins, including nine proteins of the small ribosomal subunit and 11 proteins of the large ribosomal subunit, are also present in the DmSP-II. Gene ontology analyses revealed a range of protein classes believed to be critical to sperm structure or function. These included enzymes involved in metabolism and energetics (e.g. oxidoreductases, hydrolases, and transferases), cytoskeletal and related functions, and other minor categories (e.g. proteases, isomerases, and chaperones). By far the largest category was found to be proteins with unclassified functions, a finding consistent with our analysis of the DmSP-I. Further curation of the DmSP-II revealed 40 proteins known to directly affect sperm development or function. The majority of these have been demonstrated to affect sperm motility, including the βTub85D, βTub56D and αTub84B family of proteins. A second class of proteins related to spermatid development and differentiation (e.g. *blw*, *dj*, *heph*, *Hsp83*, *jar*, and *ox*), sperm individualization (*poe* and *shi*), and germline development (*tud*) was also identified. The DmSP-II also contains four Y-linked genes (kl-3, *kl-5*, *ORY*, and *ARY*) not previously found in the DmSP-I, of which *kl-3* and *kl-5* are known to impact male fertility (Carvalho et al. 2000). The successful identification of a wide range of proteins known to have functions related to sperm and male fertility demonstrates the utility of the DmSP-II as a tool for future genetic and molecular investigations of insect sperm biology. We now turn our focus to the evolutionary insights which have been obtained through the integrated analysis of both insect and mammalian sperm proteomes in an evolutionary genomic framework.

16.3 Molecular evolution of the *Drosophila* sperm proteome

Reproductive proteins are thought to evolve rapidly owing to evolutionary pressures associated with sexual selection. However, no large-scale molecular evolutionary analysis of genes specifically encoding integral sperm proteins had been conducted prior to the characterization of the DmSP-I (Dorus et al. 2006). Our original analyses aimed to characterize the evolutionary forces acting on the proteome, through the identification of *D. simulans* orthogs and common methodologies, to estimate selective constraints on gene evolution (nonsynonymous (dN) and synonymous (dS) rates of molecular evolution). Surprisingly, this analysis revealed that most sperm proteins have evolved under purifying selection, a finding that was in stark contrast to the very rapid evolution of other male reproductive genes, most notably those expressed in the accessory gland. Among individual functional categories of sperm genes, the highest evolutionary constraint was amongst genes encoding structural proteins, central metabolic enzymes, and proteins involved in energetics. Interestingly, genes encoding DNA and RNA binding factors were found to evolve rapidly, comparable to the average evolutionary rate of accessory gland genes. Furthermore, individual DmSP genes did not show evidence of positive selection, as measured by a dN/dS ratio that significantly exceeded 1.0. Thus, it was concluded that the sperm proteome, as a whole, is evolving quite conservatively, presumably under the influence of functional and structural constraints. In retrospect, these high levels of selective constraint might be expected for several reasons, including the fact that some of the DmSP gene products perform critical cellular functions in sperm and possibly other cell types (such as in motility and primary metabolism) and that a subset of the DmSP genes are not specific to sperm and therefore may be subject to pleiotropic functional contraints.

In this chapter, we have reconfirmed these general findings by repeating our original analysis on the much larger and more comprehensive DmSP-II. Consistent with our original analysis, sperm genes were found to have an average Ka of 0.013 and an average Ka/Ks of 0.076 in a pairwise analysis with *D. simulans*. These observations confirm that purifying selection is, in fact, the predominant force acting on the sperm proteome and that the evolution of the sperm proteome is quite distinct from the evolution of genes expressed in the accessory gland (Fig. 16.1). The availability of genome annotations across the Sophophora subgenus (Clark et al. 2007) has allowed us to examine (using a maximum likelihood codon analysis) the signature of positive selection on 924 DmSPII genes (pers. comm. S. D.). This analysis identified significant evidence of positive selection on 77 sperm genes, representing approximately 8% of the genes analyzed. It is noteworthy that only 24.6% (19 of 77) of these genes are testis-specific in expression and, in most

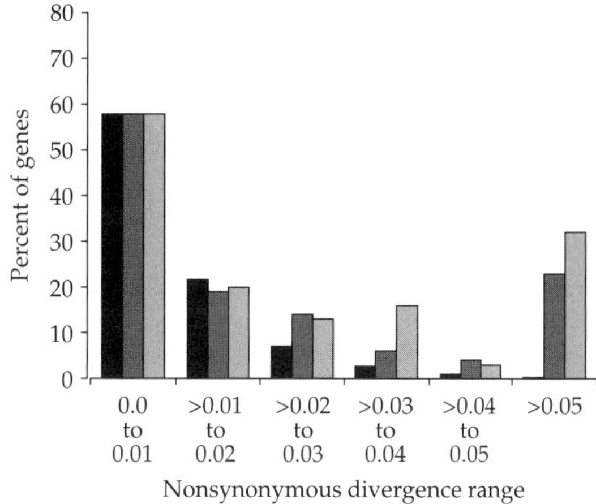

Figure 16.1 Evolutionary rates of sperm and accessory gland genes in *Drosophila*. Histogram comparing the percentages of sperm proteome (black), accessory gland (light gray) and ACP genes (dark gray) within the indicated ranges of nonsynonymous substitution rates between *D. melanogaster* and *D. simulans* orthologs. The analysis was conducted on 915 sperm proteome genes identified in the DmSP-I (Dorus et al. 2006) and DmSP-II (Wasbrough et al. 2010).

cases, the relationship between their function in sperm and positive selection is unknown. There is, however, one notable exception: *Pkd2* encodes a sperm cation channel that is required for directional sperm movement and has been implicated in sperm competition (Gao et al. 2003). This observation is potentially interesting in light of our demonstration of the rapid evolution of mouse sperm membrane proteins (discussed in Section 16.5). In conclusion, increased knowledge of the sperm proteome complemented by the availability of additional genome sequences and the application of more robust analytical methodologies has successfully identified a set of *Drosophila* sperm proteins impacted by positive selection despite the overall conservative evolutionary nature of the proteome as a whole.

16.4 Evolution of novel *Drosophila* sperm components

Despite the conservative evolution of sperm proteome genes, comparative genomic analyses have revealed that gene creation has been an influential mechanism in the molecular evolution of sperm. Duplication of genetic information is believed to be essential in the evolution of functional novelty and biological diversity, and long recognized as important to the evolution of species-specific traits (Ohno 1970). Of particular relevance to the evolution of sperm is the observation that many novel retrogenes acquire testis-enriched or testis-specific expression in Drosophila (Betran et al. 2002; Dorus et al. 2008) and mammals (Dorus et al. 2003; Emerson et al. 2004; Marques et al. 2005; Vinckenbosch et al. 2006). Despite this, the functional ramifications of this process remain to be elucidated in most cases as do the mechanisms by which new genes ultimately evolve new roles in cellular function. In this section, we review the mounting evidence that gene duplication, both via DNA- and RNA-based mechanisms, is of particular importance in sperm evolution based on our analysis of the sperm proteome in *Drosophila*. We also discuss our recent analysis of an expanded S-LAP gene family which highlights the role of gene creation in the evolution of novel gene functions in sperm (Dorus et al. 2011).

Although no comprehensive analyses of sperm gene creation or duplication were possible prior to the characterization of the DmSP-I, several examples of novel sperm genes had been documented. The first involves the evolutionary history of *ms(3)K81* ('K81') a strict paternal effect gene created through retrotransposition prior to the divergence of the *melanogaster* subgroup (Loppin et al. 2005). As a paternal effect lethal mutation, K81 flies are viable, produce motile sperm, and have no adult phenotype. Instead, the phenotype is manifested during fertilization following sperm entrance into the egg (Yasuda et al. 1995). In wild-type eggs fertilized by sperm from K81 homozygous males, paternal chromosomes systematically fail to properly separate sister chromatids during the first zygotic division leading to lethality early in embryogenesis (Loppin et al. 2005). Recent evidence suggests that K81 is critical for telomere maintenance in sperm and, thus, the K81 protein should be physically present in sperm (Dubruille et al. 2010; Gao et al. 2011). Despite this, K81 has never been empirically identified in the sperm proteome of wildtype males, possibly due to an excess of trypsin cleavage sites present in this small protein which results in short peptides difficult to identify robustly by MS (average size of predicted cleavage product < 10 amino acids). The second is the well-studied case of *Sdic*, a newly created gene encoding a protein localized to the sperm tail (Nurminsky et al. 1998). *Sdic* is an unusual case of an X-linked chimeric gene specific to *D. melanogaster* that was created through the duplication of *annexinX* (*AnnX*) and subsequent fusion with *Cdic*, a cytoplasmic dynein. This chimeric gene has also undergone a series of tandem gene duplication events resulting in a multigene cluster (Ponce and Hartl 2006). Although the precise function of *Sdic* in sperm is not known, further analysis of this gene and other novel testis-enriched genes should provide important insights into the processes involved in the integration of novel proteins as components of mature sperm and the role selection plays in their evolutionary retention. The third example is the *Drosophila* gene (*mjl*) *mojoless* (*mjl*), which was created approximately 50 million years ago through retrotransposition (Kalamegham et al. 2007). Evolutionary analysis indicate that this gene represents a retrotransposed copy of *shaggy* (*sgg*), a glycogen synthase kinase-3 encoding gene. By unknown mechanisms, *mjl* acquired male germline expression and testis function as shown by RNA interference (RNAi) experiments that resulted in male infertility. Interestingly, *mjl* partially rescues the *sgg* mutant phenotype indicating that it maintains some ancestral biochemical function despite its newly acquired role in male fertility. The final example is the tandem duplicates, *don juan* and *don juan-like*, which are coexpressed during meiotic prophase and share 42% amino acid identity. Although their precise function remains unclear, both were identified in the DmSP-I and DmSP-II (Dorus et al. 2006; Wasbrough et al. 2010), appear to localize to both the spermatid nucleus and flagellum, and have been implicated in sperm individualization (Hempel et al. 2006).

16.4.1 Novel genes in the sperm proteome

Knowledge of the DmSP-I permitted the first characterization of novel genes functioning within a specific cellular context (Dorus et al. 2008). This characterization resulted in two primary findings: (1) the creation and expansion of testis-expressed gene clusters through tandem duplication has resulted in *D. melanogaster* specific sperm genes and (2) retrotransposition has been influential in the creation of novel sperm components. Analysis of very recent, *melanogaster*-specific duplication events resulted in the identification of two particularly informative cases. First, the ancestral gene encoding *Drosophila* protamines, *Mst-35*, has been duplicated specifically in *D. melanogaster* and these duplicates have subsequently evolved under the impact of positive selection with an enrichment of evolutionary changes in the HMG box DNA binding domain. Protamines are the small basic molecules involved in genome repackaging and compaction during nuclear condensation when the genome is transitioned from histone- to protamine-based nucleosomes. It is noteworthy that protamines have been identified as the target of positive selection in a range of studies, including the proposition that differential levels of sexual selection influence protamines across various primate taxa (Wyckoff et al. 2000) and a study demonstrating that protamine promoter evolution correlates with sperm head morphology and sperm motility in rodents (Martin-Coello et al. 2009). The second case involves two tandem duplications on the X chromosome, which has resulted in a multigene cluster of *tektin* genes encoding structural components of sperm. Interestingly, two of these genes share identical sequences suggestive of a very recent duplication event (Dorus et al. 2008). The evolutionary origins of the *tektin* gene cluster is very reminiscent to that of the *Sdic* cluster. Finally, analysis of the sperm proteome also revealed four novel sperm components created through retrotransposition, including *ctp*, *Acon*, *CG8310*, and *CG32063*. It is noteworthy that two of these retrotransposition events result in X to autosome gene movement consistent with other studies of *Drosophila* retrogenes (Betran et al. 2002).

16.4.2 Expansion and diversification of S-LAP gene family

Functional analysis of the sperm proteome revealed a statistical enrichment of annotated M17 leucyl aminopeptidases relative to the genome as a whole and further analyses determined that these genes are testis-specific and encode the most abundant proteins, by mass, in *Drosophila* sperm (Dorus et al. 2011). These eight genes, now named sperm leucyl aminopeptidases (S-LAPs), underwent rapid expansion during the early evolution of the *Drosophila* clade (Fig. 16.2a). Interestingly, this most dramatic expansion of the S-LAP gene family occurred after the ancestral gene underwent a series of amino acid substitutions that radically altered critical residues within the catalytic site, most likely abolishing enzymatic activity (Fig. 16.2b.) Therefore, it appears that the S-LAP gene family has evolved a novel, but yet to be determined, sperm specific function and that this neofunctionalization may have been selectively involved in the retention of newly created S-LAP gene copies during early *Drosophila* evolution. Further functional and evolutionary studies are required to elucidate the functions of these predominant sperm proteins and the possible role of positive selection in the dramatic evolutionary expansion of this gene family.

16.5 The mouse sperm proteome: intensified selection on sperm membrane and acrosome genes

Despite the successful identification of a relatively small number of *Drosophila* sperm genes driven by positive selection, a general lack of knowledge about how and where these genes function in sperm has limited the conclusions that can be made about differential selection across the proteome. Contrary to the situation in *Drosophila*, more is known about mammalian sperm protein function and subcellular localization. Previous retrospective studies of mammalian sperm genes identified from the literature found evidence of positive selection on 35 functionally diverse sperm genes, including protamines and several cell surface proteins (Torgerson et al. 2002). The recent characterization of the mouse and rat sperm proteomes (Baker et al. 2008a, b), in conjunction with proteomes of targeted subcellular sperm compartments, have made it possible to re-examine the evolutionary impact of selection on mammalian sperm from a proteome-wide perspective (Dorus et al. 2010). This study analyzed over 1000 genes and was designed to detect candidate genes under positive selection within characterized subcellular compartments of the sperm proteome. These compartments included the flagellum accessory structure (Cao et al. 2006), and the sperm membrane and acrosome (Stein

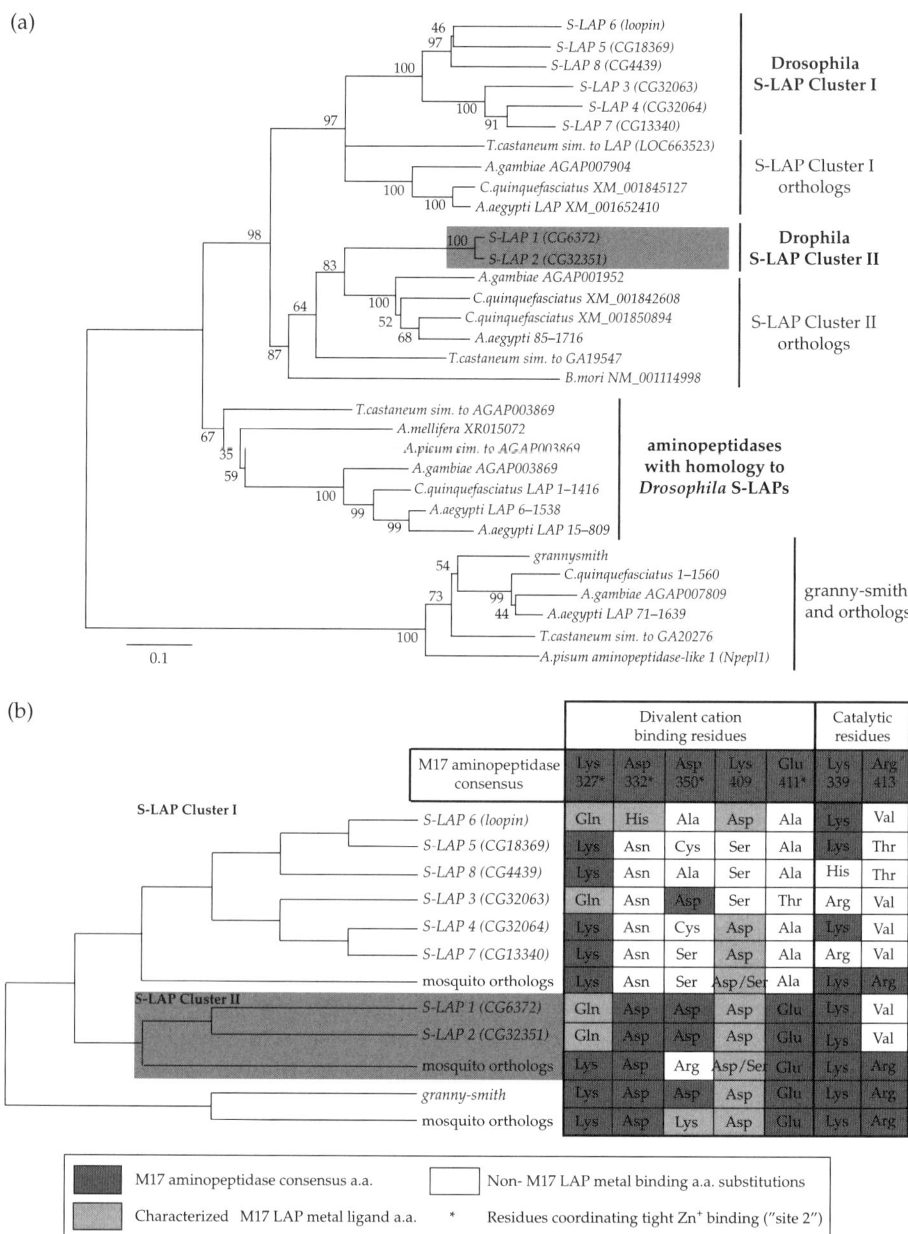

Figure 16.2 Evolutionary expansion and neofunctionalizatin of the S-LAP gene family. (a) Comparative genomic analyses identified a total of 17 related aminopeptidases in *A. gambiae*, *C. quinquefasciatus*, *A. aegypti*, *T. castaneum*, *A. pisum*, and *A. mellifera*. Bootstrap values are presented next to relevant nodes and the phylogeny is drawn to scale based on amino acid substitutions per site. S-LAP Cluster I, comprised of six genes in *Drosophila*, is related to a single copy *aminopeptidase* in mosquitos and *Tribolium*. Similarly, S-LAP Cluster II, comprised of two genes in *Drosophila*, is related to a single copy *aminopeptidase* in other insect taxa with the exception of *C. quinquefasciatus* where two gene copies are present in the genome. In contrast, *granny-smith* has 'one-to-one' orthology relationships in other insect taxa. (b) Amino acid composition at the seven residues involved in metal ion binding and catalysis. Cation binding sites, residues matching the M17 leucyl aminopeptidase consensus are highlighted in dark gray, substitutions to other metal binding amino acids in M17 leucyl aminopeptidases are highlighted in light gray and substitutions to non-M17 leucyl aminopeptidase metal binding residues in white. Catalytic residues, residues matching the M17 leucyl aminopeptidase consensus are highlighted in gray and those divergent from the consensus in white. Amino acid residues Lys327, Asp332, Asp350 and Glu411 comprise the tight, Zn^{2+}-specific binding (site 2), while residue 409 is involved in loose coordination of divalent cations (site 1).

et al. 2006) which were analyzed in relation to the remainder of the sperm proteome for which subcellular localization information was not available. This analysis identified a statistically significant twofold acceleration in evolutionary rate of genes encoding sperm cell membrane proteins compared to the remainder of the sperm proteome (Fig. 16.3a). Furthermore, maximum likelihood analyses detected the signature of positive selection on approximately 22% of sperm cell membrane proteins, representing a significant enrichment relative to other sperm proteins (Fig. 16.3b). Sperm cell membrane proteins impacted by positive selection included several well-characterized proteins involved in sperm–egg fusion, intracellular transport, and a diverse set of proteases/peptidases.

The enhancement of positive selection among proteins localized to the sperm cell membrane and acrosome, which are likely to interact with the intrauterine environment and ultimately the egg surface, focuses attention on several types of molecular interactions as possible targets of sexual selection. An essential interaction, which has been studied in great detail (reviewed in Karr et al. 2008), is the process by which sperm contact and fuse with the oocyte during fertilization. For example, this process involves molecular recognition between ligand and receptor molecules present on the sperm and zona pellucida of mammalian and other vertebrate eggs. Many of these genes tend to evolve rapidly, possibly due to coevolutionary forces, as has been described in detail in invertebrate taxa (Yang et al. 2000; Swanson et al. 2001; Galindo et al. 2003). Our analysis is generally consistent with the invertebrate data as the signature of positive selection was observed in known components of sperm–egg interactions including Zonadhesin, Zona pellucida 3 receptor, Izumo1, and the Adam gene family (Swanson and Vacquier 2002; Civetta 2003; Swanson et al. 2003; Inoue et al. 2005; Gasper

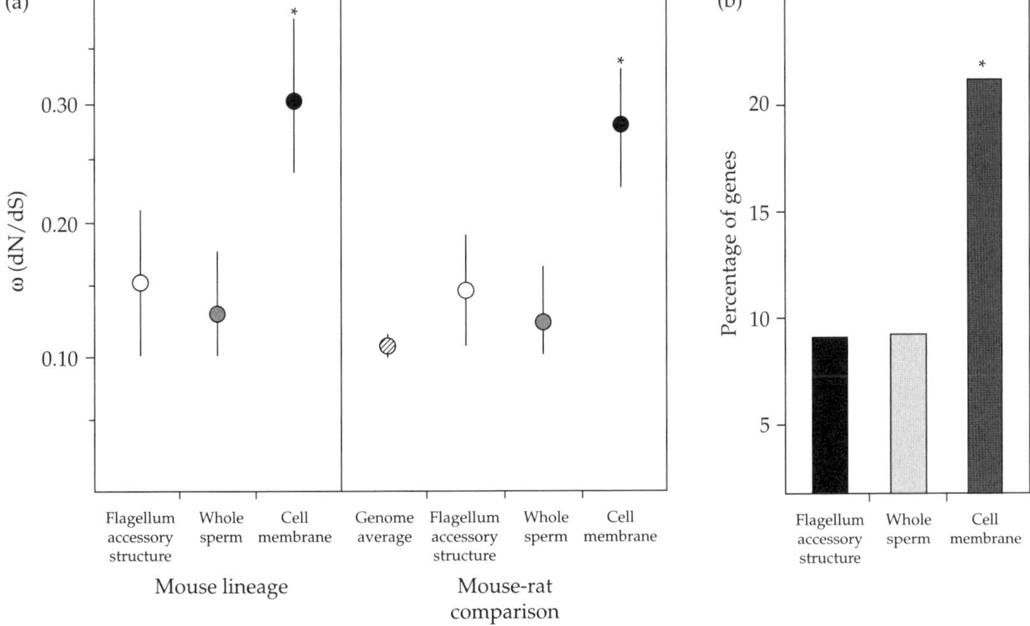

Figure 16.3 Rapid evolution and enhanced positive selection on sperm cell membrane genes. (a) Comparison of the average evolutionary rates (dN/dS) for flagellum accessory structure, whole-sperm MS, and cell membrane data sets. Average evolutionary rates using mouse–rat comparisons also include a comparison with the genomic average between orthologs (95% confidence intervals are also indicated). Significantly higher dN/dS values are observed for sperm cell membrane genes relative to other classes of sperm genes (*P <0.01). (b) Percentage of flagellum accessory structure, whole-sperm MS, and cell membrane genes displaying significant evidence for positive selection in mammals using codon-based maximum likelihood analyses. A significant enrichment of cell membrane genes under positive selection was observed (P <0.001).

and Swanson 2006). Above and beyond interactions with the oocyte, we suggest that interactions between the sperm and the surrounding environment in the male during spermatogenesis, migration, and development in the epididymis, and the female reproductive tract may drive the molecular evolution of sperm cell membrane and acrosome proteins.

16.6 Rapid evolution of immunity-related genes in mammalian sperm

In addition to reproductive genes, comparative genomics has consistently identified immunity-related genes to be among the fastest evolving in the genome, often displaying evidence of positive selection (Hughes and Yeager 1997; Tennessen 2005; Sackton et al. 2007). This has logically been attributed to the potent coevolutionary arms race between ever changing pathogens and the genes of the immune system responsible for host defense. In addition to their well-characterized roles within immune response, there have been several observations that some proteins of the immune system may also have discernable roles in fertilization. For example, the Izumo proteins which are believed to be critical to sperm–egg binding have evolved from an ancestral immunoglobulin protein. Furthermore, males with MHC haplotypes that differ from the inseminated females have been shown more likely to be successful in fertilization competition assays (Gillingham et al. 2009). Consistent with these observations, the mouse sperm proteome also contains a variety of immunity-related genes, and some of these genes also display a signature of positive selection (Table 16.1). Many of these, including *C3*, *Hc*, *Ighg1*, *Itgb2*, *Trf*, and *Mug*, are expressed at relatively low levels in the testis compared to other tissues and their rapid evolution may not be the result of selection associated with reproductive functions. However, several immunity-related sperm genes are overexpressed in the testis (*CD46*, *Irgc1*, and *CD109*) and may be under selection due to their role in reproduction. Interestingly, CD46 generally functions as a complement receptor that protects cells by inhibiting the autologous complement system in most mammals. However, CD46 has

Table 16.1 Immunity-related proteins in the mouse sperm proteome

Protein symbol	Immunity function or protein name	Evidence of selection
Cd109	Membrane antigen	Yes
Cd46	Complement regulation	Yes
Hpx	Positive regulation of immunoglobulin production	Yes
Igbp1b	Immunoglobulin (CD79A) binding protein 1b	Yes
Izumo1[a]	Sperm fusion to egg plasma membrane	Yes
Ace	Neutrophil mediated immunity	No
Apoa 1, 2 and 4	Negative regulation of cytokine secretion	No
Cd55	Innate immune response	No
DefB22	Beta-defensin 22	No
Rbpj	B cell differentiation	No
Hc	Hemolytic complement	No
Hspd1	Positive regulation of T cell activation	No
Ighg1	Complement activation, classical pathway	No
Irgc1	Immunity-related GTPase family, cinema 1	No
Il5	Positive regulation of immunoglobulin secretion	No
Was	T cell activation	No
Mug1	Murinoglobulin 1	No
C1qbp	Complement component 1	No
C3	Complement component 3	No

[a] Izumo proteins evolved from immunoglobulins but no longer have a characterized function relating to immunity

apparently evolved a specialized function in mouse sperm. CD46 localizes to the inner acrosomal membrane of the sperm and knockout mice lacking the CD46 gene exhibit accelerated acrosome reactions and improved fertilization ability (Inoue et al. 2003). A knockout phenotype resulting in improved fertilization rates *in vitro* is somewhat surprising and suggests that the adaptive evolution of this specialized immune gene may be functionally related to optimizing acrosomal reaction dynamics. Although reproductive functions have yet to be elucidated for other immunity-related sperm genes in the mouse, it is known that several complement regulatory proteins are present in the cell membrane of human spermatozoa and may protect sperm from complement mediated lysis in the female reproductive tract (Harris et al. 2006). It is therefore plausible that positive selection on mouse sperm immunity genes (especially complement proteins in the cell membrane, such as CD109 and C3) may be associated with avoiding detection by immune response mechanisms in the female reproductive tract. A preliminary analysis of *Drosophila*, where immune responses are limited to the innate system, did not identify an enrichment of immunity-related genes in the DmSP-II. This raises the possibility that the presence and function of immunity genes in mammalian sperm proteomes is an adaptation specific to those taxa possessing more complex immune systems.

16.7 Sexual selection and compartmentalized adaptation in reproductive genetic systems

Our initial evolutionary analysis of the *Drosophila* sperm proteome revealed a remarkably high level of functional constraint upon the genes encoding proteins in mature spermatozoa, leading to the proposition that sexual selection or adaptive responses to selection is compartmentalized within a restricted subset of genes with specific functions or within specific tissues of the male reproductive system (Dorus et al. 2006). Such compartmentalization of selection is clearly evident in Drosophila by the evolutionary rate disparities observed between sperm and accessory gland genes (Fig. 16.1). This was a surprising result given that: (1) sperm are believed to be a focal point of sexual selection, mediated by sperm competition (Birkhead et al. 2008), (2) testis expressed genes have been demonstrated to evolve rapidly (reviewed in Ellegren and Parsch 2007), and (3) the observation that mammalian sperm genes have been impacted by positive selection (Wyckoff et al. 2000; Torgerson et al. 2002). Purifying selection was most pronounced on sperm proteins with structural and energetic molecular functions, an observation consistent with our subsequent analysis of an expanded and more comprehensive sperm proteome. As previously mentioned, the sperm proteome is comprised, to a large extent, of metabolic/energetic and structural proteins, and more importantly, many of these sperm genes are ubiquitously or widely expressed across a variety of tissues (Wasbrough et al. 2010). It is therefore possible that some portion of the purifying selection on the proteome is due to pleiotropic constraints associated with functions in other tissues. It is noteworthy that sperm proteome genes with unannotated functions are more often testis-specific in expression and generally have higher rates of molecular evolution (Dorus et al. 2006, 2010).

Our original analysis of the sperm proteome suggested that compartmentalized adaptation in response to sexual selection was likely to occur both between male reproductive tissues (e.g. testis vs. accessory gland) as well as amongst genes expressed within a single tissue, i.e. the testis. Ultimately, a more nuanced appreciation for the rate heterogeneity of male reproductive gene evolution has been achieved through additional studies in both mammals and insects. For example, an evolutionary rate analysis across a largely comprehensive range of rodent male reproductive tissues (but not the testis) revealed marked heterogeneity between tissues and particularly rapid evolution amongst seminal vesicle expressed genes (Dean et al. 2008). A relationship between seminal vesicle gene evolution, polyandry, and predicted levels of sperm competition has been previously documented (Dorus et al. 2004) and the rapid evolution of genes encoding secreted components of the ejaculate appears to be a general taxonomical phenomenon (e.g. mammalian seminal vesicle and *Drosophila* accessory gland genes). Lastly, we have already provided evidence for subcellular adaptive compartmentalization in our discussion of the evolution of the mouse sperm proteome with respect to protein localization. This analysis strongly suggests that compartmentalized adaptation is not solely the result of heterogeneous functional constraint, but rather, can also be due to variation in the intensity (or presence) of sexual selection upon different aspects of male reproductive genetics.

Ultimately, the significance of a theory of compartmentalized adaptation is dependent on the relative importance of heterogeneous selective constraint within the relevant genetic system and the actual variation in the targets and intensity of selection. Overall, evidence seems to suggest that sexual selection does in fact vary a great deal across male reproductive architecture and it will be fascinating to determine if this is also true in genetic systems underlying other sexually selected traits (such as sexually dimorphic traits, reproductive behavior, etc.). Compartmentalized adaptation in sperm has implications beyond evolutionary genomics and may be of importance to our understanding of the development of mature sperm, a process involving both spermatogenesis in the testis and critical maturation steps in the epididymis.

16.8 Future perspectives

Although the evolutionary origins of sperm remain a mystery, recent advances in high-throughput MS has greatly increased our knowledge of the basic protein constituents of spermatozoa. Integration of this knowledge with comparative genomic and transcriptomic analysis has allowed us to measure the impact of positive selection across the proteome of a cell type that is central to sexual reproduction and a focal point of sexual selection. As a result, we now have a preliminary understanding of the heterogeneous selective forces acting across not only the sperm proteome but also across different reproductive genetics systems and tissues. The proteome thus serves as a foundational database for future functional and molecular studies of not only the spermatozoa proper, but also for proteins derived from other reproductive tissues that interact with sperm such as seminal fluid proteins, female reproductive tract proteins, and other proteins surrounding, or associated with, the ovum.

From a practical standpoint, spermatozoa can be viewed as the finished product of a complex developmental process of cellular differentiation. This view considers the testis as a factory of machinery and raw materials (the 'input' which is largely a biological 'graybox') that produces sperm (i.e. the output). Although spermatogenesis remains a highly complex graybox that remains beyond our ability to fully understand, sperm proteomes are far less complex. Therefore the sperm proteome represents a 'footprint' of spermatogenesis and we can begin to 'work backwards' from the finished product to provide guideposts and clues that will help us to focus on the most relevant processes. For example, our analysis so far has demonstrated the nonrandom distribution of DmSP genes in the genome, and further identified regions of DmSP gene clustering that should provide a useful starting point for systems-level analyses of gene regulation during spermatogenesis.

The basic discovery phase of *Drosophila* sperm proteomes is well underway and sets the stage for study of sperm proteomes of the 12 *Drosophila* species so far sequenced (Clark et al. 2007). Knowledge of the precise protein compositions of sperm across these lineages will form the basis of much larger functional and evolutionary studies that will allow us to make both intra- and interspecific inferences about sperm divergence as it relates to the various differences between life-history traits. For example, the 12 species include the *Drosophila* and *Sophophora* lineages that diverged approximately 50 million years ago and inhabit highly diverse habitats including island endemic groups (e.g. *D. grimshawi*) and world-wide human commensals (e.g. *D. melanogaster*). These large differences in evolutionary and ecological histories can eventually be juxtaposed with more refined and subtle evolutionary trajectories inherent in comparisons with more closely related species (e.g. the *melanogaster* subgroup). Ultimately such focused studies on *Drosophila* can be applied to other taxa including mammals and primates where knowledge of these systems will have direct application to human infertility.

In conclusion, sperm proteomics represents an emerging new research area that synergizes the enormous technological advances in MS (and associated computational hardware and software) with the power of developmental genetics and functional genomics. The ultimate goal is to use these new tools to probe the entire functional landscape and evolutionary history of the spermatozoa to eventually generate a systems-level understanding of sperm form and function.

References

Baker, M.A., Hetherington, L., Reeves, G.M., and Aitken, R.J. (2008a) The mouse sperm proteome characterized via IPG strip prefractionation and LC-MS/MS identification. *Proteomics* **8**: 1720–30.

Baker, M.A., Hetherington, L., Reeves, G., Müller, J., and Aitken, R.J. (2008b) The rat sperm proteome characterized via IPG strip prefractionation and LC-MS/MS identification. *Proteomics* **8**: 2312–21.

Betran, E., Thornton, K., and Long, M. (2002) Retroposed new genes out of the X in Drosophila. *Genome Res* **12**: 1854–59.

Birkhead, T, Hosken, D., and Pitnick, S. (2008) *Sperm Biology: An Evolutionary Perspective*, Burlington, MA: Academic Press.

Cao, W., Gerton, G.L., and Moss, S.B. (2006) Proteomic profiling of accessory structures from the mouse sperm flagellum. *Mol Cell Proteomics* **5**: 801–10.

Carvalho, A.B., Lazzaro, B.P., and Clark, A.G. (2000) Y chromosomal fertility factors kl-2 and kl-3 of Drosophila melanogaster encode dynein heavy chain polypeptides. *Proc Natl Acad Sci U S A* **97**: 13239–44.

Civetta, A. (2003) Positive selection within sperm-egg adhesion domains of fertilin: an ADAM gene with a potential role in fertilization. *Mol Biol Evol* **20**: 21–9.

Clark, A.G., Eisen, M.B., Smith, D.R., Bergman, C.M., Oliver, B., Markow, T.A., et al. (2007) Evolution of genes and genomes on the Drosophila phylogeny. *Nature* **450**: 203–18.

Dean, M.D., Good, J.M., and Nachman, M.W. (2008) Adaptive evolution of proteins secreted during sperm maturation: an analysis of the mouse epididymal transcriptome. *Mol Biol Evol* **25**: 383–92.

Dorus, S., Gilbert, S.L., Forster, M.L., Barndt, R.J., and Lahn, B.T. (2003) The CDY-related gene family: coordinated evolution in copy number, expression profile and protein sequence. *Hum Mol Genet* **12**: 1643–50.

Dorus, S., Evans, P.D., Wyckoff, G.J., Choi, S.S., and Lahn, B.T. (2004) Rate of molecular evolution of the seminal protein gene SEMG2 correlates with levels of female promiscuity. *Nat Genet* **36**: 1326–9.

Dorus, S., Busby, S.A., Gerike, U., Shabanowitz, J., Hunt, D.F., and Karr, T.L. (2006) Genomic and functional evolution of the Drosophila melanogaster sperm proteome. *Nat Genet* **38**: 1440–5.

Dorus, S., Freeman, Z.N., Parker, E.R., Heath, B.D., and Karr, T.L. (2008) Recent origins of sperm genes in Drosophila. *Mol Biol Evol* **25**: 2157–66.

Dorus, S., Wasbrough, E.R., Busby, J., Wilkins, E.C., and Karr, T.L. (2010) Sperm proteomics reveals intensified selection on mouse sperm membrane and acrosome genes. *Mol Biol Evol* **27**(6): 1235–46.

Dorus, S., Wilkin, E.C., and Karr, T.L. (2011) Expansion and functional diversification of a leucyl aminopeptidase family that encodes the major protein constituents of Drosophila sperm. *BMC Genomics* **12**: 177.

Dubruille, R., Orsi, G.A., Delabaere, L., Cortier, E., Couble, P., Marais, G.A., et al. (2010) Specialization of a Drosophila capping protein essential for the protection of sperm telomeres. *Curr Biol* **20**: 2090–9.

Ellegren, H. and Parsch, J. (2007) The evolution of sex-biased genes and sex-biased gene expression. *Nat Rev Genet* **8**: 689–98.

Emerson, J.J., Kaessmann, H., Betran, E., and Long, M. (2004) Extensive gene traffic on the mammalian X chromosome. *Science* **303**: 537–40.

Ficarro, S., Chertihin, O., Westbrook, V.A., White, F., Jayes, F., Kalab, P., et al. (2003) Phosphoproteome analysis of capacitated human sperm. Evidence of tyrosine phosphorylation of a kinase-anchoring protein 3 and valosin-containing protein/p97 during capacitation. *J Biol Chem* **278**: 11579–89.

Galindo, B.E., Vacquier, V.D., and Swanson, W.J. (2003) Positive selection in the egg receptor for abalone sperm lysin. *Proc Natl Acad Sci U S A* **100**: 4639–43.

Gao, G., Cheng, Y., Wesolowska, N., and Rong, Y.S. (2011) Paternal imprint essential for the inheritance of telomere identity in Drosophila. *Proc Natl Acad Sci U S A* **108**: 4932–37.

Gao, Z., Ruden, D.M., and Lu, X. (2003) PKD2 cation channel is required for directional sperm movement and male fertility. *Curr Biol* **13**: 2175–8.

Gasper, J. and Swanson, W.J. (2006) Molecular population genetics of the gene encoding the human fertilization protein zonadhesin reveals rapid adaptive evolution. *Am J Hum Genet* **79**: 820–30.

Gillingham, M.A., Richardson, D.S., Løvlie, H., Moynihan, A., Worley, K., and Pizzari, T. (2009) Cryptic preference for MHC-dissimilar females in male red junglefowl, Gallus gallus. *Proc Biol Sci* **276**: 1083–92.

Harris, C.L., Mizuno, M., and Morgan, B.P. (2006) Complement and complement regulators in the male reproductive system. *Mol Immunol* **43**: 57–67.

Hecht, N.B. (1990) Regulation of 'haploid expressed genes' in male germ cells. *J Reprod Fertil* **88**: 679–93.

Hempel, L.U., Rathke, C., Raja, S.J., and Renkawitz-Pohl. R. (2006) In Drosophila, don juan and don juan like encode proteins of the spermatid nucleus and the flagellum and both are regulated at the transcriptional level by the TAF II80 cannonball while translational repression is achieved by distinct elements. *Dev Dyn* **235**: 1053–64.

Hughes, A.L. and Yeager, M. (1997) Molecular evolution of the vertebrate immune system. *Bioessays* **19**: 777–786.

Inoue, N., Ikawa, M., Nakanishi, T., Matsumoto, M., Nomura, M., Seya, T., et al. (2003) Disruption of mouse CD46 causes an accelerated spontaneous acrosome reaction in sperm. *Mol Cell Biol* **23**: 2614–22.

Inoue, N., Ikawa, M., Isotani, A., and Okabe, M. (2005) The immunoglobulin superfamily protein Izumo is required for sperm to fuse with eggs. *Nature* **434**: 234–8.

Kalamegham, R., Sturgill, D., Siegfried, E., and Oliver, B. (2007) Drosophila mojoless, a retroposed GSK-3, has functionally diverged to acquire an essential role in male fertility. *Mol Biol Evol* **24**: 732–42.

Karr, T.L., Snook, R.R., and Swanson, W. (2008) Evolution of sperm-egg interactions. In T.R. Birkhead, D.J. Hosken, and S. Pitnick (Eds) *Sperm Biology: an Evolutionary Perspective*, pp. 305–65. London: Academic Press.

Loppin, B., Lepetit, D., Dorus, S., Couble, P., and Karr, T.L. (2005) Origin and neofunctionalization of a Drosophila paternal effect gene essential for zygote viability. *Curr Biol* **15**: 87–93.

Marques, A.C., Dupanloup, I., Vinckenbosch, N., Reymond, A., and Kaessmann, H. (2005) Emergence of young human genes after a burst of retroposition in primates. *PLoS Biol* **3**: e357.

Martin-Coello, J., Dopazo, H., Arbiza, L., Ausió, J., Roldan, E.R., and Gomendio, M. (2009) Sexual selection drives weak positive selection in protamine genes and high promoter divergence, enhancing sperm competitiveness. *Proc Biol Sci* **276**: 2427–36.

Nurminsky, D.I., Nurminskaya, M.V., De Aguiar, D., and Hartl, D.L. (1998) Selective sweep of a newly evolved sperm-specific gene in Drosophila. *Nature* **396**: 572–5.

Ohno, S. (1970) *Evolution by gene duplication*. Berlin: Springer-Verlag.

Oliva, R., de Mateo, S., and Estanyol, J.M. (2009) Sperm cell proteomics. *Proteomics* **9**: 1004–17.

Parker, G.A., Baker, R.R., and Smith, V.G. (1972) The origin and evolution of gamete dimorphism and the male-female phenomenon. *J Theor Biol* **36**: 529–53.

Pixton, K.L., Deeks, E.D., Flesch, F.M., Moseley, F.L., Björndahl, L., Ashton, P.R., et al. (2004) Sperm proteome mapping of a patient who experienced failed fertilization at IVF reveals altered expression of at least 20 proteins compared with fertile donors: case report. *Hum Reprod* **19**: 1438–47.

Ponce, R. and Hartl, D.L. (2006) The evolution of the novel Sdic gene cluster in Drosophila melanogaster. *Gene* **376**: 174–83.

Sackton, T.B., Lazzaro, B.P., Schlenke, T.A., Evans, J.D., Hultmark, D., and Clark, A.G. (2007) Dynamic evolution of the innate immune system in Drosophila. *Nat Genet* **39**: 1461–8.

Schlenke, T.A. and Begun, D.J. (2003) Natural selection drives Drosophila immune system evolution. *Genetics* **164**: 1471–80.

Stein, K.K., Go, J.C., Lane, W.S., Primakoff, P., and Myles, D.G. (2006) Proteomic analysis of sperm regions that mediate sperm-egg interactions. *Proteomics* **6**: 3533–43.

Swanson, W.J., Yang, Z., Wolfner, M.F., and Aquadro, C.F. (2001) Positive Darwinian selection drives the evolution of several female reproductive proteins in mammals. *Proc Natl Acad Sci U S A* **98**: 2509–14.

Swanson, W.J. and Vacquier, V.D. (2002) The rapid evolution of reproductive proteins. *Nat Rev Genet* **3**: 137–44.

Swanson, W.J., Nielsen, R., and Yang, Q. (2003) Pervasive adaptive evolution in mammalian fertilization proteins. *Mol Biol Evol* **20**: 18–20.

Tennessen, J.A. (2005) Molecular evolution of animal antimicrobial peptides: widespread moderate positive selection. *J Evol Biol* **18**: 1387–1394.

Torgerson, D.G., Kulathinal, R.J., and Singh, R.S. (2002) Mammalian sperm proteins are rapidly evolving: evidence of positive selection in functionally diverse genes. *Mol Biol Evol* **19**: 1973–80.

Vinckenbosch, N., Dupanloup, I., and Kaessmann, H. (2006) Evolutionary fate of retroposed gene copies in the human genome. *Proc Natl Acad Sci U S A* **103**: 3220–5.

Wasbrough, E.R., Dorus, S., Hester, S., Howard-Murkin, J., Lilley, K., Wilkin, E., et al. (2010) The Drosophila melanogaster sperm proteome-II (DmSP-II). *J Proteomics* **73**: 2171–85.

Wyckoff, G.J., Wang, W., and Wu, C.I. (2000) Rapid evolution of male reproductive genes in the descent of man. *Nature* **403**: 304–9.

Yang, Z., Swanson, W.J., and Vacquier, V.D. (2000) Maximum-likelihood analysis of molecular adaptation in abalone sperm lysin reveals variable selective pressures among lineages and sites. *Mol Biol Evol* **17**: 1446–55.

Yasuda, G.K., Schubiger, G., and Wakimoto, B.T. (1995) Genetic characterization of ms (3) K81, a paternal effect gene of Drosophila melanogaster. *Genetics* **140**: 219–29.

Yue, L., Karr, T.L., Nathan, D.F., Swift, H., Srinivasan, S., and Lindquist, S. (1999) Genetic analysis of viable Hsp90 alleles reveals a critical role in Drosophila spermatogenesis. *Genetics* **151**: 1065–79.

CHAPTER 17

Fast evolution of reproductive genes: when is selection sexual?

Alberto Civetta

17.1 Introduction

Oddity captures our attention and so it is not surprising that Darwin was intrigued by the complex and elaborate traits found only in males. He realized that an extremely long tail or bright colorations could be a problem for survival and should be selected against. However, males could use such traits to attract females to mate with them or to compete against other males for access to mating. Such reproductive advantage should be driven by a different form of selection, which he referred to as sexual selection (Darwin 1871). For a long time, biologists believed that sexual selection was restricted to secondary sexual traits used by males for premating competition and the peacock's tail became a widely used textbook example of a costly trait that had evolved under sexual selection for mating display.

More recently, we have come to the realization that the morphology of primary genitalia can be molded by pressures exerted from copulatory competition. Electron microscopic analysis of the penes of dragonflies and damselflies shows elaborate adaptations such as hammer-like structures and horn-like appendages that are used to remove or reposition sperm from rival males within the female sperm storage organs. This makes a competitor's sperm less readily available for fertilization or simply removes sperm already present in the ducts of the female storage organs (Simmons 2001). Moreover, postcopulatory competition is now well documented in a wide variety of organisms and it is known to have driven the extreme diversification of sperm morphology among species. In fact, sperm has been referred to as the most diverse cell type known (Pitnick et al. 2009). Female reproductive tract morphology is also widely divergent, highlighting the potential role of sexual selection through female cryptic choice and/or coevolution with sperm morphology (Pitnick et al. 2009). The continuity of male–male competition and female choice after mating has led to the hypothesizing and testing of sexual selection as a process molding the evolution of both pre- and postcopulatory traits (Fig. 17.1).

At the molecular level, some of the first evidence of rapid divergence of male reproductive tract genes came from studies done in the 1980s and 1990s in Dr Rama Singh's lab. Using a pioneering proteomics approach, students in Singh's lab were able to establish a common pattern of rapid divergence of proteins expressed in the testes and male accessory glands of *Drosophila* (Civetta 2003). While results from these studies supported a faster evolution of genes expressed in male reproductive tissue samples than in non-reproductive genes, a series of possible explanations for such a pattern as well as some procedural limitations were linked to these studies. Proteins expressed in male reproductive tissue might rapidly evolve as a consequence of a higher mutation rate driven by faster cell cycles linked to the constant production of sperm and ejaculatory secretions. However, proteins in the female reproductive tract showed similar patterns of divergence as male reproductive tract proteins, providing indirect support for the possibility that rapid divergence might be driven by male–female interactions (Civetta 2003). One of the major limitations of these early proteomics studies was the fact that divergence was scored as either presence or absence of spots in a gel and so the data was not informative enough to test

Rapidly Evolving Genes and Genetic Systems. First Edition. Edited by Rama S. Singh, Jianping Xu, and Rob J. Kulathinal.

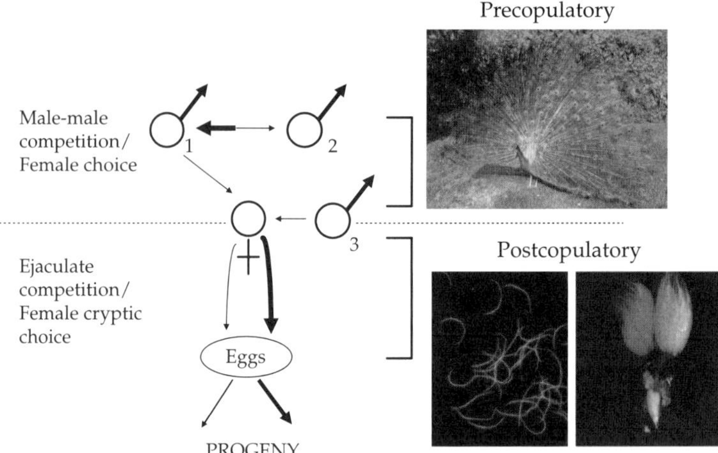

Figure 17.1 Precopulatory and postcopulatory sexual selection. Lines connecting males indicate competition, with the thicker line pointing at the winner. Lines connecting males to female indicate multiple mating, with male subscript denoting different males. Lines of different thickness connecting females to eggs and eggs to progeny represent differential ejaculate success/female cryptic choice. Pictures of precopulatory and postcopulatory traits are: A male peacock performing a mating display (taken at the Assiniboine Park Zoo, Winnipeg, MB), and *Drosophila* sperm and female reproductive tract.

between alternative hypotheses of neutrality or selection. Moreover, it was unclear how much of the divergence was driven by proteins in the somatic tissue of the reproductive tract and to what extent it was due to post-translational modifications.

With advances in DNA sequencing technologies and the development of sequence databases such as Genbank, it became possible to extend the analysis to the DNA level with the advantage that nucleotide substitutions can be classified in terms of their potential impact on gene function based on location (e.g. intron vs. coding) or the expected impact on a protein product and function (e.g. nonsynonymous vs. synonymous). DNA sequence data analysis has established a pattern of rapid divergence in the coding sequence of male genes with a role in reproduction in taxa as diverse as plants, invertebrates, and mammals (Clark et al. 2006; Panhuis et al. 2006; Turner and Hoekstra 2008). Only a handful of studies have found evidence of similar rapid divergence in genes with a role in female reproduction (Swanson et al. 2001; Galindo et al. 2003; Turner and Hoekstra 2006); however, it is likely that the scanty data are the result of the difficulty in identifying female postcopulatory receptors for sperm and ejaculate proteins. More recently, the pattern of rapid divergence of male genes has also been detected at the level of regulation of gene expression. Genes with a male-biased pattern of expression have been shown to be more abundant and evolve faster between species than do female-biased or unbiased genes (Gallach et al. 2011). This greater abundance of male-biased genes is suggestive of a greater turnover with constant loss and acquisition of new male genes. Some of the most striking examples come from genes expressed in the male accessory glands of *Drosophila* (Haerty et al. 2007), and the sperm proteome of *Drosophila* which has also been influenced by gene duplications and gene retrotranspositions into new genomic locations (Dorus et al. 2008).

In summary, male genes in general and particularly those with a role in reproduction, have been shown over and over again to have experienced rapid rates of evolutionary diversification among species resulting in a pattern that is consistent across a wide variety of taxa. In the following sections, I present a series of examples to illustrate how different hypotheses, from relaxation of selective constraints to adaptive diversification, might explain the rapid diversification of male reproductive genes. I then stress the fact that in cases where adaptive selection has been established it has not always been clear whether selection has been sexual in nature. I discuss some of the limitations and advantages of recent studies that have used genes known to affect male reproductive fitness to test the role played by sexual selection using phylogenetic information. I end by pointing out that with developments in genomics, there is a need to: (1) expand our current analytical approach from broad phylogenetic studies where only coding sequence data is tested to more focused analyses where pairs or trios of species are studied within a genomics context; (2) incorporate population genomics analytical approaches that include polymorphism information in order to validate or reject inferences of selection from interspecies comparisons; (3) perform phenotype–genotype association studies, taking advantage of available genome information, to identify candidate genes linked to sexually selected phenotypes that can later be tested in functional assays.

17.2 What has been the role of selection during the evolution of male reproductive genes?

There are a large number of studies in a wide variety of taxa that have addressed this question. Overall, most studies support the idea that rapid evolution is driven by positive or adaptive selection; however, it is worth highlighting a few exceptions. Protamines are small arginine-rich nuclear proteins important during sperm head condensation as they replace histones during the late stages of spermatogenesis (Ward 2010). It was originally proposed that rapid diversification of protamines among primates was driven by species-specific adaptations to fertilization, indicating possible sexual selection (Wyckoff et al. 2000). Interestingly, by comparing protamines in primates and other groups such as monotremes, marsupials, and ruminants, others have noticed that the protein maintains a rather constant number of arginine residues while their location changes from species to species (Rooney et al. 2000). All codons for arginine are guanine (G) rich and protamines show a large bias in G content at their second nucleotide codon position. This observation is consistent with selection for arginines, causing an increase in the frequency of G at the second codon position (Rooney et al. 2000). Moreover, high arginine content in sperm nuclear basic proteins (SNBPs), like protamines, appears to relate to the protein roles not only in DNA binding and chromatin condensation but also in changes of sperm morphology potentially linked to fertilization success as well as activation of regulatory proteins responsible for cellular and metabolic changes in fertilized eggs (Eirín-López and Ausió 2009). Thus, if the only restriction for protein function is a need for a minimum number of arginines to be retained, turnover of amino acids between species could freely occur regardless of position. Such a scenario could easily explain an elevated proportion of nonsynonymous substitutions between species as such changes would not modify the protein function. A recent study of sequence evolution that tested selection both in regulatory (promoters) and coding regions of protamines among 10 species of mice found low estimates of gene divergence by adaptive evolution at protamine 2 but not protamine 1. Interestingly, a strong association was found between species' sperm competition levels and sequence divergence within the protamine 2 promoters (Martin-Coello et al. 2009). This result highlights the need for future studies to incorporate non-coding sequence data in formal tests of sexual selection.

Genome scans have found a significantly larger proportion of genes under positive selection in the chimpanzee lineage than in humans (Arbiza et al. 2006; Bakewell et al. 2007) and it is likely that some of these genes might be driven by sexual selection linked to the more promiscuous mating system of chimpanzees than humans. However, in genome scans there is the possibility of detecting false positive signals of selection driven by error associated with lower sequence quality due to lower sequence coverage. This is a very likely possibility when comparisons are made between model organism genomes and their non-model sister species such as chimpanzee. Sequencing errors aside, a recent study by Wong found that in comparisons between squirrel (promiscuous) and owl (monogamous) monkeys, the differences found on the basis of mating systems were also influenced by mutational load, suggesting that certain lineages could be affected by mutational pressures rather than adaptive diversification (Wong 2010).

These few examples serve as a very powerful cautionary tale of the need to properly control for factors such as nucleotide or amino acid bias composition, differences in mutation rates/ generation times along compared lineages or simply errors linked to the data used to test for selection.

17.3 When is selection sexual? The phylogenetic approach

The overwhelming evidence showing positive selection at genes linked to male reproduction and/or fertilization should not be immediately used to claim sexual selection, or even worse, to support particular forms of sexual selection such as male–female conflict. It is important to test whether positive selection is, in fact, sexual selection. Traditional molecular evolution studies have looked at differences in ratios of nonsynonymous to synonymous substitutions between species at genes linked to reproduction versus other genes. In recent years, a more powerful approach to test sexual selection at the molecular level has been the combined use of phylogenetic information to test for selection along different lineages and correlation to associate lineage-specific bouts of selection with differences in mating systems among species. A handful of studies have used this approach to test sexual selection in rodents and primates (e.g. Wong 2011). The original target of study was semenogelin 2 (SEMG2), a protein that contributes to the formation of a copulatory plug in promiscuous species. The studies found a strong signal of positive selection in this gene and an association between proxies of sexual selection, such as residual testes size, and an elevated ratio of nonsynonymous to synonymous substitutions at the gene. SEMG2 has become the best example of sexual selection at the molecular level because several groups have independently confirmed the original association detected by Dorus between differences in mating systems and rate of evolution in both rodents and primates (Dorus et al. 2004). Interestingly, SEMG1 also contributes to the formation of a copulatory plug. However, studies using a phylogenetic approach have failed to detect selection linked to differences in mating system (Ramm et al. 2008; O'Connor and Mundy 2009). Recently, Wong (2011) has summarized the results of studies done in both primates and rodents, and has shown that only six out of 29 genes (21%) analyzed show any evidence of an association between positive selection and differences in mating patterns even though 26 of the 29 genes (90%) tested showed evidence of positive selection. In his review, Wong elegantly explains some of the specific problems linked to the statistical methods used to test associations between mating systems and molecular evolution. Wong also clearly points out how other factors, such as immune challenges and interactions, can lead to positive selection signals with no link to differences in sexual systems and hence no role to be attributed to sexual selection (Wong 2011). However, a problem that might have contributed to the low number of genes linked to sexual selection likely lies in the phylogenetic approach itself. Unless all or most branches in a phylogeny leading to the promiscuous species had experienced the same or similar selective pressures, selection would be undetectable. This is a problem because the power to directly test genotype–phenotype association is expected to suffer a considerable drop in studies that include few species or use short sequences for analysis (O'Connor and Mundy 2009). A perfect example of the limitations of the phylogenetic analysis is SEMG1. While SEMG1 shows no signal of sexual selection in multispecies, phylogenetic assays, a population genetics study has shown evidence of a selective sweep with low polymorphism in chimpanzees but a high rate of evolution driven by positive selection between humans and chimpanzees (Kingan et al. 2003). This population genetics study and others showing that gorillas, with a polygamous mating system, have actually lost the *SEMG1* gene (Carnahan and Jensen-Seaman 2008) demonstrate how localized branch-specific bouts of sexual selection can be easily missed in phylogenetic context studies that can detect evidence of more long-term selective bouts.

17.4 Testing sexual selection in the era of genomes

Phylogenetic approaches are powerful and have proven useful in detecting long-term selection linked to differences in mating systems. However, it is important to understand the limitations of applying phylogenetic approaches. As stated earlier, the need for a sufficient number of taxa available for phylogenetic reconstruction and, more importantly, the effect of pulling together lineages with very different evolutionary histories, can lead to missing signals of selection if sexual selection has been localized to specific lineages. Additionally, lineage-specific selective pressures that are not truly driven by mating system can lead to false positives. However, this should be less problematic in phylogenetic studies that incorporate a large number of taxa with substantive variation in mating system types. Finally, I would like to expand on a historical constraint of both phylogenetic and non-phylogenetic based tests of selection that have compared rates of nonsynonymous to synonymous substitutions. Such an approach misses information that could be derived from polymorphism data (e.g. SEMG1) or patterns of substitutions in a genome environment within which the coding sequence resides. As an example of the potential benefits and possible pitfalls of extending our current tests of selection, I will describe some of the work performed in my lab on the molecular evolution of ADAM proteins that has expanded the analysis from gene to genomes and from phylogenetics to pairs or trios of species.

The ADAM (a disintegrin and metalloprotease) gene family is ideal to address whether positive selection at reproductive genes is driven by postmating sexual selection. Sperm ADAM proteins have been linked to sperm fertilization ability, sperm migration, and aggregation (Han et al. 2009; Muro and Okabe 2011). ADAM proteins have well characterized domains and positively selected codon sites localized within the adhesion domain of sperm proteins are indicative of selection driven by molecular male × female interactions (Glassey and Civetta 2004). Moreover, we have used associations of signals of selection with differences in mating type to support sexual selection at the molecular level for two sperm surface genes, *Adam2* and *Adam18*, in primates (Finn and Civetta 2010). One limitation in our latest study was the number of species available for testing, which reduced our power to directly test genotype–phenotype associations (O'Connor and Mundy 2009; Finn and Civetta 2010). It will be straightforward to expand our test of selection to include more species given the eminent release of more primate genome sequence data. However, it is interesting to note that while *Adam2* showed evidence of selection when using either mammal sequence data or primates only, *Adam18* only showed a bout of sexual selection within primates. Similarly, signals of positive selection at other *Adam* genes detected in mammals disappeared when focusing the analysis on primates. Therefore, it is likely that the selective bout for *Adam18* is localized to a few species rather than spread over many and that the inclusion of more species in a phylogenetic based analysis will miss sexual selection at this gene.

To address the issue of localized bouts of sexual selection one could simply analyze pairs of species such as chimpanzee and humans and use genome, rather than gene, sequence comparisons. There are several reasons to use chimpanzee and humans beyond the obvious interest in understanding our own species. For example, we have found significant evidence of selection on *Adam* genes along the chimpanzee lineage in phylogenetic tests (Finn and Civetta 2010) and a genome scan found a significantly higher proportion of nonsynonymous changes for testes-expressed genes along the chimpanzee lineage than along the human lineage (Wong 2010). In the case of *Adam* genes, our prior analysis using phylogenetic approaches identified two *Adam* genes (2 and 18) that happen to be localized adjacent to each other on the primate chromosome 8. More interesting is the fact that three other sperm surface genes, for which we lacked statistical power to test selection due to the low number of primate sequences available, are immediate chromosome neighbors of *Adam2* and *Adam18* (*Adam3*, *Adam5*, and *Adam32*). This region, within chromosome 8, is a perfect target for genome sequence analysis that would incorporate both coding and non-coding sequence data. If this genome region is aligned between species, one could simply code substitutions as an event as it has been done in single gene sequence scans (Tang and Lewontin 1999). A cumulative distribution function can then be used to test for regional differences in sequence divergence without relying on coding information only. This method should allow detection of monotonic increases (rapid change) or decreases (conservation) in different genome regions and this can be used to determine whether selection signals previously detected along specific protein domains are influenced by patterns of substitutions in neighboring non-coding regions. This first pass of analysis can quickly identify whether a coding gene or protein domain is within a conserved genome environment or in a rapidly changing neighborhood. Preliminary data using this approach shows that *Adam18* is, in fact, within a rather conserved genome region between chimpanzee and humans as is evident from negative cumulative substitution scores. However, it is of interest to note that the protein adhesion domain we have previously shown to evolve under positive selection based on coding sequence analysis is in a flat area suggesting an even pattern of conserved sites and nucleotide substitutions. Thus, it is possible that the positive selection signal we have previously detected for *Adam18* is influenced by a localized mutational hotspot within the region. *Adam2* is in a noisy genomic neighborhood with multiple substitutions having arisen since the species split. Interestingly, the adhesion domain of *Adam2*, previously linked to sexual selection, is in a monotonically decreasing (conserved) genome region for substitutions thus reinforcing the likelihood that nonsynonymous substitutions within the adhesion domain of *Adam2* are not influenced by mutations (Fig. 17.2). Many reasons can explain the overall heterogeneous pattern of substitutions within the *Adam2* genome region. A general lack of selective constraints should lead to such heterogeneity but it is also possible that some informative directionality or bias in patterns of substitutions within non-coding intronic and flanking regions might be driven by selection. To test this, a careful analysis of the type of substitutions (e.g. transitions versus transversions) and localization of changes (e.g. introns, non-coding regions, putative regulatory motifs) using genome comparisons is needed. This analysis should be extended to incorporate polymorphism data from multiple closely related species to clearly identify fixed substitutions from shared polymorphisms and to add to the analysis the power of the population genomics statistical toolkit. With new sequence variation data from the 1000 Genome Project and the availability of SNP (single nucleotide polymorphism) databases, human polymorphism information should be readily available (Durbin et al. 2010) with the real challenge being the lack of polymorphism data for chimpanzees. Another potential pitfall or challenge comes from differences in sequencing coverage between genomes analyzed. For example, genome sequence comparisons between human and chimpanzee have led to the generalized suggestion that the chimpanzee genome harbors a larger proportion of genes under positive selection than the human genome (Arbiza et al. 2006; Bakewell et al. 2007). This has raised concerns that signals of selection might be driven by the lower

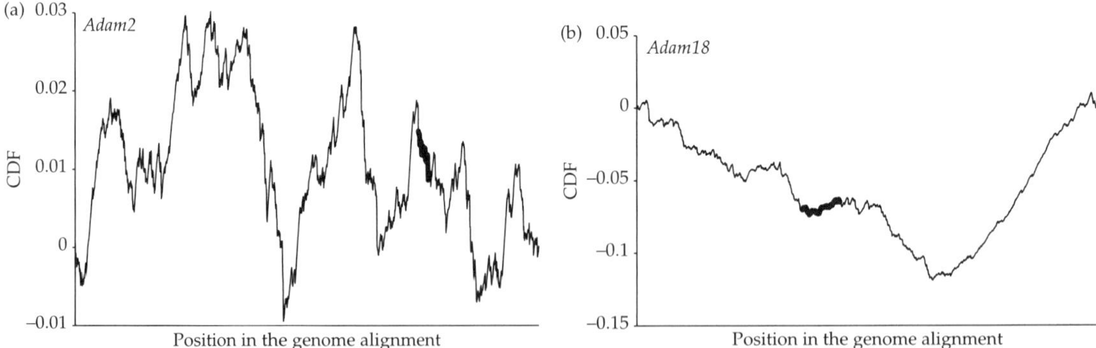

Figure 17.2 Cumulative distribution function (CDF) for substitutions between human and chimpanzee at the *Adam2* (approximately 94 kb) and *Adam18* (approximately 141 kb) genes genome regions. The CDF for the genome region harboring the protein's adhesion domain is shown as a thicker line. CDF was calculated as in Tang and Lewontin (1999).

quality, due to lower sequence coverage, of the chimpanzee compared to the human genome. In a whole-genome scan, even a very low error rate can lead to false positives. Recent studies have acknowledged this potential problem and have masked low-quality data with filters that improve alignments using synteny information and remove nucleotides with low-quality scores from the analysis (Bakewell et al. 2007; Mallick et al. 2009). Next-generation sequencing technologies are now providing biologists with unprecedented sequence coverage and filtering tools.

17.5 The need for association studies and functional assays

There are currently more than 600 eukaryotic genomes with either partial or complete sequence data available (http://www.ncbi.nlm.nih.gov/genomes/static/gpstat.html). The numbers are likely to grow exponentially with recent developments of high-throughput non-Sanger-based sequencing technologies and the continued development of next-generation sequencing platforms. While these technologies give us rapid access to sequence data, a current limitation is the lack of functional annotation even for well-studied model organisms. More problematic is the fact that gene function assignments from experimental manipulations such as gene knockouts do not necessarily identify genes that explain phenotypic variation linked to fitness. The advantage of the phenotype–genotype association approach using genome sequence information is that one can first test naturally occurring variation at a trait linked to sexual selection such as mate choice, sperm competition, or morphological adaptations for male display and competition against other males. Genes associated with such variation are then obvious candidates to test for signatures of sexual selection at the molecular level. We have a few examples of studies that have at least partially succeeded in identifying a handful of candidate genes. We have recently used different *Drosophila simulans* strains with mapped introgressions from *D. mauritiana* into their third chromosome to measure differences in sperm competitive ability when competing against a wild type *D. simulans* male. We found two broad introgressed regions (77B to 84B and 88B to 92E) within the third chromosome that resulted in second male paternity breakdown. Using genome sequence data and *D. melanogaster* gene annotations, we identified 81 broad-sense candidate genes that were subsequently narrowed down to 14 genes based on gene sequence and gene expression data analysis (Levesque et al. 2010). Interestingly four of the genes are located within the 89B cytogenetic position and so this region has become our main target for future studies using SNP-phenotype associations and gene knockdown assays.

Other SNP-phenotype association studies have identified genes potentially linked to postmating prezygotic components of sexual selection. The main suspects have always been seminal fluid proteins (SFPs) known to contribute to sperm storage ability or capable of eliciting postcopulatory responses in females such as increases in egg-laying rates or reluctance to remate (Avila et al. 2011). Some of the most powerful examples come from combinations of studies that have either tested associations between nucleotide variation at SFP genes and differences in phenotypic performance and others that have tested the effect of gene manipulations on phenotypes. For example, a gene knockout by targeted deletion of the accessory gland expressed SFP known as *Acp62F* resulted in the increase of a male's ability to place sperm in storage (Mueller et al. 2008). A population survey of sequence variation found significant associations between polymorphisms at *Acp62F* and both second male paternity success and female induced fecundity (Fiumera et al. 2007). Loss-of-function mutation and RNAi knock-down of *Acp29AB* and *Acp36DE* respectively have established the need of these proteins for accumulation and maintenance of sperm in the female storage organs (Wong et al. 2008; Avila and Wolfner 2009). Both genes have also shown evidence of associations between nucleotide polymorphisms and variation in sperm competitive ability (Clark et al. 1995; Fiumera et al. 2005).

There are also non-accessory gland protein (*Acp*) genes examples. A recent study aimed at identifying how members of the odorant binding protein (*Obp*) gene family in *Drosophila* have evolved different transcriptional profiles as a consequence of functional diversification identified the *Obp19c* transcript as enriched for GO categories linked to postmating behavior and oviposition, with high expression in ovaries and accessory glands. While the evidence from this study is still indirect, it highlights the power of an association study between olfactory behavior (phenotype) and transcripts profiles (genotype) to identify one *Obp* gene that could work to modulate female postmating behavior and physiology (Arya et al. 2010). Greenspan and Clark (2011) have recently tested associations between X-chromosome genes and differences in sperm competitive ability in *D. melanogaster* and have found four candidate genes within the X chromosome, a chromosome with a scarcity of genes for seminal proteins. Outside *Drosophila*, species of the *Allonemobius socius* complex of crickets are isolated through postmating prezygotic barriers such as male-induced egg-laying and conspecific sperm precedence. Using a combination of proteomics and gene knockdown approaches and mating designs, Marshall and collaborators have recently identified a single protein, named *ejaculate serine protease*, that is found as a single transcript in the male accessory glands and that induces egg-laying in females (Marshall et al. 2009).

While research in insect systems is amenable to this approach, other systems have been more difficult to tackle. For example, it has been rather difficult to establish direct associations between single gene knock-outs and reproductive phenotypes in vertebrates in general and mammals in particular due in part to the complexity of their reproductive biology and genetics. However, a series of recent direct gene manipulation studies have started to render a clearer picture of the precise function of some long-suspected mammalian male reproductive genes. For example, a series of null mutations in mice have established roles for a handful of genes in sperm migration from the uterus into the oviduct, sperm aggregation and binding to the zona pellucida (Han et al. 2009; Ikawa et al. 2011; Muro and Okabe 2011). We still lack data on the actual impact of natural variation at such genes on male proxies of sexual selection, but it is interesting to notice that there is some support for postcopulatory male × male competition favoring phenotypes linked to sperm migration ability and sperm aggregation. For example, longer sperm are supposed to be favored, with some metabolic limitations, due to the ability of longer sperm to swim and reach the ova faster than shorter sperm (Gomendio et al. 2011; Tourmente et al. 2011). In terms of sperm aggregation, sperm of promiscuous deer mice preferentially aggregate with sperm from the same male while sperm indiscriminately groups with unrelated sperm in monogamous species suggesting that in promiscuous species, sperm competition might have selected for cooperation among related sperm forms (Fisher and Hoekstra 2010). Therefore, it is worth testing candidate genes linked to phenotypic proxies of sexual selection in gene manipulation assays for genotype–phenotype associations in non-genetically manipulated natural populations.

17.6 Conclusions

Many studies have conclusively established a trend of rapid evolution and even adaptive selection at several genes linked to male reproductive phenotypes. Only a handful of genes showing signals of positive selection have been successfully linked to differences in proxies of sexual selection such as mating type. However, the low numbers might be a consequence of limitations on the approaches used, which require either long-term and/or phylogenetically-spread signals of selection.

Future studies of sexual selection at the molecular level could benefit from genomics and the availability of large gene-sample sizes to test bouts of sexual selection that might have been limited to single or few taxa and hence remain undetected in phylogenetic studies. Similarly, there is a need to conduct phenotypic studies of variation at traits such as mating behavior and mate choice, sperm competitive ability, and female cryptic choice as well as other proxies of sexual selection and to link differences in male competitive ability and female choice to genetic polymorphisms and divergence. Genotype–phenotype associations are now facilitated by the availability of genomes but there is still a gap in functional characterization of genes, as even fully sequenced genomes of model organisms are still poorly annotated. Genes or genome regions highlighted in phenotype–genotype association assays could be readily tested by taking advantage of gene manipulations such

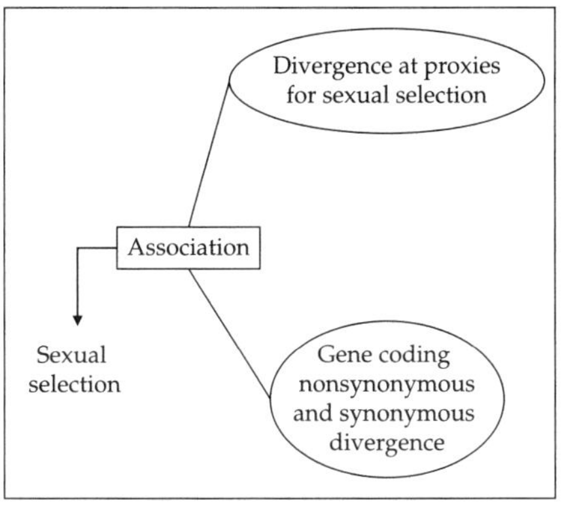

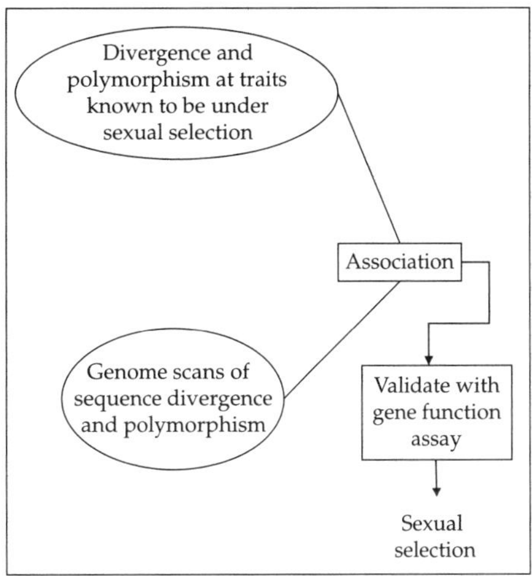

Figure 17.3 A schematic representation of differences between traditional gene-based (left panel) and genome-based (right panel) approaches to testing sexual selection at the molecular level.

as RNAi gene knock-downs, knock-outs, and the use of transgenics (Fig. 17.3). It is an exciting time for evolutionary geneticists as the availability of genome resources, bioinformatics tools, and molecular toolkits puts us in an excellent position to tackle questions linked to the genetic basis of evolutionary processes. We are also at a time when we should be able to frequently bridge what have traditionally been parallel approaches among evolutionary biologists that work either on genotypes or phenotypes, thus allowing us to decipher mechanisms of evolution from molecules to phenotypes and back.

Acknowledgments

I would like to thank Jennifer Ferguson, Scott Finn and Leanne Grieves in my lab for proofreading the original version of this chapter. This work was supported by a Natural Sciences and Engineering Research Council (NSERC) Individual Discovery Grant to Alberto Civetta.

References

Arbiza, L., Dopazo, J., and Dopazo, H. (2006) Positive selection, relaxation, and acceleration in the evolution of the human and chimp genome. *PLoS Comput Biol* **2**: e38.

Arya, G.H., Weber, A.L., Wang, P., Magwire, M.M., Negron, Y.L., Mackay, T.F., et al. (2010) Natural variation, functional pleiotropy and transcriptional contexts of odorant binding protein genes in *Drosophila melanogaster*. *Genetics* **186**: 1475–85.

Avila, F.W. and Wolfner, M.F. (2009) Acp36DE is required for uterine conformational changes in mated Drosophila females. *Proc Natl Acad Sci U S A* **106**: 15796–800.

Avila, F.W., Sirot, L.K., LaFlamme, B.A., Rubinstein, C.D., and Wolfner, M.F. (2011) Insect seminal fluid proteins: identification and function. *Annu Rev Entomol* **56**: 21–40.

Bakewell, M.A., Shi, P., and Zhang, J. (2007) More genes underwent positive selection in chimpanzee evolution than in human evolution. *Proc Natl Acad Sci U S A* **104**: 7489–94.

Carnahan, S.J. and Jensen-Seaman, M.I. (2008) Hominoid seminal protein evolution and ancestral mating behavior. *Am J Primatol* **70**: 939–48.

Civetta, A. (2003) Shall we dance or shall we fight? Using DNA sequence data to untangle controversies surrounding sexual selection. *Genome* **46**: 925–9.

Clark, A.G., Aguadé, M., Prout, T., Harshman, L.G., and Langley, C.H. (1995) Variation in sperm displacement and its association with accessory gland protein loci in Drosophila melanogaster. *Genetics* **139**: 189–201.

Clark, N.L., Aagaard, J.E., and Swanson, W.J. (2006) Evolution of reproductive proteins from animals and plants. *Reproduction* **131**: 11–22.

Darwin, C. (1871) *The descent of man, and selection in relation to sex*. London: John Murray.

Dorus, S., Evans, P.D., Wyckoff, G.J., Choi, S.S., and Lahn, B.T. (2004) Rate of molecular evolution of the seminal protein gene SEMG2 correlates with levels of female promiscuity. *Nat Genet* **36**: 1326–9.

Dorus, S., Freeman, Z.N., Parker, E.R., Heath, B.D., and Karr, T.L. (2008) Recent origins of sperm genes in Drosophila. *Mol Biol Evol* **25**: 2157–66.

Durbin, R.M., Abecasis, G.R., Altshuler, D.L., Auton, A., Brooks, L.D., Durbin, R.M., et al. (2010) A map of human genome variation from population-scale sequencing. *Nature* **467**: 1061–73.

Eirín-López, J.M. and Ausió, J. (2009) Origin and evolution of chromosomal sperm proteins. *Bioessays* **31**: 1062–70.

Finn, S. and Civetta, A. (2010) Sexual selection and the molecular evolution of ADAM proteins. *J Mol Evol* **71**: 231–40.

Fisher, H.S. and Hoekstra, H.E. (2010) Competition drives cooperation among closely related sperm of deer mice. *Nature* **463**: 801–3.

Fiumera, A.C., Dumont, B.L., and Clark, A.G. (2005) Sperm competitive ability in Drosophila melanogaster associated with variation in male reproductive proteins. *Genetics* **169**: 243–57.

Fiumera, A.C., Dumont, B.L., and Clark, A.G. (2007) Associations between sperm competition and natural variation in male reproductive genes on the third chromosome of Drosophila melanogaster. *Genetics* **176**: 1245–60.

Gallach, M., Domingues, S. and Betrán, E. (2011) Gene duplication and the genome distribution of sex-biased genes. *Int J Evol Biol* (Epub 2011 Sep 5).

Galindo, B.E., Vacquier, V.D., and Swanson, W.J. (2003) Positive selection in the egg receptor abalone sperm lysine. *Proc Natl Acad Sci U S A* **100**: 4639–43.

Glassey, B., and Civetta, A. (2004) Positive selection at reproductive ADAM genes with potential intercellular binding activity. *Mol Biol Evol* **21**: 851–9.

Gomendio, M., Tourmente, M., and Roldan, E.R. (2011) Why mammalian lineages respond differently to sexual selection: metabolic rate constrains the evolution of sperm size. *Proc R Soc B* **278**(1721): 3135–41.

Greenspan, L. and Clark, A.G. (2011) Associations between variation in X chromosome male reproductive genes and sperm competitive ability in Drosophila melanogaster. *Int J Evol Biol* (Epub 2011 May 5).

Haerty, W., Jagadeeshan, S., Kulathinal, R.J., Wong, A., Ravi Ram, K., Sirot, L.K., et al. (2007) Evolution in the fast lane: rapidly evolving sex-related genes in Drosophila. *Genetics* **177**: 1321–35.

Han, C., Choi, E., Park, I., Lee, B., Jin, S., do Kim, H., Nishimura, H., and Cho, C. (2009) Comprehensive analysis of reproductive ADAMs: relationship of ADAM4 and ADAM6 with an ADAM complex required for fertilization in mice. *Biol Reprod* **80**: 1001–8

Ikawa, M., Tokuhiro, K., Yamaguchi, R., Benham, A.M., Tamura, T., Wada, I., et al. (2011) Calsperin is a testis-specific chaperone required for sperm fertility. *J Biol Chem* **286**: 5639–46.

Kingan, S.B., Tatar, M., Rand, D.M. (2003) Reduced polymorphism in the chimpanzee semen coagulating protein, semenogelin I. *J Mol Evol* **57**: 159–69.

Levesque, L., Brouwers, B., Sundararajan, V., and Civetta, A. (2010) Third chromosome candidate genes for conspecific sperm precedence between D. simulans and D. mauritiana. *BMC Genet* **11**: 21.

Mallick, S., Gnerre, S., Muller, P., and Reich, D. (2009) The difficulty of avoiding false positives in genome scans for natural selection. *Genome Res* **19**: 922–33.

Martin-Coello, J., Dopazo, H., Arbiza, L., Ausió, J., Roldan, E.R., Gomendio, M. (2009) Sexual selection drives weak positive selection in protamine genes and high promoter divergence, enhancing sperm competitiveness. Proc. Biol. Sci. **276**: 2427–36.

Marshall, J.L., Huestis, D.L., Hiromasa, Y., Wheeler, S., Oppert, C., Marshall, S.A., et al. (2009) Identification, RNAi knockdown, and functional analysis of an ejaculate protein that mediates a postmating, prezygotic phenotype in a cricket. *PLoS One* **4**: e7537.

Mueller, J.L., Linklater, J.R., Ravi Ram, K., Chapman, T., and Wolfner, M.F. (2008) Targeted gene deletion and phenotypic analysis of the Drosophila melanogaster seminal fluid protease inhibitor Acp62F. *Genetics* **178**: 1605–14.

Muro, Y. and Okabe, M.J. (2011) Mechanisms of fertilization–a view from the study of gene-manipulated mice. *J Androl* **32**: 218–25.

O'Connor, T.D. and Mundy, N.I. (2009) Genotype–phenotype associations: substitution models to detect evolutionary associations between phenotypic variables and genotypic evolutionary rate. *Bioinformatics* **25**: i94–i100

Panhuis, T.M., Clark, N.L., and Swanson, W.J. (2006) Rapid evolution of reproductive proteins in abalone and Drosophila. *Philos Trans R Soc Lond B Biol Sci*. **361**: 261–8.

Pitnick, S., Hosken, D.J., and Birkhead, T.R. (2009) Sperm morphological diversity. In T.R. Birkhead, D.J. Hosken and S. Pitnick (Eds) *Sperm Biology: An Evolutionary Perspective*, pp. 69–149. New York: Elsevier.

Ramm, S.A., Oliver, P.L., Ponting, C.P., Stockley, P., and Emes, R.D. (2008) Sexual selection and the adaptive evolution of mammalian ejaculate proteins. *Mol Biol Evol* **25**: 207–19.

Rooney, A.P., Zhang, J., and Nei, M. (2000) An unusual form of purifying selection in a sperm protein. *Mol Biol Evol* **17**: 278–83.

Simmons, L.W. (2001) *Sperm competition and its evolutionary consequences in insects*. Princeton, NJ: Princeton University Press.

Swanson, W.J., Yang, Z., Wolfner, M.F., and Aquadro C.F. (2001) Positive Darwinian selection drives the evolution of several female reproductive proteins in mammals. *Proc Natl Acad Sci USA* **98**: 2509–14.

Tang, H. and Lewontin, R.C. (1999) Locating regions of differential variability in DNA and protein sequences. *Genetics* **153**: 485–95.

Tourmente, M., Gomendio, M, and Roldan, E.R. (2011) Sperm competition and the evolution of sperm design in mammals. *BMC Evol Biol* **11**: 12.

Turner, L.M. and Hoekstra, H.E. (2006) Adaptive evolution of fertilization proteins within a genus: variation in ZP2 and ZP3 in deer mice (Peromyscus). *Mol Biol Evol* **23**: 1656–69.

Turner, L.M. and Hoekstra, H.E. (2008) Causes and consequences of the evolution of reproductive proteins. *Int J Dev Biol* **52**: 769–80.

Ward, W.S. (2010) Function of sperm chromatin structural elements in fertilization and development. *Mol Hum Reprod* **16**: 30–6.

Wong, A. (2010) Testing the effects of mating system variation on rates of molecular evolution in primates. *Evolution* **64**: 2779–85.

Wong, A. (2011) The molecular evolution of animal reproductive tract proteins: What have we learned from mating-system comparisons? *Int J Evol Biol* Article ID 908735.

Wong, A., Albright, S.N., Giebel, J.D., Ram, K.R., Ji, S., Fiumera, A.C., and Wolfner, M.F. (2008) A role for Acp29AB, a predicted seminal fluid lectin, in female sperm storage in Drosophila melanogaster. *Genetics* **180**: 921–31.

Wyckoff, G.J., Wang, W., and Wu, C.I (2000) Rapid evolution of male reproductive genes in the descent of man. *Nature* **403**: 304–9.

CHAPTER 18

Rapid morphological, behavioral, and ecological evolution in *Drosophila*: comparisons between the endemic Hawaiian *Drosophila* and the cactophilic *repleta* species group

Patrick M. O'Grady and Therese Ann Markow

18.1 Introduction

Flies in the genus *Drosophila* have served as model systems in genetics, ecology, and evolutionary biology for over 100 years. In addition to their ease of culture and numerous attractive genetic attributes, species in this group also feature a wealth of naturally occurring diversity in morphological, behavioral and ecological characters. These three character systems evolve in concert, sometimes rapidly, as the flies interact with and adapt to their environment. For example, historical changes in the environment, such as geological processes (e.g. island formation, continental movement), long-term climatic patterns (e.g. sea level rise, temperature, and rainfall shifts) and host plant availability (e.g. cladogenesis and extinction in plant lineages), have been instrumental in laying the basis for the genera, radiations, and species groups currently seen in the Drosophilidae. These long-standing processes can have a direct impact on ecological associations and, in turn, morphological and behavioral characters associated with feeding and breeding ecology.

More recent environmental phenomena mediated by man's impact on climate and species distributions, are also having an impact on the evolution of this genus. An example of this type of perturbation is seen in Hawaiian *Drosophila* where predatory species (ants and wasps) are introduced and native species are extirpated from part of their ranges and driven higher upslope where invasive species have yet to gain a foothold (Foote and Carson 1995). Similar range reductions are predicted for alpine taxa with increasing temperatures. As temperatures increase, less heat tolerant species will be pushed out of the lower parts of their traditional ranges to higher elevations. This effect will be greatest on those *Drosophila* species that are adapted to montane environments or have seasonal shifts in species distributions that are mediated by temperature. However, desert-adapted species may be unable to undergo altitudinal or latitudinal shifts, because they specialize on cacti, whose slow growth may prevent the host distribution from shifting before the flies become extinct. Those species able to undergo rapid evolution with respect to climatic or host variables will be the most likely ones to survive.

Sexual selection on both morphological and behavioral characters can likewise play a role in driving rapid change in morphology, behavior, and ecological associations. Environmental factors are known to influence the chemical and morphological underpinnings of sexual selection in *Drosophila*, thus reproductive behaviors and the characters that execute them are also of interest in responding to environmental changes.

Rapidly Evolving Genes and Genetic Systems. First Edition. Edited by Rama S. Singh, Jianping Xu, and Rob J. Kulathinal.

18.1.1 Ecological adaptations

The ecology of the genus *Drosophila* is quite diverse. When considering ecological associations in Drosophilidae, it is useful to divide preference into two discrete categories: (1) adult feeding, oviposition site preference, and larval feeding substrates, and (2) mating site preference. While these are identical for the majority of drosophilid species, a number of taxa, such as the Hawaiian *Drosophila*, have evolved separate preferences for mating.

Lynn Throckmorton (1975) was the first to summarize ecological adaptations across the family Drosophilidae and place them in an evolutionary context. He considered most ecological adaptations in the family as a reflection of 'opportunism and versatility centering around the saprophagous leaf-mold habit' (Throckmorton 1975). He proposed that ancestral drosophilids were able to exploit a number of different types of substrate, although they may have preferred one (Throckmorton 1975). Subsequent elaboration and specialization generated broad 'feeding and oviposition guilds,' such as sap feeders, fungivores, frugivores, flower breeders, or those taxa that utilize some other decomposing vegetative structure (e.g. leaves, bark). Recent phylogenetic results (Fig. 18.1) broadly support Throckmorton's hypotheses and show support for clades of present-day drosophilid species adapted to various guilds of rotting plant material (Markow and O'Grady, 2005, 2008).

Species can be classified as either generalists or specialists depending on the breadth of feeding and oviposition substrate types they use and the phylogenetic relationships of the taxa they specialize upon. For example, a *broad generalist* would be able to capitalize on multiple resources spread across feeding guilds. *Drosophila melanogaster* is an example of a broad generalist and can be reared from flowers, fungi, fruits, and a number of other rotting plant materials. Interestingly, broad generalist species have evolved multiple times on the phylogeny and can often be seen nested within clades of more specialized taxa. *Substrate generalists* are restricted to a given feeding guild (e.g. fruit) but can utilize this resource across a wide array of unrelated plant species. Many members of the *tripunctata* radiation are generalist frugivores and can be found on many different rainforest fruits. *Substrate specialists* are a class of taxa that have adapted to a single substrate type from a single clade of host plant. Most members of the *repleta* species group fit in this category because they have evolved to necrotic parts of a single host plant family, Cactaceae. Likewise, most Hawaiian *Drosophila* are also classified as substrate specialists because individual species use one substrate type from a single clade (e.g. leaves of Araliaceae in the case of *Drosophila waddingtoni*). *True specialists* are the most narrowly defined category and can use only a single type of plant resource from a single species of host plant. *Drosophila sechellia* is a classic example, larvae develop only on rotting fruits of *Morinda citrifolia* (Louis and David 1986; Jones 2005). A small number of specialist taxa have shifted away from rotting plant material to become *parasitic* on some animal species (Throckmorton 1975; Ashburner 1981). Broad generalists, substrate generalists and substrate specialists can all be considered *polyphagous* species, while true specialists and parasitic taxa are *monophagous*.

18.1.2 Morphological adaptations

Morphology is quite variable across the genus, with the various species groups and radiations adapting to different ecological niches and microhabitats. In addition, there is also a high degree of sexual dimorphism, with males of many taxa possessing species-specific secondary sexual characteristics used in courtship and/or mating. A classic example of such a character is the sex comb of *Drosophila melanogaster*, a row of stout, peg-like bristles on the foreleg of males. Sex combs are quite variable within the broader *melanogaster* species group and have been studied extensively because of their evolutionary significance and relatively simple genetic architecture (Tanaka et al. 2009). While sex combs are restricted to the *melanogaster* and *obscura* species groups, male-specific foreleg modifications are widespread in the family and can be found in most species groups and genera (e.g. Stark and O'Grady 2009), suggesting that evolution of these characters is rapid and important to overall mating success in many species.

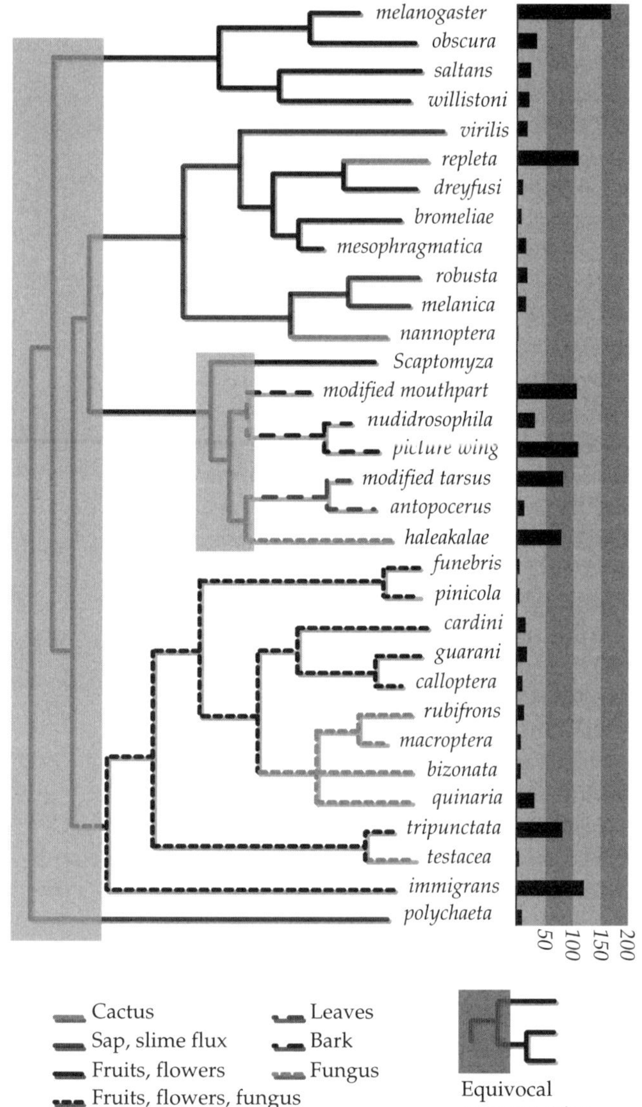

Figure 18.1 Overview of phylogenetic relationships and substrate type usage in the genus *Drosophila*. Numbers of species in each clade shown to the right of the phylogeny.

Primary sexual characters, particularly the male genitalia, also evolve rapidly (Masly et al. 2011; Richmond et al. in press). These traits are critical to species recognition and success during courtship and intromission. While male characters have been studied extensively (Hsu 1949; Ashburner et al. 2005; Coyne 1983; Vilela 1983), female characters, such as spermathecae (Pitnick et al. 1999), are also subject to rapid evolutionary change. Other characteristics of the female terminalia also evolve rapidly, but are linked more to ecology than sexual behavior. The ovipositor of many Hawaiian *Drosophila* species has diversified rapidly and has been correlated with oviposition substrate (Craddock and Kambysellis 1997; Hardy et al. 2001).

18.1.3 Behavioral adaptations

Behavioral evolution in *Drosophila* entails two broad classes of change: modifications to actual behaviors (e.g. the evolution of male guarding in some species of the *repleta* group and lek behavior in Hawaiian *Drosophila*) and changes in the structures employed in the various behaviors. Some of the structures involved in this latter category may be either external, like wing patterns used in courtship display, or internal, such as in the nervous system or reproductive organs.

18.2 Hawaiian *Drosophila* radiation

18.2.1 Phylogenetic relationships

The Drosophilidae endemic to the Hawaiian Archipelago are comprised of two large sister lineages, Hawaiian *Drosophila* and the genus *Scaptomyza* (Russo et al. 1995; Remsen and DeSalle 1998; Remsen and O'Grady 2002; O'Grady and DeSalle 2008). Hawaiian *Drosophila* contains over 400 described species (O'Grady et al. 2008) placed in eight major lineages including the *picture wing*, *modified mouthparts*, *antopocerus*, *modified tarsus*, *nudidrosophila*, *ateledrosophila*, *rustica*, and *haleakalae* species groups. Relationships among the various lineages of Hawaiian *Drosophila* (Fig. 18.1) have recently been summarized by (O'Grady et al. 2011). The sister lineage of the Hawaiian *Drosophila* is the genus *Scaptomyza*, a clade of about 300 species, over 50% of which are endemic to the Hawaiian Islands. The remainder of taxa in the genus *Scaptomyza* are found throughout the world, but particularly on islands. O'Grady and DeSalle (2008) examined phylogenetic relationships across the family Drosophilidae to demonstrate: (1) a single common ancestor of Hawaiian *Drosophila* and all *Scaptomyza* on Hawaii roughly 25 million years ago, (2) the origin of *Scaptomyza* in Hawaii, followed by the escape and global colonization of members of this genus, and (3) the subsequent escape from and colonization of the remainder of the world by members of the genus *Scaptomyza*. Molecular clock analyses place the divergence between Hawaiian *Drosophila* and the genus *Scaptomyza* at ~20 million years, a number that fits well with a *Scaptomyza* species found in Dominican amber. However, the movement of this genus out of Hawaii seems to have taken place independently within several subgenera and more likely is on the scale of the past 5–10 million years.

18.2.2 Sexual adaptations to morphology and behavior

The majority of species in Hawaiian *Drosophila* possess secondary sexual modifications in male wings, forelegs, mouthparts, and other head structures. Mating behavior in this group is among the most complex in all Diptera. Males of most species lek, guarding specific territories away from the feeding and oviposition substrate. Male–male aggression is common and a single male will often need to spar with multiple males over the course of a single day. This ranges from 10–15-minute contests occurring three to four times an hour in *D. percnosoma* to 30-second interactions taking place four to five times in a 5-minute period in *D. imparisetae* (Shelley 1987, 1989; P.M. O'Grady unpublished). The genus *Scaptomyza* provides an interesting contrast to Hawaiian *Drosophila* because they have a markedly lower degree of sexual dimorphism. While males of most Hawaiian *Drosophila* species possess extreme secondary sexual dimorphism, differences between males and females in most *Scaptomyza* species are rare and restricted to the male genital apparatus. Corresponding sexual behaviors are likewise simpler, with courtship in *Scaptomyza* being of much shorter duration and less dependent upon male display. It is possible that these two sister clades represent distinct avenues of sexual selection, one favoring 'showy' displays using energetically expensive secondary sexual characters with correspondingly complex behaviors and another employing the rapid elaboration of primary sexual characters (male genitalia) while maintaining relatively simple behaviors.

18.2.3 Ecological adaptations to morphology and behavior

O'Grady et al. (2011) examined the evolution of substrate and host plant species preference across the Hawaiian *Drosophila*. This work builds on previous studies by Heed (Heed 1968; Heed 1971), Carson (1971), Kambysellis et al. (1995), and Magnacca et al. (2008) by placing oviposition and larval feeding preference in the context of a broadly sampled phylogeny of Hawaiian *Drosophila*. This study suggested that host plant family was a much more plastic character than substrate type. This gives insight into how oviposition preference and larval feeding might evolve within this group. Based on these results, a species that has adapted to leaves of *Cheirodendron* (Arialaceae) could more readily shift to leaves of another plant, such as *Pisonia* (Nyctaginaceae), than it can change substrate types within *Cheirodendron* and oviposit on stems. While additional work is needed to fully understand this phenomenon, this constraint might be explained by differences in physical or chemical properties of leaves versus stems or by microbial communities adapted to different substrate types across plant groups.

18.3 Cactophilic *Drosophila* radiation in the New World

18.3.1 Phylogenetic relationships

While the cactophilic lifestyle has arisen more than once in the genus *Drosophila*, the largest radiation of cactus-breeding *Drosophila* belongs to the *repleta* species group which contains over 100 species. The separation between the *repleta* and the *virilis* group is estimated to have occurred approximately 10 mya (million years ago). Five subgroups are recognized within the *repleta* group: *hydei*, *mercatorum*, *fasciola*, *repleta*, and *mulleri*.

At least 60 of the *repleta* group species, particularly members of the *mulleri* supbgroup, are endemic to Mexico, primarily residing in arid or semi arid areas. Additional species are endemic to South America, where radiations of different cactus species, both *opuntia* and columnar provide niches for them. Members of the *fasciola* subgroup appear restricted to the wetter areas of the West Indies, Central and South America. *Drosophila hydei* and *D. mercatorum*, while utilizing cactus in rural areas, also are cosmopolitan, using other decaying fruit and vegetable materials associated with humans. Because cacti can be characterized chemically and their necroses can be tracked at both spatial and temporal scales, the system is an ideal one for studies of the interaction of genes and ecology in evolution.

Endemic to the Western hemisphere, the Cactaceae family contains about 1800 species. The major divisions of the family Cactaceae used by *Drosophila*: Pachycereae, Cactae, and the Opuntioidae (Fig. 18.2). Clearly, then, there are more cactus species than cactophilic *Drosophila* species, reflecting the fact that while some *Drosophila* species tend to be associated with only one species of cactus, many are cactus generalists that can utilize multiple cactus hosts. In some cases, multiple host use is restricted to hosts of the same genus, but in other cases, flies are able to utilize cacti from three different cactus lineages. The majority of cactophilic *Drosophila* utilize species of *Opuntia*, which is not surprising, given that these represent the majority of cactus species available and are basal to the other cacti. Phylogenetic analyses also indicate that opuntia-breeding is the ancestral state for cactophilic *Drosophila*.

Most well-studied of the cactophilic *Drosophila* are *D. mojavensis* and *D. arizonae* in North America. William Heed and his students and colleagues (Fellows and Heed 1972; Heed 1982; Ruiz and Heed 1988) identified the host specificities of each, and his subsequent group members characterized the yeast communities associated with each cactus species (Starmer et al. 1982). More recently, the whole genome of one cactophilic fly species, *D. mojavensis*, was published as part of the *Drosophila* Genomes Consortium effort (Drosophila 12 Genomes Consortium et al. 2007), and the genome of *D. buzzatii* currently is being sequenced by Alfredo Ruiz and his colleagues in Barcelona.

The South American *D. buzzatii* is another important focal taxon for studies of ecology, evolution and behavior (Manfrin and Sene 2006). An assortment of related species, *D. koepferi*, *D. anonieta*, *D. gouviae*, *D. buzzatii*, and *D. serido*, with different evolutionary and reproductive relationships provide a parallel system to the cactophilic drosophilids in North America.

18.3.2 Rapid evolution of ecological adaptations

Cactophilic *Drosophila* are able to utilize the moist, nutritious habitats afforded by cactus tissue only after it becomes necrotic, a condition that occurs following an injury to the plant and subsequent invasion and decomposition by bacteria and yeasts.

(a)

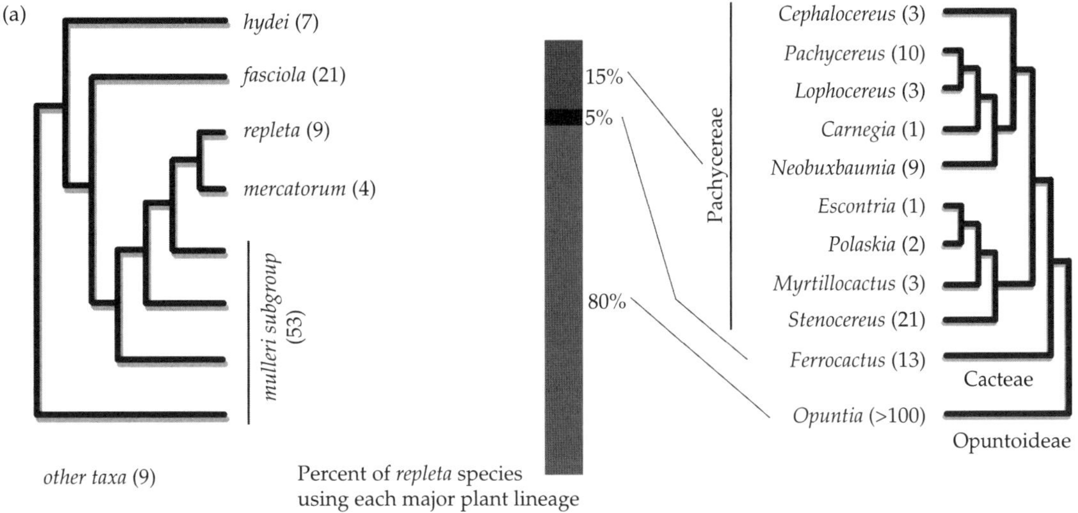

(b)

	spp.	Ecological breadth	Morphological diversity	Mating behavior			
				Visual	Auditory	Tactile	Other
repleta	>100	Low (1 family)	Primary sexual characters	No	Yes	Yes	Lek (some) Male guarding (some)
Hawaiian Drosophila	>600	High (34 families)	Secondary sexual characters	Yes	Yes	Yes	Lek (most)

(c)

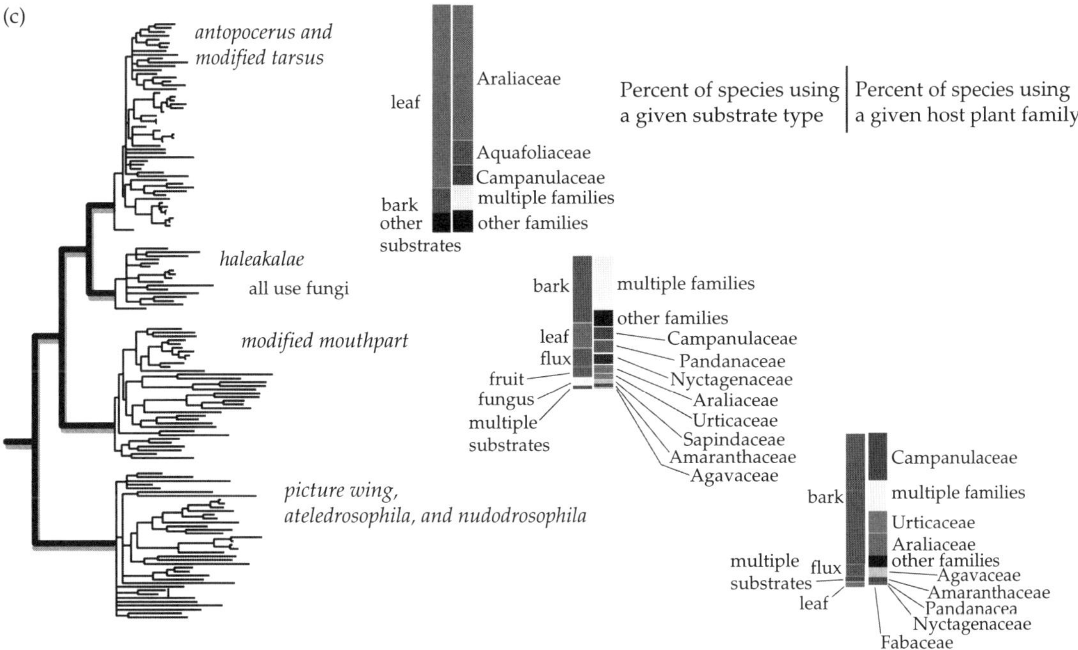

Figure 18.2 Relationship of phylogeny, ecology, and species numbers in Hawaiian *Drosophila* and the *repleta* species group. (a) Phylogeny among subgroups in the *repleta* group, with numbers of species described in each shown, and phylogeny of Cacteaceae, with numbers of species in each genus. Central bar indicates percent of repleta species using Pachycereae, Cacteae, and Opuntoideae as feeding and oviposition substrate. (b) Detail of the ecological, morphological, and behavioral diversity in the *repleta* group and Hawaiian *Drosophila*. (c) Phylogenetic relationships among the major lineages in Hawaiian *Drosophila*. Colored bars show substrate (leaf, bark, fruit, etc.) and host plant family (e.g. Araliaceae, Campanulaceae, Nyctagenaceae) use across major clades of Hawaiian *Drosophila*.

Adult flies feed and mate at or near these necrotic cacti and females deposit their eggs in the necrotic tissue, which then serves as the food for the developing larvae.

In order to exploit the cactus niche, however, the flies face several challenges. With respect to abiotic variables (Gibbs et al. 2003), many cactophilic species exhibit rapid adaptation to high temperature (Stratman and Markow 1998) and low humidity (Matzkin et al. 2007, 2009) that typifies many cactus habitats. While these resistances are physiological in nature, often related to cuticular hydrocarbon composition (Markow and Toolson 1990; Toolson et al. 1990; Etges and Jackson 2001), flies also disperse at night when the temperatures are lower and humidities higher (Markow and Castrezana 2000).

The cacti themselves, with their various toxic compounds, present additional, biotic challenges to the flies. Partial detoxification of the plant tissue is accomplished by the microbial communities, unique to each cactus species, which are responsible for the decay process. Although the flies feed on these microbes, they are still confronted with the need to process many cactus compounds. Byproducts of microbial decomposition and the chemical profiles of unaltered plant tissue constitute the specific environment that each *Drosophila* species must deal with in its own particular host cactus. Oligoarrays based upon the *D. mojavensis* genome have revealed candidate genes for changes in host use and population genetic analyses of these genes, especially alcohol dehydrogenase and glutathione S-transferase D-1, show that they are evolving rapidly among subspecies using different hosts (Matzkin 2005, 2008; Matzkin et al. 2006).

Finding the appropriate cactus host necroses is assumed to occur via olfactory cues from the volatiles specific to that cactus. Flies behaviorally have been reported to move toward material from their native host in a laboratory setting (Newby and Etges 1998). More recently, however, population genetic differentiation in olfactory receptor genes among different host-specific subspecies of *D. mojavensis* suggest the possibility of connecting particular olfactory receptors with particular hosts and their cues (Matzkin and Shumacher submitted).

18.3.3 Rapid evolution of behavioral traits

In contrast to the frequently spectacular mating behaviors exhibited by Hawaiian *Drosophila*, visual cues appear to play little of any role in mating behavior. Among the cactophilic *Drosophila*, a moving fly of any sex or species will trigger interest from a male, especially in mating chambers or vials in the laboratory, but this appears to be the primary role. Courtship consists of males closely following females and constantly licking the tip of her abdomen. At the same time, males produce species-specific auditory cues or 'songs' with their wings. Unlike the auditory courtship component of *melanogaster* group flies, *repleta* group females also sing, resulting in a dueting between members of the courting pairs. Female flies of all *repleta* species examined so far indicate their acceptance of a courting male by a characteristic wing spreading behavior without which the males will not attempt to mount. Because the male songs of close relatives are distinctive (Etges et al. 2006), they obviously have evolved quite rapidly which could explain the sexual isolation reported among different subspecies of *D. mojavensis* (Krebs and Markow 1989; Zouros and d'Entremont 1980).

Courtship and mating in at least some species of cactophilic *Drosophila* take place on the cactus, but away from the necrotic sites where feeding is taking place. For example, in *D. nigrospiracula*, males space themselves on healthy tissues in ways that suggest they are defending small territories (Markow 1988). Females land near a male and they approach each other with three different outcomes: (1) females depart rather quickly; (2) courtship takes place but the female departs; and (3) copulation occurrs (Markow 1988). *Drosophila mojavensis* males, in addition to locating themselves on healthy cactus arms, also can be found in groups on healthy tissue near the necrosis. Females arrive and there is a flurry of courtship activity some of which results in copulation. It is also noteworthy that rapid evolution in male genitalia within species and between closely related species has been reported for both South American (Soto et al. 2007) and North American (Richmond et al. in press) *repleta* group species.

Oviposition specificity also has been extensively studied in cactophilic *Drosophila* (Barker and

Starmer 1999; Fanara et al. 1999; Fanara and Hasson 2001). While *D. buzzatii* females prefer to oviposit in their native hosts compared to other cacti, its sister species, *D. koepferi*, is more of a generalist in its preferences (Soto et al. 2011). Males of both species have greater success in mating when reared on their own host (Hurtado et al. 2011) so, in the case of *D. buzzatii*, the oviposition preference supports an evolutionary association between maternal preference and offspring performance.

18.4 Conclusions: adaptive radiation versus adaptive infiltration

Adaptive radiations are characterized by high ecological and phenotypic diversity in a rapidly evolving lineage. The Hawaiian *Drosophila*, with their impressive ecological breadth, high diversity in male secondary sexual characters, and over 600 described species evolving in the past 25 million years, are a classic example of adaptive radiation in nature (Kaneshiro 1997; O'Grady et al. 2008). The *repleta* group also constitutes an impressive radiation of over 100 species evolving in the past ~30 million years, with physiological adaptations (e.g. desiccation tolerance) and diverse reproductive strategies, but lacking the ecological breadth and phenotypic diversity seen in the Hawaiian taxa. Interestingly, most species in this group are adapted to a single plant family, Cactaceae, although this family is very diverse and presents significant biological challenges to any species attempting to exploit it.

The differences in the degree to which these two groups have diversified, in terms of species numbers, the degree of ecological breadth, and the magnitude of morphological, behavioral, and physiological adaptation suggests two separate modes of radiation acting in Hawaiian *Drosophila* and the *repleta* group: one where the species spatially segregate into a number of different host plant family and substrate types, each with its own ecological requirements, and another where the species diversify on a single plant lineage but maintain separation from close relatives via a variety of differences in reproductive morphology (male genitalia and female reproductive tracts) and/or behavior (specific lek sites away from a common feeding area, male guarding in some species, courtship song characteristics and pheromone profiles).

Heed and Mangan (1986) used the term 'adaptive infiltration' when discussing the ecology of Sonoran desert *Drosophila*, three of which were in the *repleta* group. This phenomenon occurs when members of independent lineages adapt to and diversify in the same stressful environment, leading to the evolution of convergent characteristics. They applied the term narrowly and discussed only four endemic species, from two separate clades (the repleta and nannoptera groups) in this region, *D. pachea*, *D. nigrospiracula*, *D. mettleri*, and *D. mojavensis*, all of which have adapted to the harsh climate and cactophilic lifestyle. However, one can apply this term more broadly to the entire *repleta* group, the majority of which are cactophilic and occupy harsh, arid environments. While this group represents a single lineage, a complex pattern of adaptations to arid environments, some due to common descent and others due to convergence, is seen across this large group. The specific differences between taxa that have adaptively radiated compared to those that have adaptively infiltrated can be reflected in the degree of ecological breadth, physiological tolerance, rapid morphological innovation (e.g. primary and secondary sexual characters), and behavioral diversity.

Rapid evolution of behavioral and physical traits can occur under either scenario: adaptive radiation or adaptive infiltration. In the case of the Hawaiian radiation, the most obvious physical traits to have evolved are the morphological ones employed in behavior and behavioral displays. Among the repleta species, the most obvious physical traits are chemical ones, especially those used in aggregation and mating. In both cases, however, behavioral and morphological traits are linked to adapting to new ecological niches, although it is unclear whether behavioral changes precede physical changes and promote their evolution or vice versa.

References

Ashburner, M. (1981) Entomophagous and other bizarre Drosophilidae. In M. Ashburner, H. Carson, and J. Thompson (Eds) *The genetics and biology of Drosophila*, pp. 395–429. London: Academic Press.

Ashburner, M., Golic, K.G., and Hawley, R.S. (2005) *Drosophila: a laboratory handbook*. Cold Spring Harbor, NY: Cold Spring Harbor Laboratory Press.

Barker, J.S.F. and Starmer, W.T. (1999) Environmental effects and the genetics of oviposition site preference for natural yeast substrates in Drosophila buzzatii. *Hereditas* **130**(2): 145–75.

Carson, H.L. (1971) The ecology of Drosophila breeding sites. *Harold L. Lyon Arboretum Lecture No. 2*. Honolulu: University of Hawaii Press.

Coyne, J.A. (1983) Genetic basis of differences in genital morphology among three sibling species of Drosophila. *Evolution* **37**: 1101–18.

Craddock, E.M. and Kambysellis, M.P. (1997) Adaptive radiation in the Hawaiian Drosophila (Diptera: Drosophilidae): Ecological and reproductive character analyses. *Pacif Sci* **51**(4): 475–89.

Drosophila 12 Genomes Consortium, Clark, A.G., Eisen, M.B., Smith, D.R., Bergman, C.M., Oliver, B., et al. (2007) Evolution of genes and genomes on the Drosophila phylogeny. *Nature* **450**(7167): 203–18.

Etges, W.J. and Jackson, L.L. (2001) Epicuticular hydrocarbon variation in Drosophila mojavensis cluster species. *J Chem Ecol* **27**(10): 2125–49.

Etges, W.J., Over, K.F., De, O., and Ritchie, M.G. (2006) Inheritance of courtship song variation among geographically isolated populations of Drosophila mojavensis. *Anim Behav* **71**: 1205–14.

Fanara, J.J. and Hasson, E. (2001) Oviposition acceptance and fecundity schedule in the cactophilic sibling species Drosophila buzzatii and D. koepferae on their natural hosts. *Evolution* **55**(12): 2615–19.

Fanara, J.J., Fontdevila, A., Hasson, E. (1999) Oviposition preference and life history traits in cactophilic Drosophila koepferae and D. buzzatii in association with their natural hosts. *Evol Ecol* **13**(2): 173–90.

Fellows, D.P. and Heed, W.B. (1972) Factors affecting host plant selection in desert-adapted cactiphilic Drosophila. *Ecology* **53**: 850–8.

Foote, D. and Carson, H.L. (1995) Drosophila as monitors of change in Hawaiian ecosystems. In E.T. LaRoe (Ed.) *Our living resources: A report to the nation on the distribution, abundance, and health of U.S. plants, animals, and ecosystems*, pp. 368–72. Washington, DC: US Department of the Interior, National Biological Service

Gibbs, A.G., Perkins, M.C., and Markow, T.A. (2003) No place to hide: microclimates of Sonoran Desert Drosophila. *J Thermal Biol* **28**: 353–62.

Hardy, DE, Kaneshiro, KY, Val, FC, and O'Grady, PM. 2001. Review of the *haleakalae* species group of Hawaiian *Drosophila* (Diptera: Drosophilidae). Bishop Museum Bulletin in Entomology 9: 1–88. Bishop Museum Press.

Heed, W.B. (1968) Ecology of the Hawaiian Drosophilidae. *Univ Texas Publs Stud Genet* **4**(6818): 387–419.

Heed, W.B. (1971). Host plant specificity and speciation in Hawaiian Drosophila. *Taxon* **20**: 115–121.

Heed, W.B. (1982) The origin of Drosophila in the Sonoran Desert. In J.S.F Barker and W.T. Starmer (Eds) *Ecological Genetics and Evolution: The Cactus-Yeast-Drosophila Model*, pp. 65–80. Sydney: Academic Press.

Heed, W.B. and Mangan, R.L. (1986) Community ecology of the Sonoran Desert Drosophila. In M. Ashburner, H.L. Carson, and J.N. Thompson (Eds) *The Genetics and Biology of Drosophila*, 311–345. London: Academic Press.

Hsu, T.C. (1949) The external genital apparatus of male Drosophilidae in relation to systematics. *Univ Texas Publs* **4920**: 80–142.

Hurtado, J., Soto, I., Orellana, L., and Hasson, E. (2011) Mating success depends on rearing substrate in cactophilic Drosophila. *Funct Ecol* [DOI 10.1007/s10682-011-9529-z].

Jones, C.D. (2005) The genetics of adaptation in Drosophila sechellia. *Genetica* **123**: 137–45.

Kambysellis, M.P., Ho, K.F., Craddock, E.M., Piano, F., Parisi, M., Cohen, J. (1995) Pattern of ecological shifts in the diversification of Hawaiian Drosophila inferred from a molecular phylogeny. *Curr Biol* **5**(10): 1129–39.

Kaneshiro, K.Y. (1997) Perkins' legacy to evolutionary research on Hawaiian Drosophilidae (Diptera). *Pacific Sci* **51**: 450–61.

Krebs, R. and Markow, T.A. (1989) Courtship behavior and the control of reproductive isolation in Drosophila mojavensis. *Evolution* **43**: 908–13.

Louis, J. and David, J.R. (1986) Ecological specialization in the Drosophila melanogaster species subgroup: a case study of D. sechellia. *Acta Oecol* **7**: 215–29.

Magnacca, K.N., Foote, D., and O'Grady, P.M. (2008) A review of the endemic Hawaiian Drosophilidae and their host plants. *Zootaxa* **1728**: 1–58.

Manfrin, M.H. and Sene, F.M. (2006) Cactophilic *Drosophila* in South America: a model for evolutionary studies. *Genetica* **126**(1–2): 57–75.

Markow, T.A. (1988) Reproductive behavior of Drosophila melanogaster and D.nigrospiracula in the field and in the laboratory. *J Comp Psychol* **102**: 169–74.

Markow, T.A. (1991) Sexual isolation among populations of Drosophila mojavensis. *Evolution* **45**: 1525–9.

Markow, T.A. and Castrezana, S. (2000) Dispersal in cactophilic Drosophila. *Oikos* **89**: 378–86.

Markow, T. and O'Grady, P.M. (2005) Evolutionary genetics of reproductive behavior in Drosophila: connecting the dots. *Annu Rev Genet* **39**: 263–91.

Markow, T.A. and O'Grady, P.M. (2008) Reproductive ecology of Drosophila. *Funct Ecol* **22**(5): 747–59.

Markow, T.A. and Toolson, E.C. (1990) Temperature effects on epicuticular hydrocarbons and sexual isolation in D. mojavensis. In J.S.F. Barker, W.T. Starmer, and R.J. MacIntyre (Eds) *Ecological and Evolutionary Genetics of Drosophila*, pp. 315–31. New York: Plenum Press.

Masly, J.P., Dalton, J.E., Srivastava, S., Chen, L., Arbeitman, M.N. (2011) The genetic basis of rapidly evolving male genital morphology in Drosophila. *Genetics* **189**(1): 357–74.

Matzkin, L.M. (2005) Activity variation in alcohol dehydrogenase paralogs is associated with adaptation to cactus host use in cactophilic Drosophila. *Mol Ecol* **14**: 2223–31.

Matzkin, L.M. (2008) The molecular basis of host adaptation in cactophilic Drosophila: Molecular evolution of glutathione- S-transferase (Gst) in Drosophila mojavensis. *Genetics* **178**: 1073–83.

Matzkin, L.M. and Markow, T.A. (2009) Transcriptional regulation of metabolism associated with the increased desiccation resistance of the cactophilic Drosophila mojavensis. *Genetics* **182**: 1279–88.

Matzkin, L.M. and Schumacher, J.O. (submitted) Adaptive protein evolution of odorant receptors in cactophilic Drosophila.

Matzkin, L.M., Watts, T., Bitler, B.G., Machado, C.A. and Markow, T.A. (2006) Functional genomics of cactus host shifts in Drosophila mojavensis. *Mol Ecol* **15**: 4635–43.

Matzkin, L., Watts, T.D. and Markow, T.A. (2007) Desiccation resistance in four Drosophila species: sex and population effects. *Fly* **1**(5): 268–73.

Matzkin, L.M., Watts, T.D., and Markow, T.A. (2009) Evolution of stress resistance in Drosophila: Interspecific variation in tolerance to desiccation and starvation. *Funct Ecol* **23**: 521–7.

Newby, B.D. and Etges, W.J. (1998) Host preference among populations of Drosophila mojavensis (Diptera: Drosophilidae) that use different host cacti. *J Insect Behav* **11**(5): 691–712.

O'Grady, P. and DeSalle, R. (2008) Out of Hawaii: the origin and biogeography of the genus Scaptomyza (Diptera: Drosophilidae). *Biol Lett* **4**(2): 195–9.

O'Grady, P.M., Magnacca, K.N., and Lapoint, R.T. (2008) Taxonomic relationships within the endemic Hawaiian Drosophilidae. *Records Hawaii Biol Survey* **108**: 3–35.

O'Grady, P.M., Lapoint, R.T., Bonacum, J., Lasola, J., Owen, E., Wu, Y., and Desalle, R. (2011) Phylogenetic and ecological relationships of the Hawaiian Drosophila inferred by mitochondrial DNA analysis. *Molec Phylogenet Evol* **58**(2): 244–56.

Pitnick, S., Markow, T.A., and Spicer, G. (1999) Evolution of multiple kinds of female sperm storage organs in Drosophila. *Evolution* **53**: 1804–22.

Remsen, J. and DeSalle, R. (1998) Character congruence of multiple data partitions and the origin of the Hawaiian Drosophilidae. *Mol Phylogenet Evol* **9**(2): 225–35.

Remsen, J. and O'Grady, P. (2002) Phylogeny of Drosophilinae (Diptera: Drosophilidae), with comments on combined analysis and character support. *Mol Phylogenet Evol* **24**(2): 249–64.

Richmond, M.P., Johnson, S., and Markow, T.M. (in press) Evolution of reproductive morphology among recently diverged taxa in the Drosophila mojavensis species cluster. *Ecol Evol*.

Ruiz, A. and Heed, W.B. (1988) Host-plant specificity in the cactophilic Drosophila mulleri species complex. *J Anim Ecol* **57**(1): 237–49.

Russo, C.A.M., Takezaki, N., and Nei, M. (1995) Molecular phylogeny and divergence times of drosophilid species. *Mol Biol Evol* **12**(3): 391–404.

Shelley, T.E. (1987) Lek behaviour of Hawaiian Drosophila: male spacing, aggression and female visitation. *Anim Behav* **35**(5): 1394–404.

Shelley, T.E. (1989) Waiting for mates: variation in female encounter rates within and between leks of Drosophila conformis. *Behaviour* **111**(1–4): 34–48.

Soto, I.M., Carreira, V.P., Fanara, J.J., and Hasson, E. (2007) Evolution of male genitalia: environmental and genetic factors affect genital morphology in two Drosophila sibling species and their hybrids. *BMC Evol Biol* **7**: 77.

Soto, I., Goenaga, J., Hurtado, J., and Hasson, E. (2011) Oviposition and performance in natural hosts in cactophilic Drosophila. *Funct Ecol* [DOI 1007/s10682-011-9531-5].

Stark, J.B. and O'Grady, P.M. (2010) Morphological variation in the forelegs of the Hawaiian Drosophilidae. I. The AMC clade. *J Morph* **271**(1): 86–103.

Starmer, W.T., Phaff, H.J., Miranda, M., Miller, M.W., and Heed, W.B. (1982) The yeast flora associated with the decaying stems of columnar cacti and Drosophila in North America. *Evol Biol* **14**: 269–96.

Stratman, R. and Markow, T.A. (1998) Resistance to thermal stress in desert Drosophila. *Funct Ecol* **8**: 965–70.

Tanaka, K., Barmina, O., and Kopp, A. (2009) Distinct developmental mechanisms underly the evolutionary diversification in Drosophila: sex combs. *Proc Natl Acad Sci U S A* **106**: 4764–9.

Throckmorton, L.H. (1975) The phylogeny, ecology and geography of Drosophila. In R.C. King (Ed.) *Handbook of Genetics*, pp. 421–69. New York: Plenum.

Toolson, E.C., Howard, R. Jackson, L., and Markow, T.A. (1990) Epicuticular hydrocarbon composition of wild-type and laboratory-reared Drosophila mojavensis. *Ann Ent Soc Amer* **83**: 1165–76.

Vilela, C.R. (1983) A revision of the Drosophila repleta species group (Diptera, Drosophilidae). *Revta Bras Entomol* **27**: 1–114.

Zouros, E. and D'Entremont, C.J. (1980) Sexual isolation among populations of Drosophila mojavensis: response to pressure from a related species. *Evolution* **34**(3): 421–30.

CHAPTER 19

Ancient yet fast: rapid evolution of mating genes and mating systems in fungi

Timothy Y. James

19.1 Introduction

Though seldom observed directly, the life cycle of most filamentous fungi includes countless cryptic encounters between conspecific fungal cells as they grow within their substrate. Complex substrates such as soil contain hundreds of species of fungi per gram (Buée et al. 2009) and hyphal interactions are unavoidable. Most fungi are sexual, and completing the life cycle necessitates that hyphal encounters with potential mates are distinguished from enemies. Recognition of self from potential mate or enemy is governed by a set of genes known as incompatibility genes. Unlike mating in the majority of eukaryotes in which sex involves the fusion of two cells, interactions between filamentous fungal individuals are fundamentally different and may be played out in a theater of numerous redundant and synchronous interactions between nuclei and the cells that house them. Following an encounter with a compatible mate, a genetically merged and dynamic colony is formed (Rayner 1991). Hyphal fusion is seldom followed by nuclear fusion, and instead filamentous fungi postpone the formation of a diploid zygote by forming a heterokaryon in which nuclei of compatible genotype coexist in the same cell and continue to divide synchronously by mitosis as the network grows. Only immediately before meiosis, does the zygote nucleus form by karyogamy of the mated partner nuclei (Fig. 19.1).

Incompatibility genes in fungi control both sexual and competitive conspecific interactions through two distinct molecular pathways. The mating-type incompatibility genes (*MAT*) control sexual attraction and heterokaryon (see Box 19.1 for a glossary of terms) maintenance through the production of communication and signal transduction molecules (Hiscock and Kües 1999). The genes of the vegetative incompatibility pathway (*HET* genes), function to prevent illicit vegetative fusions and nuclear exchange between incompatible conspecific mycelia while allowing the network of one genotype to proliferate by branching and re-fusion to self. Vegetative incompatibility genes define 'individuality' in mycelial fungi, and vegetative incompatibility can often be observed in nature as lines in the substrate (physical boundaries) that demarcate genetically distinct individuals (Rayner 1991; Worrall 1997). In Ascomycota, a monophyletic group including most lichenized fungi, cup fungi, many molds and yeasts such as *Saccharomyces cerevisiae*, individualistic behavior occurs between haploids (homokaryons), whereas in Basidiomycota, including the rusts, smuts, and a diversity of mushroom-like groups, it occurs between heterokaryons (Fig. 19.1). In contrast to heterokaryons, filamentous homokaryons in basidiomycetes are highly promiscuous cells that are able to mate with most other conspecific homokaryons they encounter due to increased compatibility imparted by a mating system with a large number of mating types. Thus the life cycle of the basidiomycete is primarily mated or heterokaryotic, while in that of the ascomycetes, with much more limited compatibility by possession of only two mating types, sex occurs only at precisely the right condition, place, and time, and most of the life cycle exists in a primarily haploid or homokaryotic stage.

Rapidly Evolving Genes and Genetic Systems. First Edition. Edited by Rama S. Singh, Jianping Xu, and Rob J. Kulathinal.

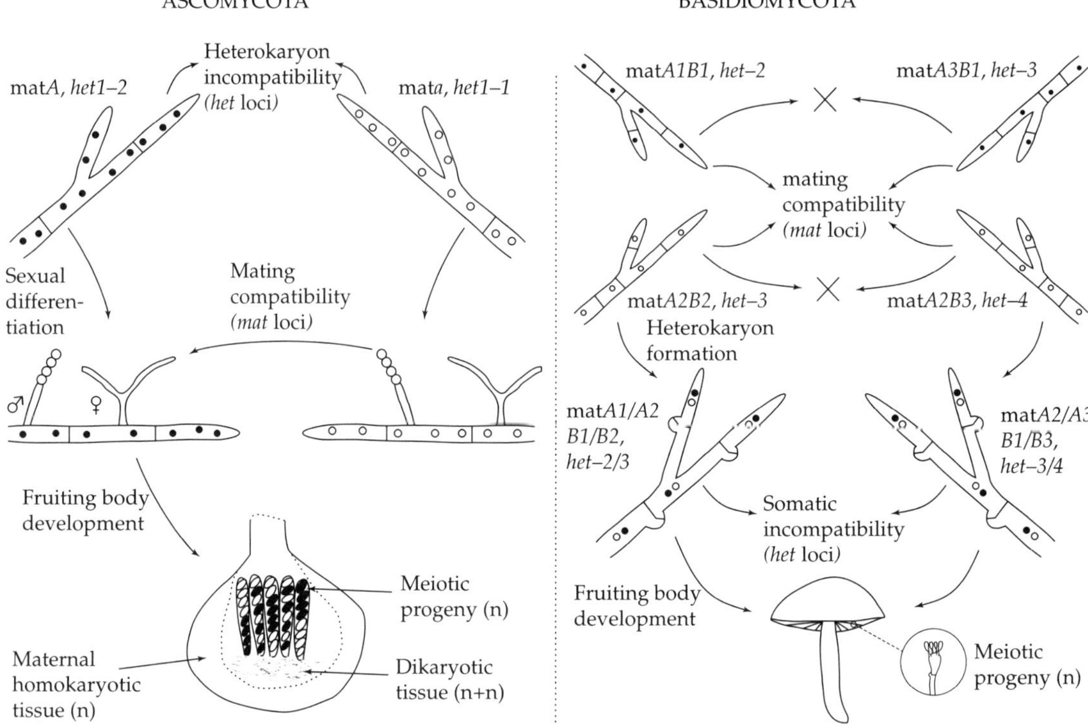

Figure 19.1 Contrasting life cycles of the filamentous Ascomycota and Basidiomycota. Major differences include the nuclear condition in which *HET* loci determine the ability to fuse (homokaryotic stage in Ascomycota, heterokaryotic stage in Basidiomycota) and genetic condition of the fruiting body (homokaryotic in Ascomycota, heterokaryotic in Basidiomycota). Shown is a tetrapolar basidiomycete with two mating-type loci (mat-*A* and mat-*B*). Both life cycles are drawn with a single *HET* locus, but most species typically have several.

Box 19.1 Glossary

Anisogamy: in fungi, occurs when a spermatium (small aerially dispersed gamete) fuses with a trichogyne (receptive female hypha).

Bipolar: a homoallelic mating incompatibility system with only a single locus. At meiosis two mating types are produced per meiocyte.

Heterokaryon: a cell type comprised of multiple nuclear genotypes maintained in an unfused state through multiple rounds of mitosis.

Homeodomain (HD) locus: a mating type locus encoding one or more transcription factors that possess a distinctive homeodomain DNA binding motif.

Homokaryon: a cell type comprised of only a single nuclear genotype.

Homothallic: a non outcrossing mating system in which a single spore is capable of giving rise to a fruiting body.

Mating type: equivalent to 'sexes' in isogamous or hermaphroditic organisms; individuals of the same mating type are sexually incompatible.

Pheromone receptor (P/R) locus: a mating type locus encoding at least one G-protein coupled transmembrane receptor and one pheromone peptide.

Tetrapolar: a homoallelic mating incompatibility system with two loci. At meiosis four mating types are produced per meiocyte.

As the result of decades of intensive research, much is known about the genes regulating hyphal fusion and mating (Hiscock and Kües 1999; Lee et al. 2010), and some of these pathways are among the best known in eukaryotes, such as the pheromone response pathway in yeast (Herskowitz 1989). Incompatibility genes in fungi have many similarities to genes controlling incompatibility and sex determination in animals and plants. The reproductive and incompatibility proteins in the plant and animal kingdoms have provided some of the clearest examples of rapid evolution. In fungi, however, little is known about the evolutionary dynamics of reproductive proteins, particularly whether incompatibility genes evolve rapidly as they do in plants and animals.

In this review I will discuss the expectations for fungal incompatibility systems generated by the plant/animal systems and provide examples of key studies that can shed light on whether these expectations hold for fungi. I will discuss evidence for accelerated evolution as well as balancing selection on incompatibility genes, loci, and mating systems in fungi. Emphasis will be placed on mushroom fungi as these species have evolved a multiallelic system that has many of the characteristics of other fast-evolving incompatibility systems such as those observed in flowering plants.

19.2 Incompatibility systems in fungi

The two types of incompatibility systems in fungi function through two distinct mechanisms leading to two distinct signal cascades. Vegetative incompatibility is a heteroallelic allorecognition system, whereas mating incompatibility is a homoallelic autorecognition system (Hiscock and Kües 1999).

In the allorecognition system, individuals that possess different alleles at one or more incompatibility loci are unable to form a heterokaryon. Depending on the species, the number of loci involved ranges from a single *HET* locus to typically several, each with two or more alleles. In *Neurospora*, which has 11 such loci, the number of potential vegetative compatibility (VC) types based on combinations of the alleles is in the thousands, and thus, random individuals pulled from the population are unlikely to be compatible. However, in species with fewer VC types, such as *Aspergillus nidulans* with six *HET* loci (Jinks et al. 1966), genetically distinct individuals may by chance or through shared ancestry possess common alleles and be able to fuse and form heterokaryons. In basidiomycetes, heterokaryons can also be formed outside of the mating cycle through somatic compatibility between genetically distinct heterokaryons, leading to the formation of higher-order heterokaryons with three or more nuclear types (Johannesson and Stenlid 2004). Currently, nothing is known of the molecular basis for vegetative incompatibility in basidiomycetes, and in ascomycetes, the genes are only known from two classes of this group, the Sordariomycetes and Eurotiomycetes. The gene products of the known *HET* loci vary widely but often include a 150 amino acid HET domain. How the HET proteins function is unknown, but the cellular result is compartmentalization and activation of the programmed cell death pathway in fused incompatible cells, which leads to boundary formation among individuals (Glass and Dementhon 2006).

The allorecognition system of fungi is analogous in function to the systems controlling tissue rejection in vertebrates (animals), for example, the major histocompatibility complex (MHC) that controls tissue and fetal rejection and the cell surface recognition molecules alr1 and alr2 controlling fusion in colonial cnidarians (Nicotra et al. 2009). Maintenance of a high diversity of alleles is necessary for these systems to function properly and failure to reject results in chimerism.

In the autorecognition system controlling mating, individuals that share alleles at *MAT* loci are the same mating type and are incompatible. Thus *MAT* loci segregate for alleles that determine a cell's mating type. Most fungi have a single *MAT* locus (bipolarity), but in many basidiomycetes, there are two *MAT* loci (tetrapolarity) and heteroallelism at both loci is required for full compatibility. MAT proteins regulate steps in mate attraction and fusion (e.g. pheromones) and once mated maintain the heterokaryotic state by nuclear signaling. The gene products of ascomycete *MAT* are primarily transcription factors, including proteins with homeodomain and high mobility group (HMG) DNA binding domains (Lee et al. 2010).

Basidiomycete *MAT* loci also encode homeodomain transcription factors, but may also encode peptide pheromones and the G protein-coupled pheromone receptors they stimulate. This is another departure from ascomycetes, wherein pheromones and receptor alleles are differentially regulated but not differentially present between the genomes of mating types. In tetrapolar basidiomycetes, one locus encodes at least one incompatible receptor and pheromone combination (P/R) and the other locus typically encodes two types of homeodomain proteins (HD1 and HD2) that are self-incompatible but can form a HD1–HD2 heterodimer in heteroallelic encounters. In bipolar basidiomycetes the *MAT* locus encodes homeodomain proteins solely or both homeodomain proteins and pheromone/receptors, but never a pheromone/receptor solely (James 2007). As with the allorecognition system, the typical model postulates that the outcome of mating interactions in filamentous fungi is largely determined post-cell fusion by coordination of nuclear migration and nuclear acceptance of the mating mycelia.

The autorecognition system is analogous in function to the genes that control sex determination in animals and plants but has more important similarities to the well-known self-incompatibility (SI) system in flowering plants where interactions occur on the cellular level independently of the sex of the parent. Here, a dichotomy can again be made between ascomycetes and basidiomycetes. Filamentous ascomycetes have anisogamous sex (Fig. 19.1) with only two mating types. Agaricomycetes (mushrooms), on the other hand, often have multiple mating types (as many as 100s) and always isogamous sex. Maintenance of an equal frequency of each of the mating types or incompatibility alleles is assured by frequency dependent selection that favors rare alleles. This parallel between a high diversity of plant SI alleles and Agaricomycetes *MAT* alleles has been long recognized and suggests similar evolutionary dynamics (Uyenoyama 2005; Newbigin and Uyenoyama 2005). Why the number of mating types in ascomycetes has been limited to two is unknown but may relate to a lower cost of mating or ancestral anisogamy (Billiard et al. 2011). This stable equilibrium of the two mating type system is similar to the sex determining systems in animals and plants with two sexes in equal frequencies (Hurst 1996). In the multiallelic mating system of basidiomycetes, however, *MAT* alleles can be lost by drift or by replacement with a slightly higher fitness allele. In the following section, I address how fitness differences between MAT alleles in the multiallelic basidiomycete system could lead to accelerated evolution.

19.3 Fungal reproductive proteins show evidence for positive and balancing selection

Non-neutral evolution takes several forms, but this review focuses on selective mechanisms that cause rapid evolution of genes within and between species, i.e., positive selection, and the selective forces that maintain polymorphism within a species, i.e., balancing selection. In this review, positive selection is used to refer to selection that increases the fitness, and therefore the frequency, of an allele relative to its ancestral allele, whereas balancing selection is used to refer to forces that prevent allele loss or fixation. Because all incompatibility systems require the maintenance of multiple alleles in order to function, balancing selection prevents allele fixation by positive selection. However, balancing selection can also act to accelerate amino acid replacement at sites under selection under certain selective regimes (e.g., negative frequency dependent selection and heterosis).

Reproductive proteins in animals and plants have often been demonstrated to evolve rapidly relative to other regions of the genome. Evidence for rapid evolution can be obtained along multiple lines. One line of evidence is based on traditional tests for positive selection by detecting an increased rate of nonsynonymous substitution over neutral expectations. These data have often shown that reproductive proteins involved in sperm–egg (Panhuis et al. 2006) or pollen–ovule interactions (Takebayashi et al. 2003) have rates of nonsynonymous substitution greater than synonymous substitution. A second line of evidence comes from studies demonstrating increased divergence in reproductive protein sequences between species relative to other proteins in animals and plants (Jagadeeshan and Singh 2005). A final line of evidence comes from comparisons of gene duplicates that have undergone neofunctionalization. For example, a testes-specific homeobox variant in mammals showed an elevated rate of substitution relative to the ubiquitously expressed copy from which it diverged (Wang and Zhang 2004).

Balancing selection extends the genealogical depth of a collection of alleles, and depending the selection coefficient may impart a very high level sequence diversity within and among species (Takahata 1990). In scenarios in which homozygous genotypes cannot be formed, such as in homoallelic incompatibility systems, the selection is very strong, and alleles are expected to be maintained for very large numbers of generations, thereby increasing the divergence among functionally distinct allelic lineages at both functionally relevant and linked neutral sites. In extreme cases, this may result in a pattern of trans-specific polymorphism wherein lineages diverged before the species in which they are found. Trans-species polymorphism has been observed for lineages of plant SI genes (Richman et al. 1996) and the complementary sex determination genes in hymenoptera, one of the largest orders of insects (Cho et al. 2006). What is unknown is whether the alleles are really functionally equivalent (in some plant SI systems they clearly are not) and whether demographic forces, mutational limitation, or both can lead to the observed differences in numbers of alleles across species.

Despite the widespread demonstration of rapid evolution of reproductive proteins, the nature of selection on the proteins is actually not often clear (Swanson and Vacquier 2002). Several general models have been put forward to explain the rapid evolution of reproductive proteins, but they are likely applicable to only a subset of proteins (Swanson and Vacquier 2002). Following this logic, I propose that at least five forces may be particularly relevant for rapid evolution and/or extreme sequence divergence of incompatibility proteins in fungi:

1. *Rare advantage:* The multiallelic MAT incompatibility systems of Agaricomycetes may maintain hundreds of alleles at a single locus (James et al. 2004). Even if a population is at equilibrium for the number of alleles they may maintain (a function of population size and mutation rate), new mutations that generate novel specificities have an initial compatibility (fitness) advantage, and this should cause increased allele turnover. This constant favoring of novel alleles could be observed as an increase in nonsynonymous substitution rates over synonymous rates because alleles containing new amino acid replacements are on average less likely to be lost than those without them during the replacement process. See Newbigin and Uyenoyama (2005) for one model of how the replacement and acceleration could work in the plant SI system. Gaps in knowledge: vanishingly little is known about the origin of new alleles, and most alleles appear to be deeply divergent in sequence and therefore, time. One explanation for the deep sequence divergence of mating type and self-incompatibility alleles hypothesizes that the absence of recombination in and near mating type loci causes the accumulation of genetic load due to a reduced efficacy of purifying selection (Uyenoyama 2005). The accumulation of genetic load then favors pairs of alleles of deeper divergence as they are less likely to display homozygosity of deleterious alleles linked to the mating specificity. Thus, new alleles are most likely to replace the closest related allele, extending the overall coalescence time of the genealogy and reducing the rare advantage effect. However, it is unclear whether this genetic load is expected at mushroom mating type loci in which recombination immediately outside of *MAT* appears to be high (James et al. 2006) and cells have the potential to purge deleterious alleles due to selection in the free-living haploid stage.
2. *Competition/sexual selection model:* because of the high density of potential mating partners, access to mates through selection on rate of nuclear migration may be a fierce arena for competition. Opportunities for 'female' strains to choose among potential nuclear donors has been observed and linked to differences among mating types (Nieuwenhuis et al. 2011). Analogies to sperm competition are clear, with the end result that proteins in control of

access to mycelia (e.g. the pheromone proteins), attractiveness to potential 'female partners' (nuclear acceptors), or proteins involved in nuclear migration such as dynein and the cytoskeletal proteins (Gladfelter and Berman 2009) may be subject to strong selection. Gaps in knowledge: how extranuclear reproductive proteins can remain specific to the nucleus in which the allele is encoded. Do the male/female roles during fungal provide an avenue for sexually antagonistic mutations to develop by divergence in male/female nuclear behavior?

3. *Constant tinkering/Red queen model:* with constant sequence diversification caused by selection and drift, the molecular interactions between proteins in the multiallelic MAT systems must be complex and dynamic. Pheromone stimulation of the receptors with seven transmembrane helices appears to involve tertiary structure, and dimerization of HD1 and HD2 proteins involves broadly defined dimerization motifs in the specificity determining regions. Are some alleles more fit than others, and are fitness improvements difficult or easy to achieve through mutation? Variance in compatibility of alleles is likely to be pronounced in fungal mating systems because MAT proteins continue to act after fertilization. This differs from some of the plant SI systems because in that model of incompatibility, proteins in the style must only identify and inhibit fertilization by pollen with the same SI type (Wheeler et al. 2009). Once fertilization is secured, there is no role for the SI system. In contrast, after fertilization in filamentous fungi, the MAT genes function to regulate and maintain a heterokaryotic state, typically through trans-acting dimerization or pheromone-receptor stimulation. Because the proteins must recognize and cooperatively interact with each and every other MAT allele, it could be speculated that not all proteins are equivalent in dimerization or activation ability, and given the variation in numbers of genes per MAT allele, alleles should be expected differ quantitatively in their degree of compatibility/recognition of other alleles (Fig. 19.2). For example at the P/R locus of *Coprinopsis cinerea*, some haplotypes encode

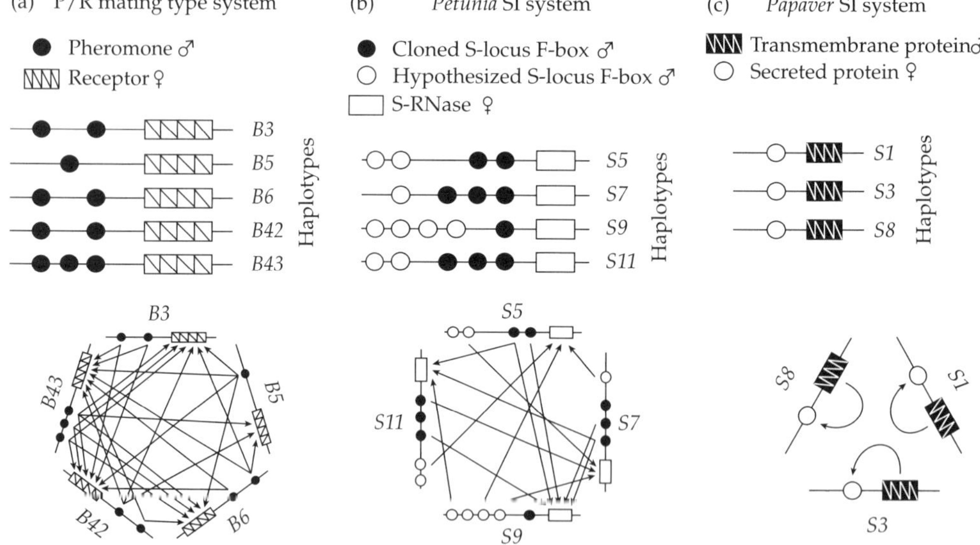

Figure 19.2 The pheromone/ pheromone receptors (P/R) in mushroom fungi use redundancy and versatile binding interactions to ensure all alleles are able to recognize and activate all other alleles (the pheromone response pathway). Positive interactions are shown using single arrows. The P/R system of mushrooms is most similar to that of *Petunia* SI in which each allele has a number of pollen-specific F-box proteins that collaboratively work to deactivate the stylar RNases of all other SI alleles (Kubo et al. 2010). The *Papaver* SI system (Wheeler et al. 2009) shows a very different interaction network, where the pollen-specific transmembrane protein need only recognize the homoallelic stylar protein to terminate pollen tube growth.

multiple pheromones (B43) that can stimulate the same receptor, while others, e.g., B5, encode only a single pheromone (Fig. 19.2). The ability of a single pheromone to stimulate multiple receptors is reminiscent of the *Petunia* SI system in which multiple S-locus F-box pollen proteins are encoded by each allele, and many are capable of deactivating multiple stylar S-RNases (Fig. 19.2). In the constant tinkering model, each allele must continuously evolve at the sites in the protein sequence that interact with partner MAT proteins in order to compensate for mutations in these partners. If the rare advantage model is also incorporated, then the global optimum is always changing because of the constant introduction of new alleles and thus there will be constant need for adaptation.

Gaps in knowledge: Because each allele interacts with a broad spectrum of other alleles (promiscuity) yet must prevent self-compatibility (Figure 19.2), it stands to reason that the fitness landscape is very rugged and exploration limited by mutation. However, mutagenesis screens have essentially never identified new or higher fitness alleles (Raper et al. 1965). Also, the biochemistry of the interactions among mating proteins is only weakly predicted by primary and secondary sequence structure.

4. *Mycoparasites:* the early-diverging fungal lineages comprising zygomycetes have a conserved use of trisporic acid as a pheromone for hyphal homing that has been exploited by related parasites for illicit invasion (Schultze et al. 2005). Diversification of pheromones and receptors and especially *HET* genes in fungi could work to prevent mycoparasites from tricking host cells into invasion as well as diminish interspecies fusions. This argues for an arms race between hosts and parasites that would facilitate rapid evolution of reproductive proteins. Gaps in knowledge: Hormones in zygomycetes are conserved and thus relatively easy for parasites to track. In the Dikarya (Ascomycota + Basidiomycota), pheromones appear to evolve rapidly, but *Candida albicans* was recently demonstrated to respond to a broad range of interspecific mating pheromones (Alby and Bennett 2011). Examining species-specific mycoparasites and their host reproductive proteins for coevolutionary patterns could test this hypothesis.
5. *Population size fluctuation model*: Given that the equilibrium number of MAT alleles is solely a function of census population size and mutation rate (Wright 1939), dramatic changes in effective population size are likely to cause departures from equilibrium that will increase the effectiveness of selection by rare advantage. For example, if a population undergoes a severe bottleneck and concomitant loss of mating type alleles followed by a recovery of population size, the number of alleles will be far below the equilibrium number and selection favoring novel alleles arising from mutation will increase. Recurrent episodes of population size change will thus speed the evolution of amino acid substitutions in specificity-determining regions of MAT alleles.

Gaps in knowledge: Fungal demography is a large unknown, and estimates of effective population size over time have never been obtained for any species. Evidence for recent range expansions exist (Kauserud et al. 2007), but whether recurrent changes in effective population size in fungi should be expected are unknown because the spatial definitions of populations are also largely unknown (James et al. 1999). In species like *Serpula lacrymans* that have undergone a recent range expansion, it would be interesting to determine whether increased rates of nonsynonymous substitutions could be detected in MAT and HET proteins.

19.4 Evidence for rapid evolution of fungal incompatibility genes and systems

It will come as no surprise that fungal incompatibility genes appear to share many of the characteristics of rapidly evolving animal and plant reproductive loci. Three aspects of rapid evolution are considered here and shown to apply to fungal incompatibility genes: rapid evolution of mating gene/protein sequences, rapid transitions among mating systems, and rapid changes in *MAT* loci.

19.4.1 Sequence evolution

Multiallelic MAT loci have been shown to evolve faster than biallelic loci based on DNA cross-hybridization of MAT genes within and among species (Specht et al. 1994; Metzenberg and Randall 1995). Sequence variation at the multiallelic loci among species was found to be much greater than biallelic loci. In the biallelic Ascomycota system, allele diversification within a species is a moot point because the two mating types alleles are 'idiomorphs,' meaning the two mating-type alleles encode entirely different gene products. Nonetheless, *MAT* genes in Ascomycota appear to evolve more quickly relative to other 'neutral' genes used to study species-level phylogeny (Barve et al. 2003).

Most of the evidence for rapid evolution and selection comes from the multiallelic MAT loci of basidiomycetes which supports the rare advantage model. Sequencing of *MAT* alleles from three model species (the split gill mushroom *Schizophyllum commune*, the inky cap mushroom *Coprinopsis cinerea*, and the corn smut *Ustilago maydis*) have revealed that alleles are invariably characterized by hyperdiverse amino acid sequences, with pairwise identities of 37–78% among alleles (Schulz et al. 1990; Stankis et al. 1992; Halsall et al. 2000). Heightened polymorphism in these systems appears to results from both positive selection and balancing selection. However, the evidence for positive selection on mating type genes has been difficult to obtain because alleles appear to be so ancient that silent mutations are saturated and insertion/deletions have made protein alignments challenging (Badrane and May 1999). Moreover, identifying and isolating these proteins from non-model organisms has remained a challenge due to their large sequence diversity that complicates isolation by PCR (polymerase chain reaction). Standard approaches for detecting positive selection include identifying proteins for which the ratio of nonsynonymous substitution rate to synonymous substitution rate (ω) is significantly greater than 1. However, this type of data is typically unrealistic in molecular evolution, because most proteins, especially incompatibility proteins, will have both conserved domains and domains that actually impart specificity, for example the dimerization (specificity) and transactivation (conserved) domains in homeodomain proteins. More sensitive tests that allow codons to behave independently within a coding sequence allow discrimination between various evolutionary models incorporating neutrality or selection (Yang et al. 2000).

As a first pass to test for non-neutral evolution in basidiomycete MAT proteins, I applied the codon models in PAML to test selection versus neutral models for homeodomain protein alignments of *Coprinellus disseminatus*, *Coprinopsis cinerea*, and *Ustilago maydis* (Table 19.1). While the average ω is considerably less than 1 for all of the proteins, likelihood ratio tests supported codon-based models that incorporated positive selection over neutral models. An additional finding from these explorations is that the number positively selected codons differed between the HD1 and HD2 genes in both *C. disseminatus* and *U. maydis*, consistent with divergent roles between the two protein types (Spit et al. 1998). Nonetheless, the data show a strong indication that for all of the species investigated, positive selection on selected codons is more likely than neutral or purifying selection only models.

Investigations of multiallelic pheromone receptors are likely to be fruitful, as these proteins are typically alignable over the seven transmembrane regions at the N-terminus even across all of Basidiomycota. Moreover, evidence that these genes may undergo positive selection has been obtained for heterothallic members of the genus *Neurospora* (Karlsson et al. 2008). Here, the genes for pheromone receptors are not mating type genes, but instead genes involved in mating and have a much lower polymorphism than in the multiallelic system. The multiallelic systems that govern some *HET* loci in fungi also show hyper-polymorphism and evidence for positive selection (Table 19.1). Both pheromones and *HET* loci are intimately involved in controlling interactions preventing illicit fusions and would be predicted to be under greater selection in more intensely competitive, highly species-rich niches such as soil.

19.4.2 Mating systems and loci

One of the most frequent transitions in the evolution of fungal mating systems has been the origin

Table 19.1 Elevated polymorphism in basidiomycete homeodomain *MAT* genes and *HET* genes from Sordariaceae and tests for positive selection using PAML (Yang et al. 2000). Model M1a is a nearly neutral model, and M2a is a variant of M1a that allows for sites with $\omega > 1$. M7 models variation in using a beta distribution with ω in the range of 0–1. M8 allows for an additional class of codons with $\omega > 1$. # selected codons with posterior probabilities > 0.5 (>0.95 in parentheses) using the empirical Bayes method. HD2 motif proteins are shown in bold.

Species	N	n	π	ω (d_n/d_s)	*P* M1a vs. M2a	*P* M7 vs. M8	# selected codons	Reference
Coprinellus disseminatus CDA1	9	1860	0.366	0.349	< 0.0001	< 0.0001	36 (4)	James et al. (2006)
Coprinellus disseminatus **CDA2**	9	1524	0.213	0.400	< 0.0001	< 0.0001	86 (23)	James et al. (2006)
Coprinopsis cinerea b1	17	1941	0.271	0.301	> 0.5	< 0.0001	18 (0)	Badrane and May (1999)
Ustilago maydis **bW**	23	966	0.133	0.547	< 0.0001	< 0.0001	106 (52)	Gillissen et al. (1992) + GenBank
Ustilago maydis bE	18	648	0.154	0.544	< 0.001	< 0.0001	6 (4)	Gillissen et al. (1992) + GenBank
Sordariaceae *het-c*	39	375	0.112	0.325	< 0.001	< 0.001	11 (3)	Wu et al. (1998)

of homothallic (non-outcorssing) mating systems from heterothallic (outcrossing) ones (Lin and Heitman 2007). There is no clear-cut evidence that the reverse transition has occurred. Our understanding of mating type gene function provides a reasonable explanation for this pattern. Early efforts to produce novel mating alleles by mutagenesis failed to create new alleles, but instead recovered a large number of self-compatible alleles that activated the mating response pathway (Raper et al. 1965). Thus, the MAT genes can become self-activating through simple changes to their molecular structure whereas the evolution of novel specificities requires multiple substitutions. Furthermore, self-compatible combinations can be readily created by recombination within the MAT locus (which normally is suppressed). Merely bringing the two-component system that is normally provided separately by two mating types together into a single genome has led to the evolution of homothallic species repeatedly in the Ascomycota (Yun et al. 1999; Nygren et al. 2011).

Within Basidiomycota, transitions from tetrapolar to bipolar have occurred numerous times (Hibbett and Donoghue 2001). These transitions appear to arise from either physical linkage of the two mating type loci or the inactivation of one of the two loci, possibly by formation of a self-compatible allele. In each of many tetrapolar to bipolar transitions investigated in Agaricomycetes, it has been the P/R mating type locus which has become self-compatible (James et al. 2011). Similar to the absence of evidence for reversal from heterothallic to homothallic, no reversals from bipolar to tetrapolar have been documented. Altogether the data on mating system transitions in fungi provide a genetic explanation for how rapid evolution of mating systems could proceed, because only a single genomic change is required, as well as an explanation for why the process is irreversible, the incompatibility mechanism is fractured in a manner that is difficult to repair.

Variation in the number of genes present within MAT haplotypes of the multiallelic mushroom species is common, and two exemplary studies are provided. These examples involve changes in gene number and organization but not type of genes involved in mating determination, which are much more static. Dynamic numbers of homeodomain genes are known from various *A* mating types of *Coprinopsis cinerea* (Kües et al. 2011). The number of complete gene copies per haplotype varies from four to seven, and among these genes are included both non-functional and non-expressed HD genes (Kües et al. 1994). Whether these result from degenerative processes or could eventually provide fodder for the future evolution of allelic diversity at the *A* mating type locus of *C. cinerea* is speculation only. Along similar lines, Fowler et al. (2004) demonstrated that the model of organization at the *Schizophyllum commune* P/R mating type locus was not as cleanly divided into two functionally redundant subloci (α and β) as previously hypothesized. Using transformation assays, individual pheromones were shown to stimulate receptors from both α and β subloci. These data, in combination with genome sequencing and phylogenetic analysis of *C. cinerea* receptors (Riquelme et al. 2005), now suggest that reorganization and recombination through time make division of mating genes into clear paralogous classes impossible.

19.5 Evidence for ancient alleles and mating systems

Also contributing to maintenance of diversity at fungal incompatibility genes are stabilizing forces that prevent allele or gene turnover. Based on what we know about the *MAT* genes in Dikarya, the loci controlling mating incompatibility have not been reinvented numerous times, unlike, for example, the repeated evolution of sex chromosomes in animal lineages such as fishes. Evidence that all Dikarya utilize the same mating response pathway through activation of a G protein-coupled pheromone receptor comes from studies that recapitulate the pheromone signaling system in yeast using P/R genes isolated from mushroom fungi (Brown and Casselton 2001). The *MAT* genes themselves are essentially the same in all filamentous ascomycetes which utilize HMG-motif and α-domain proteins as the two alleles of *MAT*. Likewise, all basidiomycetes have HD transcription factors and often P/R genes as *MAT* genes. These systems are thus ancient, having remained the same for hundreds of millions of years of evolution, each heterothallic species having inherited a set of alleles during speciation.

Genomic evidence provides further evidence of the longevity of *MAT* genes rather than the constant recruitment of new HD and HMG genes to the *MAT* locus from the many non-mating type specific copies of these genes throughout the genome. Specifically, conserved gene order near *MAT* has been detected for ascomycetes (Butler 2007) and the *MAT-A* HD genes of basidiomycetes. For example, the genes *SLA2* and *MIP* are known to be adjacent to MAT in most Pezizomycotina and Agaricomycetes, respectively. The Agaricomycete HD *MAT* genes are typically found on the largest chromosome, whereas the P/R genes are typically found on a smaller chromosome and show no evidence of conserved gene order (Kües et al. 2011). It is possible that the conserved location of *MAT* is related to a reduced recombination rate as larger chromosomes appear to have a lower rate of recombination. Alternatively, the larger chromosomes appear to be enriched for essential genes, and this location may reflect the fact that the HD MAT genes play a critical role in the basidiomycete lifecycle.

How long can independent MAT allele lineages be maintained? For biallelic species, the two lineages will be maintained as long as the system does not collapse to asexuality (a single mating type) and as long as a third mating type allele does not emerge. As mentioned above, the biallelic ascomycetes demonstrate this trans-specific inheritance of the two mating type allele lineages throughout all of the evolution of the Pezizomycotina, but instead of alleles, the two MAT variants are considered 'idiomorphs' because they encode different genes. In the basidiomycetes, MAT loci always encode for alleles instead of idiomorphs, and it has been demonstrated that for the biallelic Pucciniomycotina and Ustilaginomycotina, the same two pheromone receptor lineages have been maintained for over 370 million years of evolution (Devier et al. 2009).

Multiallelic systems found in some Ustilaginomycotina and most Agaricomycotina have high allelic and sequence diversity owing in large part to balancing selection that extends the age of the genealogy of alleles. This extension of the coalescence time relative to neutral expectations has been measured for the HD gene *b1* in *Coprinopsis cinerea* (May et al. 1999). May et al. derived an estimate of the scaling factor by which the genealogy of the *b1* gene was extended in time relative to neutral expectations as 27.8, which was larger than estimates for vertebrate MHC. This scaling factor is difficult to relate to geological time, and evidence for trans-species polymorphism in multiallelic systems would provide more data on the timing of sequence divergence at *MAT* loci relative to speciation events. One example of trans-specific polymorphism was suggested for the pheromone receptor genes of the mushroom genus *Pleurotus* (James et al. 2004). However, the absence of positional or functional information in this study makes inferring homology of the studied proteins difficult. Thus, convincing evidence of trans-specific polymorphism in multiallelic species is lacking and should be tested by using recently diverged species. A comparison among the relatively closely related species with clear homology can be drawn using the smuts *Ustilago maydis*, *U. hordei*, and *Sporisorium reilianum*. Approximate divergence times between the species are 20 million years ago (mya) for *U. maydis*–*S. reilianum* and 60 mya for the divergence of *U. hordei* from the other two species based on back of the envelope calculations using 18S rRNA sequence data (Berbee and Taylor 1993). Reciprocal monophyly is observed for each region of the HD2 gene *bW* analyzed separately though diversity is much greater in the specificity-determining region based on branch lengths (Fig. 19.3). One exception is the specificity determining region of the bipolar smut *U. hordei*, which is additionally unusual in that its mating system has been reduced to a biallelic system. Clearly, the alleles in multiallelic species are turning over faster than 20 mya, but the timescale investigated is very limited. Sister species and species complexes need to be investigated.

How frequently new genes are recruited to become *HET* loci is largely unknown as most molecular information concerns a single order of filamentous ascomycetes, Sordariales. However, investigation of the draft genomes of aspergilli (Eurotiomycetes) identified the presence of most of the characterized *HET* genes from *Neurospora crassa* and *Podospora anserina* (Pál et al. 2007). Moreover, within the family Sordariaceae, the *het-c* locus shows clear evidence of ancient trans-specific polymorphism, with three *het-c* lineages found in most of the species studied (Wu et al. 1998). These data show that the heterokaryon incompatibility systems, like *MAT* genes are not merely allelic incompatibilities that arise commonly through genetic drift or isolation but are complex systems that have been long maintained.

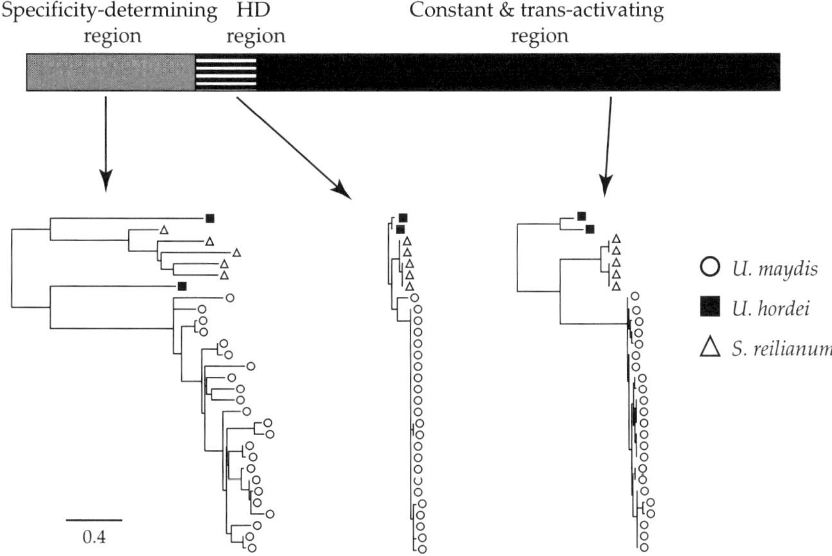

Figure 19.3 The HD2 protein bW of Ustilaginaceae shows variable and constant domains. Phylogenies were constructed for three separate regions of the protein and shown are the phylograms with branch lengths at a constant scale. The specificity-determining region displays the highest level of polymorphism based on branch lengths, but with the exception of the two *U . hordei* alleles, all other alleles are reciprocally monophyletic between species.

19.6 Conclusions

Because fungi lack motility, their interactions with other fungal conspecific and interspecific cells are defining moments during the lifetime of a mycelium. However, individuality is not imparted upon the mycelium, rather, genotypes are more dynamic than other multicellular organisms because nuclear interactions and movements can vary across the mycelium. Critical for these interactions is the pheromone signaling pathway that has been conserved throughout Dikarya. Evidence is presented here that the genes involved in incompatibility in fungi experience rapid evolution and heightened polymorphism due to the effects of selection. Mating-type genes are subject to simple balancing and (mostly) symmetrical selection making them subject to many of the phenomena witnessed in similar systems involved in mate and tissue recognition such as MHC and plant SI. Both *HET* and *MAT* genes are also shown to undergo positive selection at specific codons. Like the animal and plant systems, the forces responsible for positive selection are not entirely clear. Another major unanswered question in the evolution of the multiallelic mating type system is how novel mating type alleles are generated. How many changes are required and are recombination events a source of new alleles? In summary, the multiallelic incompatibility loci are marvels of evolution that are complicated, yet tractable, systems whose study could shed great insight into principals of receptor–ligand signaling, sexual selection, and perhaps even reproductive isolation and speciation.

References

Alby, K. and Bennett, R.J. (2011) Interspecies pheromone signaling promotes biofilm formation and same-sex mating in *Candida albicans*. *Proc Natl Acad Sci U S A* **108**: 2510–15.

Badrane, H. and May, G. (1999) The divergence-homogenization duality in the evolution of the b1 mating type gene of *Coprinus cinereus*. *Mol Biol Evol* **16**: 975–86.

Barve, M.P., Arie, T., Salimath, S.S., Muehlbauer, F.J., and Peever, T.L. (2003) Cloning and characterization of the mating type (MAT) locus from *Ascochyta rabiei* (teleomorph: *Didymella rabiei*) and a MAT phylogeny of legume-associated *Ascochyta* spp. *Fungal Genet Biol* **39**: 151–67.

Berbee, M.L. and Taylor, J.W. (1993) Dating the evolutionary radiations of the true fungi. *Can J Bot* **71**: 1114–27.

Billiard, S., Lopez-Villavicencio, M., Devier, B., Hood, M.E., Fairhead, C., and Giraud, T. (2011) Having sex, yes, but with whom? Inferences from fungi on the evolution of anisogamy and mating types. *Biol Rev* **86**: 421–42.

Brown, A.J. and Casselton, L.A. (2001) Mating in mushrooms: increasing the chances but prolonging the affair. *Trends Genet* **17**: 393–400.

Buée, M., Reich, M., Murat, C., Morin, E., Nilsson, R. H., Uroz, S., et al. (2009) 454 Pyrosequencing analyses of forest soils reveal an unexpectedly high fungal diversity. *New Phytol* **184**: 449–56.

Butler, G. (2007) The evolution of MAT: the Ascomycetes. In J. Heitman, J. Kronstad, J.W. Taylor, and L.A. Casselton (Eds) *Sex in Fungi*, pp. 3–18. Washington, DC: ASM Press.

Cho, S.C., Huang, Z.Y., Green, D.R., Smith, D.R. and Zhang, J.Z. (2006) Evolution of the complementary sex-determination gene of honey bees: Balancing selection and trans-species polymorphisms. *Genome Res.* 16: 1366–1375.

Devier, B., Aguileta, G., Hood, M.E., and Giraud, T. (2009) Ancient trans-specific polymorphism at pheromone receptor genes in basidiomycetes. *Genetics* **181**: 209–23.

Fowler, T.J., Mitton, M.F., Rees, E.I., and Raper, C.A. (2004) Crossing the boundary between the $B\alpha$ and $B\beta$ mating-type loci in *Schizophyllum commune*. *Fungal Genet Biol* **41**: 89–101.

Gillissen, B., Bergemann, J., Sandmann, C., Schroeer, B., Bolker, M., and Kahmann, R. (1992) A two-component regulatory system for self/non-self recognition in *Ustilago maydis*. *Cell* **68**: 647–57.

Gladfelter, A. and Berman, J. (2009) Dancing genomes: fungal nuclear positioning. *Nat Rev Microbiol* **7**: 875–86.

Glass, N.L. and Dementhon, K. (2006) Non-self recognition and programmed cell death in filamentous fungi. *Curr Opin Microbiol* **9**: 553–8.

Halsall, J.R., Milner, M.J., and Casselton, L.A. (2000) Three subfamilies of pheromone and receptor genes generate multiple *B* mating specificities in the mushroom *Coprinus cinereus*. *Genetics* **154**: 1115–23.

Herskowitz, I. (1989) A regulatory hierarchy for cell specialization in yeast. *Nature* **342**: 749–57.

Hibbett, D.S. and Donoghue, M.J. (2001) Analysis of character correlations among wood decay mechanisms, mating systems, and substrate ranges in homobasidiomycetes. *Syst Biol* **50**: 215–42.

Hiscock, S.J. and Kües, U. (1999) Cellular and molecular mechanisms of sexual incompatibility in plants and fungi. *Int Rev Cytol* **193**: 165–295.

Hurst, L.D. (1996) Why are there only two sexes? *Proc Roy Soc B-Biol Sci* **263**: 415–422.

Jagadeeshan, S. and Singh, R.S. (2005) Rapidly evolving genes of *Drosophila*: Differing levels of selective pressure in testis, ovary, and head tissues between sibling. *Mol Biol Evol* **22**: 1793–801.

James, T.Y. (2007) Analysis of mating-type locus organization and synteny in mushroom fungi- beyond model species. In J. Heitman, J. Kronstad, J.W. Taylor, and L.A. Casselton (Eds) *Sex in fungi: molecular determination and evolutionary implications*, pp. 317–31. Washington, DC: ASM Press.

James, T.Y., Lee, M., and Van Diepen, L.T.A. (2011) A single mating-type locus composed of homeodomain genes promotes nuclear migration and heterokaryosis in the white-rot fungus *Phanerochaete chrysosporium. Eukaryot Cell* **10**: 249–61.

James, T.Y., Liou, S.R., and Vilgalys, R. (2004) The genetic structure and diversity of the A and B mating-type genes from the tropical oyster mushroom, *Pleurotus djamor*. *Fungal Genet Biol* **41**: 813–25.

James, T.Y., Porter, D., Hamrick, J.L. and Vilgalys, R. (1999) Evidence for limited intercontinental gene flow in the cosmopolitan mushroom, *Schizophyllum commune. Evolution* **53**: 1665–1677.

James, T.Y., Srivilai, P., Kües, U., and Vilgalys, R. (2006) Evolution of the bipolar mating system of the mushroom *Coprinellus disseminatus* from its tetrapolar ancestors involves loss of mating-type-specific pheromone receptor function. *Genetics* **172**: 1877–91.

Jinks, J.L., Caten, C.E., Simchen, G., and Croft, J.H. (1966) Heterokaryon incompatibility and variation in wild populations of *Aspergillus nidulans*. *Heredity* **21**: 227–39

Johannesson, H. and Stenlid, J. (2004. Nuclear reassortment between vegetative mycelia in natural populations of the basidiomycete *Heterobasidion annosum. Fungal Genet Biol* **41**: 563–70.

Karlsson, M., Nygren, K., and Johannesson, H. (2008) The evolution of the pheromonal signal system and its potential role for reproductive isolation in heterothallic *Neurospora*. *Mol Biol Evol* **25**: 168–78.

Kauserud, H., Svegarden, I.B., Saetre, G.P., Knudsen, H., Stensrud, O., Schmidt, O., Doi, S., Sugiyama, T. and Hogberg, N. (2007) Asian origin and rapid global spread of the destructive dry rot fungus *Serpula lacrymans. Mol Ecol* **16**: 3350–3360.

Kubo, K., Entani, T., Takara, A., Wang, N., Fields, A.M., Hua, Z.H., et al. (2010) Collaborative non-self recognition system in S-RNase-based self-incompatibility. *Science* **330**: 796–9.

Kües, U., Tymon, A.M., Richardson, W.V.J., May, G., Gieser, P.T., and Casselton, L.A. (1994) *A* mating-type factors of *Coprinus cinereus* have variable numbers of specificity genes encoding two classes of homeodomain proteins. *Mol Gen Genet* **245**: 45–52.

Kües, U., James, T.Y., and Heitman, J. (2011) Mating type in Basidiomycetes: Unipolar, bipolar, and tetrapolar patterns of sexuality. In S. Pöggeler and J. Wöstemeyer (Eds) *Evolution of fungi and fungal-like organisms, The Mycota XIV*, pp. 97–160. Berlin: Springer Verlag.

Lee, S.C., Ni, M., Li, W., Shertz, C., and Heitman, J. (2010) The evolution of sex: a perspective from the fungal kingdom. *Microbiol Mol Biol Rev* **74**: 298–340.

Lin, X. and Heitman, J. (2007) Mechanisms of homothallism in fungi and transitions between heterothallism and homothallism. In J. Heitman, J. Kronstad, J.W. Taylor, and L.A. Casselton (Eds) *Sex in Fungi*, pp. 35–57. Washington, DC: ASM Press.

May, G., Shaw, F., Badrane, H., and Vekemans, X. (1999) The signature of balancing selection: fungal mating compatibility gene evolution. *Proc Natl Acad Sci U S A* **96**: 9172–7.

Metzenberg, R.L. and Randall, T.A. (1995) Mating type in *Neurospora* and closely related ascomycetes: some current problems. *Can J Bot* **73**: S251–S257.

Newbigin, E. and Uyenoyama, M.K. (2005) The evolutionary dynamics of self-incompatibility systems. *Trends in Genetics* **21**: 500–505.

Nicotra, M.L., Powell, A.E., Rosengarten, R.D., Moreno, M., Grimwood, J., Lakkis, F.G., et al. (2009) A hypervariable invertebrate allodeterminant. *Curr Biol* **19**: 583–9.

Nieuwenhuis, B.P.S., Debets, A.J.M., and Aanen, D.K. (2011) Sexual selection in mushroom-forming basidiomycetes. *Proc Roy Soc B Biol Sci* **278**: 152–7.

Nygren, K., Strandberg, R., Wallberg, A., Nabholz, B., Gustafsson, T., Garcia, D., et al. (2011) A comprehensive phylogeny of *Neurospora* reveals a link between reproductive mode and molecular evolution in fungi. *Mol Phylogenet Evol* **59**: 649–63.

Pál, K., Van Diepeningen, A.D., Varga, J., Hoekstra, R.F., Dyer, P.S., and Debets, A.J.M. (2007) Sexual and vegetative compatibility genes in the aspergilli. *Stud Mycol* **59**: 19–30.

Panhuis, T.M., Clark, N.L., and Swanson, W.J. (2006) Rapid evolution of reproductive proteins in abalone and *Drosophila*. *Phil Trans Soc B Biol Sci* **361**: 261–8.

Raper, J.R., Boyd, D.H., and Raper, C.A. (1965) Primary and secondary mutations at incompatibility loci in *Schizophyllum*. *Proc Natl Acad Sci U S A* **53**: 1324–32.

Rayner, A. D. M. (1991) The challenge of the individualistic mycelium. *Mycologia* **83**: 48–71.

Richman, A.D., Uyenoyama, M.K., and Kohn, J.R. (1996) Allelic diversity and gene genealogy at the self-incompatibility locus in the Solanaceae. *Science* **273**: 1212–16.

Riquelme, M., Challen, M.P., Casselton, L.A., and Brown, A.J. (2005) The origin of multiple *B* mating specificities in *Coprinus cinereus*. *Genetics* **170**: 1105–119.

Schultze, K., Schimek, C., Wostemeyer, J., and Burmester, A. (2005) Sexuality and parasitism share common regulatory pathways in the fungus *Parasitella parasitica*. *Gene* **348**: 33–44.

Schulz, B., Banuett, F., Dahl, M., Schlesinger, R., Schafer, W., Martin, T., et al. (1990) The b alleles of *U. maydis*, whose combinations program pathogenic development, code for polypeptides containing a homeodomain related motif. *Cell* **60**: 295–306.

Specht, C.A., Stankis, M.M., Novotny, C.P., and Ullrich, R.C. (1994) Mapping the heterogeneous DNA region that determines the 9 Aα mating-type specificities of *Schizophyllum commune*. *Genetics* **137**: 709–14.

Spit, A., Hyland, R.H., Mellor, E.J.C., and Casselton, L.A. (1998) A role for heterodimerization in nuclear localization of a homeodomain protein. *Proc Natl Acad Sci U S A* **95**: 6228–33.

Stankis, M.M., Specht, C.A., Yang, H., Giasson, L., Ullrich, R.C., and Novotny, C.P. (1992) The *Aα* mating locus of *Schizophyllum commune* encodes two dissimilar multiallelic homeodomain proteins. *Proc Natl Acad Sci U S A* **89**: 7169–73.

Swanson, W.J. and Vacquier, V.D. (2002) Reproductive protein evolution. *Annu Rev Ecol Syst* **33**: 161–79.

Takahata, N. (1990) A simple genealogical structure of strongly balanced allelic lines and trans-species evolution of polymorphism. *Proc Natl Acad Sci U S A* **87**: 2419–23.

Takebayashi, N., Brewer, P.B., Newbigin, E., and Uyenoyama, M.K. (2003) Patterns of variation within self-incompatibility loci. *Mol Biol Evol* **20**: 1778–94.

Uyenoyama, M.K. (2005) Evolution under tight linkage to mating type. *New Phytol* **165**: 63–70.

Wang, X.X. and Zhang, J.Z. (2004) Rapid evolution of mammalian X-linked testis-expressed homeobox genes. *Genetics* **167**: 879–88.

Wheeler, M.J., De Graaf, B.H.J., Hadjiosif, N., Perry, R.M., Poulter, N.S., Osman, K., et al. (2009) Identification of the pollen self-incompatibility determinant in Papaver rhoeas. *Nature* **459**: 992–118.

Worrall, J.J. (1997) Somatic incompatibility in basidiomycetes. *Mycologia* **89**: 24–36.

Wright, S. (1939) The distribution of self-sterility alleles in populations. *Genetics* **24**: 538–552.

Wu, J., Saupe, S.J., and Glass, N.L. (1998) Evidence for balancing selection operating at the *het-c* heterokaryon incompatibility locus in a group of filamentous fungi. *Proc Natl Acad Sci U S A* **95**: 12398–403.

Yang, Z.H., Nielsen, R., Goldman, N., and Pedersen, A.M.K. (2000) Codon-substitution models for heterogeneous selection pressure at amino acid sites. *Genetics* **155**: 431–49.

Yun, S.-H., Berbee, M.L., Yoder, O.C., and Turgeon, B.G. (1999) Evolution of the fungal self-fertile reproductive life style from self-sterile ancestors. *Proc Natl Acad Sci U S A* **96**: 5592–7.

PART IV

Pathogens and their Hosts

CHAPTER 20

Rapid evolution of innate immune response genes

Brian P. Lazzaro and Andrew G. Clark

20.1 The evolution of immunity

The immune system is a central mediator of inherently antagonistic interactions between hosts and pathogens. Genes in the immune system often evolve more rapidly than genes in other physiological systems (e.g. Murphy 1991; Schlenke and Begun 2003), presumably as a consequence of this antagonism. The mode of immune system evolution, however, can depend on a multitude of factors, including whether the pathogens are generalists or specialists, the prevalence and diversity of infectious agents in hosts' natural environments, and pleiotropic functions of immune genes. Even within the immune system, there is every reason to expect that selective pressures will vary across functionally distinct components.

Host immune systems are generally defined in terms of the physiological process of recognizing and eliminating potentially pathogenic infection. In order to be effective, any immune system must therefore possess mechanisms for surveillance, for signal transduction and stimulation of appropriate antipathogen activity, and for sequestration and killing of the pathogen. For the pathogen, surviving the immune response is essential. Pathogens, therefore, may experience strong selective pressure to evade recognition, subvert or suppress signal transduction, and/or resist host killing mechanisms. Pathogen success on any of these fronts, however, imposes renewed selective pressure on the host to evolve re-established immunity. Thus, hosts and pathogens may reciprocally adapt to each other, serially evolving under positive Darwinian selection but without achieving any substantial change in the relationship status quo. This particular coevolutionary model is sometimes termed a 'coevolutionary arms race' or 'Red Queen dynamics,' the former referring to serial escalation that maintains parity between antagonists and the latter referring to the Lewis Carroll character's assertion to Alice that in Wonderland 'it takes all the running you can do, to keep in the same place' (van Valen 1973; Dawkins and Krebs 1978).

This chapter provides an overview of the evolutionary dynamics of insect antimicrobial and antiviral immune systems, emphasizing the fruit fly, *Drosophila*. Insects have no analog to the charismatic antibody-mediated acquired immunity that allows vertebrates to generate hyperdiversity and memory of previous infection through somatic recombination and clonal expansion (Murphy et al. 2007). Instead, insects rely solely on 'innate' immunity. Innate immune systems, which are also central components of vertebrate defense, are hardwired into the genome and therefore might be more sensitive to host–pathogen coevolutionary dynamics. Innate immune responses to microbes include defensive phagocytosis and the production of broad-spectrum antimicrobial peptides (reviewed in Lemaitre and Hoffmann 2007). Innate immunity to RNA viruses and transposable elements is mediated by RNA interference (RNAi), a cellular mechanism for recognizing and degrading double-stranded RNA (dsRNA), and subsequently single-stranded RNA homologous to the activating dsRNA (van Mierlo et al. 2011). Components of both the antimicrobial immune system and antiviral RNAi have been shown to evolve rapidly and adaptively in *Drosophila* and other insects.

Rapidly Evolving Genes and Genetic Systems. First Edition. Edited by Rama S. Singh, Jianping Xu, and Rob J. Kulathinal.

20.2 Orthology and gene family evolution in antimicrobial immunity

Insect immune responses to microbes can include both defensive phagocytosis and production of secreted antimicrobial peptides (AMPs). The mechanistic basis for the systemic production of AMPs has been well studied and appears from comparative genomic analyses to be highly conserved across invertebrates (reviewed in Lazzaro 2008). There are two primary signaling pathways used to activate AMP production, named the Toll pathway and the Imd pathway after key constituent genes. The Imd pathway has homology to the mammalian tumor necrosis factor pathway, and mammalian Toll-like signaling pathways are named for their homology to their insect counterpart. Nearly all the core signaling proteins in both the Toll and Imd pathways are conserved as strict orthologs across available sequenced invertebrate genomes.

The Imd and Toll pathways can each be stimulated by host recognition of microbial cell wall components. This recognition is achieved by peptidoglycan recognition proteins (PGRPs) and the Gram-negative binding proteins (GNBPs; misleadingly named because their recognition spectrum is not restricted to Gram-negative bacteria). PGRPs and GNBPs each exist as multigene families of roughly four to 15 members in most insects and mammals. These gene families remain evolutionarily stable over short time periods, but family members undergo considerable duplication and deletion over longer evolutionary timescales (e.g. Evans et al. 2006; Sackton et al. 2007; Waterhouse et al. 2007; Zhou et al. 2007). Fig. 20.1 shows the distribution of turnover rates of genes involved in recognition, signaling, and effector classes, showing that the signaling class has the lowest rate. Genes encoding antimicrobial peptides show extremely high rates of gene family expansion and contraction. While genes encoding some peptides, such as cecropins and defensins, are nearly ubiquitous in insects, most peptide gene families are much more taxonomically restricted (e.g. Evans et al. 2006; Sackton et al. 2007; Waterhouse et al. 2007; Zhou et al. 2007). Peptides in the Defensin class are the most taxonomically widespread, being found in insects, mammals, and plants. The most distantly

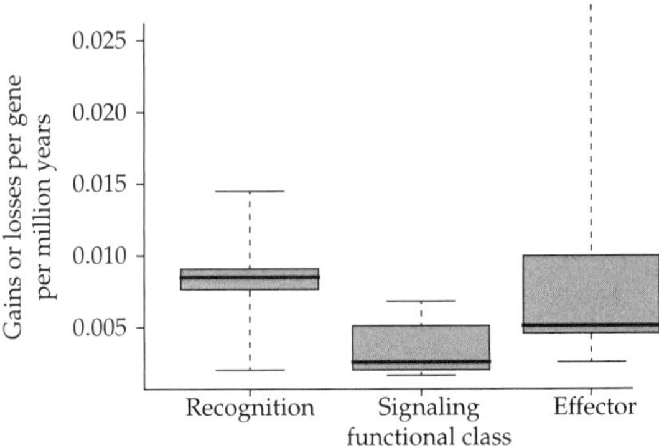

Figure 20.1 Rates of turnover of copy number of different classes of innate immune genes, as inferred from the 12 *Drosophila* genome sequences. On the *Drosophila* phylogeny, the gene copy number of each class was determined in the 12 species, and a maximum likelihood procedure was used to estimate the rate of change in copy number along the branches of the phylogeny. The clear conservativeness of copy number of the signaling genes stands in contrast to both the recognition and effector (antimicrobial) peptides. Redrawn from Sackton et al. (2007).

related Defensins may, however, be the product of convergent evolution (Broekaert et al. 1995) to a similar tertiary structure.

There are exceptions to the pattern of strong conservation of the Toll and Imd pathways and diversification of PGRPs, GNBPs, and AMPs. For example, the pea aphid genome sequence indicates that aphids have lost key genes in the Imd pathway and are completely without PGRPs (Gerardo et al. 2010). Even though the honeybee has intact Toll and Imd pathways, the bee exhibits reduced copy number in most multigene families, resulting in a nearly two-thirds reduction in the complement of identifiable immune system genes (Evans et al. 2006). It is unclear whether these insects are actually immunocompromised. The gene losses may be offset by indirect protection from infection through hygienic hive behavior in the case of the honeybee or protection by secondary symbionts in the case of the aphid. Alternatively, the immunological functions lost with the deletions of these genes may be regained through other, yet-unidentified genetic mechanisms. These questions cannot be addressed by comparative genomic analysis, but can only be answered with careful functional study.

The evolutionary examination of genes involved in defensive phagocytosis has been less thorough.

Nonetheless, there are clear indications that, like PGRPs and GNBPs, phagocytic receptors duplicate and delete at rates that are significantly higher than the genome average (e.g. Sackton et al. 2007). For example, receptors in the eater/nimrod/hemese class are highly diversified in *Drosophila* (Sackton et al. 2007; Zou et al. 2007; Somogyi et al. 2008). Scavenger receptors are also diversified across distinct insect taxa. *Class C scavenger receptors* have expanded from one progenitor to four genes in the melanogaster group of *Drosophila* (Lazzaro 2005; Sackton et al. 2007), and the *Class B scavenger receptor* family is greatly expanded in *Tribolium castaneum* (Zou et al. 2007). The *Tep* gene family encodes protease-activated opsonins that tag microbes and other pathogens for phagocytosis and immunological elimination. *Tep* genes are highly diversified in mosquitoes (Waterhouse et al. 2007) and experience rapid gene family evolution across insect taxa (Evans et al. 2006; Zou et al. 2007; Gerardo et al. 2010).

In summary, the rate of gene copy number evolution varies greatly across different functional components of the immune system, but is relatively consistent across insect taxa. Core signaling genes in the Imd and Toll pathways tend to be maintained as strict orthologs across insect taxa. In contrast, PGRP and GNBP recognition proteins that activate these pathways and the AMPs that are among their downstream targets are highly diversified across insects. This diversification in recognition and effector proteins may arise as a consequence of different species' ecological exposure to distinct suites of microbes. Alternatively, these genes may be subject to 'threshold' evolution, where gene copies can duplicate and delete nearly neutrally provided some minimum capacity for microbial recognition and clearance is retained. Whichever model is more correct, rates of evolution for these gene families are considerably higher than for most genes in the genome.

20.3 Molecular evolution of the antimicrobial immune system

Despite their strict maintenance of orthology across very distantly related taxa, signaling genes in the Toll and Imd pathways evolve surprisingly rapidly at the amino acid level. Genes in these pathways are among the most divergent in the immune system in comparisons between *D. melanogaster* and the mosquitoes *Anopheles gambiae* and *Aedes aegypti* (Waterhouse et al. 2007), and several individual signaling genes exhibit significant evidence of adaptive evolution within *Drosophila* (Begun and Whitley 2000; Schlenke and Begun 2003; Jiggins and Kim 2007; Sackton et al. 2007). This observation has been interpreted in light of the capacity of some pathogens to subvert or block host immune signaling (Begun and Whitley 2000; Schmid-Hempel 2008). The essential requirement of these pathways for antimicrobial immunity and their highly conserved orthology may be the very features that expose them to pathogen manipulation. Whereas recognition proteins and AMPs are comprised of diverse and varied gene families, the two signaling pathways are a 'bottleneck' at which pathogens can choke off the immune response. The ubiquitous orthology of these pathways may further serve to make them attractive targets for interference by generalist pathogens. The adaptive evolutionary signature in these pathways may be amplified by the correlated amino acid substitutions within and among proteins that maintain pathway function while escaping pathogen manipulation (DePristo et al. 2005).

Compared to other functional classes of genes in the innate immune system, genes encoding receptors display the strongest signature of positive selection (Fig. 20.2). Genes encoding opsonins and receptors for phagocytosis tend to evolve under positive selection at the amino acid level. In particular, *Tep* genes have been shown to evolve adaptively in *Drosophila* (Jiggins and Kim 2006; Sackton et al. 2007), *Anopheles* (Little and Cobbe 2005), and the cladoceran crustacean *Daphnia* (Little et al. 2004), with selected sites predominantly found in and around the domain that is proteolytically cleaved for TEP activation. The expanded class C scavenger receptor family in the *melanogaster* species group also evolves unusually quickly at the amino acid level (Lazzaro 2005), as do several other scavenger receptors and bacteria-binding phagocytosis receptors in the nimrod class (Sackton et al. 2007). In contrast, there is little indication of adaptive amino-acid level evolution in PGRP and GNBP

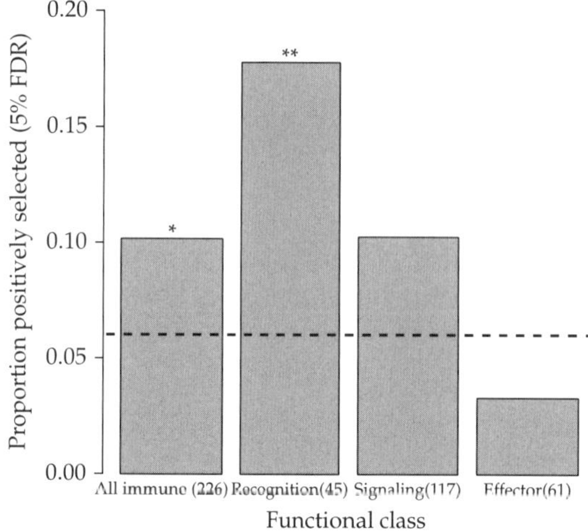

Figure 20.2 Rates of amino-acid substitution are accelerated in recognition and signaling proteins (as estimated by the maximum likelihood fits to the codon substitution model as implemented in PAML), resulting in a greater proportion of genes in these classes showing evidence for positive selection ($K_A/K_S > 1$). The dotted line represents the genome-wide average proportion of positively selected genes. Redrawn from Sackton et al. (2007).

recognition proteins that activate Toll and Imd signaling (Schlenke and Begun 2003; Jiggins and Kim 2006; Sackton et al. 2007). The observation of adaptive evolution in signaling genes but not in the recognition factors that activate signaling seems to be generalizable across invertebrates (e.g. Little et al. 2004; Little and Cobbe 2005; Bulmer and Crozier 2006), although individual genes may vary in the degree to which they are selected in different taxa (e.g. Levine and Begun 2007; Sackton et al. 2007). The distinct evolutionary trajectories of phagocytosis receptors versus PGRPs and GNBPs may stem from differences in binding affinity. Opsonins and phagocytic receptors bind to a diversity of pathogen molecules, some of which may be evolutionarily very labile. In contrast, GNBPs and PGRPs that activate the immune system are targeted to highly conserved microbial cell wall compounds like peptidoglycan and β-glucans.

Despite their rapid gene family turnover, AMP genes in *Drosophila* show little indication of rapid evolution at the amino acid level (e.g. Lazzaro and Clark 2003; Jiggins and Kim 2005; Sackton et al. 2007). This contrasts with the observation that AMP gene duplication is frequently associated with adaptive amino acid diversification in vertebrates (Tennessen et al. 2005). AMP gene duplication has also been coupled with amino acid divergence in termites and mosquitoes (Bulmer and Crozier 2004; Dassanayake et al. 2007), so the data from *Drosophila* may represent a departure from the norm. Amino acid diversification may result in altered antimicrobial activity (Tennessen 2005; Yang et al. 2011), and both gene family expansion and amino acid diversification may be driven by adaptation to commonly encountered microbes. Drosophila species may associate less with specific coevolving microbes, obviating the need for amino-acid level adaptation in AMP genes. The limited survey work that has been conducted suggests that most microbes associated with *D. melanogaster* in the field are generalist opportunists (Corby-Harris et al. 2007; P. Juneja and B.P. Lazzaro unpublished data), and the *Drosophila* antimicrobial immune system may be adapted to management of these more persistent but less threatening challenges (Hultmark 2003).

20.4 The evolution of defense against viruses and transposable elements

Genes responsible for defense against viruses and transposable elements (TEs) can exhibit exceptionally fast evolutionary rates. Double-stranded RNA (dsRNA) associated with RNA viruses and active transposons are targeted for silencing and degradation by RNAi machinery in plant, insect, and mammalian cells. Three *Drosophila* genes that are required for processing and silencing of transposon- and virus-derived dsRNA (*Dicer-2*, *R2D2*, and *Argonaut-2*) are among the fastest evolving 3% of genes in the *D. melanogaster* genome (Fig. 20.3). These genes exhibit highly elevated Ka/Ks ratios and McDonald–Kreitman test statistics that indicate strong positive selection across the *melanogaster* subgroup species (Obbard et al. 2006). *Dicer-2* and *Argonaut-2* in particular appear to have been recent targets of selective sweeps, resulting in significantly reduced genetic diversity at these loci in *D. melanogaster* and related species (Obbard et al. 2006, 2011). Modeling of the selective process suggests multiple recurrent, recent, and independent

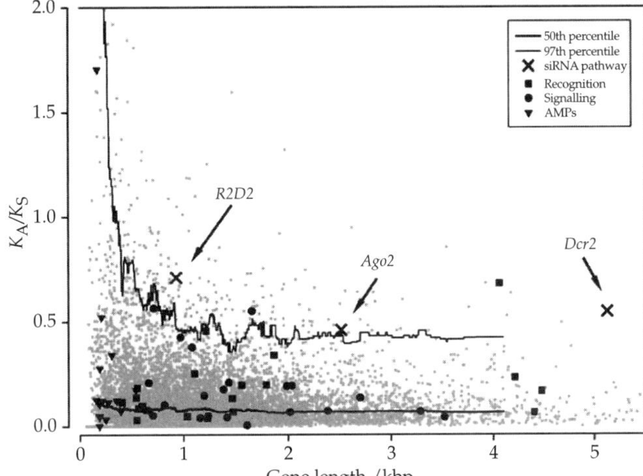

Figure 20.3 Rates of adaptive evolution of genes involved in immune response, expressed as the ratio of nonsynonymous (K_A) to synonymous (K_S) rates of nucleotide substitution. Three genes involved in antiviral response, *R2D2*, *Ago2*, and *Dcr2* are among the top 3% most rapidly evolving genes in *Drosophila*. From Obbard et al. (2006).

sweeps at *Argonaut-2* in *D. melanogaster*, *D. simulans*, and *D. yakuba* (Obbard et al. 2011).

RNAi is an effective defense against RNA viruses, and several viruses have mechanisms for suppressing or subverting the host defensive RNAi of plants, mammals, insects, and worms (reviewed in Li and Ding 2006). Viral suppression of RNAi (VSR) can occur through a variety of mechanisms which may spur molecular arms races between hosts and viruses. These could include competitive binding and sequestration of processed siRNAs (which would dampen the host RNAi response), competitive binding of full-length dsRNA (which would prevent access by endogenous RNAi machinery), and direct inhibition of host RNAi proteins (Li and Ding 2006; Obbard et al. 2009). The sites that are putatively evolving adaptively in host RNAi genes tend not to be restricted to known functional domains, but are distributed throughout the proteins (Obbard et al. 2006; Kolaczkowski et al. 2011; Obbard et al. 2011). Putatively adaptive substitutions occur in a domain critical for RNA-binding by *D. melanogaster* Dicer-2 (Kolaczkowski et al. 2011), and are particularly prevalent on molecular surfaces of other genes (Obbard et al. 2006, 2011; Kolaczkowski et al. 2011), which could perhaps indicate coevolution with viral genes that physically interact with host RNAi machinery or correlated compensatory coevolution among physically correlated amino acid residues or interacting RNAi proteins (as in DePristo et al. 2005; Callahan et al. 2011). An arms race between hosts and viruses implies not only rapid evolution in the antiviral machinery, but also rapid evolutionary turnover in viral VSRs. This condition is satisfied by the rapid molecular evolutionary origin and elimination of VSRs across viral taxa. VSRs are often encoded by overlapping genes that differ in reading frame, which arise when an existing gene sequence becomes translated in an alternative reading frame, a process known as overprinting (Li and Ding 2006). This results in gene sets that vary in age, with structurally novel proteins arising instantaneously in viral lineages and resulting in remarkable VSR functional diversity. As would be expected under an arms race model, VSRs show elevated rates of protein divergence relative to other viral genes (Obbard et al. 2009).

RNAi genes more conventionally associated with germline silencing of transposable elements (TEs) also show evidence of recent and recurrent adaptation (Obbard et al. 2009; Kolaczkowski et al. 2011). TE silencing in the germline is executed by the PIWI-interacting, or piRNA, pathway. Active TEs can be severely deleterious to host lineages and are strongly selected against (reviewed in Lee and Langley 2010). Theory predicts, and empiricism bears out, that piRNAs which silence transposons should be adaptive through reducing the deleterious effects of TE mobilization (Lu and Clark 2010).

piRNA pathway genes show strong evidence of adaptive evolution and are among the 5% most rapidly evolving genes in *D. simulans* (Obbard et al. 2009). Several piRNA genes also show evidence of recent selective sweeps in *D. melanogaster* (Kolaczkowski et al. 2011).

Because there is mechanistic overlap between antiviral and anti-transposon RNAi functions, it is difficult to definitively declare that rapid evolution of RNAi genes is due to coevolution specifically with viruses or with TEs at the exclusion of the other. Several piRNA components appear to have additional antiviral functions and some VSRs may affect piRNA pathway genes (reviewed in Obbard et al. 2009, 2011; Kolaczkowski et al. 2011), which could result in the rapid evolution of piRNA genes without invoking transposon-driven selection. At the same time, the antiviral RNAi genes *Dicer-2* and *Argonaut-2* are recruited for anti-TE function in an RNAi mechanism so far thought to be unique to *Drosophila* (Obbard et al. 2009). Unlike transposons, however, viruses have a known mechanism for suppressing host RNAi, and have themselves the capacity to rapidly evolve in response to evolutionary change in the host. These factors suggest that antagonistic host-virus coevolution may be a more probable driver of rapid evolution in RNAi genes than is host–TE coevolution.

20.5 Concluding remarks

Immune systems tend to evolve rapidly and adaptively, and the innate immune system of insects and invertebrates is no exception. The precise nature of evolution in immune system genes unsurprisingly depends on gene function, but not all immune genes evolve in a manner that is necessarily intuitive. For example, pathogen recognition proteins that activate antimicrobial immune signaling show little evidence of adaptive amino acid evolution, suggesting that these genes tend not to coevolve. At the same time, the gene families encoding these recognition proteins recurrently show taxon-specific expansion and deletion, potentially indicating adaptation to the spectrum of microbes encountered by distinct species. Cell-surface and secreted proteins that bind microbes for phagocytosis, on the other hand, show both rapid gene family diversification and pervasive adaptive amino acid evolution. The difference in evolutionary profiles between the two classes of pathogen-recognition proteins is likely a function of differences in the ligands which they recognize. Perhaps surprisingly, intracellular signaling proteins that activate systemic antimicrobial immunity tend to evolve rapidly at the amino acid level, even though these are not expected to have obvious contact with pathogens and show virtually no diversification at the gene family level across very distantly related taxa. This has been interpreted to result from coevolution with pathogens that interfere with these highly conserved signaling processes. Although antimicrobial peptides are ubiquitous components of innate immune systems, specific peptide families are typically highly taxonomically restricted and vary across taxa in the rate at which they evolve at the amino acid level. Finally, RNAi genes that defend against viruses and transposable elements evolve extraordinarily quickly, probably in reflection of tight coevolution with highly specific and quickly evolving viruses that have high mutation rates and short replication times.

Also surprisingly, there is little indication that genes in the innate immune system maintain polymorphism through balancing natural selection. Instead, where there is evidence of adaptive evolution, the data reveal rapid directional selection more consistent with coevolutionary arms races. This may partially stem from the fact that serial directional selection is experimentally easier to detect than balanced polymorphism, but also is indicative of the nature of invertebrate immune system evolution.

References

Broekaert, W.F., Terras, F.R., Cammue, B.P., and Osborn, R.W. (1995) Plant defensins: novel antimicrobial peptides as components of the host defense system. *Plant Physiol* **108**: 1353–8.

Begun, D.J. and Whitley, P. (2000) Adaptive evolution of relish, a Drosophila NF-κB/IκB protein. *Genetics* **154**: 1231–8.

Bulmer, M.S. and Crozier, R.H. (2004) Duplication and diversifying selection among termite antifungal peptides. *Mol Biol Evol* **21**: 2256–64.

Bulmer, M.S. and Crozier, R.H. (2006) Variation in positive selection in termite GNBPs and Relish. *Mol Biol Evol* **23**: 317–26.

Callahan, B., Neher, R.A., Bachtrog, D., Andolfatto, P., and Shraiman, B.I. (2011) Correlated evolution of nearby residues in Drosophilid proteins. *PLoS Genet* **7**(2): e1001315.

Corby-Harris, V., Pontaroli, A.C., Shimkets, L.J., Bennetzen, J.L., Habel, K.E., and Promislow, D.E. (2007) Geographical distribution and diversity of bacteria associated with natural populations of *Drosophila melanogaster. Appl Environ Microbiol* **73**: 3470–9.

Dassanayake, R.S., Silva Gunawaradene, Y.I., and Tobe, S.S. (2007) Evolutionary selective trends of insect/mosquito antimicrobial defensin peptides containing cysteine-stabilized alpha/beta motifs. *Peptides* **28**: 62–75.

Dawkins, R. and Krebs, J.R. (1979) Arms races between and within species. *Proc Roy Soc Lond B Biol Sci* **205**: 489–511.

DePristo, M.A., Weinreich, D.M., and Hartl, D.L. (2005) Missense meanderings in sequence space: a biophysical view of protein evolution. *Nat Rev Genet* **6**(9): 678–87.

Evans, J.D., Aronstein, K., Chen, Y.P., Hetru, C., Imler, J.L., Jiang, H., (2006) Immune pathways and defence mechanisms in honey bees *Apis mellifera. Insect Mol Biol* **15**: 645–56.

Gerardo, N.M., Altincicek, B., Anselme, C., Atamian, H., Barribeau, S.M., de Vos, M., et al. (2010) Immunity and other defenses in pea aphids, *Acyrthosiphon pisum. Genome Biol* **11**: R21.

Hultmark, D. (2003) Drosophila immunity: paths and patterns. *Curr Opin Immunol* **15**: 12–19.

Jiggins, F.M. and Kim, K.W. (2005) The evolution of antifungal peptides in Drosophila. *Genetics* **171**: 1847–59.

Jiggins, F.M. and Kim, K.W. (2006) Contrasting evolutionary patterns in Drosophila immune receptors. *J Evol Biol* **63**: 769–80.

Jiggins, F.M. and Kim, K.W. (2007) A screen for genes evolving under positive selection in Drosophila. *J Evol Biol* **20**: 965–70.

Kolaczkowski, B., Hupalo, D.N., and Kern, A.D. (2011) Recurrent adaptation in RNA interference genes across the Drosophila phylogeny. *Mol Biol Evol* **28**(2): 1033–42.

Lazzaro, B.P. (2005) Elevated polymorphism and divergence in the class C scavenger receptors of *Drosophila melanogaster* and *D. simulans. Genetics* **169**(4): 2023–34.

Lazzaro, B.P. (2008) Natural selection on the Drosophila antimicrobial immune system. *Curr Opin Microbiol* **11**(3): 284–9.

Lazzaro, B.P. and Clark, A.G. (2003) Molecular population genetics of inducible antibacterial peptide genes in *Drosophila melanogaster. Mol Biol Evol* **20**: 914–23.

Lee, Y.C. and Langley, C.H. (2010) Transposable elements in natural populations of *Drosophila melanogaster. Philos Trans R Soc Lond B Biol Sci* **365**(1544): 1219–28.

Lemaitre, B. and Hoffmann, J. (2007) The host defense of *Drosophila melanogaster. Annu Rev Immunol* **25**: 697–743.

Levine, M.T. and Begun, D.J. (2007) Comparative population genetics of the immunity gene, Relish: is adaptive evolution idiosyncratic? *PLoS ONE* **2**: e442.

Little, T.J. and Cobbe, N. (2005) The evolution of immune-related genes from disease carrying mosquitoes: diversity in a peptidoglycan and a thioester-recognizing protein. *Insect Mol Biol* **14**: 599–605.

Little, T.J., Colbourne, J.K., and Crease, T.J. (2004) Molecular evolution of Daphnia immunity genes: polymorphism in a gram-negative binding protein gene and an alpha-2-macroglobulin gene. *J Mol Evol* **59**: 498–506.

Lu, J. and Clark, A.G. (2010) Population dynamics of PIWI-interacting RNAs (piRNAs) and their targets in Drosophila. *Genome Res* **20**(2): 212–27.

Murphy, P.M. (1991) Molecular mimicry and the generation of host defense protein diversity. *Cell* **72**: 823–6.

Murphy, K.M., Travers, P., and Walport, M. (2007) *Janeway's Immunobiology*, 7th edition. London: Garland Science.

Obbard, D.J., Jiggins, F.M., Halligan, D.L., and Little, T.J. (2006) Natural selection drives extremely rapid evolution in antiviral RNAi genes. *Curr Biol* **16**: 580–5.

Obbard, D.J., Gordon, K.H.J., Buck, A.H., and Jiggins, F.M. (2009) The evolution of RNAi as a defence against viruses and transposable elements. *Philos Trans R Soc Lond B Biol Sci* **364**: 99–115.

Obbard, D.J., Jiggins, F.M., Bradshaw, N.J., and Little, T.J. (2011) Recent and recurrent selective sweeps of the antiviral RNAi gene *Argonaute-2* in three species of Drosophila. *Mol Biol Evol* **28**(2): 1043–56.

Sackton, T.B., Lazzaro, B.P., Schlenke, T.A., Evans, J.D., Hultmark, D., and Clark, A.G. (2007) Dynamic evolution of the innate immune system in Drosophila. *Nat Genet* **39**: 1461–8.

Schlenke, T.A. and Begun, D.J. (2003) Natural selection drives *Drosophila* immune system evolution. *Genetics* **164**: 1471–80.

Schmid-Hempel, P. (2008) Parasite immune evasion: a momentous molecular war. *Trends Ecol Evol* **23**: 318–26.

Somogyi, K., Sipos, B., Pénzes, Z., Kurucz, E., Zsámboki, J., Hultmark, D., et al. (2008) Evolution of genes and repeats in the Nimrod superfamily. *Mol Biol Evol* **25**: 2337–47.

Tennessen, J.A. (2005) Molecular evolution of animal antimicrobial peptides: widespread positive selection. *J Evol Biol* **18**: 1387–94.

van Mierlo, J.T., van Cleef, K.W., and van Rij, R.P. (2011) Defense and counterdefense in the RNAi-based antiviral immune system in insects. *Methods Mol Biol* **721**: 3–22.

Van Valen, L. (1973) A new evolutionary law. *Evol Theory* **1**: 1–30.

Waterhouse, R.M., Kriventseva, E.V., Meister, S., Xi, Z., Alvarez, K.S., Bartholomay, L.C., et al. (2007) Evolutionary dynamics of immune related genes and pathways in disease-vector mosquitoes. *Science* **316**: 1738–43.

Yang, W., Cheng, T., Ye, M., Deng, X., Yi, H., Huang, Y., et al. (2011) Functional divergence among silkworm antimicrobial peptide paralogs by the activities of recombinant proteins and the induced expression profiles. *PLoS ONE* **6**(3): e18109.

Zou, Z., Evans, J.D., Lu, Z., Zhao, P., Williams, M., Sumathipala, N., et al. (2007) Comparative genomic analysis of the *Tribolium* immune system. *Genome Biol* **8**: R177.

CHAPTER 21

Rapid evolution of the plague pathogen

Ruifu Yang, Yujun Cui, and Dongsheng Zhou

21.1 Introduction

Plague, one of the most devastating infections in the human history, is a zoonotic infection that spreads to humans from natural rodent reservoirs, commonly via bites of infected fleas. *Yersinia pestis*, the causative agent of plague, is a multihost and multivector pathogen, involving more than 200 species of wild rodents as hosts and over 80 species of fleas as vectors (Anisimov et al. 2004). Different hosts and vectors have their own specific ecological niches to inhabit. During its expansion and adaptation to new niches, *Y. pestis* could undergo rapid genetic changes in its genome in response to novel natural selective forces.

To fully understand its evolution, a large collection of *Y. pestis* strains from different time periods and regions around the world would be needed. However, due to its high virulence and its potential use in bioterrorism, obtaining sufficient samples of this organism has been difficult. Until recently, most genetic data about *Y. pestis* have come from a relatively small number of strains from Western Europe (Pourcel et al. 2004) and there had been a significant lack of data on strains from China and the former Soviet Union. This is unfortunate as most of the existing plague foci are located in these regions and is where *Y. pestis* is thought to have originated.

Two recent developments are changing our understanding of the evolution of this pathogen. The first is due to Dr Anisimov's effort in analyzing the strains and populations of *Y. pestis* from the former Soviet Union. The second is the recently published works on genome diversities of *Y. pestis* in China by our laboratory, thus helping bridging part of this gap (Song et al. 2004; Cui et al. 2008; Li et al. 2008, 2009; Zhou and Yang 2009). However, different studies have traditionally used different strain typing techniques. There is an urgent need to study the same batch of strains of *Y. pestis* representing different regions around the world using the same standardized typing schemes in order to fulfill the systematic understanding of the genome diversities of this deadly bacterium. The recent emergence and widespread application of high-throughput DNA sequencing technologies are making this possible. For example, using whole-genome sequencing and single nucleotide polymorphism (SNP) analysis, a relatively comprehensive set of strains of *Y. pestis* from different parts of the world was recently analyzed and the histories of its spread were proposed (Morelli et al. 2010).

Plague is a typical zoonosis pathogen and its long-term existence in the natural foci is impacted by the interactions between the hosts, the vectors, the pathogen genotype, and the environment (Zhou et al. 2004a). The plague pathogen is postulated to have appeared in the Mongolian bobak (*Marmota sibirica* Radde, 1862) populations in Central Asia during the Pleistocene (Suntsov and Suntsova 2000). The late Pleistocene cooling, which induced deep-freezing of the ground in southern Siberia, Mongolia, and Manchuria, likely played a significant role for its emergence. The main ecological factors of plague pathogen evolution likely included the species-specific behavior of the Mongolian bobak. The bobaks hibernate in the arid petrophytic landscapes in winter, bringing the larvae of the flea *Oropsylla silantiewi* Wagn. with them. After its emergence, *Y. pestis* expanded to other hosts, including synanthropic rats and gerbils, pikas, voles, and other kinds of marmots, to

Rapidly Evolving Genes and Genetic Systems. First Edition. Edited by Rama S. Singh, Jianping Xu, and Rob J. Kulathinal.

become a multihost pathogen. During its expansion to different hosts, *Y. pestis* likely modified its genome to adapt itself to new niches and gradually diversified to contain strains with different features. The diagnostic features include the capacity of the strains to ferment glycerol, reduce nitrate, and their virulence properties to humans. Indeed, these diversified strains are linked to the types of rodent hosts, geographical landscapes, and flea vectors.

Molecular phylogenetic analyses have demonstrated that *Y. pestis* split from *Y. pseudotuberculosis* very recently, from about 6500–20,000 years ago (Achtman et al. 2004). At present, the detailed mechanisms on how *Y. pestis* speciated from its ancestor remain incompletely understood. However, several processes are known to have played a role in its evolution. These include plasmid acquisition, horizontal gene transfers that include both gene acquisition and loss, and neutral or positive selection. In this chapter, we present the diversity data of *Y. pestis* published by different groups using different methods and correlate these data with the evolution of the *Y. pestis* genome and its pathogenicity. Together, these data demonstrate the rapid evolution of the plague pathogen since its recent emergence.

21.2 Plasmid acquisition in *Y. pestis*

There are different kinds of plasmids in strains and populations of *Y. pestis* from different regions around the world. Most of the strains contain three plasmids, i.e. pPCP1 (also designated pYP, pPla, or pPst, 9.5 kb or ~6 Mdal), pMT1 (also designated pFra, pTox or pYT, ~100 kb or ~65 Mdal) and pCD1 (also designated pYV, pCad, pLcr, or pVW, ~70 kb or ~45 Mdal). The first two plasmids, pPCP1 and pMT1, are unique to *Y. pestis* and their acquisitions are thought to play key roles in the evolution of the plague microbe. All pathogenic *Yersinia* strains contain the virulence-associated pCD1 plasmid, which encodes finely-tuned type III secretion machinery consisting of antiphagocytic factors. pPCP1 encodes the plasminogen activator and the bacteriocin pesticin. pMT1 is responsible for the synthesis of fraction 1 antigen and phospholipase D. The plasminogen activator is involved in the dissemination of the plague bacterium from the site of the initial fleabite to other sites, while phospholipase D (previously termed as a murine toxin) plays a major role in the survival of the plague bacteria in fleas.

Of the three plasmids, the greatest size variations are found in the pMT1 plasmid among the strains, followed by plasmid pCD1, and then pPCP1 (Dong and Yu 1994). Filippov et al. (1990) investigated plasmid content in 242 *Y. pestis* strains from various natural plague foci of the former Soviet Union and other countries. Of these strains, 172 (71%) were shown to carry three plasmids described previously. Twenty of them (8%) harbored additional cryptic plasmids, most often one about 20 MDa in size. Plasmid pPCP1 displayed considerable constancy of its molecular mass among the strains. On the contrary, significant size variations in pCD1 (45–49 MDal) and, especially, pMT1 (60–190 MDa) were found. Molecular masses of these plasmids showed a significant association with host origins.

There are also other plasmids in *Y. pestis* and many of them are geography specific. For example, a 6-kb cryptic plasmid (pYC; 5919 bp) was found in the *Y. pestis* isolates from Yunnan province in southwestern China. There is evidence that this plasmid is increasing in frequency in the southern regions of Yunnan province (Dong et al. 2000). Song et al. (2004) reported a 21,742-bp plasmid pCRY from an avirulent strain of *Y. pestis* from *Microtus brandti* in inner Mongolia.

There are quite a number of studies on plasmid variation among strains of *Y. pestis* from different regions in the former Soviet Union (Anisimov et al. 2004). Variations in plasmid contents and sizes have been used to classify strains of *Y. pestis* into different groups of plasmid variants (called plasmidovars), which are often associated, to a high but not exclusive degree, with strain source and phenotype, as shown earlier. These results indicate the potential utility of this method for epidemiological investigations and for contributing to the determination of the pathogenic potential of isolates.

Among the diverse genome evolution mechanisms, plasmid acquisition is probably the quickest and most economical way of evolving new capacities and novel genotypes. While scientists have obtained many profiles of *Y. pestis* plasmids and correlated them to strain origins, most research has focused on plasmids pPCP1 and pMT1 and the role of other region-specific plasmids in evolution has not been fully understood. By comparative analysis, pMT1 has more than 50% nucleotide sequence identity to a cryptic plasmid pHCM2 in *Salmonella typhi* (Prentice et al. 2001), indicating a recent shared common ancestry and/or the potential transfer of the plasmid between *Y. pseudotuberculosis* or *Y. pestis* and *Salmonella typhi* in fleas or hosts. Plasmid pCRY, encoding a cryptic type IV secretion system, was found to share a high nucleotide sequence similarity to plasmids harbored by members of the Enterobactericeae, such as p307 in *E. coli*, pGSH500 in *Klebsiella pneumoniae*, pYVe439-80 in *Y. enterocolitica*, and pCP301 in *Shigella flexneri* (Song et al. 2004). This data suggests frequent recent transfers of plasmids between *Y. pseudotuberculosis* or *Y. pestis* and Enterobactericeae members in fleas or hosts. The abundant evidence of plasmid transfers indicates that plasmid acquisitions have played key roles in the evolution of *Y. pestis*.

21.3 The impact of phages on genome structure

Bacteria are thought to be the most abundant organisms on our planet, but the number (10^7–10^8/mL) of viruses in the environments exceeds greatly that (10^6/mL) of bacteria. In the environment, phages play indispensable roles in ecology, and they can significantly shape the host bacterial genome and promote their spread and evolution. In some human or animal pathogenic bacteria, phages often bring virulence-associated genes to increase and/or diversify their host bacterial pathogen's virulence properties. For example, the Shiga-toxin-producing *Escherichia coli* O104: H4 responsible for the recent food-poisoning outbreak in Germany evolved by acquiring a phage carrying genes encoding the Shiga toxin 2 from enterohemolytic *E. coli* by a strain of the enteroaggregative *E. coli* (Rohde et al. 2011). Although we can not decipher exactly how phages shape bacterial genomes, comparisons of phage or phage-like sequences in bacterial genomes by bioinformatics analysis often yield surprising patterns.

21.4 Prophages in the *Y. pestis* genome

Four fragments of prophage sequences were annotated in seven publicly available *Y. pestis* complete genomes. Of these four prophage sequences, only one (prophage2) was found in the genome of *Y. pseudotuberculosis*, the most closely related species of *Y. pestis*. This result suggested that the other three prophages were likely acquired after speciation of *Y. pestis* from *Y. pseudotuberculosis* (Table 21.1). Prophages 1 and 3 were found in all seven *Y. pestis* genomes, suggesting that they were likely acquired at the early stage of *Y. pestis* evolution. Prophage4, also designated DFR13, encodes a filamentous

Table 21.1 Annotated prophages in *Y. pestis* genomes

Strain ID[a]	Accession number	Length (base pairs)					Percentage[b]
		Prophage1	Prophage2	Prophage3	Prophage4	Total	
CO92	NC_003143	9,931	17,246	46,368	8,758	82,303	1.64%
91001	NC_005810	9,645	17,264	10,331	0	37,240	0.66%
KIM	NC_004088	9,644	17,281	39,338	0	66,263	1.44%
Antiqua	NC_008150	9,929	17,281	33,300	0	60,510	1.29%
Nepal516	NC_008149	9,929	17,275	46,363	0	73,567	1.62%
Angola	NC_010159	9,432	17,282	50,290	0	77,004	1.71%
Pestoides_F	NC_009381	9,928	17,270	44,450	0	71,648	1.59%
IP32953	NC_006155	0	17,624	0	0	17,624	1.60%[c]

[a] IP32953 is an *Y. pseudotuberculosis* strain and the other seven are all *Y. pestis* strains

[b] Percentages were calculated as the total length of prophages divided the genome size of each strain

[c] This value included three IP32953-specific prophages that are not shown in this table

phage (*Ypfϕ*) and contributes to the pathogenicity of *Y. pestis* in mice. It has been shown that prophage 4 forms an unstable episome in strains of biovar Antiqua and biovar Mediavalis. However, this prophage is stably integrated into the chromosome of strains of biovar Orientalis as tandem repeats (Derbise et al. 2007). Population genetic studies have shown that prophage 4 is widely distributed in biovar Orientalis stains of *Y. pestis*, whereas very few strains in biovar Antiqua, Medievalis, and Microtus contained prophage4. For example, only eight of the 325 *Y. pestis* strains of other three biovars were found to contain prophage4 but all 52 biovar Orientalis strains tested were positive for this prophage (Li et al. 2008). These results suggested that prophage4 was acquired by the *Y. pestis* ancestor, but only fixed in the Orientalis lineages, likely due to selective advantages it conferred for this group of strains. This hypothesis is supported by the active expression of genes on prophage4. However, the genes in the other three prophages were not expressed and may be degenerating to defective bacteriophages or phage remnants (Table 21.2).

The sizes of prophages 1 and 2 are comparable across the seven completely sequenced *Y. pestis* genomes and their sequences are highly similar among the strains with low diversities (Table 21.1 and Fig. 21.1). The data indicate that these two phages may have played indispensable roles in *Y. pestis* physiology and genome evolution. However, sequences of prophage3 were highly polymorphic among the seven strains. The length of prophage3 ranged from 10–50kbp. In the Angola strain, the prophage3 sequence was split into two parts separated by 186 kbp. The large DNA insert fragment contained four insertion sequence (IS) elements, including an IS100, an IS1541, and two IS285, suggesting that the IS-introduced genomic changes can contribute to significant strain variation among prophage3 sequences. The observation that prophage3 sequences could be partly lost or interrupted in strains of *Y. pestis* indicated that while this phage might have played some roles in *Y. pestis* evolutionary history, it is most likely nonessential and dispensable.

The comparative genomics results of prophages in *Y. pestis* revealed evidence for an arms race and coevolution between phages and host bacteria. Virulent phages can lyse susceptible host bacteria without leaving their signatures on surviving bacterial genomes. However, for lysogenic phages such as prophages 1 and 2 in *Y. pestis*, the phage genomes can integrate into the genome of the host bacterium, and maintain its structure and function if it can benefit the host bacterial population. Signatures of this 'win–win' model have been observed through analyzing prophage-like sequences in the target bacterial genomes. For phages that are not harmful but also with no apparent benefit for their host, although they can temporarily integrate into the genome of the host bacterium, they will be jettisoned eventually during evolution, such as what has occurred for prophage3 in *Y. pestis*.

21.5 CRISPRs diversity and the battle between phage and *Y. pestis*

Clustered, regularly interspaced short palindromic repeats (CRISPRs), a family of repeat elements that typically consist of non-contiguous direct repeats

Table 21.2 Characteristics of *Y. pestis* prophages

ID	Start–end (in the CO92 genome)		Note	Alternate designations[a]
	CDS	Position		
Prophage1	YPO1087–1098	1234752–1244683	Phage remnant	GI09
Prophage2	YPO1239–1252	1398814–1410182	Defective Mu-like bacteriophage	GI10
Prophage3	YPO2084–2140	2363016–2409384	Defective lambdoid bacteriophage	DFR12
Prophage4	YPO2271–2281	2554162–2562920	Filamentous phage	DFR13, *Ypfϕ*

[a] GI, genome island

GI and DFR designations here follow those in Zhou et al. (2004b), and *Ypfϕ* is from Derbise et al. (2007)

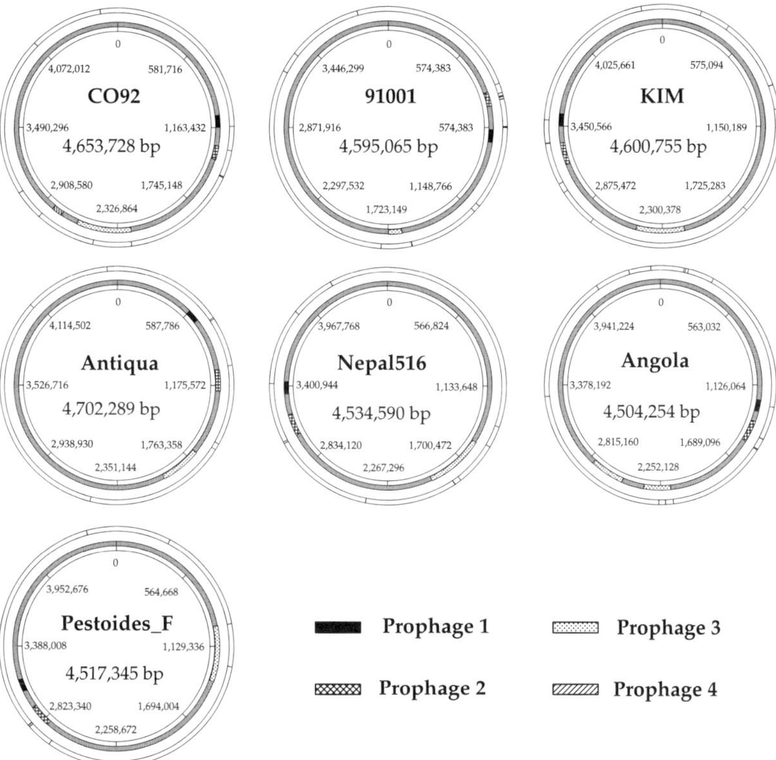

Figure 21.1 The distributions of prophages among *Y. pestis* genomes. The bars on the outside circle of each genome indicate the positions of dispersed prophage-associated genes. The second circles from outside show the positions and lengths of the four prophages described in Tables 21.1 and 21.2. The lengths of the prophages were magnified ten times relatively to other genomic regions for easy visualization. Therefore, the prophages and some adjacent genes seemed overlapped. The innermost circles indicate the genomic locations relative to the origin of replication, at '0.'

(DR, 24–47 bp) separated by stretches of similarly-sized unique sequences, are widely distributed in the genomes of both Bacteria and Archaea (Horvath and Barrangou 2010). CRISPRs that are associated with Cas (CRISPR-associated) proteins and leader sequences (the non-coding sequences flanking the CRISPRs on one side and acting as a promoter) constitute a prokaryotic immune system against bacteriophage attacks via a RNA interference (RNAi)-like mechanism (Marraffini and Sontheimer 2010). The DR sequences of CRISPRs are usually conserved among strains of the same species. DR sequences of different species can be organized into multiple clusters. The sequences of some clusters have the tendency to form stable, conserved RNA secondary structures, which may be involved in implementing the functions of CRISPRs (Kunin et al. 2007). The unique parts, also called 'spacers', were acquired from invading foreign genetic elements and preserved in CRISPR loci. These 'spacers' can be used as templates to provide resistance to phages or to conjugative plasmids that contain sequences homologous to the spacers (Garneau et al. 2010). The spacer regions in CRISPRs seem to be one of the most rapidly diversified regions in prokaryotic genomes, and even in the same species, different spacer arrays often exist in different isolates (Cui et al. 2008). The highly polymorphic CRISPRs in different isolates has provided a powerful set of markers for microevolutionary studies, and laid a foundation for 'spoligotyping,' which had been widely used in genotyping strains of *Mycobacterium tuberculosis*.

Three types of CRISPRs loci, designated as YPa, YPb, and YPc, have been identified in the chromosome of *Y. pestis* (Pourcel et al. 2005). Our laboratory collected 125 strains of *Y. pestis* from 26 natural plague foci of China, the former Soviet Union, and Mongolia. We sequenced and analyzed their CRISPRs loci variation. In the tested 375 CRISPRs loci (125 strains × 3 loci each), a total of 131 spacers were found, and 77 (59%) of them had sequences homologous to a prophage (YPO2096 ~2135 in CO92 genome), whereas 22 spacers (17%) were homologous to other non-viral regions in the *Y. pestis* chromosome. There were 83, 37, and 11 spacers respectively from the YPa, YPb, and YPc loci, suggesting that the three CRISPRs of *Y. pestis* had different activities and the YPa locus was the most active of the three. The lengths of spacers varied between 29–34 bp, with most at 32 bp (79%). The largest CRISPRs locus found until now in *Y. pestis* contained 14 spacers, while some contain only one spacer. Interestingly, the YPb and YPc loci of the Angola strain (GenBank accession number: NC_010159) contained only one truncated DR and the leader sequence, and no spacer was observed at the two loci (Cui et al. 2008). The diversity of CRISPRs has provided good markers for genotyping *Y. pestis*. For example, plague natural foci L and M (one in Inner Mongolia and the other in Qinghai province) are geographically distinct in China. However, the *Y. pestis* isolates from these two regions could not be distinguished by their phenotypes, biochemical features, or by other conventional phenotyping methods (Zhou et al. 2004b; Li et al. 2008). We found that there was a 120-bp difference (the length of two spacers as well as the DR sequences) at the YPc locus between strains of these two foci. This size difference could be easily used to differentiate them by conventional polymerase chain reaction (PCR)-gel electrophoresis method (Cui et al. 2008).

The highly variable spacer compositions in CRISPRs loci have provided high-resolution markers for strain genotyping. In addition, the variation patterns of CRISPRs also recorded the interactions between prokaryotes and phages. In a study on 109 CRISPR loci from 61 *Y. pestis* strains, Pourcel et al. (2005) generalized the following rules of CRISPR evolution: (1) extrachromosomal gene elements can be recognized rapidly and inserted into CRISPR loci as new spacers to enhance the host cell's immunity; (2) the addition of new spacers is always in the leader sequence side of the CRISPR region; and (3) one or more existing spacers can be deleted randomly, and 'useful' spacers will likely be kept by natural selection. The patterns of sequence variation at CRISPR loci would record part of foreign genetic elements (mostly coming from phages) with which the prokaryotic organisms had interacted during their evolutionary history. As the addition of new spacers is polarized, information preserved in spacer arrays will also show the directional evolutionary traces within one species. Based on these three rules, we predicted and tested the following evolutionary scenario of *Y. pestis* in China. Firstly, since the CRISPRs are shaped by interactions between bacteria and their local ecological niches (especially the phage profiles in different foci), the distribution of spacers and their compositions in *Y. pestis* strains should be strongly correlated with the distribution of natural plague foci. Secondly, based on the genotype information at the CRISPRs, the transmission route of an ancient lineage of *Y. pestis*, the microtus strains, could be inferred. This hypothetical route encircled the Takla Makan Desert and ZhunGer Basin. Our predictions were largely confirmed. The route of expansion shown in this study started from Tajikistan, with one population passing through the Kunlun Mountains, and moved to the Qinghai-Tibet Plateau. Another population headed north via the Pamirs Plateau, the Tianshan Mountains, the Altai Mountains, and the Inner Mongolian Plateau. Other *Y. pestis* genotypes and lineages might emerge from along these routes. The inferred patterns were supported by results using other molecular typing methods, such as MLVA and DFR (Pourcel et al. 2005; Li et al. 2008, 2009), indicating that CRISPRs can be used as high-resolution evolutionary research and genotyping tools.

21.6 Gene acquisition, loss, and inactivation

Darwinian (positive) selection changes allele frequencies in populations due to their effects on population fitness. The selective pressures can be observed not only on genotypes but also phenotypes of populations (Zhou and Yang 2009). Over time, this process will result in the adaptation of organisms to specialize in particular niches, and eventually lead to the emergence of new species. Acquired DNA fragments from other sources will be retained in the bacterial genome if they confer an advantage to the host cells. If the acquired fragment confers no beneficial function or is deleterious, their inactivation and/or deletion will eventually occur. Selective pressures on inactivation, loss, and acquisition of functional genes all follow the principles of Darwinian evolution (Zhou and Yang 2009). The selection of small allelic variants (e.g. gene inactivation) may confer small evolutionary changes, while the accumulation of large genomic changes (gene loss or acquisition) can alter phenotypes in 'quantum leaps'.

Evidences of gene acquisition, loss, and inactivation have been demonstrated in the genomes of *Y. pestis* (Parkhill et al. 2001; Deng et al. 2002; Song et al. 2004). Compared to its progenitor *Y. pseudotuberculosis* (Chain et al. 2004), about 10% of the genes of *Y. pestis* are inactivated or absent (Parkhill et al. 2001; Deng et al. 2002; Song et al. 2004). Since *Y. pestis* spends its life almost exclusively in a flea–reservoir–flea cycle, the organism could accumulate the inactivated genes encoding enteropathogenic determinants required for its progenitor *Y. pseudotuberculosis* (Simonet et al. 1996). In addition, gene loss and inactivation are closely linked to flea-borne transmission (Sun et al. 2008) and increased virulence (Montminy et al. 2006) in *Y. pestis* (Zhou and Yang 2009). In comparison to its closest relative *Y. pseudotuberculosis*, *Y. pestis* has acquired two unique virulence plasmids (pPCP1 and pMT1) (Parkhill et al. 2001; Chain et al. 2004) and probably two chromosomal regions through horizontal gene transfer (Wang et al. 2007). Indeed, pPCP1 and pMT1 encode a set of determinants for virulence and transmission, such as plasminogen activator (Pla), murine toxin (Ymt), and F1 capsule. In summary, gene acquisition, loss, and inactivation all greatly promoted the recent emergence and rapid diversification of *Y. pestis* within natural plague foci (Zhou et al. 2004b; Li et al. 2008).

21.7 Rearrangements and copy number variants

Chromosomal rearrangements and copy number variants (CNVs) have been shown to play key roles in genome evolution and genetic diseases of eukaryotes (Volker et al. 2010). However, how these variations promote bacterial evolution remains largely unknown. Using pulsed field gel electrophoresis (PFGE) and rRNA gene cluster digests by rare endonuclease enzymes, certain large genome structural variants could be detected (Tsuru et al. 2006). PFGE, which separates large DNA fragments (e.g. derived through the digestion of the bacterial chromosome with restriction endonucleases that cleave infrequently) (Smith and Condemine 1990) can facilitate detecting chromosomal rearrangements. The method has been highly effective in molecular epidemiological studies of bacterial isolates, and can be superior to other methods in discriminating among isolates in many common pathogens such as *Escherichia coli* and *Staphylococcus aureus* (Arbeit et al. 1990). For example, the method has been used to evaluate the clonal relatedness among bacterial isolates and to investigate outbreaks (Bercovier et al. 1979). It has also proved effective for qualitative evaluation of intraspecific genetic variation, permitting identification of individual isolates within a given species by comparing their macrorestriction patterns (Filippov et al. 1995).

Using the PFGE techniques, Lucier and Brubaker (1992) found that the *Spe*Idigested DNA patterns of eight *Y. pestis* strains were closely related to their respective biovars. Similar results were obtained by Rakin and Heesemann (1995) using the I-*Ceu*I digested macrorestriction patterns for nine *Y. pestis* strains and by Huang et al. (2002) using the restriction enzyme *Spe*I.

The choice of an appropriate macrorestriction enzyme is critical for PFGE analysis. Huang et al. compared three different enzymes (*Not*I, *Sfi*I, and *Spe*I) for generating genomic restriction patterns. *Not*I digestion generated closely grouped bands of high molecular weight, similar to that reported by Lucier and Brubaker (1992), with many comigrating fragments. In contrast, *Sfi*I digestion produced many closely grouped smaller-molecular-weight bands. However, *Spe*I produced a relatively wider range of DNA fragments that could easily be resolved by PFGE separation conditions used by Huang et al. Therefore, *Spe*I digestion seems the best of the three for PFGE analysis of *Y. pestis*.

For evaluating the extent of variability of the pulsotype by PFGE within one strain, Guiyoule et al. (2004) randomly picked eight and four colonies from the stock culture of strain Saigon 55–1239 and Kenya 169, respectively, and their genomes were subjected to PFGE. Five and three different *Spe*I restriction patterns were observed for each of these two strains, indicating a high heterogeneity of the pulsotypes within a given strain. They also found similar phenomena by using other restriction enzymes, such as *Xba*I or *Not*I. These results pointed to the hypervariability of chromosome structures in *Y. pestis* and the potential problems of relying only on PFGE for strain typing of *Y. pestis*. In our lab, while the potential PFGE variability among colonies of the same strain has not been investigated, we did find different FAFLP (fluorescent amplified fragment length polymorphism) profiles among three subcultures of the same strain (Pei et al. 2004).

The number of transposable elements, including ISs, can impact genome plasticity by mediating genetic rearrangements in bacterium (Bickhart et al. 2009). For example, IS can mediate the spontaneous deletion of a chromosomal region of *Y. pestis* KIM6+. This region contains the pigmentation locus, genes for pesticin sensitivity, and the HMWP2 virulence locus (Fetherston and Perry 1994). Whole-genome sequencing of *Y. pestis* revealed abundant insertion sequences in their genome with different copy numbers of IS element (Parkhill et al. 2001; Deng et al. 2002; Song et al. 2004). For IS100 in the first three sequenced *Y. pestis* strains, their copy numbers in individual strains were (number of IS) is: CO92 (44) > KIM (35) > 91001 (30); for IS 1541: CO92 (62) > KIM (49) > 91001 (43); for IS 285: 91001 (23)> CO92 (21) > KIM (19); and for IS1661: KIM (10)> CO92 (9) > 91001(8).

Copy number variants associated with tandem repeats has been instrumental in mammalian genetics for the construction of genetic maps and is one of the main bases of DNA fingerprinting in forensic applications. Tandem repeats are usually classified into satellites (spanning megabases of DNA, associated with heterochromatin), minisatellites (repeat units in the range 6–100 bp, spanning hundreds of base pairs) and microsatellites (repeat units in the range 1–5 bp, spanning a few tens of nucleotides). Tandem repeat polymorphisms of mini- and microsatellites are a highly significant source of very informative markers for the identification of pathogenic bacteria, especially for the recently emerged, highly monomorphic species such as *Y. pestis* (Kim et al. 2002; Banu et al. 2004). Tandem repeat variants probably contribute to a pathogen's adaptation to its host, and they can also account for bacterial phenotypic variations depending on their location in the genome. For example, if a tandem repeat is located within the regulatory region of a gene, it can play an on/off switch role in gene expression at the transcriptional level (Le Fleche et al. 2001). Similarly, if they are located within coding regions with repeat unit length of three bases they may cause the antigen variations for that bacterium, but if the repeat unit length is not a multiple of three bases, they can induce a truncated translation product.

VNTR sequences are common in the *Y. pestis* genome and occur frequently in protein-coding gene regions, at an average of 2.18 arrays per 10 kbp. These sequences are distributed evenly throughout both the genome and the two virulence plasmids, pCD1 and pMT1 (Klevytska et al. 2001). Strain typing based on tandem-repeat variation may be a powerful complement to the existing phylogenetic tools for *Y. pestis*. For example, Adair et al. (2000) identified a tetranucleotide repeat sequence, $(CAAA)_N$, in the genome of *Y. pestis*, and demonstrated that this region had nine different alleles and a high gene diversity value of 0.82 (calculated as 1 minus the sum of the squared allele frequencies) within a set of 35 diverse *Y. pestis* strains.

Multiple-locus VNTR analysis (MLVA) was shown capable of both distinguishing closely related strains and successfully classifying more distant relationships. Pourcel et al. (2004) and Klevytska et al. (2001) examined representative strains of *Y. pestis* using 25 and 42 VNTR loci respectively, and they found vast differences in gene diversity among these loci. Pourcel et al. (2004) grouped 180 *Y. pestis* into 61 different genotypes. These genotypes were distributed in the three biovars, with biovar *Medievalis* showing a very high heterogeneity. They also proposed the seven most informative VNTR markers for rapid characterization of new strains. Compared to other genotyping methods, MLVA is easily standardized for establishing databases. This is very important because

Y. pestis is one of the most dangerous bioterrorism agents and international exchange of *Y. pestis* strains is almost impossible. As a result, web-based comparison and identification of VNTR genotypes of *Y. pestis* would be ideal.

Le Flèche et al. (2001) created a database (http://minisatellites.u-psud.fr) of tandem repeats for pathogenic bacteria based on publicly available bacterial genomes and illustrated its application by the characterization of minisatellites from two important human pathogens, *Y. pestis* and *Bacillus anthracis*. They found that *Y. pestis* contains 64 minisatellites with each repeat unit at least 9 bp long, and with each unit repeated at least seven times.

Denoeud et al. (2004) then presented an Internet-based resource to help develop and perform tandem repeats-based bacterial strain typing. The tools are accessible through the Web link provided earlier. There are four parts to the web page: the 'Tandem Repeats Database' enables the identification of tandem repeats across entire genomes. The 'Strain Comparison Page' identifies tandem repeats that differ among genome sequences of strains from the same species. The 'Blast in the Tandem Repeats Database' facilitates the search for a known tandem repeat and the prediction of amplification product sizes across large taxonomic groups. The 'Bacterial Genotyping Page' is a service for strain identification at the subspecies level.

21.8 Neutral versus adaptive evolution

Y. pestis has encountered a diverse array of habitats, e.g. from the Qinghai-Tibet plateau to deserts and grasslands across many parts of China. The pathogens from different plague foci have their own characteristics, suggesting adaptation to local niches.

Genetic variations, including base substitution, gene loss, gene acquisition, duplication, insertion, and genome rearrangement, can occur randomly. A deleterious genetic variant has a negative effect on the phenotype, and thus decreases the fitness of the organism, which will be removed by purifying selection. A neutral one has no harmful or beneficial effect on the organism. Being fixed by genetic drift, neutral genetic variations occurs in a population at a steady rate due to random sampling and chance, and forms the basis for the molecular clock hypothesis. The genetic or phenotypic changes due to drift are not driven by environmental or adaptive pressures, and may be neutral to reproductive success. According to the theory of neutral evolution, the majority of sequences in the genome evolve under purifying selection and genetic drift, with only a small fraction of genetic variation actually being beneficial and fixed by positive (Darwinian) selection. Positive selection makes gene variants more or less common depending on their contributions to reproductive success. A beneficial genetic variation has a positive effect on the phenotype, and thus increases the fitness of the organism. Indeed, a single event of fixed beneficial genetic variation could lead to dramatic evolution of phenotype differences.

Genetic variations beneficial to mammalian blood-borne infection or vector-borne transmission by *Y. pestis* would be stabilized by vertical inheritance under positive selection. Darwinian adaptive evolution could select for *Y. pestis* to diverge from *Y. pseudotuberculosis* to a new emerging pathogen that was not only able to parasitize insects in part of their life cycles, but also being highly virulent to rodents and humans, causing pandemics of a systemic and often fatal disease. Survival of *Y. pestis* in nature primarily depends on rodents and fleas, while fleas parasitize rodents and act as vectors for bacterial transmission. Natural environments in various plague foci will have distinct sets of rodents and fleas. Positive selection and the inheritance of beneficial mutations bring about adaptive change, by which favorable genotypes and phenotypes become more common in the reproducing populations of an organism. For example, the pseudogenization of rcsA driven by positive selection allowed the formation of *Y. pestis* biofilms, which enhanced the transmission of the bacteria (Zhang 2008). Similarly, gene acquisition, loss, and inactivation could promote parallel diversification of *Y. pestis* in different plague foci, which is reflected by expansion of various plague foci. Overall, the complex interactions between the environment, the hosts, and *Y. pestis* could all contribute to the microevolution of *Y. pestis* (Zhou et al. 2004b; Tong et al. 2005; Li et al. 2008).

21.9 Conclusions

The evolution of *Y. pestis* is an important topic for genome studies and will remain so in the foreseeable future. Such analyses will help us reveal the evolutionary mechanisms. Correlating studies of genetic variations with functions could facilitate the identification of selective forces and sequencing large collections of bacterial strains from different plague foci could help us reveal the nature of their mechanisms. Plague outbreaks have impacted humanity significantly, with three plague pandemics in recent recorded history. As was shown recently, samples from buried plague victims' bones can be analyzed for *Y. pestis* DNA to obtain an ancient pandemic strain's genetic background to help us understand *Y. pestis* evolution (Papagrigorakis et al. 2006; Bos et al. 2011; Schuenemann et al. 2011). These and other studies will continue to shed more light on the patterns and rates of *Y. pestis* evolution through the ages and across host groups, vector types, and geographic regions.

References

Achtman, M., Morelli, G., Zhu, P., Wirth, T., Diehl, I., Kusecek, B., et al. (2004) Microevolution and history of the plague bacillus, Yersinia pestis. *Proc Natl Acad Sci U S A* **101**(51): 17837–42.

Adair, D.M., Worsham, P.L., Hill, K.K., Klevytska, A.M., Jackson, P.J., Friedlander, A.M., et al. (2000) Diversity in a variable-number tandem repeat from Yersinia pestis. *J Clin Microbiol* **38**(4): 1516–19.

Anisimov, A.P., Lindler, L.E., and Pier, G.B. (2004) Intraspecific diversity of Yersinia pestis. *Clin Microbiol Rev* **17**(2): 434–64.

Arbeit, R.D., Arthur, M., Dunn, R., Kim, C., Selander, R.K., and Goldstein, R. (1990) Resolution of recent evolutionary divergence among Escherichia coli from related lineages: the application of pulsed field electrophoresis to molecular epidemiology. *J Infect Dis* **161**(2): 230–5.

Banu, S., Gordon, S.V., Palmer, S., Islam, M.R., Ahmed, S., Alam, K.M., et al. (2004) Genotypic analysis of Mycobacterium tuberculosis in Bangladesh and prevalence of the Beijing strain. *J Clin Microbiol* 2004, **42**(2): 674–82.

Bercovier, H., Alonso, J.M., Bentaiba, Z.N., Brault, J., and Mollaret, H.H. (1979) Contribution to the definition and the taxonomy of Yersinia enterocolitica. *Contrib Microbiol Immunol* **5**: 12–22.

Bickhart, D.M., Gogarten, J.P., Lapierre, P., Tisa, L.S., Normand, P., and Benson, D.R. (2009) Insertion sequence content reflects genome plasticity in strains of the root nodule actinobacterium Frankia. *BMC Genomics* **10**: 468.

Bos, K.I., Schuenemann, V.J, Golding, G.B, Burbano, H.A, Waglechner, N., Coombes, B.K., et al. (2011) A draft genome of Yersinia pestis from victims of the Black Death. *Nature* **478**: 506–10.

Chain, P.S., Carniel, E., Larimer, F.W., Lamerdin, J., Stoutland, P.O., Regala, W.M., et al. (2004) Insights into the evolution of Yersinia pestis through whole-genome comparison with Yersinia pseudotuberculosis. *Proc Natl Acad Sci U S A* **101**(38): 13826–31.

Cui, Y., Li, Y., Gorge, O., Platonov, M.E., Yan, Y., Guo, Z., et al. (2008) Insight into microevolution of Yersinia pestis by clustered regularly interspaced short palindromic repeats. *PLoS One* **3**(7): e2652.

Deng, W., Burland, V., Plunkett, G., 3rd, Boutin, A., Mayhew, G.F., Liss, P., et al. (2002) Genome sequence of Yersinia pestis KIM. *J Bacteriol* **184**(16): 4601–11.

Denoeud, F. and Vergnaud, G. (2004) Identification of polymorphic tandem repeats by direct comparison of genome sequence from different bacterial strains: a web-based resource. *BMC Bioinformatics* **5**(1): 4.

Derbise, A., Chenal-Francisque, V., Pouillot, F., Fayolle, C., Prevost, M.C., Medigue, C., et al. (2007) A horizontally acquired filamentous phage contributes to the pathogenicity of the plague bacillus. *Mol Microbiol* **63**(4): 1145–57.

Dong, X. and Yu, D. (1994) Plasmids in Yersinia pestis: functions and their role in epidemiology. *Yu Fang Yi Xue Qing Bao Za Zhi* **10**(3): 138–44.

Dong, X.Q., Lindler, L.E., and Chu, M.C. (2000) Complete DNA sequence and analysis of an emerging cryptic plasmid isolated from Yersinia pestis. *Plasmid* **43**(2): 144–8.

Fetherston, J.D. and Perry, R.D. (1994) The pigmentation locus of Yersinia pestis KIM6+ is flanked by an insertion sequence and includes the structural genes for pesticin sensitivity and HMWP2. *Mol Microbiol* **13**(4): 697–708.

Filippov, A.A., Solodovnikov, N.S., Kookleva, L.M., and Protsenko, O.A. (1990) Plasmid content in Yersinia pestis strains of different origin. *FEMS Microbiol Lett* **55**(1–2): 45–8.

Filippov, A.A., Oleinikov, P.V., Motin, V.L., Protsenko, O.A., and Smirnov, G.B. (1995) Sequencing of two Yersinia pestis IS elements, IS285 and IS100. *Contrib Microbiol Immunol* **13**: 306–9.

Garneau, J.E., Dupuis, M.E., Villion, M., Romero, D.A., Barrangou, R., Boyaval, P., et al. (2010) The CRISPR/Cas bacterial immune system cleaves bacteriophage and plasmid DNA. *Nature* **468**(7320): 67–71.

Guiyoule, A., Grimont, F., Iteman, I., Grimont, P.A., Lefevre, M., and Carniel, E. (1994) Plague pandemics investigated by ribotyping of Yersinia pestis strains. *J Clin Microbiol* **32**(3): 634–41.

Horvath, P. and Barrangou, R. (2010) CRISPR/Cas, the immune system of bacteria and archaea. *Science* **327**(5962): 167–70.

Huang, X.Z., Chu, M.C., Engelthaler, D.M, and Lindler, L.E. (2002) Genotyping of a homogeneous group of Yersinia pestis strains isolated in the United States. *J Clin Microbiol* **40**: 1164–73.

Kim, W., Hong, Y.P., Yoo, J.H., Lee, W.B., Choi, C.S., and Chung, S.I. (2002) Genetic relationships of Bacillus anthracis and closely related species based on variable-number tandem repeat analysis and BOX-PCR genomic fingerprinting. *FEMS Microbiol Lett* **207**(1): 21–7.

Klevytska, A.M., Price, L.B., Schupp, J.M., Worsham, P.L., Wong, J., and Keim, P. (2001) Identification and characterization of variable-number tandem repeats in the Yersinia pestis genome. *J Clin Microbiol* **39**(9): 3179–85.

Kunin, V., Sorek, R., and Hugenholtz, P. (2007) Evolutionary conservation of sequence and secondary structures in CRISPR repeats. *Genome Biol* **8**(4): R61.

Le Fleche, P., Hauck, Y., Onteniente, L., Prieur, A., Denoeud, F., Ramisse, V., et al. (2001) A tandem repeats database for bacterial genomes: application to the genotyping of Yersinia pestis and Bacillus anthracis. *BMC Microbiol* **1**(1): 2.

Li, Y., Dai, E., Cui, Y., Li, M., Zhang, Y., Wu, M., et al. (2008) Different region analysis for genotyping Yersinia pestis isolates from China. *PLoS One* **3**(5): e2166.

Li, Y., Cui, Y., Hauck, Y., Platonov, M.E., Dai, E., Song, Y., et al. (2009) Genotyping and phylogenetic analysis of Yersinia pestis by MLVA: insights into the worldwide expansion of Central Asia plague foci. *PLoS One* **4**(6): e6000.

Lucier, T.S. and Brubaker, R.R. (1992) Determination of genome size, macrorestriction pattern polymorphism, and nonpigmentation-specific deletion in Yersinia pestis by pulsed-field gel electrophoresis. *J Bacteriol* **174**(7): 2078–86.

Marraffini, L.A. and Sontheimer, E.J. (2010) CRISPR interference: RNA-directed adaptive immunity in bacteria and archaea. *Nat Rev Genet* **11**(3): 181–90.

Montminy, S.W., Khan, N., McGrath, S., Walkowicz, M.J., Sharp, F., Conlon, J.E., et al. (2006) Virulence factors of Yersinia pestis are overcome by a strong lipopolysaccharide response. *Nat Immunol* **7**(10): 1066–73.

Morelli, G., Song, Y., Mazzoni, C.J., Eppinger, M., Roumagnac, P., Wagner, D.M., et al. (2010) Yersinia pestis genome sequencing identifies patterns of global phylogenetic diversity. *Nat Genet* **42**(12): 1140–3.

Papagrigorakis, M.J., Yapijakis, C., Synodinos, P.N., and Baziotopoulou-Valavani, E. (2006) DNA examination of ancient dental pulp incriminates typhoid fever as a probable cause of the Plague of Athens. *Int J Infect Dis* **10**: 206–14.

Parkhill, J., Wren, B.W., Thomson, N.R., Titball, R.W., Holden, M.T., Prentice, M.B., et al. (2001) Genome sequence of Yersinia pestis, the causative agent of plague. *Nature* **413**(6855): 523–7.

Pei, D., Pang, X., Song, Y., Zhai, J., Chen, Z., Liu, H., et al. (2004) Fluorescent amplified fragment length polymorphism for genotyping Yersinia pestis. *Chi J End* **23**(3): 210–14.

Pourcel, C., Andre-Mazeaud, F., Neubauer, H., Ramisse, F., and Vergnaud, G. (2004) Tandem repeats analysis for the high resolution phylogenetic analysis of Yersinia pestis. *BMC Microbiol* **4**(1): 22.

Pourcel, C., Salvignol, G., and Vergnaud, G. (2005) CRISPR elements in Yersinia pestis acquire new repeats by preferential uptake of bacteriophage DNA, and provide additional tools for evolutionary studies. *Microbiology* **151**(Pt 3): 653–63.

Prentice, M.B., James, K.D., Parkhill, J., Baker, S.G., Stevens, K., Simmonds, M.N., et al. (2001) Yersinia pestis pFra shows biovar-specific differences and recent common ancestry with a Salmonella enterica serovar Typhi plasmid. *J Bacteriol* **183**(8): 2586–94.

Rakin, A. and Heesemann, J. (1995) The established Yersinia pestis biovars are characterized by typical patterns of I-CeuI restriction fragment length polymorphism. *Mol Gen Mikrobiol Virusol* **3**: 26–9.

Rohde, H., Qin, J., Cui, Y., Li, D., Loman, N.J., Hentschke, M., et al. (2011) Open-source genomic analysis of Shiga-toxin-producing E. coli O104: H4. *N Engl J Med* **365**(8): 718–24.

Schuenemann, V. J., Bos, K., Dewitte, S., Schmedes, S., Jamieson, J., Mittnik, A., et al. (2011) From the cover: Targeted enrichment of ancient pathogens yielding the pPCP1 plasmid of Yersinia pestis from victims of the Black Death. *Proc Natl Acad Sci U S A* **108**: E746–52.

Simonet, M., Riot, B., Fortineau, N., and Berche, P. (1996) Invasin production by Yersinia pestis is abolished by insertion of an IS200-like element within the inv gene. *Infect Immun* **64**(1): 375–9.

Smith, C.L. and Condemine, G. (1990) New approaches for physical mapping of small genomes. *J Bacteriol* **172**(3): 1167–72.

Song, Y., Tong, Z., Wang, J., Wang, L., Guo, Z., Han, Y., et al. (2004) Complete genome sequence of Yersinia pestis strain 91001, an isolate avirulent to humans. *DNA Res* **11**(3): 179–97.

Sun, Y.C., Hinnebusch, B.J., and Darby, C. (2008) Experimental evidence for negative selection in the evolution of a Yersinia pestis pseudogene. *Proc Natl Acad Sci U S A* **105**(23): 8097–101.

Suntsov, V.V. and Suntsova, N.I. (2000) Ecological and geographical aspects of the plague agent Yersinia pestis speciation. *Dokl Biol Sci* **370**(1-6): 74–6.

Tong, Z., Zhou, D., Song, Y., Zhang, L., Pei, D., Han, Y., et al. (2005) Pseudogene accumulation might promote the adaptive microevolution of Yersinia pestis. *J Med Microbiol* **54**(Pt 3): 259–68.

Tsuru, T., Kawai, M., Mizutani-Ui, Y., Uchiyama, I., and Kobayashi, I. (2006) Evolution of paralogous genes: Reconstruction of genome rearrangements through comparison of multiple genomes within Staphylococcus aureus. *Mol Biol Evol* **23**(6): 1269–85.

Volker, M., Backstrom, N., Skinner, B.M., Langley, E.J., Bunzey, S.K., Ellegren, H., et al. (2010) Copy number variation, chromosome rearrangement, and their association with recombination during avian evolution. *Genome Res* **20**(4): 503–11.

Wang, X., Han, Y., Li, Y., Guo, Z., Song, Y., Tan, Y., et al. (2007) Yersinia genome diversity disclosed by Yersinia pestis genome-wide DNA microarray. *Can J Microbiol* **53**(11): 1211–21.

Zhang, J. (2008) Positive selection, not negative selection, in the pseudogenization of rcsA in Yersinia pestis. *Proc Natl Acad Sci U S A* **105**(42): E69; author reply E70.

Zhou, D. and Yang, R. (2009) Molecular Darwinian evolution of virulence in Yersinia pestis. *Infect Immun* **77**(6): 2242–50.

Zhou, D., Han, Y., Song, Y., Huang, P., and Yang, R. (2004a) Comparative and evolutionary genomics of Yersinia pestis. *Microbes Infect* **6**(13): 1226–34.

Zhou, D., Han, Y., Song, Y., Tong, Z., Wang, J., Guo, Z., et al. (2004b) DNA microarray analysis of genome dynamics in Yersinia pestis: insights into bacterial genome microevolution and niche adaptation. *J Bacteriol* **186**(15): 5138–46.

CHAPTER 22

Evolution of human erythrocyte-specific genes involved in malaria susceptibility

Wen-Ya Ko, Felicia Gomez, and Sarah A. Tishkoff

22.1 Introduction

Malaria is a mosquito-borne blood infection caused by apicomplexan parasites of the genus *Plasmodium*. It is a severe infectious disease prevalent in tropical and subtropical areas including sub-Saharan Africa, South Asia, and the Americas, resulting in 190–311 million clinical cases and 1 million deaths per year (World Health Organization 2008). Some of the earliest theoretical evolutionary perspectives of malaria infection in human populations were initiated by J.B.S. Haldane, who questioned the unreasonably high mutation rate inferred for mutations causing thalassemia under the assumption of mutation-selection balance for a recessive lethal mutation (Haldane 1949). Haldane proposed the possibility of heterozygote advantage as an alternative hypothesis to account for the high incidence of thalassaemia in a population which recently immigrated to the US from Sicily, a region that was once under severe malaria threat. Since Haldane's malaria hypothesis, studies on identifying the candidate loci and their variants affecting malaria susceptibility have been carried out extensively using case–control or population approaches. Although the genetic basis of malaria susceptibility appears to be extremely complex, a number of candidate genes have been identified that appear to play a role in malaria susceptibility. These include genes coding for erythrocyte-specific structural proteins and metabolic enzymes, and receptors expressed on the surface of red blood cells (RBCs) and endothelium cells in blood vessels, as well as genes that play a role in innate and adaptive immunity (Kwiatkowski 2005; Weatherall 2008). Thus, in addition to immunity-related genes, genetic variation in many erythrocyte-specific genes has been playing a central role during the course of an evolutionary arms race between human and malaria-causing *Plasmodium* parasites.

The effects of malaria pressure on the human genome are profound. However, the evolutionary trajectory may vary among these genes because each gene has its own distinct function and may play a different role against parasite invasion. Indeed, whereas overdominance selection (i.e. heterozygote advantage) is well recognized as the underlying force for maintaining high frequency of alleles such as the HbS allele (causing 'sickle hemoglobin') of the β-hemoglobin (*HBB*) locus in many malaria-endemic populations, evidence of directional positive selection was demonstrated at several genes coding for glycosylated membrane receptors exploited by the *Plasmodium* parasites for entrance into RBCs including *DARC* and *GYPB* (Hamblin and Di Rienzo 2000; Ko et al. 2011). In many other cases, ambiguous results were obtained for identifying the causal alleles and genes relevant to malaria resistance or for determining the form of natural selection for a causal allele that confers resistance to malaria infection (e.g. $ICAM^{Kilifi}$ allele; see Fernandez-Reyes et al. 1997; Flint et al. 1998).

The complex genetic basis underlying malaria resistance perhaps can be better unraveled from an evolutionary genetics perspective. Here, we briefly review the evolution of several human genes that play major roles in malaria infection by *Plasmodium* parasites. Evolutionary dynamics and types of

Rapidly Evolving Genes and Genetic Systems. First Edition. Edited by Rama S. Singh, Jianping Xu, and Rob J. Kulathinal.

natural selection on these genes are greatly affected by their roles interacting with parasites at different invasion stages and the innate deleterious effects of mutations that confer protective effects on malaria infection. In addition, because human populations are highly substructured, in many genes multiple mutations favored by selection have arisen independently in different ethnic groups or geographic areas. In the cases of multigene families, genetic exchange between duplicated genes (e.g. gene conversion) appears to be an important mechanism for creating novel haplotypes that may be important for the evolutionary arms race between host and parasite.

22.2 Adaptive evolution in erythrocyte-specific genes

22.2.1 Genetic variants causing erythrocytic structural, regulatory, or enzymatic deficiency: candidates for heterozygote advantage

Haldane's malaria hypothesis has led to several classical studies on heterozygote advantage against malaria. All of these genetic disorders appear to be erythrocyte specific. Erythrocytes are non-nucleus-containing blood corpuscles filled with hemoglobins, oxygen-transport metalloproteins, which are evolutionarily conserved across most vertebrates. Hemoglobin is comprised of two pairs of α-globin and β-globin that are encoded by two α-globin genes (*HBA1* and *HBA2*) and one β-globin (*HBB*) gene, respectively (Fig. 22.1a). Studies of sickle-cell anemia were the first to confirm Haldane's hypothesis (Allison 1954). Hemoglobin S (HbS) is a common allele of *HBB* and is present at high frequency across a broad range of sub-Saharan Africa and parts of the Middle East (Fig. 22.1b). The geographic distribution and prevalence of the HbS allele correlates with the geographic range of endemic malaria (Piel et al. 2010). The HbS allele causes an amino acid change from glutamic acid to valine at residue 6 of the β-globin chain. Individuals homozygous for HbS (HbSS) typically develop sickle cell anemia, a fatal disorder due to severe erythrocyte malformation into a sickle-shape which can result in severe anemia and various life-threatening clinical symptoms including splenic sequestration, vaso-occlusive, aplastic, and haemolytic crisis (Rees et al. 2010). Heterozygous individuals (HbAS) usually have no obvious clinical abnormality. Case–control studies of malaria patients have shown that HbAS carriers benefit from an ~10-fold increase in protection from severe malaria infection with estimated selection coefficient (s) of ~0.1 (Hedrick 2011).

Unlike sickle-cell anemia which is due to structural abnormality of erythrocytes, thalassemia results from genetic mutations that cause insufficient or no production of one of the α-globin or β-globin chains. Thalassemias are the most widespread hemoglobinopathies and Mendelian disorders in humans, causing a serious global health concern. More than 200 mutations are known to cause thalassemia. These mutations occur either at α-globin or β-globin genes and encompass a broad spectrum of mutations including point mutations, small insertions or deletions, or gene deletions. Furthermore, many of these mutations have been confirmed as protective against malaria (Weatherall and Clegg 2002). In general, homozygous thalassemias often cause severe clinical symptoms or lethality, whereas individuals heterozygous for thalassemia only suffer mild forms of anemia, similar to the case of the hemoglobin S allele, which has highest fitness if present in heterozygous form.

In addition to sickle-cell disease and thalassemia anemia, the geographic range of deficiency of the enzyme glucose-6-phosphate dehydrogenase (G6PD) is also correlated with the prevalence of malaria, in agreement with Haldane's malaria hypothesis (Cappellini and Fiorelli 2008). G6PD is a rate-limiting enzyme in the pentose phosphate metabolic pathway for producing pentose (5-carbon sugar) and nicotinamide adenine dinucleotide phosphate (NADPH), which leads to the production of glutathione, an important antioxidant for preventing cellular damage by peroxides and free radicals. This pentose phosphate pathway is critical in erythrocyte metabolism particularly for the *Plasmodium*-infected erythrocytes because parasites inside the RBCs will break down hemoglobins for their own nutrition and reproduction requirement, resulting in the release of vast toxic materials such as iron which is a source of oxidative stress. G6PD deficiency is also a common genetic

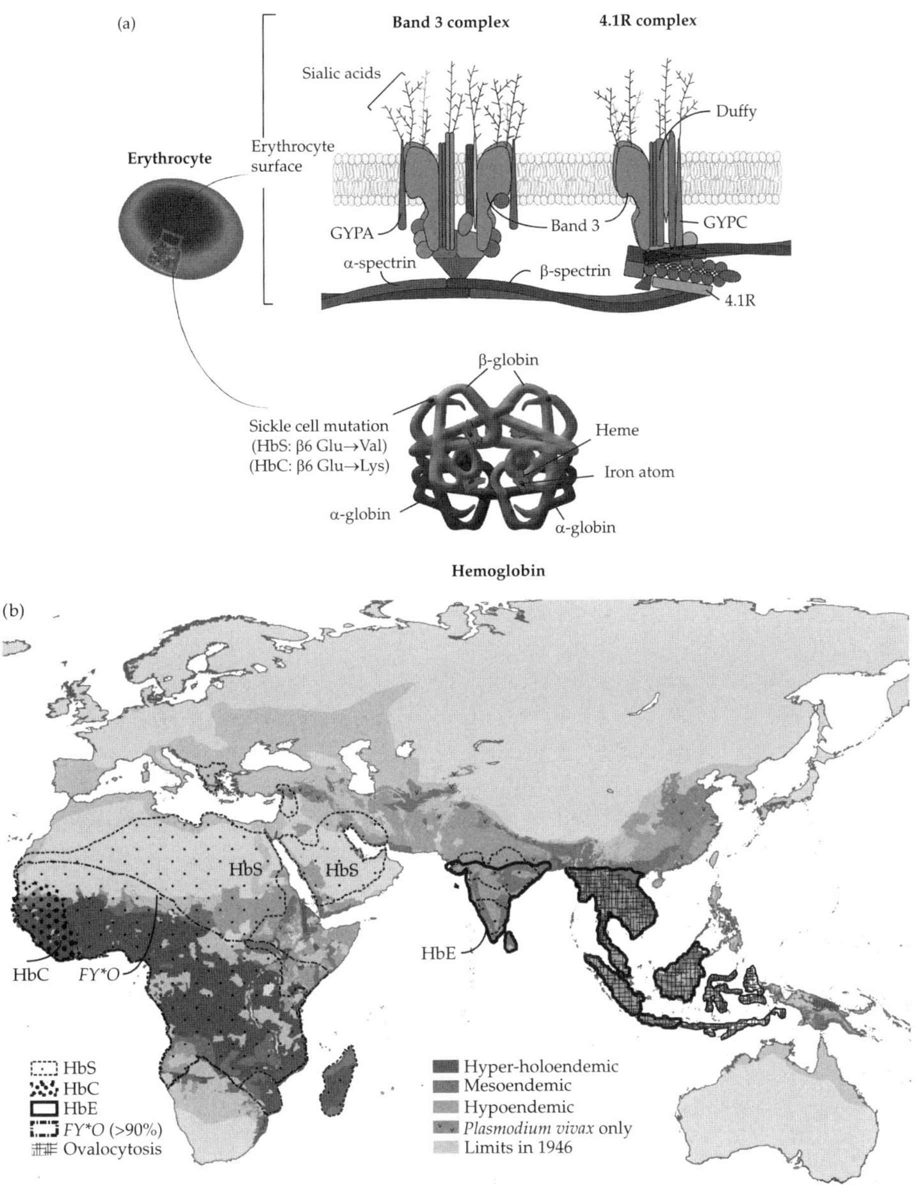

Figure 22.1 (a) Diagrammatic representations of the hypothetical models of two multiple protein complexes (i.e. Band 3 and 4.1R complexes) proposed by Salomao et al. (2008) in the membrane of erythrocyte surface and the molecular structure of hemoglobin are represented. The protein complexes demonstrate the interaction between cytoskeletal proteins, α- and β-spectrins encoded by *SPTA1* and *SPTB*, respectively, and the interaction between cytoskeletal proteins and certain integral and transmembrane proteins. Only proteins with genetic variants identified as associated with malaria susceptibility are labeled. Sialic acids, a family of monosaccharides, are heavily distributed on several major sialoglycoproteins such as glycophorin A (GYPA), B (GYPB), C (GYPC), and Duffy expressed on the erythrocyte surface. These sialoglycoproteins involve direct ligand–receptor interaction during the merozoite invasion of erythrocytes by *Plasmodium* parasites. Sialic acids on these receptors are essential for the binding to parasite ligands whereas the protein backbone is also important for binding specificity. Hemoglobin consists of two α-globins encoded by *HBA1* and *HBA2* and two β-globins encoded by *HBB*. Numerous genetic variants identified on these globin genes confer some protective effects against malaria, but also result in various hemoglobinopathies such as thalassemias and sickle-cell diseases. (b) The global map of the spatial limits and endemic levels of malaria (except for the Americas) and the geographic distributions of the HbS, HbC, HbE, and *FY*O* alleles, and of Southeast Asian ovalocytosis. The endemicity classes of malaria: dark gray, hyperendemic and holoendemic (area in which childhood infection prevalence is > 50%); medium gray, mesoendemic (area with infection prevalence between 11–50%); and light gray, hypoendemic (area with infection prevalence $\leq$10%). The spatial limit for malaria transmission in 1946 is also shown. Geographic distributions of alleles are according to López et al. (2010) and Howes et al. (2011). The figures are adapted from Snow et al. (2005), Salomao et al. (2008), and Schechter (2008).

disorder in humans, affecting 400 million people worldwide. More than 140 different genetic variants that cause G6PD deficiency have been discovered (Cappellini and Fiorelli 2008). The observations of long-range linkage disequilibrium (LD) on the chromosomal region surrounding the G6PD-A- variant, a derived allele common in sub-Saharan Africa, and high levels of genetic differentiation between the haplotypes of G6PD-A- and ancestral alleles are consistent with a genetic signature of recent adaptive evolution (Tishkoff et al. 2001, Sabeti et al. 2002; Verrelli et al. 2002). This pattern is attributed to the rapid increase in frequency of the G6PD-A- carrying haplotypes and, consequently, recombination has had little chance to break the genetic linkage between the beneficial mutation and adjacent neutral variants, resulting in long-range LD. However, it remains unclear whether the G6PD-A- allele has been maintained due to balancing selection in heterozygous females or due to directional selection in both females and males.

Hereditary ovalocytosis is another blood abnormity that causes the oval shape of RBCs rather than the typical biconcave disc shape and ovalocytic erythrocytes have been shown to confer some resistance to *Plasmodium* infection (Gallagher 2005). Ovalocytosis can be caused by a number of genetic variants from multiple genes including spectrin $\alpha 1$ (*SPTA1*) and spectrin β (*SPTB*) genes coding for cytoskeletal proteins, and erythrocyte transmembrane proteins band 4.1 (*EPB41*) and band 3 (*SLC4A1*) that are also associated with cytoskeletal structural integrity (Fig. 22.1a). Given the genetic variability underlying this disease, many individuals who inherit ovalocytosis are asymptomatic or suffer only mild forms of hemolytic anemia. The incidence of hereditary ovalocytosis appears to be higher in populations of African or Mediterranean descent and Malayan natives; these populations originate from regions where malaria is or was endemic (Gallagher 2005). In fact, *in vitro* experiments have suggested that several amino acid variants of the α-chain of spectrin present in Africa (e.g. $Spa^{1/65}$ and $Spa^{1/46}$) might provide some protective effects against the invasion of *P. falciparum* (Dhermy et al. 2007). A nine-amino acid deletion at band 3 (encoded by *SLC4A1*) that causes ovalocytosis in Southeast Asia and Southwest Pacific populations also confers some protective effects against malaria for heterozygous carriers and might have evolved adaptively (Wilder et al. 2009a).

22.2.2 Positive selection on erythrocyte-surface receptors

In the genus *Plasmodium*, five species are considered to be human parasites (i.e. *P. vivax, ovale, malariae, knowlesi*, and *falciparum*). Among them, *P. falciparum* is the most widespread and life-threatening parasite that causes severe symptoms of malaria. Although there is some controversy about the origin of *P. falciparum* as a human parasite and the time estimation of the most recent common ancestry of all *P. falciparum* strains in humans, recent studies have shown that *P. falciparum* forms a monophyletic group (the *Laverania* clade) with several other species including *P. billbrayu, billcollinsi, gaboni*, and *reichenow* that parasitize African great apes (Prugnolle et al. 2011). These findings indicate that coevolution between the *Laverania Plasmodium* and the great hominoid genomes might have been long-standing. However, it is generally thought that selective pressure of malaria in humans has greatly increased in the last 10,000 years due to the development of agriculture and increased human population density which exacerbate the spread of malaria (Tishkoff et al. 2001).

Coevolution between host and parasite can be characterized as continuous processes of adaptation and counter-adaptation in both species. Erythrocyte invasion by *Plasmodium* parasites depends on distinct molecular interactions between the merozoite ligands and several host receptors expressed on the erythrocyte surface. Signatures of the evolutionary arms race between humans and *Plasmodium* parasites have been observed at genes coding for several erythrocyte-surface receptors. Among them, Glycophorin A (GYPA) and Glycophorin B (GYPB) are two major glycoproteins that can be recognized by erythrocyte-binding antigen 175 (EBA-175) and erythrocyte-binding ligand 1 (EBL-1), respectively, expressed by *P. falciparum*. In addition, Glycophorin C (not homologous to GYPA and GYPB), which codes for the Gerbich blood group antigens, is another highly glycosylated surface-protein that can be recognized by EBA-140 of *P. falciparum*. Accelerated rates of protein evolution were observed at these genes, providing strong evidence of positive selection among the hominoids including humans (Baum et al. 2002; Wang et al. 2003; Wilder et al. 2009b; Ko et al. 2011). Interestingly, two forms of selection were observed separately at different parts of the GYPA extracellular domain. While rapid protein evolution was identified at the peptide encoded by exons 3–4 (i.e. positive selection), balanced polymorphisms were observed at the O-sialoglycan-rich NH2 terminal peptide in some African populations living in areas with high exposure to *P. falciparum* (Ko et al. 2011).

The Duffy glycoprotein encoded by the *FY* gene is another membrane-receptor expressed on the surfaces of RBCs. The Duffy glycoprotein can be exploited by another *Plasmodium* parasite, *P. vivax*, during erythrocyte invasion. The *FY*O* allele differs from the ancestral *FY*B* allele by a point mutation (T – > C) in the GATA box of the promoter region of *FY*, resulting in no expression of Duffy on the erythrocyte surface and complete resistance to *P. vivax* infection in its homozygous form. The *FY*O* allele has reached near fixation in most sub-Saharan African populations, but is rare outside Africa. Signatures of positive selection at the *FY* locus have been observed in several studies based on the unusually high levels of population differentiation and homozygosity at this locus, particularly for the *FY*O* allele (Hamblin and Di Rienzo 2000).

22.3 Evolutionary response of the human genome to malaria infection

The evolutionary dynamics of host–parasite interaction depend on a number of genetic features such as the genetic mating system, pleiotropy and epistatic effects, dominance relationship between selective and non-selective alleles, and generation-time difference between host and parasite. The effect of host–parasite interaction on genetic variation in both host and parasite genomes can be commonly categorized into stable polymorphisms, dynamic polymorphisms with temporal or spatial fluctuations in allele frequency, and selective sweeps (Woolhouse et al. 2002). *Plasmodium* protozoans have likely coevolved with our hominid ancestors for a long evolutionary time. Nonetheless, the selective pressure by *P. falciparum* infection that causes the most virulent forms of malaria in humans appears to be relatively recent, around 10,000 years ago (Tishkoff et al. 2001; Hedrick 2011). Although the genetic basis underlying malaria susceptibility is complex, several general patterns appear to be consistent across these erythrocyte-specific genes. While most membrane-receptor genes have accelerated rates of protein evolution, structural/enzymatic genes appear to evolve relatively slowly among humans and the great apes. However, a substantial amount of genetic variation has been observed in each of these genes due to over-dominance selection (e.g. *G6PD*, *HBA1*, and *HBB*; see Fig. 22.2a). Fig. 22.2b also illustrates the proportions of the coding region that are under purifying selection ($d_N/d_S < 1$ where d_N/d_S is the ratio of nonsynonymous to synonymous substitution rates), positive selection ($d_N/d_S > 1$), and relaxation of functional constraint ($d_N/d_S = 1$). It is interesting to note that the majority of the coding region was estimated to be under strong functional constraint ($d_N/d_S \approx 0$) for each of these structural/enzymatic genes and, indeed, many variants in each gene are found to be associated with distinct, but usually harmful, phenotypic outcomes (e.g. sickle-cell anemia and thalassemia). In addition, many of these variants appear to be specific to a geographic area and have risen adaptively in geographically diverse populations (Flint et al. 1998).

22.3.1 Maintenance of deleterious mutations due to selective pressure of malaria

According to molecular evolutionary theory, all new mutations fall primarily into three fitness categories. A vast majority of mutations are harmful to the carriers by reducing either survival or fertility. These deleterious mutations are usually quickly eliminated from populations by purifying selection and, therefore, have little or no contribution to genetic diversity in populations. Secondly, a considerable proportion of mutations have little or no effect on individual fitness. The evolutionary dynamics of these 'effectively' neutral mutations in a population are governed solely by

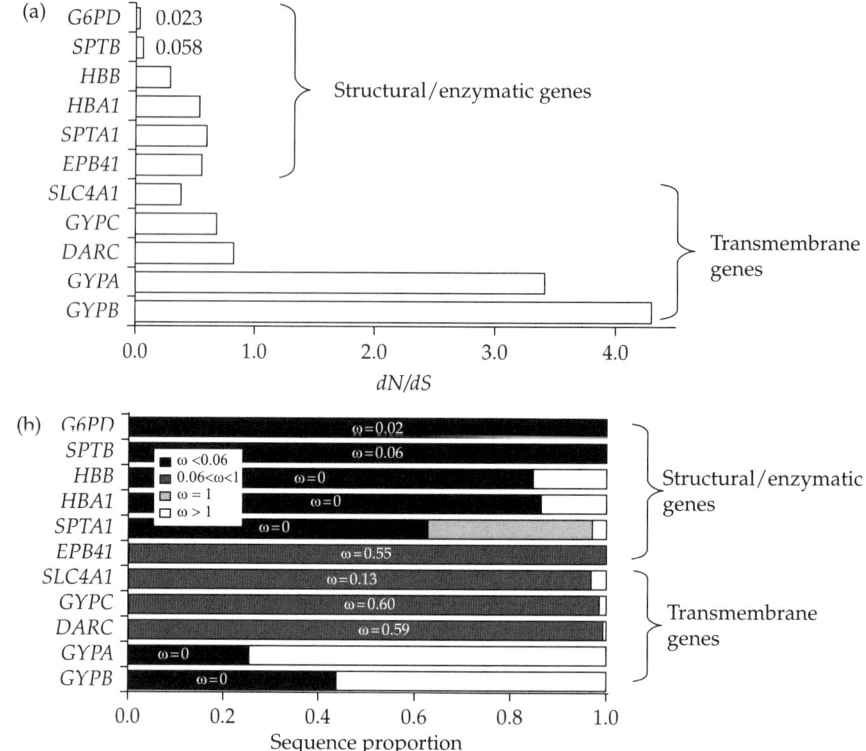

Figure 22.2 d_N/d_S analysis for gene-specific molecular evolution in hominid species (i.e. human, chimpanzee, gorilla, and orangutan) was performed for eleven erythrocyte-related genes using CODEML packaged in PAML. A maximum likelihood approach was used to estimate the ratio of nonsynonymous to synonymous substitution rate ($d_N/d_S = \omega$) assuming constant ω over a gene tree, but heterogeneous ω among sites (i.e. the M2 model in CODEML; see Yang et al. 2000). The M2 model allows three classes of sites: the conserved sites ($\omega < 1$), the neutral sites ($\omega = 1$), and the adaptive sites ($\omega >$ 1pug) The overall d_N/d_S estimates across all species are shown in (a) and the proportions of coding sequence undergoing purifying selection ($\omega < 1$), neutral evolution ($\omega = 1$), and positive selection ($\omega > 1$) are presented in (b). The transmembrane genes tend to have greater d_N/d_S estimates than the structual/enzematic genes ($P = 0.05$ for the Mann–Whitney U test after exclusion of *GYPB* due to non-independent evolution between *GYPA* and *GYPB*; see Ko et al. 2011). The coding sequences for these genes were obtained for each of the hominidae species from the Ensemble Genome Browser if they are annotated. The estimates of d_N/d_S in (a) are labeled for the values smaller than 0.1.

genetic drift. Finally, there are advantageous mutations that confer fitness advantages to the carriers and, consequently, these mutations have better chances to be passed on to the next generation than their alternative alleles (Ohta 1992). Understanding the distribution of fitness effects (DFE) of new mutations in humans is fundamental to studies that seek to identify the genetic basis of disease. Despite some discrepancy among studies, Eyre-Walker and colleagues (2006) estimated the DFE of amino acid changing mutations and showed that greater than 60% of mutations have deleterious effects with selection coefficient > 0.001 (i.e. deleterious and mildly deleterious mutations), and one-quarter of these mutations are strongly selected against (selection coefficient ≥ 0.1). The DFE for mutations could also differ greatly between genes that are subjected to different levels of functional constraint. Generally speaking, those structural/enzymatic genes that are candidates for overdominance selection tend to have lower d_N/d_S estimates in comparison with the membrane receptor genes, reflecting stronger functional constraint for the former class of genes that are important for the structure or metabolism of erythrocytes (Fig. 22.2). For example, whereas about 140 different G6PD variants that cause reduced enzyme activity have been identified, affecting 400 million people worldwide

(Cappellini and Fiorelli, 2008), the protein evolution of G6PD has been slow and 97% of the *G6PD* gene is estimated to be under purifying selection (Fig. 22.2b).

Because malaria is a very strong selective pressure, deleterious mutations that occurred in these structural/enzymatic genes might have risen adaptively if they confer any protection from malaria despite the fact that their harmful effects often cause different forms of hemoglobinopathies with varying levels of severity. In other words, a severe infectious disease like malaria can alter the distribution of fitness effects of new mutations by allowing a considerable amount of deleterious mutations to evolve adaptively in populations if they confer protective effects. In environments where malaria is absent, these deleterious mutations are not expected to segregate in populations at appreciable frequencies. Whether these mutations will be maintained in populations by overdominance selection or be driven to fixation by positive selection may depend on their original deleterious effects (s) when malaria is absent and the dominance effects (h) of new mutations relative to the ancestral alleles. Under the assumption of Mendelian inheritance given a selection scheme of 1, $1 - hs$, and $1 - s$ for A_1A_1, A_1A_2, and A_2A_2 genotypes, respectively, a mutation (A_2) that is lethal or harmful when homozygous (i.e. $1 - s \approx 0$ for A_2A_2), but offers protective effects against malaria when heterozygous, is a candidate for overdominance selection ($1 - hs > 1$ for A_1A_2). A typical example is the HbS allele prevalent in sub-Saharan African populations (Fig. 22.1b). While the heterozygous carriers of HbS are healthy and protected from severe forms of malaria, HbS homozygotes have severe and lethal symptoms of sickle-cell anemia. As a result, HbS heterozygosity is favored by selection in malaria-endemic populations and balanced polymorphism is maintained.

In contrast, HbC and HbE are two other structural variants of β-globin that have risen recently in West Africa and Southeast Asia, respectively. The homozygous carriers of HbC and of HbE appear to suffer only mild forms of sickle-cell anemia in comparison with the HbS homozygote, and the heterozygotes are generally asymptomatic. The protective effect of the HbC allele was found to be greater in homozygotes than in heterozygotes (Kwiatkowski 2005). The HbC and HbE alleles apparently have milder deleterious effects than the HbS allele. If the less severe form of sickle-cell anemia has only little or no effect on survival and if HbC or HbE homozygosity confers greater protection from malaria than the heterozygous forms (e.g.$w_{HbC/HbC} > w_{HbC/HbA} > w_{HbA/HbA}$, where w is viability), these alleles are expected to eventually reach fixation. If the protective effects against malaria cannot overcome the deleterious effects of anemia in the homozygous form (i.e. heterozygosity has the highest relative fitness value), the polymorphism will be maintained in populations. Strong deleterious mutations that confer protective effects but are completely dominant are not expected to rise in a population. This is because for completely dominant mutations, both heterozygous and homozygous carriers are lethal. One exception is the gene family of hemoglobins α 1 (HBA1) and α 2 (HBA2), which both code for α-globin proteins. In the case of a single-locus system, a homozygous mutation (e.g. HbSS) often leads to severe or lethal symptoms whereas heterozygous carriers usually only suffer from mild symptoms. However, in the case of α-hemoglobin, which is coded by two genes, homozygous individuals for a deleterious mutation at one α-hemoglobin gene only suffer from mild anemia because the other wild-type α-hemoglobin gene can still produce normal α-globin. This appears to be an effective genetic system to battle against malaria by allowing for mutations from a broad spectrum of fitness (s) and dominance (h) effects because heterozygous and homozygous carriers of a strongly deleterious and completely dominant mutation conferring protection from malaria are able to survive. Natural selection can act effectively in such a system because a dominant allele is expected to increase in frequency faster at the initial stage than a recessive mutation (i.e. spend less time at low allele frequency) and, therefore, has a lower probability of loss by genetic drift (Hartl and Clark 1997).

Many variants of α- or β- hemoglobin, *G6PD*, or the other erythrocytic genes that play a role in malaria resistance have only mildly deleterious effects (e.g. HbC and HbE). Therefore, both positive and overdominance selection are possible mechanisms responsible for their rapid increases in allele frequency in a population. However, since many alleles associated with malaria resistance have recent origins within the last 10,000 years (Weatherall and Clegg 2002), genetic signatures may be indistinguishable between overdominance and positive selection.

22.3.2 Effects of population substructure on genetic variation in malaria-endemic human populations

Although Haldane's malaria hypothesis was originally proposed to account for the unusual high frequency of thalassemia homozygotes under the simple assumption of Mendelian inheritance at a single locus (Haldane 1949), it is now clear that the genetic basis of thalassemia encompasses more than 200 mutations occurring at the two α-hemoglobin loci and at the β-hemoglobin locus. A large number of variants have been observed for G6PD deficiency and for pyruvate kinase deficiency. Similarly, various mutations causing ovalocytosis have also been found at spectrin α1, spectrin β, band 4.1 or band 3. A common feature among these genetic traits is that most mutations are region specific. For example, HbS is a common allele across sub-Saharan Africa and some parts of the Middle East, but is rarely found in Southeast Asia, while HbE is distributed in the opposite geographic pattern. HbC has a recent origin (<5000 years old; Wood et al. 2005) and is only restricted to some populations in West Africa. The *FY*O* allele is nearly fixed in sub-Saharan African populations, but is rare outside Africa, Madagascar, and the Arabian Peninsula while the *FY*A* allele is common in eastern Asian and Pacific populations (Howes et al. 2011).

The evolutionary dynamics of these mutations perhaps can be better understood conceptually based on Sewall Wright's fitness landscape metaphor and the shifting-balancing theory which emphasizes the importance of population subdivision and the effects of genetic drift in the process of evolutionary adaptation (Wright 1932, 1988). Fig. 22.3 illustrates a map of the fitness landscape with multiple fitness peaks and valleys, indicating higher and lower fitness, respectively. A circle represents a local population and its diameter is proportional to the field of possible gene combinations within this population as defined by Wright. Expansion or contraction of the circle is governed jointly by mutation rates, random genetic drift, and natural selection. An increase in mutation rates or a decrease in stabilizing selection enlarges the population circle, indicating an expansion of genetic variation from a selective peak whereas changes in the reverse direction result in a greater concentration around the peak (Fig. 22.3a,b). While selection is a driving force that brings the population toward a selective peak, genetic drift moves the population circle in a random direction (Fig. 22.3d,e). Although genetic drift often moves a population circle away from a fitness peak, it also provides opportunities for the population to cross over a saddle toward a neighboring fitness hill ('peak shift'). In addition, an environmental change could also cause a peak shift by sliding the entire contour map (Fig. 22.3c). A species is often subdivided into many local populations that occupy different locations on a fitness map (Fig. 22.3f). These subpopulations are genetically differentiated by random drift, but allow occasional migrants to exchange genetic variants between subpopulations. When a local population acquires a beneficial mutation, this allele will rise adaptively to high frequency and may spread to neighboring populations due to migration that would allow the recipient population to advance toward a high fitness peak. The process, referred to as 'intergroup selection' by Wright, continues repeatedly with other surrounding populations until all populations that are under the similar selection scheme have reached the fitness peak.

Human populations are, indeed, highly substructured. In particular, Africa and Southeast Asia, both regions with high levels of malaria also have high levels of population substructure. Tishkoff et al. (2009) identified 14 major ancestral population clusters in Africa in which each population cluster contains several ethnical groups that share similar genetic, cultural, and linguistic properties. Nine population clusters have been identified in Southeast Asia (Durbin et al. 2010). Wright's scheme provides useful insights into the evolutionary mechanism that underlie the geographic distribution of malaria-resistance alleles observed among different ethnic groups. The abundance of muta-

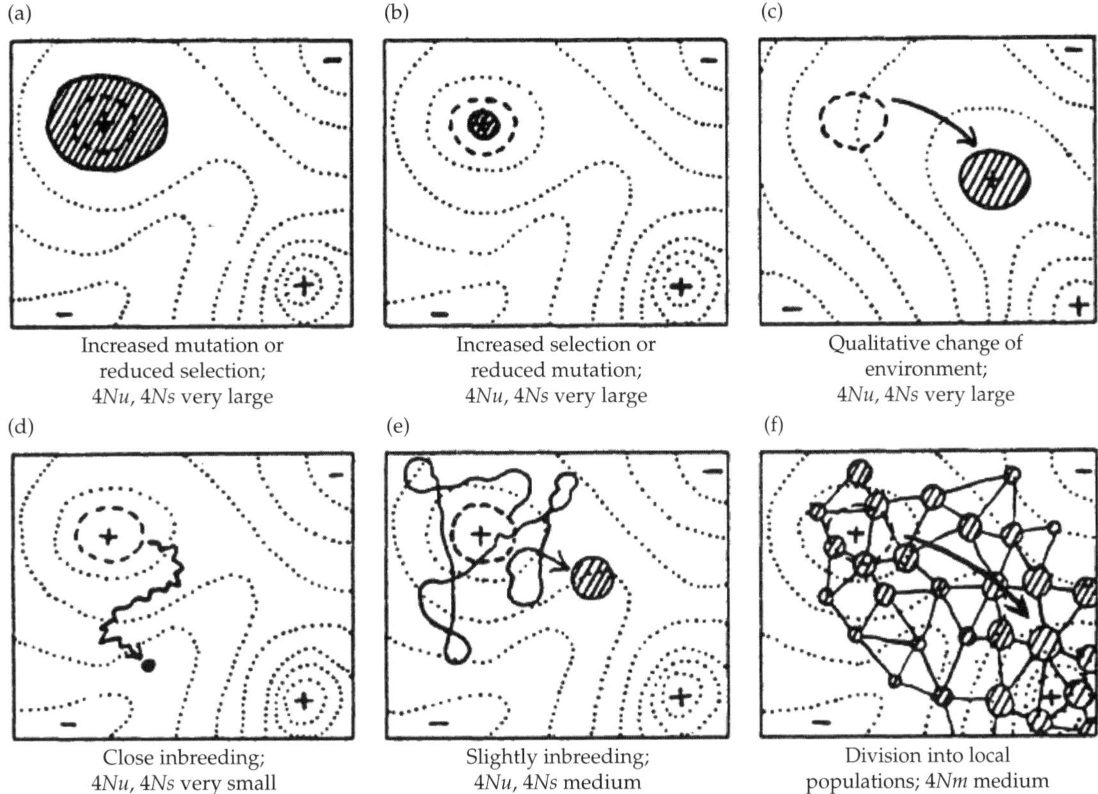

Figure 22.3 Diagrammatic representation of fitness landscape and effects of population subdivision on evolutionary adaptation. Wright (1988) envisioned that organisms exist on a fitness landscape characterized by multiple fitness peaks and valleys. An increase in population scaled mutation rates (4Nu) or a decrease in stabilizing selection (4Ns) enlarge a population circle whose diameter is proportional to the field of genetic combinations (a). A change in a reverse direction results in a contraction of the population circle (b) and an environmental change can be viewed as a landscape shift (c). A population circle is expected to be largely shrinking and moving away from the fitness peak when the population size is very small because the effects of genetic drift greatly exceed the joint effects of 4Nu and 4Ns on the evolutionary trajectory of a population (d). Panel (e) illustrates the joint outcome of genetic drift, mutation rates, and selection when the effects of these three forces are at the same order. The evolutionary trajectory of the population is partly random with a loose control by selection, which might result in a 'peak shift.' When a species is subdivided into many local populations with sufficient rates of migration (4Nm), migration events could overcome the antagonistic effects of stabilizing selection that prevents the population moving away from a local fitness center, and, subsequently, triggers a 'peak shift.' Wright considered population subdivision as an effective mechanism for species adaptation since many local populations can generate greater genetic combinations and expand the exploration area on a fitness landscape. The figure is adapted from (Wright 1932, 1988).

tions observed in each of the hemoglobinopathies described earlier results not only from the change in the distribution of fitness effects of mutations but also from population subdivision that allows unique mutations to arise independently in each subpopulation. Although infrequent, migration events between subpopulations would allow a beneficial allele, such as G6PD-A-, to spread over a large geographic region, as observed in sub-Saharan Africa (Tishkoff et al. 2001). Interestingly, multiple origins of the HbS allele have been suggested because there are five major distinct haplotypes (four in Africa: Central Africa Republic, Benin, Cameron, and Senegal, and another one in India and Saudi Arabia; see Flint et al. (1998)). This hypothesis of multiple origins and convergent adaptation has recently gained some support from theoretical modeling (Ralph and Coop 2010). Alternatively, Flint et al. (1998) preferred a single origin of HbS and argued that the original mutation was redistributed onto different haplotype backgrounds by gene conversion followed by a random loss of either the original or the gene-conversion derived haplotypes in subpopulations.

22.3.3 Effects of gene conversion between homologous sequences on genetic variation at loci associated with malarial susceptibility

Gene families encompass a great proportion of genes in the human genome and are characterized by their high sequence similarity if they arose recently through gene duplications. In a multigene family, gene conversion and unequal crossing over are two common genetic mechanisms that homogenize genetic material between homologous loci but can introduce novel variation at individual loci. Studies on human major compatibility complex (MHC) have shown that gene conversion and unequal crossing over are responsible for the unusually high levels of polymorphism observed at class I and class II loci of MHC that are targets of natural selection for pathogen recognition (Ohta 2010). Several erythrocyte-specific genes described previously are also members of multigene families. For example, glycophorins A, B, and E are duplicated genes, which are similar to each other at the nucleotide level (> 95%). Various kinds of copy number variants have been identified for these loci (Blumenfeld and Huang 1995). Additionally, high levels of gene conversion have been observed at the *GYPA*, *GYPB*, and *GYPE* genes. In particular, the N allele of the MN blood group polymorphism at *GYPA* was generated by gene conversion and caused two amino acid changes at the extracellular domain that interacts with erythrocyte-binding antigen 175 (EBA-175) expressed by *P. falciparum* (Ko et al. 2011). A newly identified haplotype of *GYPB* causing three amino acid changes also contains a gene-conversion-derived mutation at the extracellular peptide that can be recognized by erythrocyte-binding ligand 1 (EBL-1) of *P. falciparum*. Signatures of adaptive evolution have been detected at both loci, suggesting that gene conversion is indeed an effective mechanism for creating novel haplotypes upon which natural selection could act (Ko et al. 2011). Evolution of other erythrocyte-specific genes that are also members of gene families (e.g. hemoglobin α spectrin α, and spectrin β gene families), however, remain to be explored.

22.4 Future perspectives

Many classic studies identifying genetic variants involved in malaria susceptibility, such as the sickle-cell causing allele (HbS) and G6PD-A- allele, have been well established in the past decades, perhaps due to their strong deleterious effects that result in obvious phenotypic or clinical abnormality. However, our recent understanding of the distribution of fitness effects of mutations in humans suggests that a considerable proportion of new mutations are likely to be mildly deleterious or effectively neutral with obscure clinical symptoms. Many mutations that fall into this fitness class might have also evolved adaptively in human populations due to severe selection pressure caused by malarial infection. These variants are also expected to contribute to genetic burdens in humans and, therefore, the factors contributing to the maintenance of these variants at high frequency are of importance. Recent technological advances in areas of whole-genome genotyping and next-generation sequencing will enable better identification of causal genetic variants that underlie phenotypic adaptation and human disease. For example, with the data available from the 1000 genome project, Genovese et al. (2010) was able to identify two *APOL1* variants as the risk alleles responsible for the higher rates of kidney disease in African Americans. Further functional and sequence analyses have suggested that these two alleles confer some resistance to trypanosome infection and might have risen adaptively in the Yoruba population of Nigeria. Similar approaches can be taken for identifying variants that played a role in malaria resistance but cause deleterious effects on human health. However, since human populations are highly substructured, particularly in Africa, it will be of great importance to conduct a fine-scale investigation across diverse ethnic groups for discovering regional-specific novel variants underlying malaria susceptibility.

References

Allison, A.C. (1954). Protection afforded by sickle-cell trait against subtertian malarial infection. *Br Med J* **1**: 290–4.

Baum, J., Ward, R.H., and Conway, D.J. (2002). Natural selection on the erythrocyte surface. *Mol Biol Evol* **19**: 223–9.

Blumenfeld, O.O. and Huang, C.H. (1995). Molecular genetics of the glycophorin gene family, the antigens for MNSs blood groups: multiple gene rearrangements and modulation of splice site usage result in extensive diversification. *Hum Mutat* **6**: 199–209.

Cappellini, M. and Fiorelli, G. (2008). Glucose-6-phosphate dehydrogenase deficiency. *Lancet* **371**: 64–74.

Dhermy, D., Schrevel, J., and Lecomte, M. (2007). Spectrin-based skeleton in red blood cells and malaria. *Curr Opin Hematol* **14**: 198–202.

Durbin, R.M., Abecasis, G.R., Altshuler, D.L., Auton, A., Brooks, L.D., Durbin, R.M., et al. (2010). A map of human genome variation from population-scale sequencing. *Nature* **467**: 1061–73.

Eyre-Walker, A., Woolfit, M., and Phelps, T. (2006). The distribution of fitness effects of new deleterious amino acid mutations in humans. *Genetics* **173**: 891–900.

Fernandez-Reyes, D., Craig, A., Kyes, S., Peshu, N., Snow, R.W., Berendt, A., et al. (1997). A high frequency African coding polymorphism in the N-terminal domain of ICAM-1 predisposing to cerebral malaria in Kenya. *Hum Mol Genet* **6**: 1357–60.

Flint, J., Harding, R., Boyce, A., and Clegg, J. (1998). The population genetics of the haemoglobinopathies. *Baillièr's Clin Haematol* **11**: 1–51.

Gallagher, P. (2005). Red cell membrane disorders. *Hematology* **2005**: 13–18.

Genovese, G., Friedman, D.J., Ross, M.D., Lecordier, L., Uzureau, P., Freedman, B.I., et al. (2010). Association of trypanolytic ApoL1 variants with kidney disease in African Americans. *Science* **329**: 841–5.

Haldane, J.B.S. (1949). The rate of mutations of human genes. *Proc Eighth Intl Congress Genet* 267–73.

Hamblin, M. and Di Rienzo, A. (2000). Detection of the signature of natural selection in humans: evidence from the Duffy blood group locus. *Am J Hum Genet* **66**: 1669–79.

Hartl, D.L. and Clark, A.G. (1997). *Principles of population genetics*. Sunderland, MA: Sinauer Associates.

Hedrick, P.W. (2011). Population genetics of malaria resistance in humans. *Heredity* **107**: 283–304.

Howes R., Patil, A., Piel, F., Nyangiri, O., Kabaria, C., Gething, P., et al. (2011). The global distribution of the Duffy blood group. *Nature Communications* **2**, 266.

Ko, W., Kaercher, K., Giombini, E., Marcatili, P., Froment, A., Ibrahim, M., et al. (2011). Effects of natural selection and gene conversion on the evolution of human glycophorins coding for MNS blood polymorphisms in malaria-endemic African populations. *Am J Hum Genet* **88**: 741–54.

Kwiatkowski, D.P. (2005). How malaria has affected the human genome and what human genetics can teach us about malaria. *Am J Hum Genet* **77**: 171–92.

López, C., Saravia, C., Gomez, A., Hoebeke, J., Patarroyo, M. (2010). Mechanisms of genetically-based resistance to malaria. *Gene* **467**: 1–12.

Ohta, T. (1992). The nearly neutral theory of molecular evolution. *Annu Rev Ecol Syst* **23**: 263–86.

Ohta, T. (2010). Gene conversion and evolution of gene families: An overview. *Genes* **1**: 349–56.

Piel, F.B., Patil, A.P., Howes, R.E., Nyangiri, O.A., Gething, P.W., Williams, T.N., et al. (2010). Global distribution of the sickle cell gene and geographical confirmation of the malaria hypothesis. *Nature Comm* **1**: 140.

Prugnolle, F., Durand, P., Ollomo, B., Duval, L., Ariey, F., Arnathau, C., et al. (2011). A fresh look at the origin of Plasmodium falciparum, the most malignant malaria agent. *PLoS Pathog* **7**: 1–8.

Ralph, P. and Coop, G. (2010). Parallel adaptation: One or many waves of advance of an advantageous allele? *Genetics* **186**: 647–68.

Rees, D., Williams, T., and Gladwin, M. (2010). Sickle-cell disease. *Lancet* **376**, 2018–31.

Sabeti, P.C., Reich, D.E., Higgins, J.M., Levine, H.Z.P., Richter, D.J., Schaffner, S.F., et al. (2002). Detecting recent positive selection in the human genome from haplotype structure. *Nature* **419**: 832–7.

Salomao, M., Zhang, X., Yang, Y., Lee, S., Hartwig, J., Chasis, J., et al. (2008). Protein 4.1 R-dependent multiprotein complex: new insights into the structural organization of the red blood cell membrane. *Proc the Natl Acad Sci U S A* **105**: 8026–31.

Schechter, A. (2008). Hemoglobin research and the origins of molecular medicine. *Blood* **112**: 3927–38.

Snow, R.W., Guerra, C.A., Noor, A.M., Myint, H.Y., and Hay, S.I. (2005). The global distribution of clinical episodes of Plasmodium falciparum malaria. *Nature* **434**: 214–17.

Tishkoff, S., Varkonyi, R., Cahinhinan, N., Abbes, S., Argyropoulos, G., Destro-Bisol, G., et al. (2001). Haplotype diversity and linkage disequilibrium at human G6PD: recent origin of alleles that confer malarial resistance. *Science* **293**: 455–61.

Tishkoff, S.A., Reed, F.A., Friedlaender, F.R., Ehret, C., Ranciaro, A., Froment, A., et al. (2009). The genetic structure and history of Africans and African Americans. *Science* **324**: 1035–44.

Verrelli, B., McDonald, J., Argyropoulos, G., Destro-Bisol, G., Froment, A., Drousiotou, A., et al. (2002). Evidence for balancing selection from nucleotide sequence analyses of human G6PD. *Am J Hum Genet* **71**: 1112–28.

Wang, H.Y., Tang, H., Shen, C.K.J., and Wu, C.I. (2003). Rapidly evolving genes in human. I. The glycophorins and their possible role in evading malaria parasites. *Mol Biol Evol* **20**: 1795–804.

Weatherall, D. (2008). Genetic variation and susceptibility to infection: the red cell and malaria. *Brit J Haematol* **141**: 276–86.

Weatherall, D. and Clegg, J. (2002). Genetic variability in response to infection: malaria and after. *Genes Immunity* **3**: 331–7.

Wilder, J.A., Stone, J.A., Preston, E.G., Finn, L.E., Ratcliffe, H.L., and Sudoyo, H. (2009a). Molecular population genetics of SLC4A1 and Southeast Asian ovalocytosis. *J Hum Genet* **54**: 182–7.

Wilder, J.A., Hewett, E.K., and Gansner, M.E. (2009b). Molecular evolution of GYPC: evidence for recent structural innovation and positive selection in humans. *Mol Biol Evol* **26**: 2679–87.

Wood, E., Stover, D., Slatkin, M., Nachman, M., and Hammer, M. (2005). The β-globin recombinational hotspot reduces the effects of strong selection around HbC, a recently arisen mutation providing resistance to malaria. *Am J Hum Genet* **77**: 637–642.

Woolhouse, M.E., Webster, J.P., Domingo, E., Charlesworth, B., and Levin, B.R. (2002). Biological and biomedical implications of the co-evolution of pathogens and their hosts. *Nat Genet* **32**: 569–77.

World Health Organization. (2008). *World malaria report 2008*. Geneva: WHO.

Wright, S. (1932). The roles of mutation, inbreeding, crossbreeding and selection in evolution. *Proc 6th Int Cong Genet* **1**: 356–66.

Wright, S. (1988). Surfaces of selective value revisited. *Am Nat* **131**: 115–23.

Yang, Z., Nielsen, R., Goldman, N., and Pedersen, A. (2000). Codon-substitution models for heterogeneous selection pressure at amino acid sites. *Genetics* **155**: 431–49.

PART V

From Gene Expression to Development to Speciation

CHAPTER 23

The rapid evolution of gene expression

Carlo G. Artieri

23.1 Introduction

Evolution is, like all subjects involving natural history, a very slow process occurring over a vast span of geological time. This is often frustrating to the evolutionary biologist, whose research cannot help but study the aftermath of processes long past. The desire to overcome such challenges may explain our fascination with those systems that are rapidly evolving, changing over durations comprehensible in the scale of human history (e.g. millennia rather than millions of years). Such systems have the potential to allow us to observe evolution as it is occurring. In addition, our interest in rapid evolution may also be biased as we may stubbornly hold on to the notion that the lineage leading to our very own species is the product of a bout of accelerated change. Somewhat disappointingly, when methods became available that allowed us to compare divergence in the sequence of our genome with that of our closest relative, the chimpanzee, it was found that our two species are no more different than *morphologically identical* sibling species of fruit flies (King and Wilson 1975). Given the obvious differences in 'anatomy and way of life' differentiating us from our nearest relatives, King and Wilson suggested that changes in the manner in which genes are regulated (expressed) could account 'for the major biological differences between humans and chimpanzees.' Therefore, relative to coding sequences, the divergence of gene expression levels and patterns may itself be an agent of rapid phenotypic evolution.

Whereas our grasp of the various factors influencing nucleotide sequence evolution has grown considerably in the many decades since DNA sequences have been available for comparison, our understanding of the evolutionary 'forces' acting upon the mechanisms underlying gene regulation itself lags behind. Clearly this is in no small part due to the relative infancy of our ability to easily interrogate the regulatory state of biological systems, let alone those of multiple closely related species. Nevertheless, the unique challenges associated with the study of regulatory evolution should not be understated. A small sample of such difficulties may include the following:

1. Because of material and technical constraints, expression studies in many organisms have ignored tissue and developmental stage-specific expression variation focusing on the entire organism as a single transcriptome 'pool'
2. The evolution of these complex transcriptomes involves both quantitative (expression level) and qualitative (presence/absence of transcripts) components. Furthermore, many loci are capable of expressing alternatively spliced variants (isoforms) that add an additional intralocus dimension to the notion of divergence.
3. Our knowledge of the fitness consequences associated with changes in gene expression output or alternative splicing is restricted to a small number of cases, and unlike coding sequences we lack a well agreed upon neutral standard against which we can test hypotheses about potential selective pressures.
4. Complicating the evaluation of selection, gene expression occurs within the context of potentially large, complex, and tightly regulated networks, whose details we are only beginning to explore.

Rapidly Evolving Genes and Genetic Systems. First Edition. Edited by Rama S. Singh, Jianping Xu, and Rob J. Kulathinal.

5. Finally, only a handful of the regulatory mechanisms controlling these networks are well understood and we remain incapable of predicting most *cis*-regulatory positions either from primary sequence or based on the structure of the transcription factors that bind to them, i.e. we lack a regulatory 'code'

Encouragingly, we have begun to overcome these challenges using novel experimental and theoretical approaches on an increasing number of comparable datasets from closely related species. One significant conclusion from many of these studies is that gene sequences and their levels of expression evolve in qualitatively (if not often quantitatively) similar manners, suggesting that they are subject to similar selective regimes and amenable to similar techniques of evolutionary analysis (Khaitovich et al. 2005; Artieri and Singh 2010a). In this chapter, I shall draw upon recent studies in order to highlight those significant advances that have been made in our understanding of some of the general factors responsible for determining the rate of interspecific divergence in gene expression levels and patterns as well as indicate some areas of research that remain to be explored. I shall focus on studies employing whole-transcriptome profiling techniques (e.g. microarrays and RNA-Seq) as patterns and processes responsible for broad evolutionary patterns are not adequately explored by case studies of single genes. Finally, this review concerns the selective pressures influencing the evolution of gene expression itself rather than the evolutionary details of the underlying molecular mechanisms of its regulation.

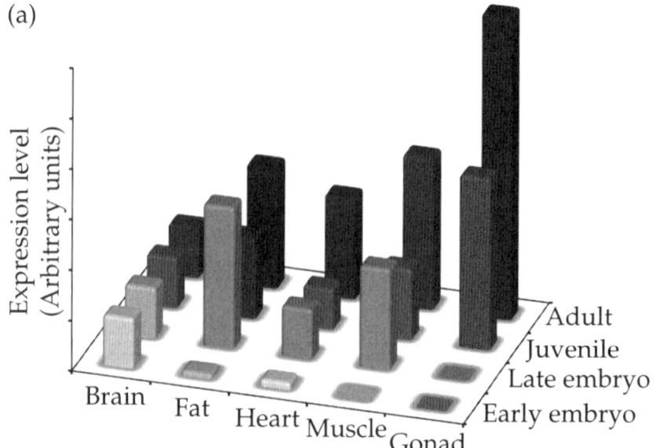

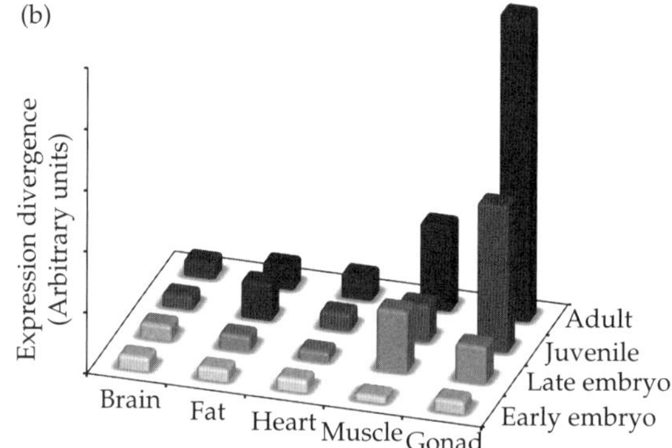

Figure 23.1 Each organism contains many separate transcriptomes. (a) Unlike its coding sequence, the expression level of any particular gene varies widely depending on the tissue (x axis) and developmental stage (y axis) in which it is sampled. (b) Similarly, divergence in expression levels between species is also heterogenous among tissues and stages, leading to situations where rapid evolution (such as in the adult gonad in this hypothetical example) may be missed by profiling only a single tissue, or alternatively 'averaging' over multiple transcriptomes by sampling from pools of whole organisms. Not shown are the potential external environmental influences leading to changes in gene expression which is itself a phenotypic trait that can vary between species, leading to genotype × environment interactions that complicate the study of interspecific differences in expression level.

23.2 One genome harbors many transcriptomes

With few exceptions, each cell of a multicellular organism harbors its own copy of an identical genome. However, from an identical genome, each cell deploys its own subset of the totality of potentially expressed loci, which is contingent on its spatial location within an organism, its temporal position within the overall developmental state of the organism, and the organism's external environment (Fig. 23.1a). I use the term 'transcriptome' to denote all RNA transcripts expressed within the cell, irrespective of their function and protein coding potential. Whereas developmental stage and external environmental influences are relatively straightforward to control in the laboratory, even expression experiments performed on dissected tissues involve a mixture of a variety of cell types, and by extension, transcriptomes. In the case of research with small organisms such as *Drosophila* or *Caenorhabdi-*

tis, pools obtained from a population of homogenous individuals (or tissues) have traditionally been used in order to extract sufficient biological material from which to conduct expression profiling. Thus comparisons between and among such samples are subject to the possibility that what is actually being measured is a mixture of different transcriptomes of unequal abundance.

The technical considerations associated with transcriptome heterogeneity are surely significant, yet often overlooked. For instance, is it correct to say that a testis-specific gene is expressed at a lower level in one species if that sample is known to possess smaller testes (relative to body weight) as compared to the other tissues under investigation? While there are certainly fewer transcripts of the RNA in question, the differences are unlikely to be solely based on regulatory divergence at this locus, but rather reflect differing allometric abundances of the tissue in which it is expressed (though the differing testes sizes may themselves reflect regulatory divergence during a point 'upstream' in the course of development). Similarly, a large number of genes have been classified as 'sex-biased' due to being more abundant in samples extracted from one sex as compared to the other. However, a recent large-scale developmental study performed in *Drosophila* suggested that many 'female-biased genes' (FBGs) identified from whole-fly extractions represent genes that are highly expressed in early embryonic development and reflect measurements derived from embryos contained within the adult female reproductive tract (Graveley et al. 2011). Such observations become especially important in comparisons between and among different species, where the ratios of tissue and/or cellular composition may be quite different. With the emergence of new technologies that sharply reduce the amount of starting material required to conduct gene expression profiling experiments, experimental design will likely shift towards comparisons of homogenous cellular extractions. This will certainly be more informative than the present 'averaging over multiple transcriptomes' effect that certainly reduces our ability to detect real differences (and increases the likelihood that we detect nonregulatory differences) that exist among less abundant cell types.

23.3 Transcriptome divergence is complex

The degree of divergence in DNA sequence can be represented via a single numerical value, such as classical genetic distance (D) or the ratio of nonsynonymous substitutions per nonsynonymous site to synonymous substitutions per synonymous site in the case coding sequences (d_N/d_S) (Graur and Li 2000). Consequently, estimates of the divergence time between the species being compared allow us to identify unique instances, or even categories of genes that are evolving rapidly relative to a genomic standard. Comparisons of transcriptomes among various species are substantially more complicated as divergence rates of expression profiles are tissue, stage, and cell dependent (Fig. 23.1b). Traditionally, the vast majority of comparative gene expression studies have focused on *quantitative* measures of expression or more specifically, comparisons of the abundance of particular transcripts among samples. While techniques for semiquantitative measurements of the abundance of individual RNAs have been available for decades, it is the advent of high-density DNA microarrays in the mid 1990s that paved the way for large-scale statistical comparisons of whole transcriptomes (Schena et al. 1995). Unfortunately, the reliance of most microarray platforms on sequenced, annotated genomes (at least for the purpose of meaningful interspecific comparisons) has limited the number of species groups among which expression divergence has been studied. Nevertheless, important insights have been gleaned. For instance, it is clear that stabilizing selection (i.e. selection limiting divergence) is acting on the majority of loci, as divergence in levels of expression is not proportional to the amount of evolutionary time separating the species under study. Rather, divergence saturates rapidly (Bedford and Hartl 2009). Furthermore, those genes that do diverge at rates greater than expected by a model of stabilizing selection often do so in predictable ways (see later in chapter).

Gene expression may also diverge in a *qualitative* manner. For instance, studies performed in the field of developmental biology have long showed that some loci are expressed in highly restricted temporal and spatial patterns and that these patterns may differ among species. Such differences can be seen as qualitative as the switch between complete lack of expression of a locus in one species to expression in another likely involves fundamentally different evolutionary mechanisms when compared to changes in the level of expression of a locus (e.g. the *de novo* gain or loss of a regulator in the former vs. the modification of the activity of a promoter or regulatory element in the latter) (e.g. Sucena and Stern 2000). These qualitative differences are being revealed by novel, more-sensitive techniques for measuring expression levels, chief among which is RNA-Seq. RNA-Seq involves direct, short-read sequencing of cDNA generated from reverse transcribed RNA (Wang et al. 2009), in which the number of reads generated from any particular transcript is proportional to both its length and, more importantly, its abundance in the sample under study. RNA-Seq thus provides a digital picture of the transcriptome, with increased dynamic range as compared to microarrays, and, due to rapid technology improvements, very small input sample requirements—going even so far as allowing transcriptome profiling at single-cell resolution (e.g. Tang et al. 2010). Beyond its aforementioned benefits, perhaps the most novel contribution of RNA-Seq is its ability to profile the state of alternative splicing among a sample's expressed transcripts. As has been appreciated for some time (reviewed in Xing and Lee 2006), individual genomic loci are often able to selectively ligate subsets of their total potential exons in order to create different alternative transcripts—via still somewhat poorly understood effectors. In providing direct access to sequence information, RNA-Seq allows the identification of splice junction spanning reads (tracts of short sequence that cannot be mapped directly to the genome without adding an intronic 'gap') and thus comparison of their abundance between samples. Importantly for the purposes of understanding transcriptome divergence, isoforms themselves can vary among samples in both quantitative (abundance) and qualitative ways (presence/absence). Given that studying alternative transcripts in any comprehensive fashion within well-annotated model organisms has been challenging, it is not surprising that more traditional studies have generally focused on studying isoform evolution in individual or small numbers of genes.

23.4 Factors affecting the rate of evolution of gene expression

23.4.1 Spatial heterogeneity

Not all tissues are subject to the same selective pressures or constraints and average rates of coding sequence evolution for genes expressed in different tissues can vary wildly. For instance, a large number of studies conducted in a wide variety of taxa have shown persuasively that genes expressed in male reproductive tissues are rapidly diverging in their coding sequences relative to the genomic average—see Meisel (2011) for a recent discussion. One of the first studies to extend the tissue-specificity of evolutionary rates to the level of gene expression in a whole-transcriptome manner was that of Khaitovich and colleagues (2005), who compared divergence in both coding sequence and expression level in five paired tissues from human and chimpanzee: brain, heart, kidney, liver, and testis. Notably, they observed parallel patterns of divergence at loci expressed in these tissues: genes expressed in the liver and testis were rapidly diverging in both coding sequence and expression level, whereas genes expressed in the brain were more conserved at both levels. Such patterns were not restricted to those genes uniquely expressed in a single tissue: genes detectably expressed in multiple tissues remained more conserved in expression level when profiled in brain tissue than in liver tissue, for instance. In addition, the authors noted a further pattern influencing rates of divergence: broadly expressed genes (those detectably expressed in multiple tissues) were less divergent than genes showing a more restricted pattern of expression. This as well as subsequent studies have suggested that such conservation reflects the action of negative selection acting to minimize deleterious pleiotropic effects caused by mutations affecting genes with broadly dispersed function.

Many of these observations have been confirmed in subsequent studies (e.g. Blekhman et al. 2008). However the parallelism between rates of expression level and coding sequence divergence has not been observed in all cases. While most studies have found a positive, albeit often weak, correlation between both levels of divergence, some earlier studies suggested that expression levels and coding sequences diverge independently—e.g. see the discussion in Artieri et al. (2007). Few of these studies have directly compared levels of divergence among different tissues or developmental stages; however, there is some evidence that rates of divergence at both coding and expression levels are primarily determined by the stage(s) where the locus is (1) most highly expressed and (2) functionally relevant (Jordan et al. 2005).

23.4.2 Temporal heterogeneity

Studies exploring how expression patterns vary over development—the second 'axis' in Fig. 23.1—have been motivated by a classical observation predating Darwin's formulation of the theory of evolution: Karl Ernst Von Baer noted that the early developmental stages of organisms are more morphologically conserved than later stages; subsequently named 'Von Baer's third law' (see Gould 1978 for extensive review). Though Von Baer's third law holds generally among several different phyla, modern analyses have revealed that the very earliest stages of development (e.g. stages prior to and including gastrulation) can be quite divergent among even closely related species (Raff 1996). These observations have led to a revised 'developmental hourglass' model of phenotypic divergence, wherein the most conserved developmental stage(s) among organisms, termed the 'phylotypic period,' occur during mid-embryogenesis (reviewed extensively in Raff 1996) (Fig. 23.2). Furthermore, an explicit molecular mechanism has been suggested for this model: the most spatially organized portions of embryonic development, which take place during the period of embryogenesis when organogenesis begins, have highly-integrated and thus tightly-regulated biological networks. These periods may be more conserved due to selection against mutations creating deleterious pleiotropic effects across highly connected regulatory networks (a proposal termed 'developmental constraint'). As development progresses it becomes more 'modular' causing mutations to be more likely

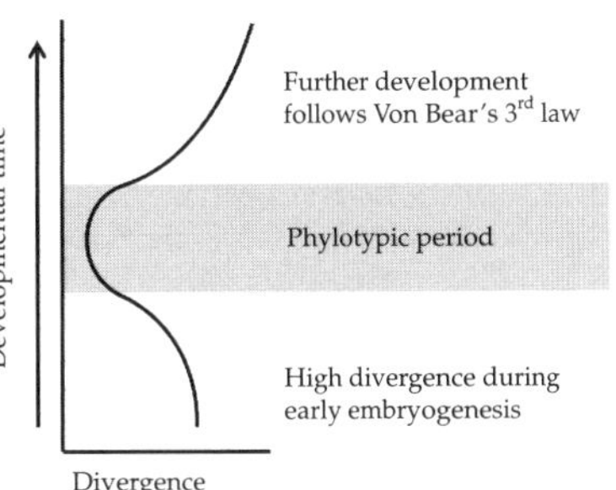

Figure 23.2 The 'developmental hourglass' model of embryonic divergence (adapted from Raff 1996). Empirical observation has shown that the earliest stages of embryogenesis can be quite divergent among species relative to a later stage of conservation known as the phylotypic period. Once through this period (the 'waist' of the hourglass), subsequent stages are allowed to accumulate divergence over development in a manner consistent with Von Baer's third law (Kalinka et al. 2010).

to affect only specific regions of organismal morphology and thus increasing the apparent rate of evolution of later developmental stages.

An initial foray with the aim of exploring whether patterns of divergence of gene expression conformed to the hourglass shape of embryonic morphological divergence was conducted in a pair of elegant interspecific expression studies over precisely staged developmental time courses (Kalinka et al. 2010; Irie and Kuratani 2011). Kalinka and colleagues compared the levels of expression of ~3000 genes over the course of embryogenesis across six species of *Drosophila* spanning 40 million years of evolution. The authors observed that expression levels were more conserved during the extended germband stage, which is widely reported to be the arthropod phylotypic period. Furthermore, genes conforming to the hourglass pattern of divergence show an overrepresentation of functions associated with cellular and organismal development as well as regulation of gene expression—including the well-studied and highly conserved homeobox or HOX genes—supporting the notion that the need for tight regulation of developmentally important effectors leads to stronger purifying selection. Irie and Kuratani (2011) took a much broader approach and compared expression divergence of orthologs among a broad phylogenetic selection of vertebrates: mouse, chicken, frogs, and zebrafish. As discussed earlier, they also found evidence that the stage meeting the majority of the criteria of the vertebrate phylotypic stage showed the greatest degree of expression level conservation among species, with earlier and later stages showing significantly more divergence.

A handful of studies have explored coding sequence divergence of genes expressed beyond the phylotypic stage and have found evidence that divergence increases as development progresses (see Artieri and Singh 2010b). Unfortunately, similar studies comparing levels of gene expression over the course of development have been lacking. The most comprehensive study of this type was conducted by Artieri and Singh (2010a), who compared expression levels of ~2250 genes in males of three species of *Drosophila*: *D. melanogaster*, *D. simulans*, and *D. sechellia*. *D. simulans* and *D. sechellia* shared a common ancestor approximately 0.5–1.0 million years ago (mya), and both form a clade with *D. melanogaster* ~2.5–5.0 mya. Expression was sampled over four developmental time points representing three of the four major phases of development in holometabolous insects (late third instar larvae, early pupation, late pupation, and newly-emerged adult). While the authors did not observe a monotonic, linear increase in the number of genes with significantly different levels of expression over the developmental interval measured, in all comparisons where there was a significant difference in the number of genes differentially expressed between stages, it was the later stage that showed more differentially expressed genes, supporting the notion that gene expression is more conserved in earlier stages.

23.5 Beyond comparisons of expression levels

Given the challenges associated with performing direct comparisons of expression levels across species, some studies have focused instead on how patterns of expression vary between treatments, sexes, or developmental time points *within* species, and then comparing how such transcriptional patterns differ *between* species (see Lu et al. (2009) for a general review). For instance, Papatsenko and colleagues (2011) compared patterns of expression of orthologs generated from within-species embryonic expression time courses in *D. melanogaster* and mosquito (*Anopheles gambiae*), two species of Diptera who shared a common ancestor ~200 mya. Rather than ask whether orthologs retained similar levels of expression, they used statistical models to determine whether they showed conserved profiles of up- and downregulation over their respective species' embryonic developmental time course. They identified clusters of genes whose expression patterns differed between the species and whose likely biological function was associated with known anatomical differences in the development of embryonic membranes in *Drosophila* and *Anopheles*. Another example of using within-species comparisons to draw between-species evolutionary conclusions involves a large-scale analysis of sex-biased gene expression conducted in seven species of *Drosophila* spanning ~60 million years of evolution (Zhang et al. 2007). Zhang and colleagues found that while the majority of fly species possessed a greater proportion of their adult transcriptome that was male-biased in expression, two of the species were primarily female-biased (*D. pseudoobscura* and *D. mojavensis*). Nevertheless, in all species, the magnitude of male bias was significantly greater than that of female bias, indicating that regardless of the actual number of genes in each category of bias, male biased genes were closer to being sex-specific. Interestingly, a significantly greater number of orthologs could be identified among female-biased genes as compared to male-biased genes, suggesting that the rate of 'birth' of new genes, via such processes as gene duplication, retrotransposition, or *de novo* generation are fundamentally higher among male-biased genes.

The two examples discussed here highlight how rapid evolution of the transcriptome can be studied beyond raw comparisons of the expression levels of orthologs. Identification of a heterochronic shift in the time of activation of a particular cluster of genes during embryogenesis significantly aids in the explanation of what would have simply been interpreted as a difference in expression level if interrogated during a single time-point. Similarly, while levels of expression of a locus may not change significantly when comparing females between species, the finding that it is significantly male-biased in only one of the two species may allow us to draw conclusions about the selective pressures underlying this potentially rapid evolutionary shift (see Ellegren and Parsch (2007) for a discussion of sex-biased expression). In addition, the observation that certain classes of genes are more prone to lineage-specificity indicate that the contents of the transcriptome itself is changing, with evidence that this is occurring at an accelerated rate in males as compared to females, and that simple comparison of rates of both sequence and expression divergence of orthologs is insufficient to understand how biological complexity evolves (see Singh and Artieri (2010) for a more detailed discussion of this phenomenon).

23.6 Open questions and future directions

Despite the relative infancy of our ability to explore expression divergence among species, we are beginning to elucidate some of the factors that drive or facilitate the rapid evolution of gene expression. As a general consideration, it appears that genes with more spatially restricted patterns of expression evolve more rapidly at both coding and expression levels than do genes that are broadly expressed. In addition, there is evidence that this is the case for coding sequences among genes that show restricted vs. broad patterns of temporal expression as well (Artieri et al. 2009), though whether these observations apply equally to patterns of divergence in expression level remains an open question. It is clear that certain tissues show accelerated patterns of expression divergence as compared to others, and that furthermore there is compelling evidence that, in most cases, these patterns parallel those seen in the evolution of coding sequences (Khaitovich et al. 2005). Many cases of tissue-specific rapid evolution will likely reflect the particular idiosyncrasies of the selective pressures acting on any particular species, though a long-standing hypothesis suggests that such pressures may often be correlated with changes in morphology (e.g. King and Wilson 1975). As illustrated in this chapter, the paucity of available whole-transcriptome expression studies comprehensively exploring regulatory changes in a tissue-specific manner between species has made it difficult to determine whether there is a systematic association between rapid evolutionary shifts in expression and morphological change, or whether the numerous discussions of this phenomenon within the literature represent ascertainment bias (Fraser 2011).

Novel techniques for profiling expression are producing data from an increasing number of species, tissues, and developmental stages, which will allow us to address some of the open questions regarding the causes and mechanisms of rapid expression divergence. Perhaps the most pressing question from an evolutionary perspective concerns whether instances of divergence represent adaptive rather than neutral processes. Divergence of expression level does not *ipso facto* indicate that adaptive evolution has occurred—equally plausible is the possibility that selection has been relaxed along one or more of the lineages. RNA-Seq's ability to identify single nucleotide polymorphisms and hence measure allele-specific expression can be exploited in hybrids of closely related species in order to detect instances where groups of functionally associated genes share a greater than expected degree of allelic expression bias in the direction of the same parental lineage (Fraser 2011). Such instances uncover adaptive evolutionary change in the *cis*-regulatory elements (i.e. DNA regulatory elements) controlling the expression of these genes. Alternatively, adaptation may occur via *trans*-regulatory divergence (e.g. evolutionary change in transcription factors). Such divergence can be detected using RNA-Seq on interspecific hybrids, where alleles that show differences in expression discordant from expression levels in the pure species indicate *trans*-regulatory divergence between species (McManus et al. 2010); however, confidently inferring adaptive evolution in such *trans* loci will require their identification followed by appropriate tests of selection.

A second open question in the field of gene expression evolution has been the extent and significance of alternative splicing. RNA-Seq data have revealed that a much larger than previously appreciated number of loci produce alternative isoforms (e.g. Graveley et al. 2011), though whether all of these isoforms are themselves functional or are translated into functional proteins is unknown. Tissue- and developmental stage-biased alternative transcripts may provide a fertile ground for rapid intralocus divergence, However, the lack of comprehensive alternative isoform annotations in all but a few key model organisms means that it remains quite tedious to identify, let alone compare orthologous isoforms on a whole genome scale. Recent advances in both reference genome assisted and *de novo* (i.e. without a sequenced genome) transcriptome assembly as well as comparative annotation pipelines are allowing us to more fully interrogate this potential dimension of regulatory evolution.

A third direction of active research involves determining the regulatory function(s) and evolutionary signatures of the multitude of expressed non-coding RNAs (ncRNAs; Saxena and Carninci 2010). Some classes of ncRNAs, such as microRNAs (miRNAs), are known to have extensive post-transcriptional regulatory roles, including destabilization of mRNAs as well as translational inhibition by targeting ~7nucleotide 'seed' sequences in the 3′ untranslated regions of their targets (Chen and Rajewski 2007). Though miRNAs themselves appear to be remarkably conserved over long phylogenetic distances, their relatively short seed sequences allow rapid turnover of their targets, and therefore these ncRNAs may themselves be an agent of rapid evolution of gene expression. Other classes of ncRNAs such as long non-coding RNAs (lncRNAs) remain much less well understood, leaving unanswered their potential contributions to transcriptome divergence.

Finally, whereas the majority of this chapter has focused on the evolution of independent loci (or groups of loci), it is the functional interactions among these loci that determine when, where and under what external circumstances genes are expressed. It is all but certain that the specific position of a given locus within regulatory networks plays a large role in determining its rate of divergence. Early evidence suggests that genes belonging to functional network clusters can coevolve in terms of expression level (Fraser 2011), thus opening the possibility that large networks of genes could evolve rapidly in a coordinated manner. As organismal phenotype is ultimately determined via the action of such networks, such a possibility may help to explain the puzzlingly rapid morphological evolution that launched this discussion (King and Wilson 1975). Alternatively, phenotypic outcomes of selection on organismal level traits could occur via expression changes of one or a few loci within the same regulatory network. Similar outcomes could potentially be accomplished via changes in different network members, thus increasing the likelihood of cases of convergent evolution by expanding the mutational target size of traits. Ultimately, our understanding of how divergence of gene expression relates back to divergence in organismal phenotype will certainly require an understanding not only of the structure of complex coregulated networks of loci, but also of the molecular mechanisms by which these loci regulate one another.

Despite the relative infancy of the field, the studies discussed in this chapter have highlighted many broad factors with the potential to influence the rate of divergence of various elements of the transcriptome. In the past, technological constraints have hampered our ability to study the (many) transcriptomes of organisms in a manner that takes into account their heterogeneous nature. Yet those data that we do possess have made it clear that this heterogeneity plays a crucial role in determining patterns and rates at which gene expression evolves. Future analyses will no doubt seek to combine both 'axes' of transcriptome divergence—how expression patterns of tissues and even cells diverge between species in the context of development—in order to identify those genes and networks that are rapidly evolving as well as the evolutionary factors facilitating this divergence.

References

Artieri CG. and Singh RS. (2010a) Molecular evidence for increased regulatory conservation during metamorphosis, and against deleterious cascading effects of hybrid breakdown in Drosophila. *BMC Biol* **8**: 26.

Artieri CG. and Singh RS. (2010b) Demystifying phenotypes: The comparative genomics of evo-devo. *Fly* **4**: 18 20.

Artieri CG, Haerty W, and Singh RS. (2007) Association between levels of coding sequence divergence and gene misregulation in Drosophila male hybrids. *J Mol Evol* **65**: 697–704.

Artieri CG, Haerty W, and Singh RS. (2009) Ontogeny and phylogeny: molecular signatures of selection, constraint, and temporal pleiotropy in the development of Drosophila. *BMC Biol* **7**: 42.

Bedford T. and Hartl DL. (2009) Optimization of gene expression by natural selection. *Proc Nat Acad Sci U S A* **106**: 1133–8.

Blekhman R, Oshlack A, Chabot AE, Smyth GK, and Gilad Y. (2008) Gene regulation in primates evolves under tissue-specific selection pressures. *PloS Genet* **4**: e1000271.

Chen K. and Rajewsky N. (2007) The evolution of gene regulation by transcription factors and microRNAs. *Nat Rev Genet* **8**: 93–103.

Ellegren H. and Parsch J. (2007) The evolution of sex-biased genes and sex-biased gene expression. *Nat Rev Genet* **8**: 689–98.

Fraser HB. (2011) Genome-wide approaches to the study of adaptive gene expression evolution. *Bioessays* **33**: 469–77.

Gould SJ. (1978) *Ontogeny and Phylogeny*. Cambridge, MA: Belknap Press.

Graur D. and Li, WH. (2000) *Fundamentals of Molecular Evolution*, 2nd edition. Sunderland, MA: Sinauer Associates.

Graveley BR, Brooks AN, Carlson JW, Duff MO, Landolin JM, Yang L, et al. (2011) The developmental transcriptome of Drosophila melanogaster. *Nature* **471**: 473–9.

Irie N. and Kuratani S. (2011) Comparative transcriptome analysis reveals vertebrate phylotypic period during organogenesis. *Nat Commun* **2**: 248.

Jordan IK, Mariño-Ramírez L, Koonin EV. (2005) Evolutionary significance of gene expression divergence. *Gene* **345**: 119–26.

Kalinka AT, Varga KM, Gerrard DT, Preibisch S, Corcoran DL, Jarrells J, et al. (2010) Gene expression divergence recapitulates the developmental hourglass model. *Nature* **468**: 811–14.

Khaitovich P, Hellmann I, Enard W, Nowick K, Leinweber M, Franz H, et al. (2005) Parallel patterns of evolution in the genomes and transcriptomes of humans and chimpanzees. *Science* **309**: 1850–4.

King MC and Wilson AC. (1975) Evolution at two levels in humans and chimpanzees. *Science* **188**: 107–16.

Lu Y, Huggins P, and Bar-Joseph Z. (2009) Cross species analysis of microarray expression data. *Bioinformatics* **25**: 1476–83.

Majewski J. and Pastinen T. (2011) The study of eQTL variations by RNA-seq: from SNPs to phenotypes. *Trends Genet* **27**: 72–9.

McManus CJ, Coolon JD, Duff MO, Eipper-Mains J, Graveley BR, and Wittkopp PJ. (2010) Regulatory divergence in Drosophila revealed by mRNA-seq. *Genome Res* **20**: 816–25.

Meisel RP. (2011) Towards a more nuanced understanding of the relationship between sex-biased gene expression and rates of protein-coding sequence evolution. *Mol Biol Evol* **28**: 1893–900.

Papatsenko D, Levine M, and Goltsev Y. (2011) Clusters of temporal discordances reveal distinct embryonic patterning mechanisms in Drosophila and anopheles. *PLoS Biol* **9**: e1000584.

Raff RA. (1996) *The Shape of Life: Genes, Development, and the Evolution of Animal Form*. Chicago, IL: University of Chicago Press.

Saxena A. and Carninci P. (2011) Whole transcriptome analysis: what are we still missing? *Wiley Interdiscip Rev Syst Biol Med* **3**: 527–43.

Schena M, Shalon D, Davis RW, and Brown PO. (1995) Quantitative monitoring of gene expression patterns with a complementary DNA microarray. *Science* **270**: 467–70.

Singh RS and Artieri CG. (2010) Male sex drive and the maintenance of sex: evidence from Drosophila. *J Hered* **101**(Suppl 1): S100–6.

Sucena E. and Stern DL. (2000) Divergence of larval morphology between Drosophila sechellia and its sibling species caused by cis-regulatory evolution of ovo/shaven-baby. *Proc Natl Acad Sci U S A* **97**: 4530–34.

Tang, F., Barbacioru, C., Nordman, E., Li, B., Xu, N., Bashkirov, V.I., et al. (2010) RNA-Seq analysis to capture the transcriptome landscape of a single cell. *Nat Protoc* **5**: 516–35.

Wang Z, Gerstein M, and Snyder M. (2009) RNA-Seq: a revolutionary tool for transcriptomics. *Nat Rev Genet* **10**: 57–63.

Xing Y and Lee C. (2006) Alternative splicing and RNA selection pressure—evolutionary consequences for eukaryotic genomes. *Nat Rev Genet* **7**: 499–509.

Zhang Y, Sturgill D, Parisi M, Kumar S, and Oliver B. (2007) Constraint and turnover in sex-biased gene expression in the genus Drosophila. *Nature* **450**: 233–7.

CHAPTER 24

Rate variation in the evolution of development: a phylogenetic perspective

Artyom Kopp

24.1 Introduction

The words 'rapid evolution' imply a quantitative comparison. Some characters can change more rapidly than others, or the same character can evolve at different rates in different groups of organisms (Fig. 24.1). In either case, rapid evolution is a testable quantitative hypothesis; in fact, a large body of phylogenetic literature is devoted to formulating and testing hypotheses about variation in evolutionary rates. Rigorous quantitative methods have been developed for both discrete and continuously variable traits (Pagel et al. 2004; O'Meara et al. 2006). Both types of methods incorporate explicit models of character evolution and use probabilistic reconstruction of phylogenetic relationships and ancestral character states so that the marginal estimates of evolutionary rates account for both sources of uncertainty. The quantitative power of these approaches makes it possible to compare the rates of evolution between different characters, pinpoint the nodes in the phylogeny where the rates of character evolution change, and test for coevolution of different traits or correlation between phenotypes and lineage diversification or external ecological changes. There are many examples where the application of phylogenetically structured tests has changed our inference of evolutionary scenarios from comparative data (Garland et al. 2005).

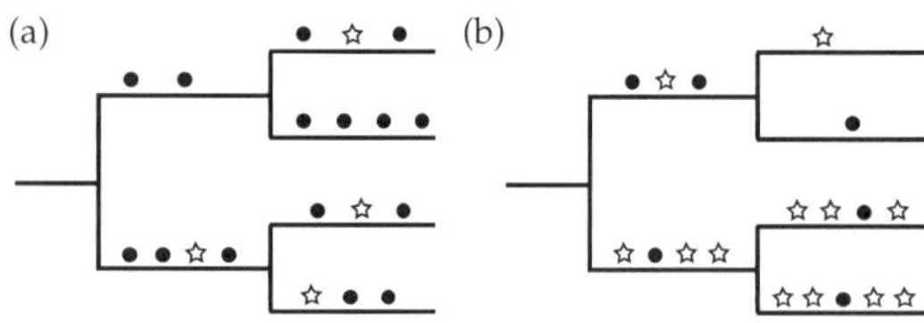

Figure 24.1 Different scenarios of variation in evolutionary rates. Circles and stars represent different groups of characters, for example, the presence or absence of particular genes or gene interactions in developmental pathways. Symbols above each branch indicate evolutionary changes in character states that have occurred on that branch. (a) The closed-circle characters show rapid evolution in the entire lineage, relative to the star-shaped characters. (b) The closed-circle characters evolve at a roughly constant rate, while the star-shaped characters show accelerated evolution in the bottom clade.

Evolutionary developmental biology ('evo-devo') has lagged behind other areas in the application of phylogenetic approaches. Most evo-devo studies examine a small number of taxa and make no use of quantitative comparative methods, relying instead on a simple parsimony framework to look at character evolution. This is not a sign of innumeracy or conservatism on the part of developmental biologists; rather, it reflects great difficulties in obtaining suitable data. Development, and especially the structure of developmental gene networks, is perhaps the hardest level of biological organization to compare and quantify on a sufficiently large scale. In contrast, molecular sequences lend themselves naturally to quantitative measures, so that comparing evolutionary rates and patterns among different characters and clades is a matter of relatively straightforward mathematics (Li 2006). Similarly, a variety of well-established methods exist for quantifying variation in morphological and other terminal phenotypes (Zelditch 2004). Importantly, both molecular sequences and morphological traits can be determined from one or

Rapidly Evolving Genes and Genetic Systems. First Edition. Edited by Rama S. Singh, Jianping Xu, and Rob J. Kulathinal.

a few dead specimens, and thus can be compared in large numbers of non-model taxa.

Development, the process that connects molecular sequences to adult phenotypes, is far less accessible. Developmental traits demand experimental analysis, especially if the structure of genetic pathways is to be determined. Such analysis requires each taxon to be cultured in the lab, and is so laborious that large taxon samples are usually impractical. At the genetic level, the tools for experimental analysis simply do not exist in most species. It is not surprising, therefore, that few studies have attempted to apply quantitative phylogenetic methods to the evolution of development. We can document rapid evolution for many phenotypic traits, but in most cases we know little about their developmental underpinnings. For example, genome-wide analyses of gene expression and the pervasive pattern of hybrid male sterility indicate that testis development and spermatogenesis evolve at very high rates, but our understanding of the genetic basis of these changes is fragmentary at best (Coyne and Orr 2004; Ellegren and Parsch 2007). At the opposite side of the spectrum, some genetic pathways have been characterized in great detail (Busser et al. 2008), but we lack sufficient taxon samples to look at the tempo and mode of their evolution. Only recently, and only for a few traits, are the conflicting requirements of in-depth mechanistic analysis and large comparative datasets beginning to be met.

As evo-devo is becoming a mature field, quantitative phylogenetic methods hold great promise as a unifying framework despite the difficulties in their application. Numbers and phylogenetic trees are the common currency of evolutionary biology, allowing examples drawn from different traits and organisms to be integrated and compared. Thus, tree-based quantification of developmental changes is essential for any effort to draw general lessons from diverse case studies and perhaps identify some overarching rules of developmental evolution. Do some developmental pathways evolve more rapidly than others? Is the rate of their evolution uniform, or accelerated in particular clades or during certain periods in their history? Are evolutionary changes equally likely throughout the pathway, or concentrated at specific nodes? Most importantly, what are the reasons for rate variation between developmental pathways and evolutionary lineages? Can differences in the rate of evolution be explained by external forces, or do rapidly evolving pathways and nodes share some intrinsic topological or molecular features? All these questions are inherently quantitative and comparative, and can only be addressed from a phylogenetic perspective.

In this chapter, I review several recent studies that examine the evolution of development using phylogenetic methods. I look at the challenges that stand in the way of this approach, and preview the types of general questions that will be opened for investigation as tree-based analyses take hold in the evo-devo field.

24.2 Examples of rate variation in the evolution of development

24.2.1 Same clade, different pathways: evolution of vulval development in rhabditid nematodes

A detailed understanding of cell–cell interactions during vulval development in the model nematode *Caenorhabditis elegans* (Sternberg 2005) has opened the way for investigating the evolution of developmental processes in the larger lineage of rhabditid worms (Kiontke et al. 2007). In *C. elegans*, the vulva is composed of the progeny of three ventral ectoblast cells: P5.p, P6.p, and P7.p. (The prefix 'P' denotes the embryonic cell lineage that gives rise to the vulva and some other tissues, while the suffix '.p' indicates the posterior daughter of the previous cell division.) P6.p is the 'primary' cell that produces eight terminally differentiated daughter cells, while the 'secondary' ectoblasts, P5.p and P7.p, produce seven progeny each. The adjacent ventral ectoblasts, P3.p, P4.p, and P8.p, are the 'tertiary' cells that have the potential to become vulval progenitors if exposed to the proper inductive signal but retain epidermal fates in wild-type *C. elegans*. Thus, P3.p–P8.p comprise a six-cell vulval 'competence group.' The inductive signal in *C. elegans* is provided by the anchor cell (AC) in the somatic gonad. In the absence of this signal, all ventral ectoblasts retain the epidermal fate and no vulva develops.

Although the vulva forms from the ventral ectoblasts in all rhabditid nematodes, there are many differences among species in its development. In some species, including the well-studied *Pristionchus pacificus*, vulval induction requires signaling from multiple somatic gonad cells rather than from the AC alone (Sommer 2005; Kiontke et al. 2007). In other genera, such as *Mesorhabditis* and *Diplogastrellus*, at least some species show gonad-independent vulval development, indicating that no inductive signal is required at all (Kiontke et al. 2007). The composition and fates of the vulval competence group also differ widely within rhabditida. In some species, P3.p and P4.p undergo programmed cell death and do not contribute to any adult structures, while in others P3.p, P4.p, and P8.p survive but are not competent to develop into vulval cells even in the presence of an inductive signal (Kiontke et al. 2007).

Phylogenetic analysis of over 40 developmental characters in 51 rhabditid species reveals that different aspects of vulval development show different rates and patterns of evolution (Kiontke et al. 2007). Some traits, such as which ectoblasts contribute to the adult vulva, have not undergone any changes within this lineage. Other characters, such as the composition of the competence group and the source of the inductive signal, show multiple evolutionary changes among the same species (Fig. 24.2). Developmental characters can vary in the directionality as well as the frequency of transitions. For example, the size of the vulval competence group shows a directional trend toward reduction (double arrowheads in Fig. 24.2) while the source of the inductive signal has changed in both directions (single arrowheads in Fig. 24.2) (Kiontke et al. 2007).

The study of Kiontke et al. (2007) represents one of the most extensive phylogenetic analyses of development to date, and demonstrates how comparative methods can be used to quantify the rate of evolutionary change in development and reveal quantitative trends that are not obvious otherwise. A large taxon sample essential for the phylogenetic approach was made possible in this case by the relative ease of analyzing cell interactions by laser ablation—an experimental technique that does not rely on genetic tools and can be implemented in non-model species. Naturally, this approach has limited depth, as it does not reveal the genetic pathways underlying cell fate decisions. Genetic analysis shows that these pathways have diverged considerably between *C. elegans* and *P. pacificus* (Sommer 2009), hinting at an even more rapid evolution of development than is apparent from cellular-level analysis.

24.2.2 Same pathway, different clades: evolution of sex combs and pigmentation in *Drosophila*

In *Drosophila*, as in nematodes, the power of an experimental model species combines with high phenotypic diversity among its relatives to enable comparative studies of development. *D. melanogaster*, the workhorse of developmental genetics, belongs to a speciose clade (the *melanogaster* species group) that is notable for its diversity of sex-specific morphological traits. The best studied of these traits are sex combs and color patterns. The sex comb is a group of modified mechanosensory bristles that develops on the front legs of males and is used during courtship and mating. This structure is a recent evolutionary innovation; it is absent in most *Drosophila* species but has undergone dramatic diversification in the lineage that includes the *melanogaster* species group. Sex comb development is controlled by a sex- and segment-specific developmental pathway centered on the HOX gene *Sex combs reduced* (*Scr*) and the sex determination gene *doublesex* (*dsx*) (Fig. 24.3a) (Tanaka et al. 2011). The diversity of sex comb structures in the *melanogaster* group reflects frequent evolutionary changes in the regulation of *Scr* and *dsx*.

In many species of the *melanogaster* group, the pigmentation of wings and posterior abdominal segments is also sexually dimorphic (Kopp et al. 2000; Prud'homme et al. 2006). The genetic control of color patterns is better understood in the abdomen, where it involves an interaction between *dsx* and another HOX gene, *Abdominal-B* (*Abd-B*) (Fig. 24.3b) (Kopp et al. 2000; Williams et al. 2008). These and other transcription factors differentially regulate enzymes in the melanin synthesis pathway to produce different combination of pigments in different regions of the cuticle (Wittkopp et al. 2003).

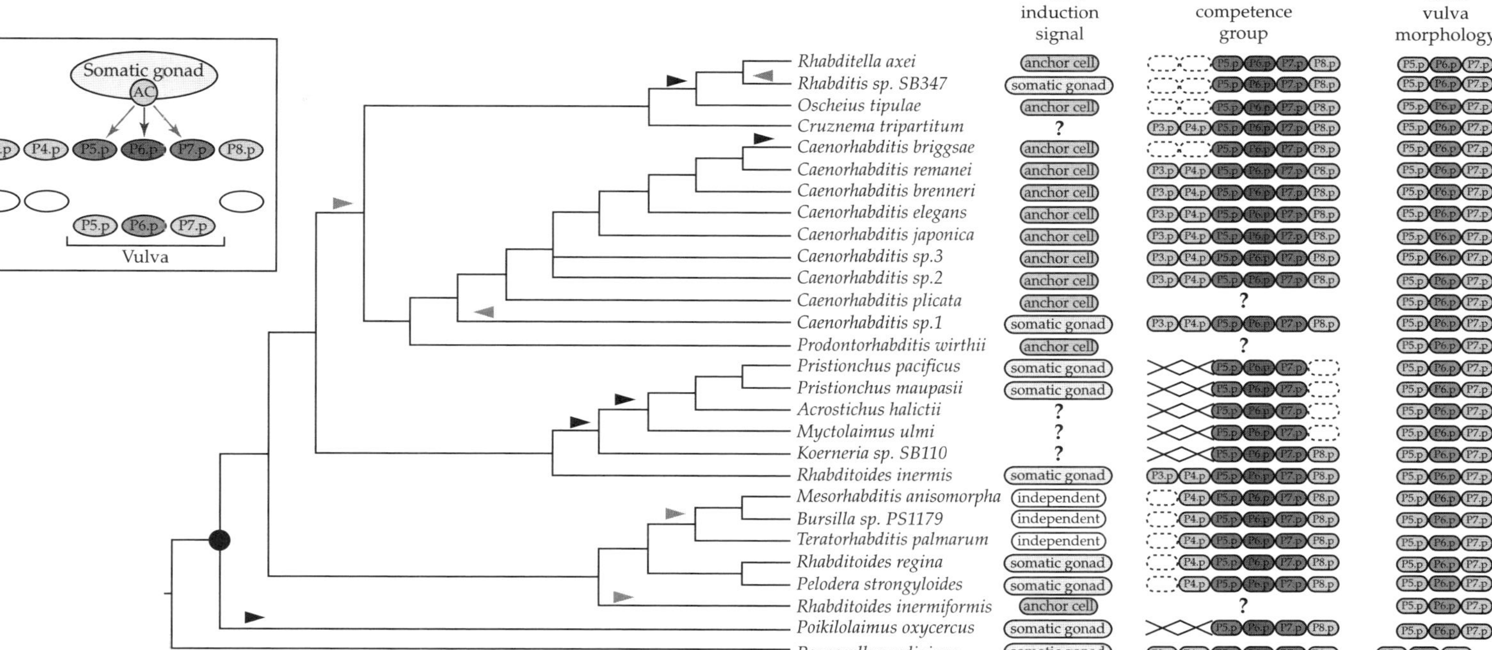

Figure 24.2 Evolution of vulval development in rhabditid nematodes (based on Kiontke et al. (2007)). The inset summarizes vulval development in *C. elegans*. P3.p–P8.p (middle row) form a six-cell equivalence group, but only P5.p–P7.p contribute to the adult vulva (lower row). Vulval development is induced by signaling (arrows) from the anchor cell (AC) located in the somatic gonad (top row). In response to this signaling, P6.p becomes the primary vulval cell (dark gray in middle and lower rows) while P5.p and P7.p become secondary cells (medium gray, middle row; and light gray, lower row). In wild-type *C. elegans*, P3.p, P4.p, and P8.p (light gray middle row) do not receive the inductive signal and thus retain the epithelial fate (white). The main panel shows the phylogeny of rhabditid nematodes with experimentally determined vulval competence groups in the centre column, the source of inductive signal in the left column, and the composition of the adult vulva in the right column. Arrowheads at internal nodes (grey arrowheads refer to the first column; black arrowheads to the middle column) indicate inferred evolutionary transitions in developmental character states. The direction of each arrowhead shows the direction of evolutionary change. For example, a grey arrowhead pointing to the right shows a change in the source of the inductive signal from multiple somatic gonad cells (the inferred ancestral state) to anchor cell alone or gonad-independent development, while a grey arrowhead pointing to the left indicates a reversion toward the ancestral state.

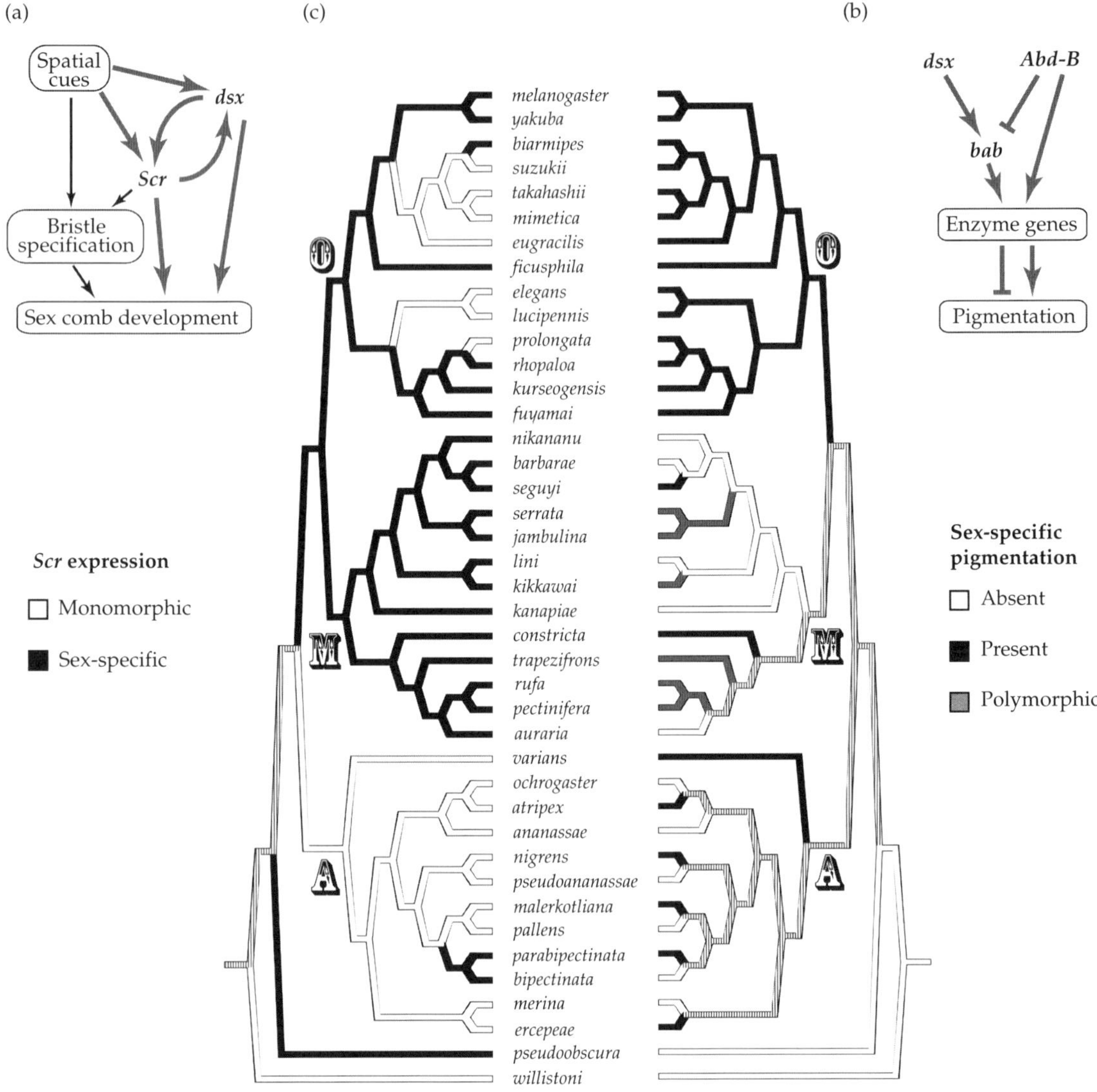

Figure 24.3 Evolution of sex comb development and sex-specific abdominal pigmentation in the *melanogaster* species group. (a) Genetic control of sex comb development (simplified from (Tanaka et al., 2011). (b) Genetic control of abdominal pigmentation (simplified from Wittkopp et al. (2003)). (c) Phylogeny of the *melanogaster* species group and the distribution of sex-specific characters in different clades (taxon sample reduced from Barmina and Kopp (2007) and Jeong et al. (2006)). *O* (Oriental), *M* (*montium*), and *A* (*ananassae*) labels at internal nodes mark the three major clades within the *melanogaster* species group. *D. pseudoobscura* and *D. willistoni* are outgroup species. Hatched branches indicate ambiguous state reconstructions.

Sex-specific pigmentation of posterior abdominal segments has been gained and lost many times due to changes both in the upstream transcription factors and in the pigment synthesis enzymes (Jeong et al. 2006; Williams et al. 2008). Importantly, genetic analysis shows that changes in different loci were responsible for the evolution of similar color patterns in different species (Kopp 2009).

Extensive taxon sampling in some of these studies allows us to examine the evolution of developmental traits using quantitative phylogenetic methodology. Barmina and Kopp (2007) used

Bayesian character reconstruction to show that sex-specific expression of *Scr* in the presumptive sex comb region has been gained and lost multiple times in the *melanogaster* species group, and that these gains and losses are tightly correlated with parallel transitions between different sex comb morphologies. Other phylogenetic hypotheses (such as a single gain of sex-specific *Scr* expression followed by repeated losses, or multiple gains with no losses) could be rejected with good statistical support. A similar approach was used to show that each of two alternative cellular mechanisms of sex comb development has likely evolved more than once (Tanaka et al. 2009).

Interestingly, evolutionary changes in these developmental pathways are not distributed evenly on the phylogeny (Fig. 24.3c). Comparison among the three major clades within the *melanogaster* species group (the *montium* and *ananassae* subgroups and the Oriental lineage) is particularly intriguing. Sex-specific expression of *Scr* has been gained and lost repeatedly in the Oriental lineage but is completely conserved in its sister clade, the *montium* subgroup. Conversely, sex-specific abdominal pigmentation shows rapid evolution in the *ananassae* and *montium* subgroups, but is static in the Oriental lineage (Fig. 24.3c). These comparisons suggest that the rate of evolution of the same developmental pathway can vary between sister clades, prompting justifiable questions about the causes of such variation (see following section).

24.2.3 Same clade, same pathway, different genes: evolution of embryonic development and sex determination in insects

Evolutionary rates can vary within as well as between developmental pathways. A classical example comes from the gene network that controls insect sex determination. In all insects studied to date, sexual dimorphism in most somatic tissues is controlled by the sex-specific splicing of the *dsx* transcription factor: a male-specific isoform is produced in males and promotes male phenotypes, and a female-specific isoform is produced in females and promotes female-specific traits (Verhulst et al. 2010). Sex-specific splicing of *dsx* is in turn controlled by the RNA-binding protein encoded by *transformer* (*tra*). The *tra/dsx* developmental nexus is conserved in all holometabolous insects, a clade that spans at least 275 million years of divergence (Verhulst et al. 2010). In contrast, the more upstream parts of this pathway, involving the primary sex-determining signals and the upstream regulators of *tra*, evolve with extraordinary speed. Insect sex determination mechanisms include XY and ZW sex chromosome systems, dominant male- or female-determining genes, haplodiploidy, environmental sex determination, and others (Verhulst et al. 2010). In fact, the primary sex determination signal can vary even within species (Hediger et al. 2010). At the same time, most sexually dimorphic phenotypes are lineage-specific, indicating that the downstream parts of the sex determination pathway, such as the downstream targets of *dsx*, are also turning over at a high rate.

A similar 'hourglass' pattern, where the top and bottom tiers of developmental pathways diverge more rapidly than the middle, is seen in embryonic development. Analysis of over 3000 genes in the embryos of six *Drosophila* species shows that expression divergence is lowest, and selective constraint highest, at the extended germband stage, while the earlier and later developmental stages show greater variation in gene expression (Kalinka et al. 2010). The hourglass divergence pattern is most pronounced for genes involved in gene regulation and cell differentiation; genes that function in metabolism, immunity, stress response, and other non-developmental processes do not tend to follow this pattern. Analysis of specific developmental pathways also reveals within-pathway rate variation consistent with the hourglass model. In the segmentation hierarchy, which is among the best characterized pathways in any organism, the bottom tier of genes (segment polarity genes, which establish segmental boundaries and initiate patterning within segments) show highly conserved expression across insect orders, while the more upstream tiers such as the gap genes show major differences even among Dipteran families (Peel et al. 2005; Jaeger 2011).

On microevolutionary timescales, genetic analysis of various traits shows that convergent phenotypic changes in different species are often, though not always, due to changes in the same loci, leading to the suggestion that some genes are pre-disposed to greater evolutionary changes by virtue of their position in developmental pathways (Gompel and Prud'homme 2009; Kopp 2009; Stern and Orgogozo 2009). Overall, it appears that within-pathway variation in evolutionary rates is a common pattern.

24.3 Technical and conceptual challenges to quantifying the evolution of development

The most obvious obstacle to the application of phylogenetic methods in developmental biology is, of course, the sheer difficulty of characterizing development in many different taxa. Although new technologies are rapidly enabling comparative analyses of genomes and genome-wide expression profiles in non-model taxa, the study of development will always require experimentation. This inevitably limits the taxonomic scale of evo-devo studies and restricts the use of quantitative comparative methods to relatively few traits. As the examples in this chapter show, the most promising traits for this analysis are those that vary among the relatives of well-established model taxa, especially where additional experimental models can be developed (Sommer 2009).

Even in a fantasy universe where developmental pathways could be studied rapidly and cheaply in any number of taxa, there would be significant challenges to applying quantitative comparative methods to the evolution of development. How does one code developmental characters on phylogenetic trees? Since the blueprint for development is contained in regulatory interactions among genes, one possible approach is to treat each interaction as a discrete character with two states (present/absent) and estimate the rate of evolutionary transitions between these states. This approach was used, for example, by Barmina and Kopp (2007) to reconstruct the evolution of sex-specific *Scr* regulation and link it to the diversification of *Drosophila* sex combs. Under this model, each pathway can be represented as a group of binary characters, and phylogenetic methods developed for other discrete traits can be used to quantify the rates of their evolution and ask whether different pathways evolve at different rates, whether the rate of evolution of the same pathway varies across lineages, whether different parts of the same pathway show evidence of correlated changes, and so on.

For larger pathways, it may be practical to quantify the rate of change using network-based metrics. One possible approach was recently proposed by Shou et al. (2011), who assumed that a network edge (e.g. a regulatory interaction) is orthologous between two species if the nodes it connects (e.g. transcription factor and its target) are orthologous, and estimated the rate of network rewiring from the fraction of conserved orthologous edges. Their analysis showed, for example, that transcriptional and other regulatory networks evolve faster than protein–protein interaction or metabolic networks (Shou et al. 2011); a similar approach could be used to compare different regulatory pathways.

However, treating regulatory interactions as discrete characters is a simplification. In reality, gene interactions are quantitative and we can expect their changes to be quantitative as well, especially on microevolutionary timescales. Quantitative changes in regulatory interactions can lead to qualitatively different phenotypes; the vulval specification pathway reviewed earlier is particularly illuminating in this respect. Computational models show that changes in the quantitative tuning of the same signaling network, without any changes in network topology, can produce a wide variety of cellular phenotypes including species-specific cell behaviors observed in different *Caenorhabditis* species (Giurumescu et al. 2009; Hoyos et al. 2011). Needless to say, measuring quantitative changes in regulatory interactions is even harder than ascertaining their presence or absence in each taxon.

The second conceptual obstacle to quantifying the evolution of developmental pathways is the inevitable vagueness in defining the pathways themselves. Since most genes play multiple roles in development, gene networks that control different phenotypes are in fact interconnected. In focusing on a particular trait, where do we draw the boundaries of the relevant developmental pathway? These decisions are somewhat arbitrary and affected by historical and experimental contingencies. In the future, formal network analysis may complement the traditional approach that proceeds from traits to genes to pathways. As large transcriptional networks are beginning to be elucidated using genome-scale approaches, objective mathematical algorithms can be used to decompose these networks into modules that show relatively tight interconnection within and sparser links to other modules (Newman and Girvan 2004; Guimera and Nunes Amaral 2005). If such network modules can be linked to particular phenotypes, they may offer a more objective way of defining trait-specific pathways and characterizing their evolution.

24.4 Future directions: the promise of phylogenetic approaches to the evolution of development

Despite the difficulties in applying formal phylogenetic methods to the evolution of development, this approach can play an important role as evo-devo becomes a mature field. Accumulation of empirical data from a variety of organisms and traits is spurring efforts to synthesize the lessons from individual case studies and identify general rules of developmental evolution (Carroll 2008; Stern and Orgogozo 2008; Kopp 2009). Phylogenetic techniques can contribute to this synthesis in two major ways: by providing an objective framework for documenting systematic trends in the evolution of development, and by helping to identify the causative factors that may explain these trends.

Description is always the first step toward explanation: there is little point in seeking a causative process if the pattern it seeks to explain is poorly supported. As this chapter illustrates, there are many cases where some pathways or clades appear to evolve more rapidly than others. Mapping character changes on phylogenetic trees using quantitative methods is essential for identifying these differences and testing their significance. Parsimony approaches that dominate the evo-devo literature offer only limited utility in this regard. Since development is clearly susceptible to homoplastic change (Barmina and Kopp 2007; Kiontke et al. 2007; Tanaka et al. 2009), explicit models of character evolution that incorporate transition probabilities and branch length information can offer substantial improvement in quantifying the rate of evolution and testing for correlation among characters. Bayesian and likelihood-based methods can assess statistical support for competing scenarios of character evolution (Pagel et al. 2004; O'Meara et al. 2006). When applied to developmental pathways, these methods can estimate the minimum number of evolutionary changes and the probability of transitions between different types of development (Barmina and Kopp 2007; Feng et al. 2011).

If phylogenetic analysis confirms the apparent pattern, a bigger question awaits: what factors explain the differences in evolutionary rates within and between developmental pathways, or between different clades or periods in evolution? This is where phylogenetic methods can yield some of the biggest breakthroughs in evo-devo, as they provide formal statistical tools for identifying correlations between evolutionary patterns and candidate explanatory factors. These factors can be either external to development, such as ecological opportunities or sexual selection, or intrinsic, i.e. differences in the structure of the developmental pathways themselves. To offer a few examples: do the pathways downstream of sex-determining genes show faster evolution than other developmental pathways? Does development evolve more rapidly during periods of rapid speciation? Does the rate of evolution correlate with topological features of the pathway? Does the origin of a novel regulatory interaction precipitate a burst of rapid evolution elsewhere in the same pathway? Is there evidence of coevolution among different regulatory links, and can this coevolution explain convergent phenotypic changes?

Although correlation is not the same as causation, comparative analysis plays an essential role in the development of evolutionary theories that seek to explain the patterns of species richness and phenotypic diversity. Evaluating support for potential cause–effect relationships in comparative data has become one of the central directions in modern phylogenetics. For example, phylogenetic analysis shows that the transition from solitary to cooperative breeding in birds is more likely to occur in species with less promiscuous mating systems, supporting a causal link between genetic relatedness and cooperation (Cornwallis et al. 2010). Similarly, a phylogenetic study of centrarchid fishes showed that piscivory reduces the rate of evolution of skull and jaw morphology, suggesting that the proximity of an evolving lineage to an adaptive peak limits the rate of its phenotypic diversification (Collar et al. 2009). Importantly, tree-based tests can help resolve causal relationships among multiple correlated variables. For instance, it was possible to disentangle habitat use, mating system, and body size in ungulates to show that polygyny consistently leads to the evolution of sexual size dimorphism (Perez-Barberia et al. 2002). In all these examples, comparing the fit of different models of character evolution led to the identification of historical trends that were not immediately obvious, and helped distinguish between competing evolutionary explanations for these trends. We can hope that, as empirical data continue to accumulate, the application of quantitative phylogenetic methods in evo-devo will lead to similar advances in defining the general patterns and rules of developmental evolution.

References

Barmina, O., and Kopp, A. (2007) Sex-specific expression of a HOX gene associated with rapid morphological evolution. *Dev Biol* **311**: 277–86.

Busser, B.W., Bulyk, M.L., and Michelson, A.M. (2008) Toward a systems-level understanding of developmental regulatory networks. *Curr Opin Genet Dev* **18**: 521–9.

Carroll, S.B. (2008) Evo-devo and an expanding evolutionary synthesis: a genetic theory of morphological evolution. *Cell* **134**: 25–36.

Collar, D.C., O'Meara, B.C., Wainwright, P.C., and Near, T.J. (2009) Piscivory limits diversification of feeding morphology in centrarchid fishes. *Evolution* **63**: 1557–73.

Cornwallis, C.K., West, S.A., Davis, K.E., and Griffin, A.S. (2010) Promiscuity and the evolutionary transition to complex societies. *Nature* **466**: 969–72.

Coyne, J.A. and Orr, H.A. (2004) *Speciation*. Sunderland, MA: Sinauer Associates.

Ellegren, H., and Parsch, J. (2007) The evolution of sex-biased genes and sex-biased gene expression. *Nat Rev Genet* **8**: 689–98.

Feng, C.M., Xiang, Q.Y., and Franks, R.G. (2011) Phylogeny-based developmental analyses illuminate evolution of inflorescence architectures in dogwoods (Cornus s. l., Cornaceae). New Phytol **191**(3): 850–69.

Garland, T., Jr., Bennett, A.F., and Rezende, E.L. (2005) Phylogenetic approaches in comparative physiology. *J Exp Biol* **208**: 3015–35.

Giurumescu, C.A., Sternberg, P.W., and Asthagiri, A.R. (2009) Predicting phenotypic diversity and the underlying quantitative molecular transitions. *PLoS Comput Biol* **5**: e1000354.

Gompel, N., and Prud'homme, B. (2009) The causes of repeated genetic evolution. *Dev Biol* **332**: 36–47.

Guimera, R., and Nunes Amaral, L.A. (2005) Functional cartography of complex metabolic networks. *Nature* **433**: 895–900.

Hediger, M., Henggeler, C., Meier, N., Perez, R., Saccone, G., and Bopp, D. (2010) Molecular characterization of the key switch F provides a basis for understanding the rapid divergence of the sex-determining pathway in the housefly. *Genetics* **184**: 155–70.

Hoyos, E., Kim, K., Milloz, J., Barkoulas, M., Penigault, J.B., Munro, E., and Felix, M.A. (2011) Quantitative variation in autocrine signaling and pathway crosstalk in the Caenorhabditis vulval network. *Curr Biol* **21**: 527–38.

Jaeger, J. (2011) The gap gene network. *Cell Mol Life Sci* **68**: 243–74.

Jeong, S., Rokas, A., and Carroll, S.B. (2006) Regulation of body pigmentation by the Abdominal-B Hox protein and its gain and loss in Drosophila evolution. *Cell* **125**: 1387–99.

Kalinka, A.T., Varga, K.M., Gerrard, D.T., Preibisch, S., Corcoran, D.L., Jarrells, J., et al. (2010) Gene expression divergence recapitulates the developmental hourglass model. *Nature* **468**: 811–14.

Kiontke, K., Barriere, A., Kolotuev, I., Podbilewicz, B., Sommer, R., Fitch, D.H., et al. (2007) Trends, stasis, and drift in the evolution of nematode vulva development. *Curr Biol* **17**: 1925–37.

Kopp, A. (2009) Metamodels and phylogenetic replication: a systematic approach to the evolution of developmental pathways. *Evolution* **63**: 2771–89.

Kopp, A., Duncan, I., Godt, D., and Carroll, S.B. (2000) Genetic control and evolution of sexually dimorphic characters in Drosophila. *Nature* **408**: 553–9.

Li, W.-H. (2006) *Molecular evolution*. Sunderland, MA: Sinauer Associates.

Newman, M.E., and Girvan, M. (2004) Finding and evaluating community structure in networks. *Phys Rev E Stat Nonlin Soft Matter Phys* **69**: 026113.

O'Meara, B.C., Ane, C., Sanderson, M.J., and Wainwright, P.C. (2006) Testing for different rates of continuous trait evolution using likelihood. *Evolution* **60**: 922–33.

Pagel, M., Meade, A., and Barker, D. (2004) Bayesian estimation of ancestral character states on phylogenies. *Syst Biol* **53**: 673–84.

Peel, A.D., Chipman, A.D., and Akam, M. (2005) Arthropod segmentation: beyond the Drosophila paradigm. *Nat Rev Genet* **6**: 905–16.

Perez-Barberia, F.J., Gordon, I.J., and Pagel, M. (2002) The origins of sexual dimorphism in body size in ungulates. *Evolution* **56**: 1276–85.

Prud'homme, B., Gompel, N., Rokas, A., Kassner, V.A., Williams, T.M., Yeh, S.D., et al. (2006) Repeated morphological evolution through cis-regulatory changes in a pleiotropic gene. *Nature* **440**: 1050–3.

Shou, C., Bhardwaj, N., Lam, H.Y., Yan, K.K., Kim, P.M., Snyder, M., et al. (2011) Measuring the evolutionary rewiring of biological networks. *PLoS Comput Biol* **7**: e1001050.

Sommer, R.J. (2005) Evolution of development in nematodes related to C. elegans. *WormBook* **Dec 14**: 1–17.

Sommer, R.J. (2009) The future of evo-devo: model systems and evolutionary theory. *Nat Rev Genet* **10**: 416–22.

Stern, D.L., and Orgogozo, V. (2008) The loci of evolution: how predictable is genetic evolution? *Evolution* **62**: 2155–77.

Stern, D.L., and Orgogozo, V. (2009) Is genetic evolution predictable? *Science* **323**: 746–51.

Sternberg, P.W. (2005) Vulval development. *WormBook* **Jun 25**: 1–28.

Tanaka, K., Barmina, O., and Kopp, A. (2009) Distinct developmental mechanisms underlie the evolutionary diversification of Drosophila sex combs. *Proc Natl Acad Sci U S A* **106**: 4764–9.

Tanaka, K., Barmina, O., Sanders, L.E., Arbeitman, M.N., and Kopp, A. (2011) Evolution of sex-specific traits through changes in HOX-dependent doublesex expression. *PLoS Biol* **9**: e1001131.

Verhulst, E.C., van de Zande, L., and Beukeboom, L.W. (2010) Insect sex determination: it all evolves around transformer. *Curr Opin Genet Dev* **20**, 376–83.

Williams, T.M., Selegue, J.E., Werner, T., Gompel, N., Kopp, A., and Carroll, S.B. (2008) The regulation and evolution of a genetic switch controlling sexually dimorphic traits in Drosophila. *Cell* **134**: 610–23.

Wittkopp, P.J., Carroll, S.B., and Kopp, A. (2003) Evolution in black and white: genetic control of pigment patterns in Drosophila. *Trends Genet* **19**: 495–504.

Zelditch, M. (2004) *Geometric morphometrics for biologists: a primer*. Amsterdam: Elsevier Academic Press.

CHAPTER 25

Natural hybridization as a catalyst of rapid evolutionary change

Michael L. Arnold, Jennafer A.P. Hamlin, Amanda N. Brothers, and Evangeline S. Ballerini

25.1 Introduction

That natural hybridization has played an important role in many clades in terms of both genomic and organismal evolution is now well established (Anderson and Stebbins 1954; Arnold 1997, 2006; Mallet 2005; Baack and Rieseberg 2007; Soltis and Soltis 2009). Sexual reproduction involving individuals from divergent evolutionary lineages can result in small- or large-scale genomic reorganization (e.g. gene silencing, multiplication of repetitive elements; see Arnold 2006; Baack and Rieseberg 2007), adaptive trait transfer (or origin), hybrid speciation and adaptive radiations (Anderson 1949; Arnold 1997, 2006; Seehausen 2004; Soltis and Soltis 2009). Furthermore, the evolutionary outcomes from natural hybridization are expected to be rapid because the effect from combining genomes from divergent lineages mimics a large-scale mutation event in which many unique genotypes are produced simultaneously (Stebbins 1959). Thus, evolutionary events resulting from hybridization should occur over a short time span due to the production of extreme genotypic and phenotypic novelty upon which natural selection can then act.

The outcomes from natural hybridization listed previously should not be considered discrete relative to one another, but rather viewed much more like interacting terms in a mathematical formula. For example, natural hybridization-mediated species diversification may very well depend upon genomic reorganization (Arnold 2006; Soltis and Soltis 2009). Likewise, the transfer of genetic material between divergent lineages may cause changes in the pattern of expression of the introgressed genes due to their new genomic surroundings (Baack and Rieseberg 2007). Yet, the categories of possible outcomes are useful constructs for discussing the role played by genetic exchange in the evolution of various species complexes. In the next section, we will focus on the potential outcomes of natural hybridization affecting adaptive evolution and diversification. In particular, we will discuss findings that test whether genetic exchange has led to adaptive trait introgression, hybrid speciation, and/or adaptive radiation in various plant and animal assemblages. In each case, we will specifically address the expectation that such evolutionary innovations will occur over relatively short time spans.

25.2 Adaptive trait introgression: when strange is really good

Like introgressive hybridization (or simply, 'introgression'), the concept of adaptive trait introgression originated with Edgar Anderson and his colleagues. Anderson and Hubricht (1938) defined introgression in the following manner: '... through repeated back-crossing of the hybrids to the parental species there is an infiltration of the germplasm of one species into that of another.' Though introgression has been seen across the breadth of eukaryotic clades, like point mutations we would assume that much of this transfer would be non-adaptive. However, Anderson and Stebbins (1954) proposed that when adaptive effects do occur, it is because:

Rapidly Evolving Genes and Genetic Systems. First Edition. Edited by Rama S. Singh, Jianping Xu, and Rob J. Kulathinal.

> introgressive hybridization elements of an entirely foreign genetic adaptive system can be carried over into a previously stabilized one, permitting the rapid reshuffling of varying adaptations and complex modifier systems. Natural selection is presented not with one or two new alleles but with segregating blocks of genic material belonging to entirely different adaptive systems.

It is important to note that these authors were not mainly pointing to the role of introgression in the transfer of previously evolved adaptations, but rather their emphasis was on the origin of new adaptive systems through these genetic exchange events (Anderson and Stebbins 1954).

From the earlier description we conclude that 'adaptive trait introgression' can reflect either the transfer of adaptations present in one of the hybridizing lineages or, alternatively, the development of novel adaptations due to recombination between divergent genomes. One of the best examples of such transfer is the species complex of Darwin's finches (Grant and Grant 2010). It is important to note that not only is adaptive trait introgression apparent among Darwin's finch species, but the speed at which adaptive evolution was observed to occur (i.e. over a few generations; see Grant and Grant 2010) is instructive for the present discussion. Likewise, our own work with the plant complex known as the Louisiana Irises has identified both rapid adaptive trait introgression as well as the origin of novel adaptive traits within two generations of hybridization (Arnold and Martin 2010). In the next two sections, we will focus on one example each from an animal and fungal clade that reflect the rapid nature of the transfer and origin of adaptive traits via introgressive hybridization.

25.2.1 Adaptive trait transfer in *Canis*: wolves in dogs' clothing

The evolution of animal pigmentation is often described as adaptive, and is at least partially controlled by interactions of various genes of the melanocortin pathway (e.g. Hoekstra et al. 2006). Though mutations in the *Melanocortin 1 receptor* (*Mc1r*) gene are commonly found to cause variations in the pigmentation of animals, Anderson et al. (2009) defined an alternative melanocortin pathway component, the *K* locus, as causal in coat color variation in North American gray wolves (*Canis lupus*). This variation ranges from pale wolves in open tundra regions to darker individuals in forested areas.

The habitat associations of paler- and darker-colored wolves were suggestive of selective constraints leading to these different pelage types in areas with more or less ambient light, respectively. The molecular variability at the *K* locus supported the hypothesis of positive selection favoring different alleles in different habitats, in that there was extremely low haplotype diversity at this locus, and yet a high frequency of alleles that cause darker pelage in the forested areas (Anderson et al. 2009). Significantly, findings from Anderson et al. (2009) implicated introgression from domestic dogs as the source of the alleles causing darker coat color in North American wolves (Fig. 25.1).

The geographical distribution and ecological associations of the melanistic variant support the hypothesis that gene flow was from domestic dogs into the wild forms. In particular, the *K* locus variant that causes dark pelage is found across dog breeds including ancient lineages from Asia and Africa. This contrasts sharply with the occurrence of this variant in wild canid populations only in North America except for Italian wolves that reside in an area known for recent hybridization between wolves and domestic dogs (Anderson et al. 2009). Likewise, the introgression event from domestic dogs into North American gray wolves would have occurred sometime since the former migrated with humans to North America ca. 12,000 ybp (years before present; Fig. 25.1) (Anderson et al. 2009).

A test of the adaptive nature of the pelage trait introgression was provided by haplotype analyses in domestic dogs and North American and Italian wolves and coyotes. Haplotypes associated with dark pelage in all domestic and wild canid samples clustered into a well-defined group regardless of species of origin (Anderson et al. 2009). In contrast, non-melanistic alleles more often than not grouped together by species. This phylogenetic signal is consistent with the origin of these alleles in domestic dogs with subsequent introgression into not only North American wolves, but also Italian wolves and

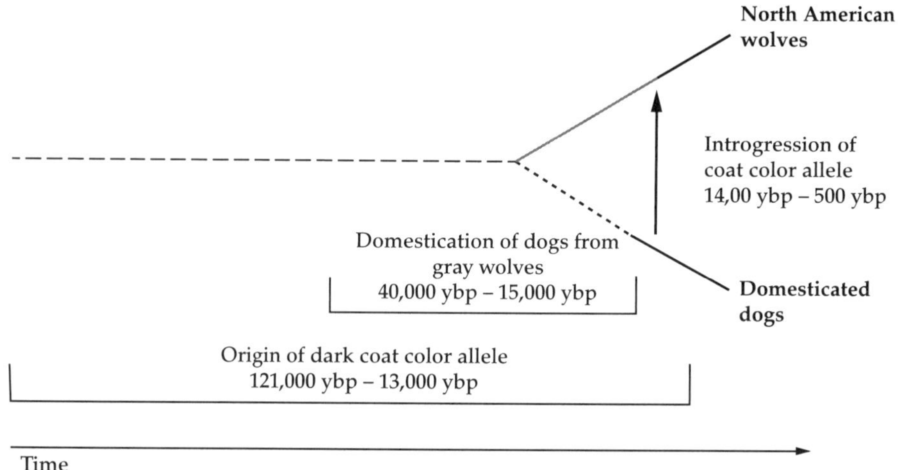

Figure 25.1 Hypothesized series of events leading to the adaptive introgression of the melanistic coat color allelic variants (i.e. *K* locus) from domestic dogs into North American gray wolves. The black, dashed lines before and after the divergence of the wolf and domestic dog lineages reflect the uncertainty of when these allelic variants arose. The gray section of the wolf lineage indicates that, if the melanistic *K* locus variants arose before the divergence of dogs and wolves from a common ancestor, they were lost in the wolf lineage. The dark segment at the tip of the wolf lineage indicates the approximate timing of the introgression of the melanistic alleles from dogs into wolves, after dogs were brought into North America by migrating humans (Anderson et al. 2009).

coyotes (Anderson et al. 2009). The limited time for introgression, along with the molecular signature of a selective sweep at this locus (Anderson et al. 2009), suggest both the rapid and adaptive nature of this introgression event affecting North American wild canids (Fig. 25.1).

25.2.2 Adaptive trait origin in *Saccharomyces cerevisiae*: Hybrids make the best wine

The evolutionary history of the human-associated yeast, *Saccharomyces cerevisiae*, has included whole-genome duplications, duplications of already existing genes, introgressive hybridization with other *Saccharomyces* species, and the acquisition of genes from unrelated organisms via horizontal transfer (Gordon et al. 2009; Novo et al. 2009). Many, if not most, of these genetic transfers likely occurred over the past several thousand years during which humans have utilized *Saccharomyces* species to produce various fermentation-based products such as wine (Novo et al. 2009). During this brief time period, numerous adaptive shifts have occurred. The ecological settings that *S. cerevisiae* occupies include a wide array of fermentative environments as well as those of a human commensal or pathogen.

The acquisition of genetic material from other lineages—either through introgression or lateral exchanges—has been identified as a key step in the adaptive process leading to the broad ecological amplitude of *S. cerevisiae*. For example, adaptation of EC1118, one of the commercial yeast strains associated with wine production, to the harsh environmental setting of fermentation (e.g. anaerobiosis, depletion of nutrients and increasing alcohol concentrations), has been partially attributed to the acquisition of foreign genes through lateral transfer and introgressive hybridization (Novo et al. 2009). This strain was found to possess three large stretches of DNA not found in other *S. cerevisiae* isolates, two of which most likely originated from another fungal genus while the third was likely introduced by introgressive hybridization with a *Saccharomyces* species.

The three regions of inserted DNA contained 34 genes, 20 of which possess inferred functions in the metabolism and transport of sugar or nitrogen (Novo et al. 2009). Wine production occurs in a nitrogen-limited environment and requires the conversion of huge amounts of sugar into alcohol.

The putative functions of these genes acquired from other species are thus consistent with adaptive trait origin via introgression and horizontal transfer. Furthermore, the inserted regions detected in the EC1118 winemaking strain are differentially distributed across other strains of *S. cerevisiae*. Specifically, the majority of strains carrying the inserted regions are involved in wine production. That the inserted regions are distributed across large geographic distances and in numerous winemaking strains was taken as evidence of continuous genetic restructuring of these *S. cerevisiae* variants by rapid, genetic exchange-mediated adaptive evolution (Novo et al. 2009).

25.3 Hybrid speciation: when opposites attract

Both homoploid and allopolyploid hybrid speciation have been detected in plant and animal clades (Arnold 1997; 2006; Rieseberg 1997; Soltis and Soltis 2009). The former involves the combination of divergent genomes resulting in hybrids with more or less the same chromosome number as the hybridizing taxa, while the latter involves hybridization accompanied by increases in chromosome number by multiples of the parental haploid sets (for reviews, see Rieseberg 1997; Soltis and Soltis 2009). Like adaptive trait introgression and natural hybridization-mediated adaptive radiations, hybrid speciation is, by definition, a sympatric process that is expected to occur rapidly. Indeed, in cases of allopolyploid speciation, the process is considered instantaneous given that reproductive isolation is expected to be immediately present between the polyploid and its diploid progenitors. Though sympatric speciation had been *persona non grata* for many decades following the neo-Darwinian synthesis (Arnold 1997, 2006), it is now understood that a large proportion of lineage diversification has likely occurred in the presence of at least intermittent gene flow between diverging taxa (Pinho and Hey 2010).

To illustrate some of the mechanisms associated with divergence-with-gene-flow, we will consider one case of homoploid hybrid speciation in animals and one of allopolyploidy in plants. As stated previously, these categories are somewhat arbitrary, especially in the case of homoploid hybrid speciation. Thus, examples of homoploid hybrid speciation, including the one discussed in the following section, could also be referred to the category of 'adaptive trait origin via introgression' since the possession of novel adaptations is the mechanism by which at least partial reproductive isolation from the parental lineages may arise (e.g. Gompert et al. 2006).

25.3.1 Homoploid hybrid speciation: hybrid butterflies (quickly) change their spots

The neotropical butterfly genus, *Heliconius*, is renowned for its aposematic wing color patterns that are a paradigm of Müllerian mimicry (Merrill et al. 2011). The mimicry in wing markings provides a greater level of protection for all of the associated *Heliconius* species due to the cumulative effects on predator behavior (Merrill et al. 2011). Furthermore, sister taxa often belong to 'mimicry rings' in which the variation in wing patterning also results in some degree of reproductive isolation from closely related species (Merrill et al. 2011). Indeed, associations between wing color variation and both pre- and postzygotic reproductive isolation have been detected, with loci affecting wing coloration and reproductive isolation being clustered within the *Heliconius* genome (Merrill et al. 2011). This genetic linkage would impede the breaking up, through recombination and segregation, of the loci that affect ecological traits (i.e. aposematic markings) from those that result in pre- and postzygotic reproductive isolation (Salazar et al. 2010; Merrill et al. 2011).

In addition to being a model of Müllerian mimicry and ecological speciation, the *Heliconius* species complex has recently become a focus of hypotheses surrounding the process of natural hybridization, specifically, homoploid hybrid speciation. *Heliconius heurippa* has been identified as a hybrid derivative of *H. melpomene* and *H. cydno*. The original definition of this species was based upon its admixed morphological and genomic characteristics, both of which could be recreated through experimental, introgressive hybridization between *H. melpomene* and *H. cydno* (Salazar et al. 2010). The admixed morphology of *H. heurippa* can be explained by the introgression of the alleles causing the red-banded *H. melpomene* phenotype onto the color background of *H. cydno* (Salazar et al. 2010). Consistent with this model of introgression, 670 single nucleotide polymorphisms from 29 unlinked genes grouped *H. heurippa* most closely with *H. cydno*. In contrast, 344 SNPs from genes associated with the red-band markings placed *H. heurippa* most closely with *H. melpomene* (Salazar et al. 2010).

The molecular signature of the hybrid speciation event resulting in *H. heurippa* supports the hypothesis of adaptive transfer, with some of the genetic loci putatively affecting pattern formation of the forewing and thus an important ecological trait (i.e. predation avoidance; Salazar et al. 2010). Furthermore, as with the other aposematic species, the origin of a novel pattern of wing coloration (in this case through introgressive hybridization) has resulted in some measure of reproductive isolation between the homoploid hybrid and other *Heliconius* species (Salazar et al. 2010). Finally, the origin of the wing color patterning present in the homoploid hybrid, *H. heurippa*, would have likely arisen rapidly given that only three generations of experimental, introgressive hybridization are necessary to construct this phenotype (Salazar et al. 2010).

25.3.2 Allopolyploid speciation: *Tragopogon* hybrid polyploids form again, and again, and again ... in less than 100 years ...

Whole-genome duplication events have been detected in the evolutionary history of all eukaryotic lineages (see Arnold (2006) for a review and references). In particular, it is now accepted that the majority of plant clades likely underwent multiple rounds of allopolyploidy (Soltis and Soltis 2009; Symonds et al. 2010). Indeed, even the evolutionary history of the model 'diploid' organism, *Arabidopsis thaliana*, included at least two whole-genome duplication events (Bowers et al. 2003). These findings indicate the degree to which plant speciation reflects a reticulate, rather than a purely divergent, evolutionary pattern (Soltis and Soltis 2009). This reticulate pattern of evolution is illustrated nowhere better than in cases of recent, allopolyploid speciation: such 'neoallopolyploids' have most often formed multiple times from hybridization events between the same diploid parents (see Soltis and Soltis (2009) for a review and references).

One of the best-known examples of a neoallopolyploid complex comes from the recent work of Pam Soltis and Doug Soltis on the plant genus *Tragopogon*. Using the two species, *T. mirus* and *T. miscellus*, as a model system, they tested hypotheses concerning the formation and subsequent evolution of allopolyploids. From their studies of *Tragopogon*, and their reviews of findings for other neoallopolyploid complexes (Soltis and Soltis 2009; Symonds et al. 2010), several general evolutionary trends associated with allopolyploid speciation have been identified. As already stated, most allopolyploids are the product of multiple origination events from hybridization between the same diploid parent species (Fig. 25.2). Associated with

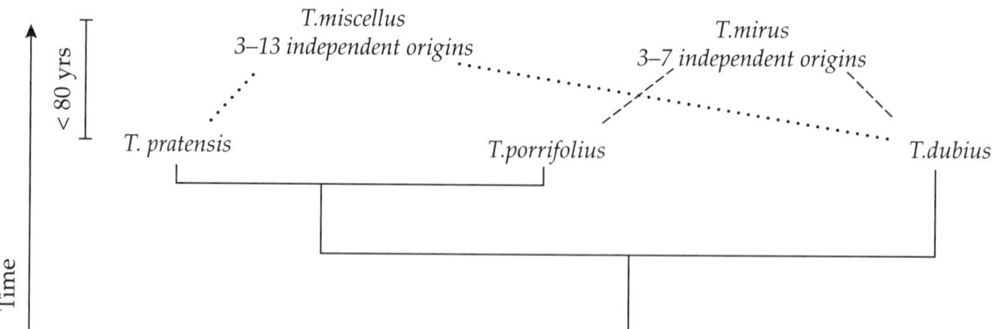

Figure 25.2 The evolutionary origin of the *Tragopogon* allopolyploids, *T. miscellus* and *T. mirus*. These two hybrid species formed sometime over the past 80 years. *T. miscellus* and *T. mirus* arose from the hybridization of *T. dubius* × *T. pratensis* and *T. dubius* × *T. porrifolius*, respectively. Both allopolyploids originated numerous times, through independent hybridization events (Soltis and Soltis 2009; Symonds et al. 2010).

their derivation from multiple origins, these neoallopolyploids possess elevated genetic variation. Yet, this elevated genetic variability does not necessarily reflect current patterns of variability in the diploid progenitors. For example, the allelic variation detected in *T. mirus* and *T. miscellus* was described as a '... snapshot of historical population structure in diploid progenitors, rather than modern diploid genotypes' (Symonds et al. 2010).

In addition to the detection of multiple reticulate events forming the *Tragopogon* allopolyploids, reflecting a web-like rather than a simple tree-like evolutionary history, the formation of these allopolyploids occurred within approximately the past 80 years (Fig. 25.2). The dating of the hybrid speciation events was substantiated by extensive sampling of natural *Tragopogon* populations over the past 100 years thereby revealing when the parental species first came into contact thus allowing the formation of the allopolyploids (see Symonds et al. (2010) for a review). The limited time period, during which the *Tragopogon* species formed, as well as the theoretical expectations of instantaneous speciation through the production of allopolyploid offspring *per se*, indicates the rapidity of this natural hybridization-mediated evolutionary process (Fig. 25.2; Soltis and Soltis 2009; Symonds et al. 2010).

25.4 Natural hybridization and adaptive radiations: hybrid speciation on steroids

As already discussed, the concept of natural hybridization-generated adaptive evolution and speciation was proposed and expanded upon during the mid-1900s (Anderson 1949; Anderson and Stebbins 1954; Stebbins 1959). However, it was not until the work of Seehausen (2004) that an explicit model was developed to predict how natural hybridization might act as a catalyst for adaptive radiations. Specifically, this model rested upon the burgeoning genetic and ecological data from cases of natural hybridization, along with the recently developed conceptual framework of ecological speciation. Because the large amount of morphological variability generated from crosses between divergent lineages has been referred to as a 'hybrid swarm' (e.g. Anderson 1949), Seehausen (2004) named his model the 'hybrid swarm theory.'

Tests of the generality of the hybrid swarm model of adaptive radiation require a combination of data that inform several predictions (Seehausen 2004). First, a period of hybridization preceding numerous radiations must be demonstrated. Second, a majority of the extant diversity must be shown to have derived from a period of crossing between divergent lineages. Third, a portion of the diversity originating during natural hybridization must be functional (e.g. adaptive morphological variation in hybrid derivatives should reflect genetic variation inherited from both of the progenitors). Fourth, natural hybridization must increase the likelihood of adaptive radiations (Seehausen 2004).

Some of the best-characterized examples of adaptive radiations associated with natural hybridization come from the spectacularly diverse African rift lake cichlids. We have chosen a recently published example from the cichlids (Joyce et al. 2011), along with Alpine lake whitefish (Hudson et al. 2011), and a Hawaiian plant clade (i.e. the 'silverswords'; Barrier et al. 1999, 2001) to illustrate tests of various components of the hybrid swarm model.

25.4.1 Hybridization and adaptive radiations of Lake Malawi cichlids: from hybrid swarm to 800 species, in one lake?!

East African Great Lake cichlids are well-known examples of truly explosive adaptive radiations. Significantly, the cichlid diversifications are assumed to have occurred over a very short period of time following the invasion of the African rift lakes by riverine lineages (Seehausen 2006). The various endemic species flocks are often so morphologically and ecologically diverse that they have been separated into hundreds of related species, yet the assemblages from any given lake have most often been inferred to be monophyletic (Seehausen 2006). This latter inference, however, may reflect a lack of appropriate data for testing for the alternative signatures of paraphyly and monophyly (Joyce et al. 2011). In particular, sequences from mitochondrial DNA (mtDNA) have often been the sole data used to define phylogenetic history. Sequences from a uniparentally inherited, non-recombining genome may not provide a sensitive assay for evolutionary pattern and process when an adaptive radiation was founded upon multiple lineages, particularly when hybridization was involved (Arnold 1997, 2006). For example, Joyce et al. (2011) highlighted the lack of a robust test of the hypothesis of monophyly for the adaptive radiation of cichlids in Lake Malawi. The inference of monophyly was drawn without assays of key riverine taxa that could have played a role in the founding of the Lake Malawi radiation, and was made using only mtDNA sequence data. Thus, the evolutionary origin of possibly the largest radiation of cichlids in any rift lake (i.e. up to 800 species formed) was not well defined.

To identify the likely progenitor(s) of the Lake Malawi cichlid adaptive radiation, Joyce et al. (2011) collected both nuclear (2045 polymorphic AFLP loci) and mitochondrial sequence (control region) data from representative taxa (in terms of ecological associations and morphological characteristics) from the six previously-defined mtDNA clades as well as 17 cichlid populations from river systems of varying distances from Lake Malawi. The genetic screens of the riverine and Lake Malawi samples detected recombination among lineages belonging to multiple cichlid clades indicating a hybrid origin for this adaptive radiation. Current phylogenetic methodologies are used to detect putative examples of reticulate evolution by testing for non-congruence between trees generated from different data sets. For example, the most diverse clade within the lake, that of the 'rock-dwelling mbuna' appears to possess nuclear and mitochondrial genomes from highly divergent progenitor lineages (Joyce et al. 2011). As predicted by the hybrid swarm model, the Lake Malawi cichlids reflect introgressive hybridization during the earliest stages of an adaptive radiation leading to the rapid origin of hundreds of species with admixed genomes.

25.4.2 Hybridization and adaptive radiations in Alpine lake whitefish: Swiss fish diversify after the last big thaw

The *Coregonus lavaretus* clade (Alpine lake whitefish) is made up of numerous species that arose from a series of adaptive radiations since the last glacial maxima in Europe (< 15,000 ybp; Hudson et al. 2011). The recency of the radiations indicates the rapid, parallel nature of the speciation events in the series of Alpine lakes in which these taxa occur (Fig. 25.3). Furthermore, the differentiation among coexisting species, at both random genetic loci and in phenotypes associated with feeding ecology, reflects both the rapidity of differentiation and the adaptive nature of the radiations in the various lacustrine habitats (Hudson et al. 2011).

The radiation of the Alpine lake whitefish has resulted in endemic flocks of up to six species in each lake, each flock consisting of sympatric forms that vary in ecological adaptations that are found across different lakes (Hudson et al. 2011). Of particular interest for the topic of this chapter, the evolutionary history of this assemblage is postulated to have included introgressive hybridization. Using >1000 AFLP loci and sequence data from the mitochondrial control region, Hudson et al. (2011) were able to test for phylogenetic and population genetic patterning within and among 36 bodies of water associated with the *Coregonus* radiation. These analyses provided a number of inferences concerning the genetic and evolutionary trajectories for the endemic lake assemblages. The overall signal was one of cytonuclear discordance. This discordance reflected strongly contrasting patterns of genetic variability for the nuclear AFLP and cytoplasmic mtDNA data sets. The nuclear data supported a monophyletic origin for the entire radiation, with major lake systems falling into separate clades, while the mtDNA sequence data revealed the presence of two divergent lineages scattered across the various species flocks (Fig. 25.3; Hudson et al. 2011).

Taken as a whole, the genetic and ecological data supported the hybrid swarm model in the diversification of the *Coregonus* flocks. Based upon nuclear genotype data, flocks from different lakes were placed within separate clades yet had similar adaptations; this observation was consistent with the *in situ* origination of the species within lakes due to environmental selection. However, as predicted by the hybrid swarm model (Seehausen 2004), the flocks apparently derived from multiple lineages as reflected by the admixture of divergent

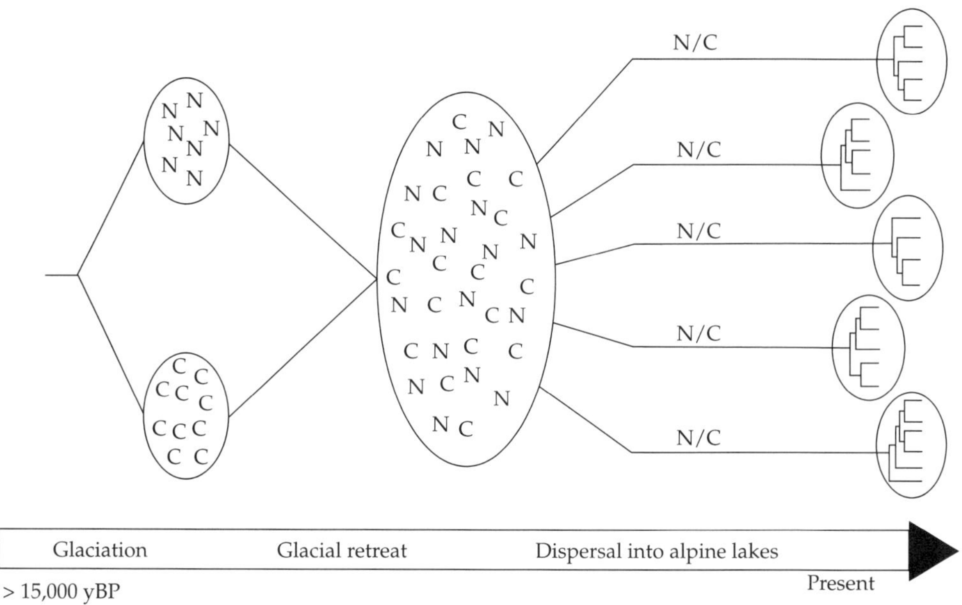

Figure 25.3 Inferred evolutionary history of the Alpine lake whitefish species complex. The origin of the individual species flocks apparently involved hybridization between multiple divergent lineages residing in glacial refugia, resulting in the formation of hybrid swarms. The signature of hybridization was lost from the nuclear genomes of the resulting lineages due to recombination. However, the hybrid nature of the founders of the individual flocks is reflected by the presence of multiple, divergent mtDNA haplotypes both within and among different lakes. Remarkably, the adaptive radiation that followed hybridization between the divergent whitefish lineages occurred within the space of < 15,000 years (Hudson et al. 2011).

mtDNA haplotypes within lakes (Fig. 25.3; Hudson et al. 2011). Not only was it likely that the various Alpine lake whitefish flocks originated rapidly and sympatrically, but there is ample evidence that these flocks were also derived from a genetically enriched, hybrid stock.

25.4.3 Hybridization and adaptive radiations in Hawaiian silverswords: allopolyploids in an island paradise

The Hawaiian silversword assemblage, like the East African rift lake cichlids, is known as a paradigm of the process of adaptive radiation. This complex: (1) consists of 30 species belonging to three genera endemic to six of the eight main Hawaiian islands, (2) is distributed across widely varying ecological settings from lava flows to bogs, (3) possesses species with radically different growth forms including cushion plants, shrubs, trees, and lianas, and (4) is most closely related to North American tarweeds, but unlike the diploid tarweed progenitors, silverswords possess polyploid genomes (Barrier et al. 1999, 2001).

At a minimum, for the adaptive radiation of the silverswords to be consistent with the hybrid swarm model of Seehausen (2004), the derivation of the ancestral lineage(s) that invaded the Hawaiian archipelago would necessarily need to have been allopolyploid (i.e. hybrid) derivatives of the diploid tarweeds. Barrier et al. (1999) tested this hypothesis by using the sequence variation found at two floral homeotic genes to construct phylogenetic relationships among silversword and tarweed lineages. Their findings identified two to three tarweed lineages that apparently contributed to the extant silversword complex thereby confirming an allopolyploid hybrid speciation event at the base of this adaptive radiation (Barrier et al. 1999). In a later study of the molecular evolutionary patterns of the same floral homeotic genes, Barrier et al. (2001) detected a signature of accelerated evolution at these regulatory loci following the formation of the hybrid lineages. Specifically, they found an increase in nonsynonymous versus synonymous base pair substitutions in these homeotic genes following allopolyploidization (Barrier et al. 2001). These observations suggested a role for the interaction between hybridization and regulatory gene evolution in catalyzing adaptive radiations (Barrier et al. 2001). Furthermore, the allopolyploid origin for this adaptive radiation indicates that the expected rapidity of the ecological diversification of this plant clade (Seehausen 2004) was preceded by an instantaneous hybrid speciation event.

25.5 Conclusions and future prospects

Outcomes of natural hybridization can lead to lineage diversification and adaptive evolutionary change. The earlier examples also suggest that genetic exchange-mediated evolution will often be rapid (Figs. 25.1–25.3). This would be expected given that the process involves invasion of open, novel ecological settings (Seehausen 2004; Arnold 2006). Rapid stabilization of hybrid lineages might also be predicted given the necessity for some measure of reproductive isolation from parental, and other hybrid, genotypes (Rieseberg 1997).

It is expected that the rapid nature of some, if not most, outcomes of natural hybridization reflects the balance between the formation of hybrid genotypes/phenotypes and natural selection for or against these hybrids across varying environmental settings. Assays of hybrid and parental fitness across generations and habitats is still lacking for most systems (Arnold and Martin 2010). However, the number of studies that give direct estimates of fitness have increased, and as predicted (Anderson 1949; Anderson and Stebbins 1954; Arnold 1997, 2006) demonstrate varying hybrid fitness across environments and time (e.g. Grant and Grant 2010; Whitney et al. 2010). In addition to the deficit of examples of direct estimates of fitness, there is also little understanding of the adaptive phenotypic traits that are being created through recombination between divergent lineages. Though some correlations have been noted, for example, between important ecological traits and adaptive radiations (Hudson et al. 2011), there is a general lack of detailed descriptive or experimental data to test for causality between natural hybridization and the origin of adaptations and adaptive radiations. However, when such tests have been possible, signatures of rapid, natural hybridization-mediated adaptive evolution have been detected (e.g. Rieseberg et al. 2003; Martin et al. 2006; Grant and Grant 2010; Whitney et al. 2010).

Acknowledgments

During the writing of this review, A.N.B. and E.S.B. were supported by National Science Foundation grants DEB-0949479/0949424 (a collaborative grant between M.L.A. and N.H. Martin, Texas State University-San Marcos) and DEB-1049757 (M.L.A.), and by funds from the Office of the Vice President for Research at the University of Georgia. J.A.P.H. was supported by the National Science Foundation PIRE grant, OISE-0730218 (R. Mauricio, PI).

References

Anderson, E. (1949) *Introgressive Hybridization*. New York: John Wiley and Sons, Inc.

Anderson, E. and Hubricht, L. (1938) Hybridization in Tradescantia. III. The evidence for introgressive hybridization. *Am J Bot* **25**: 396–402.

Anderson, E. and Stebbins, G.L., Jr. (1954) Hybridization as an evolutionary stimulus. *Evolution* **8**: 378–88.

Anderson, T.M., vonHoldt, B.M., Candille, S.I., Musiani, M., Greco, C., Stahler, D.R., et al. (2009) Molecular evolutionary history of melanism in North American Gray Wolves. *Science* **323**: 1339–43.

Arnold, M.L. (1997) *Natural Hybridization and Evolution*. Oxford: Oxford University Press.

Arnold, M.L. (2006) *Evolution Through Genetic Exchange*. Oxford: Oxford University Press.

Arnold, M.L. and Martin, N.H. (2010) Hybrid fitness across time and habitats. *Trends Ecol Evol* **25**: 530–6.

Baack, E.J. and Rieseberg, L.H. (2007) A genomic view of introgression and hybrid speciation. *Curr Opin Gen Dev* **17**: 513–18.

Barrier, M., Baldwin, B.G., Robichaux, R.H., and Purugganan, M.D. (1999) Interspecific hybrid ancestry of a plant adaptive radiation: Allopolyploidy of the Hawaiian silversword alliance (Asteraceae) inferred from floral homeotic gene duplications. *Mol Biol Evol* **16**: 1105–13.

Barrier, M., Robichaux, R.H., and Purugganan, M.D. (2001) Accelerated regulatory gene evolution in an adaptive radiation. *Proc Natl Acad Sci U S A* **98**: 10208–13.

Bowers, J.E., Chapman, B.A., Rong, J., and Paterson, A.H. (2003) Unraveling angiosperm genome evolution by phylogenetic analysis of chromosomal duplication events. *Nature* **422**: 433–8.

Gompert, Z., Fordyce, J.A., Forister, M.L., Shapiro, A.M., and Nice, C.C. (2006) Homoploid hybrid speciation in an extreme habitat. *Science* **314**: 1923–5.

Gordon, J.L., Byrne, K.P., and Wolfe, K.H. (2009) Additions, losses, and rearrangements on the evolutionary route from a reconstructed ancestor to the modern Saccharomyces cerevisiae genome. *PLoS Genet* **5**: e1000485.

Grant, P.R. and Grant, B.R. (2010) Natural selection, speciation and Darwin's finches. *Proc Cal Acad Sci* **61** (supp II): 245–60.

Hoekstra, H.E., Hirschmann, R.J., Bundey, R.A., Insel, P.A., and Crossland, J.P. (2006) A single amino acid mutation contributes to adaptive beach mouse color pattern. *Science* **313**: 101–4.

Hudson, A.G., Vonlanthen, P., and Seehausen, O. (2011) Rapid parallel adaptive radiations from a single hybridogenetic ancestral population. *Proc R Soc B* **278**: 58–66.

Joyce, D.A., Lunt, D.H., Genner, M.J., Turner, G.F., Bills, R., and Seehausen, O. (2011) Repeated colonization and hybridization in Lake Malawi cichlids. *Curr Biol* **21**: R108.

Mallet, J. (2005) Hybridization as an invasion of the genome. *Trends Ecol Evol* **20**: 229–37.

Martin, N.H., Bouck, A.C., and Arnold, M.L. (2006) Detecting adaptive trait introgression between Iris fulva and I. brevicaulis in highly selective field conditions. *Genetics* **172**: 2481–9.

Merrill, R.M., Schooten, B.V., Scott, J.A., and Jiggins, C.D. (2011) Pervasive genetic associations between traits causing reproductive isolation in Heliconius butterflies. *Proc R Soc B* **278**: 511–18.

Novo, M., Bigey, F., Beyne, E., Galeote, V., Gavory, F., Mallet, S., et al. (2009) Eukaryote-to-eukaryote gene transfer events revealed by the genome sequence of the wine yeast Saccharomyces cerevisiae EC1118. *Proc Natl Acad Sci USA* **106**: 16333–8.

Pinho, C. and Hey, J. (2010) Divergence with gene flow: models and data. *Annu Rev Ecol Evol Syst* **41**: 215–30.

Rieseberg, L.H. (1997) Hybrid origins of plant species. *Annu Rev Ecol Syst* **28**: 359–89.

Rieseberg, L.H., Raymond, O., Rosenthal, D.M., Lai, Z., Livingstone, K., Nakazato, T., et al. (2003). Major ecological transitions in wild sunflowers facilitated by hybridization. *Science* **301**: 1211–16.

Salazar, C., Baxter, S.W., Pardo-Diaz, C., Wu, G., Surridge, A., Linares, M., et al. (2010) Genetic evidence for hybrid trait speciation in Heliconius butterflies. *PLoS Genet* **6**: e1000930.

Seehausen, O. (2004) Hybridization and adaptive radiation. *Trends Ecol Evol* **19**: 198–207.

Seehausen, O. (2006) African cichlid fish: a model system in adaptive radiation research. *Proc R Soc B* **273**: 1987–98.

Soltis, P.S. and Soltis, D.E. (2009) The role of hybridization in plant speciation. *Annu Rev Plant Biol* **60**: 561–88.

Stebbins, G.L., Jr. (1959) The role of hybridization in evolution. *Proc Amer Phil Soc* **103**: 231–51.

Symonds, V.V., Soltis, P.S., and Soltis, D.E. (2010) Dynamics of polyploid formation in Tragopogon (Asteraceae): Recurrent formation, gene flow, and population structure. *Evolution* **64**: 1984–2003.

Whitney, K.D., Randell, R.A., and Rieseberg, L.H. (2010) Adaptive trait introgression of abiotic tolerance traits in the sunflower Helianthus annuus. *New Phytol* **187**: 230–9.

CHAPTER 26

Rapid evolution of pollinator-mediated plant reproductive isolation

Annika M. Moe, Wendy L. Clement, and George D. Weiblen

26.1 Plant–insect diversification

When considering the tree of life, the fact that some lineages are much more taxonomically rich than others suggests that rates of species diversification are highly variable. Explaining patterns of species diversity according to changes in diversification rate is limited by our power to reconstruct patterns of speciation and extinction through time, but this has not deterred speculation on the rate of evolution in mega-diverse groups such as flowering plants and insects (Sanderson and Donoghue 1994; Farrell 1998). The role of specialized interactions between insect herbivores and their host plants has been especially popular in explaining insect diversity by coevolutionary processes (Ehrlich and Raven 1964; Farrell et al. 1992). That reproductive isolation of herbivore populations may arise due to specialization on novel plant hosts is illustrated by the apple maggot fly, *Rhagoletis pomonella,* which broadened its host range over the past 300 years from native hawthorn (*Crataegus*) to include the introduced apple (*Malus*) in North America (Feder et al. 1994). Yet examples of herbivore speciation as a consequence of adaptation to different host plants, such as soapberry bugs (Carroll and Boyd 1992) and pea aphids (Peccoud et al. 2010), rarely consider the rate of host plant diversification. The role of herbivores in affecting plant diversification may be intensified when herbivores also provide pollination services, and directly affect reproduction of the host plant. In this chapter we focus on conditions in which insect pollinators acting as agents of reproductive isolation could influence the rate of speciation in flowering plants.

26.2 Pollination and reproductive isolation

While examining orchids, Darwin hypothesized that coevolution between flowering plants and pollinators might be responsible for their correlated patterns of diversity (Darwin 1862). Overall patterns of angiosperm diversity suggest that elevated diversification rates might be associated with biotic pollination (Jesson 2007). However, critical evaluation of this hypothesis leads to the conclusion that biotic pollination is 'neither a necessary nor sufficient condition for large numbers of species' (Gorelick 2001). Recent meta-analysis (Vamosi and Vamosi 2010) attributed episodes of angiosperm diversification primarily to geography, or 'space to diversify,' and only secondarily to biotic pollination. Given the limitations of such broad comparisons and correlative methods in identifying evolutionary processes, we focus on particular systems in which (and mechanisms by which) pollinators are implicated in accelerated plant diversification.

The role of pollinators as agents of selection on floral traits and the idea that pollinator specialization on divergent floral forms could result in the reproductive isolation of plant varieties gained broad acceptance during the 20th century (Grant 1949; Kiester et al. 1984; Johnson et al. 1998). Nonetheless, there are few specific cases of increased plant diversification attributed to pollinator interactions (Hodges et al. 2004; Sargent 2004).

Rapidly Evolving Genes and Genetic Systems. First Edition. Edited by Rama S. Singh, Jianping Xu, and Rob J. Kulathinal.

Pollinator foraging behavior has repeatedly been shown to play a role in plant reproductive isolation through constancy of floral visitation (Kephart and Theiss 2004). Modes of pollen transfer have also been implicated. For example, floral symmetry restricts the approach and movement of pollinators such that pollen placement may be precise and further reduces the likelihood of interbreeding among dissimilar floral forms (Sargent 2004). Floral mechanical means of reproductive isolation has been documented in species-rich groups such as orchids (Sun et al. 2011), gingers (Kay 2006), and louseworts (Yang et al. 2007). However, it is difficult to discern whether such mechanisms are causes or consequences of diversification if we admit the possibility of selection for traits reducing the likelihood of interbreeding when hybrids are less fit (e.g. reinforcement). In this chapter, we describe a system in which reinforcement appears unlikely and that also happens to meet conditions for rapid, pollinator-mediated speciation. Simple models predict pollinator-mediated plant diversification in the case of obligate mutualisms with highly host-specific pollinating seed predators where plant and pollinator reproduction are closely coupled (Kiester et al. 1984). The fig–fig wasp mutualism not only meets these criteria, but molecular phylogenetic studies (Datwyler and Weiblen 2004) have established the evolutionary historical context necessary for comparing diversification rates.

26.3 *Ficus* versus Castilleae

The recently discovered sister group relationship of figs (*Ficus*) to the tribe Castilleae (Moraceae) provides the opportunity to compare extant diversity and infer conditions associated with changes in diversification rate. Sister-group comparison may provide insights on such conditions given that these lineages share similar floral traits, modes of pollination, and time since divergence from their most recent common ancestor at least 65 million years ago (Zerega et al. 2005). Wind pollination is the inferred ancestral condition of the mulberry family (Moraceae) with a hypothesized shift to biotic pollination in the most recent common ancestor of *Ficus* and Castilleae (Datwyler and Weiblen 2004). Associated with the transition from wind to insect pollination were morphological changes in the position of flowers within inflorescences, particularly the arrangement of bracts that encircle the flowers of each inflorescence (Clement and Weiblen 2009). Whereas *Ficus* is one of the largest and most ubiquitous plant genera in tropical and subtropical forests with more than 800 species worldwide, Castilleae spans the same habitats and geographic range but comprises 11 genera and only ~60 species in total. This tenfold difference in species richness cannot be solely attributed to the shift from wind to biotic pollination in the common ancestor. But rather, we must consider other ecological or evolutionary differences among the descendants. Along the continuum of generalization and specialization in plant–pollinator interactions, brood-site pollination mutualisms involving insects are the most extremely specialized (Waser and Ollerton 2006). Although comparatively little is known about pollination syndromes in Castilleae, recent ecological studies (Sakai 2001; Zerega et al. 2004; Clement 2008) have uncovered an insect brood-site pollination syndrome similar to that of figs throughout the geographic distribution of the tribe. From what we do know of pollination syndromes in Castilleae, the evolution of a specialized brood-site pollination syndrome from a more conventional and generalized mode of insect pollination does not appear to account for the richness of *Ficus* relative to Castilleae. In seeking to explain the tenfold difference in numbers of species between sister clades, we describe their pollination ecology in detail and conditions affecting the evolution of reproductive isolation in particular.

Ficus is characterized by a completely enclosed inflorescence, or syconium, which is the site of the obligate mutualism with pollinating fig wasps (Agaonidae, Hymenoptera). The urn-shaped receptacle bearing numerous unisexual flowers is closed at the apex by involucral bracts that form a narrow passage, or ostiole, which is only accessible to certain agaonid wasps (Fig. 26.1). The agaonid life cycle begins and ends in syconia, where galled pistillate flowers nourish larvae, and mating occurs in the fig cavity immediately upon eclosion. Female wasps collect pollen from staminate flowers, emerge from ripening syconia, and search for receptive syconia in which to oviposit. Floral

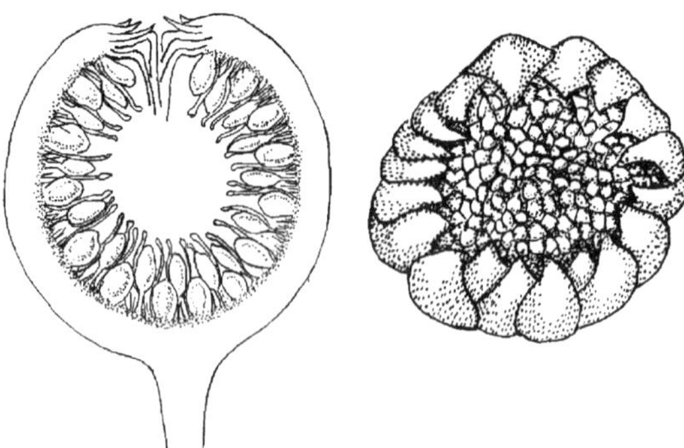

Figure 26.1 Inflorescence morphology of *Ficus* and Castilleae (Moraceae). The fig, or syconium (left), completely encloses the flowers within a hollow receptacle accessible only through a bract-lined opening, or ostiole. The staminate inflorescence of *Antiaropsis decipiens* (Castilleae, right) is discoid and has dozens of tightly packed flowers surrounded by involucral bracts. The illustration for *Ficus* first appeared in Whitfeld and Weiblen (2010) in *Harvard Papers in Botany* **15**: 1–10 and is reprinted here with permission of the editors of *Harvard Papers in Botany.* The illustration for *Antiropsis* first appeared in (Zerega et al. 2004) (© University of Chicago 2004).

volatiles and agaonid chemosensory antennae are involved in locating and choosing hosts. Access to potential brood sites requires passage through the ostiole where involucral bracts strip the wings and antennae of agaonids such that the syconium generally entombs each floral visitor. As agaonids lay eggs in a fraction of the pistillate flowers, pollination may be either active or passive but in either case, the development of seeds and galls is assured.

The specialized nature of fig/pollinator interactions has made the system a focal point for studies of coevolution (Herre 1989; Ganeshaiah et al. 1995; Weiblen 2004; Ma et al. 2009). Each *Ficus* species is associated with one or several pollinating fig wasp species (Herre et al. 2008) and approximately parallel patterns of phylogenetic diversity among more than 800 taxa has drawn much speculation on processes of diversification (Weiblen and Bush 2002; Machado et al. 2005; Jackson et al. 2008; Jousselin et al. 2008). Whether diversification is the result of cospeciation, host switching, or hybridization, the intertwining of fig and pollinator life cycles is implicated. When life cycles are linked, specificity has the potential to influence reproductive isolation of diverging populations in several ways. Among the possibilities are chemosensory responses of pollinators to fig volatiles (Grison-Pige et al. 2002), navigation of ostiolar bracts according to pollinator head shape (van Noort and Compton 1996), gall and seed formation as mediated by the interaction of the ovipositor with floral morphology (Weiblen 2004), variation in larval performance among hosts, and pollen compatibility. We will argue that interactions which are lacking in the Castilleae brood-site pollination mutualism could affect the rapid evolution of reproductive isolation in pollinators and figs simultaneously.

Compared to syconia, the inflorescences of thrips-pollinated Castilleae are discoid or urn-shaped (Sakai 2001; Clement and Weiblen 2009) but the receptacle does not completely enclose the flowers (Fig. 26.1). Inflorescences are unisexual with either stigmas or stamens protruding beyond the involucral bracts of pistillate and staminate inflorescences, respectively (Datwyler and Weiblen 2004). Although studies of fig pollination are more numerous, reports of Castilleae pollination from each major tropical region involve thrips (Thysanoptera) (Sakai 2001; Zerega et al. 2004; Clement 2008). Thrips feed on pollen at all life stages and only incidentally pollinate in the course of foraging, but they are known to breed in flowers and can be highly host-specific (Mound 2005). In the case of Castilleae, pistillate inflorescences provide no reward such that pollination involves deceit by floral mimicry. Thrips lay eggs prior to anthesis in the relatively short-lived staminate inflorescences, where nymphs later feed on pollen and eventually pupate in fallen litter. Unlike fig wasps, thrips feed as adults and move between plants while foraging and seeking opportunities for mating and oviposition. Thrips are predominantly associated with staminate inflorescences of Castilleae but the similar appearance and odor of pistillate inflorescences attracts occasional thrips where passive pollination of exposed stigmas is sufficient to affect fertilization (Zerega et al. 2004; Clement 2008). The fact that individual adult thrips have an opportunity to visit flowers in multiple inflorescences whereas fig wasps are limited to visiting a single inflorescence per generation is a key difference between pollination syndromes, possibly affecting the evolution of reproductive isolation and the rate of host plant speciation.

Differing extinction rates provide an alternative explanation for the relatively greater richness of *Ficus* but there is little reason to expect that Castilleae are more extinction-prone. The groups share identical habitats and pan-tropical geographic distributions. The complete enclosure of flowers within the syconium that severely limits opportunities for pollination favors fig species as more likely candidates for extinction than Castilleae. We argue that a higher rate of speciation in *Ficus*, due to particular conditions of the pollination syndrome, promotes the rapid evolution of reproductive isolation and explains why figs outnumber their sister group in species by ten to one. Species-specificity and floral constancy of pollinators are often invoked as reproductive isolating mechanisms in plants (Waser and Ollerton 2006). The discovery of thrips as primary pollinators of Castilleae, comprising not only greater than 95% of visitors to inflorescences but also exhibiting one-to-one host species-specificity in Panama and Papua New Guinea (Sakai 2001; Zerega et al. 2004), indicates a degree of specialization that appears rather similar to fig pollination. A closer examination of life history differences between these brood-site pollination syndromes is needed to identify conditions beyond species-specificity that favors more rapid evolution of pollinator-mediated reproductive isolation in figs than Castilleae.

Here we elaborate on two conditions that appear likely to have accelerated fig diversification relative to Castilleae. The first involves the nature of the reward for pollination services. Although both systems provide brood sites, fig wasps are seed predators whereas thrips are pollen feeders. Fig pollination enhances pollinator fitness by provisioning seed resources to offspring in galled pistillate flowers whereas pollination of Castilleae does not directly contribute to thrips fitness. A fig that hosts the offspring of a particular pollinator also achieves fitness through the attraction of that pollinator such that gene flow in fig populations is closely coupled with the reproductive consequences of wasp host choice. In dioecious Castilleae, however, thrips oviposition and pollen feeding only occur on non-pollinated plants such that the female component of plant fitness is not positively associated with foraging for brood sites. Whether or not fitness consequences of host choice by floral visitors are closely coupled with pollination may influence the rate by which plant reproductive isolation evolves if an additional condition is met.

The second condition for rapid plant speciation has to do with the number of floral visits per pollinator generation. Because fig wasps generally visit only a single inflorescence per generation, host choice has more immediate fitness consequences for fig wasps than for thrips. Once a fig wasp has located a host fig and entered the ostiole, her reproductive success is completely dependent on the suitability of that particular host whereas thrips have the option of bet hedging with visits to multiple inflorescences. The existence of 'tomb blossoms,' especially in functionally dioecious *Ficus* species having 'female' figs in which pollinators absolutely fail to achieve fitness, may impose intense selection on wasps to discern host quality prior to passing the ostiole. Recent manipulative pollination experiments with functionally dioecious figs documented the complete failure of pollinators to reproduce in sympatric, close relatives of a preferred host species (Moe 2011). Whereas visiting the wrong host imposes an absolute fitness cost to a fig wasp, thrips visiting suboptimal hosts, such as the pistillate infloresences of Castilleae that provide no reward, at least affords the possibility of locating a more suitable host with subsequent foraging. Positive selection for highly discriminatory host choice in response to the volatile chemical attractants of receptive figs is consistent with extremely species-specific patterns of fig/pollinator association (Bronstein 1987; Weiblen et al. 2001), the low incidence of pollinator sharing among sympatric fig species (Weiblen et al. 2001; Moe et al. 2011), and the apparent rarity of natural hybrids in at least some fig lineages (Parrish et al. 2003; Moe 2011). Such selection on fig wasps sets the stage for the rapid evolution of reproductive isolation in the host species.

26.4 A pollinator-mediated model for fig speciation

The obligate association of mutualistic partners (Fig. 26.2) sets conditions such that the discriminatory behavior of floral visitors is sufficient for plant reproductive isolation in the absence of postzygotic mechanisms such as pollen incompatibility, hybrid inferiority, or infertility. Crossing experiments recently demonstrated that closely related fig species are interfertile and that hybrid seedlings grow at rates comparable to non-hybrids (Moe 2011). There are also numerous anecdotal reports of fig hybrids in nature (Parrish et al. 2003; Machado et al. 2005; Moe 2011) However, manipulative experiments bypassing the host recognition phase of the pollinator life cycle found that the offspring of pollinators which successfully galled novel host species did not develop to maturity (Moe 2011). These observations together suggest that reproductive isolation among fig populations depends less upon postzygotic mechanisms and more upon pollinator fitness consequences of attraction to fig volatile chemistry. Given that wasp generation times are at least an order of magnitude shorter than those of their host trees (Fig. 26.2), the wasp chemosensory apparatus and associated behaviors are also likely to evolve more rapidly than postzygotic isolating mechanisms in host figs. We argue that the evolution of prezygotic reproductive isolation is driven by the wasp olfactory response that, according to the results of manipulative experiments, is selected for attraction to figs similar to that of the natal fig. Variation in volatile chemical bouquets introduced into a fig population through mutation, migration, hybridization, or even non-genetic factors such as local soil conditions or microbial interactions has the potential to become a target for discrimination that could effectively achieve reproductive isolation among variant subpopulations within a few generations (Fig. 26.3). The divergence of fig subpopulations, either by genetic drift or local adaptation, is expected to lag behind that of pollinator host race formation according to differences in generation time but is nonetheless expected to outpace that of Castilleae where selection for pollinator discriminatory behavior is neither so intense nor positively associated with pollination.

The proposed model finds obvious application in fig/pollinator associations characterized by extreme host specificity and congruent cophylogenetic patterns (Weiblen and Bush 2002) but it also applies to alternative modes of speciation. Pollinator sharing among fig species (Molbo et al. 2003), incongruent fig and pollinator phylogenies (Machado et al. 2005), and cyto-nuclear discordance indicative of fig hybridization (Renoult et al. 2009) suggest that host-switching may also be an important mechanism of diversification in this system (Herre et al. 2008). Hybridization has the potential

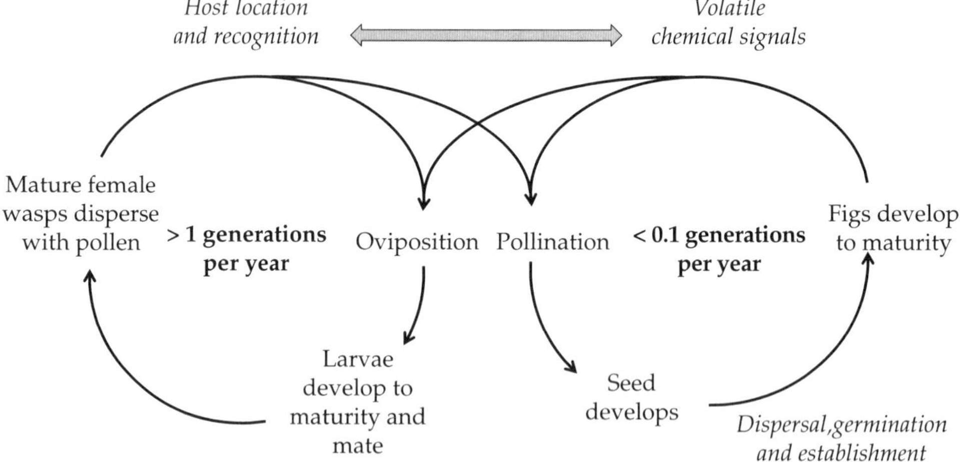

Figure 26.2 Intersecting life cycles of figs and fig wasps. The evolutionary dynamics of reproductive isolation are likely to involve genes affecting the production of volatile chemical signals by figs and their recognition by wasps. Genes affecting the success of oviposition, pollination, seed and larval development are also likely to be important. Due to the interplay of relatively shorter wasp generation times with longer fig life cycles, selection acting on wasp genetic systems and/or the evolution of reproductive isolation among wasp populations may increase the rate of speciation in figs.

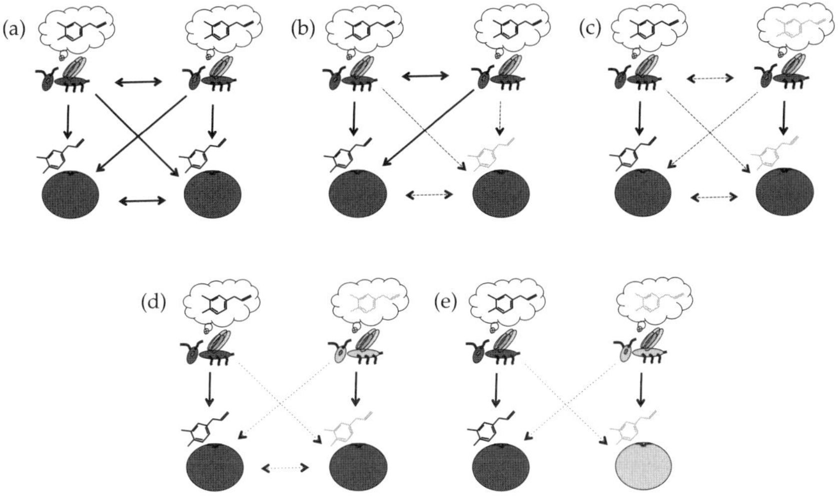

Figure 26.3 A model for the rapid evolution of pollinator-mediated reproductive isolation in figs. Horizontal arrows represent gene flow. Vertical and diagonal arrows represent plant/pollinator interactions. Solid arrows indicate frequent events whereas dashed arrows indicate rare events. (a) Host population with individuals having volatile bouquets with equal probability of attracting pollinators. (b) Variant fig arises with initial low probability of attracting pollinators. (c) Once colonized, the probability of the new variant attracting specific pollinators increases, given that wasps prefer volatiles similar to that of their birth fig. (d) Preference of pollinators for different variants leads to assortative mating, reproductive isolation, and speciation of pollinator host races. (e) Host plant speciation lags behind pollinator speciation due to longer generation times.

to generate novel profiles of chemical attractants and new targets for discriminatory pollinators such that a hybrid fig, once colonized by a pollinator whose offspring prefer similar hosts, could found a new lineage capable of exploiting ecological niches inaccessible to the parental species within a few generations (Gross and Rieseberg 2005). Evidence from fig wasp phylogeography (Haine et al. 2006; Moe and Weiblen 2010) suggests allopatric speciation where geographic variation in fig volatile profiles could also contribute to the rapid evolution of pollinator-mediated reproductive isolation. A next step in validating the proposed model and investigating its generality would be to examine the molecular evolution of genes affecting the fig wasp chemosensory apparatus and fig volatile chemistry in diverse fig/pollinator lineages and geographic contexts.

26.5 Future directions: plant–pollinator interactions and rapid evolution

The coupling of plant and pollinator life cycles may accelerate plant speciation under certain conditions. Given the continuum of variation from specialized to generalized animal-pollinated systems, it seems unreasonable to expect diversification in biotic pollination systems to be elevated relative to abiotic pollination overall. Comparisons of highly specialized systems similar to fig pollination are needed to gain further insights on conditions for rapid, pollinator-mediated plant diversification. Speciation in figs appears to be accelerated through: (1) the linkage of plant and pollinator reproduction, (2) severe pollinator fitness consequences for 'mistakes' such that highly discriminatory behavior is selected, and, (3) substantially shorter generation times in pollinators than in host plants. It will be necessary to examine these criteria in other brood-pollination mutualisms involving yucca and yucca moths (Pellmyr 2003), senita cactus and senita moths (Fleming and Holland 1998), and phyllanthoid euphorbs and *Epicephala* moths (Kato et al. 2003). For instance, *Yucca* is not more rich in species than its wind-pollinated sister group (Smith et al. 2008). Although plant and pollinator reproduction are also linked in yucca pollination, moths have generation times comparable to their hosts and moths may visit flowers of multiple plants such that the consequences of suboptimal choices are not as severe as for fig wasps. Such comparisons may shed light on whether the evolution of pollinator-mediated reproductive isolation has matched plant speciation by polyploidy or hybridization in rapidity and extent.

References

Bronstein, J.L. (1987) Maintenance of species-specificity in a neotropical fig-pollinator wasp mutualism. *Oikos* **48**: 39–46.

Carroll, S.P., and Boyd, C. (1992) Host race radiation in the soaberry bug - natural history with the history. *Evolution* **46**: 1052–69.

Clement, W.L. (2008) *Phylogeny and pollination ecology of Castilleae (Moraceae): Investigating the Evolutionary History of the Fig's Closest Relatives*. Minnesota, MN: Plant Biology, University of Minnesota.

Clement, W.L. and Weiblen, G.D. (2009) Morphological evolution in the mulberry family (Moraceae). *Syst Bot* **34**: 530–52.

Darwin, C. (1862) *On the various contrivances by which British and foreign orchids are fertilised by insects, and on the good effects of intercrossing*. London: John Murray.

Datwyler, S.L. and Weiblen, G.D. (2004) On the origin of the fig: Phylogenetic relationships of moraceae from ndhF sequences. *Am J Bot* **91**: 767–77.

Ehrlich, P.R. and Raven, P.H. (1964) Butterflies and plants: A study in coevolution. *Evolution* **18**: 586–603.

Farrell, B.D. (1998) 'Inordinate fondness' explained: why are there so many beetles? *Science* **281**: 555–9.

Farrell, B.D., Mitter, C., and Futuyma, D.J. (1992) Diversification at the insect-plant interface. *BioScience* **42**: 34–42.

Feder, J.L., Opp, S.B., Wlazlo, B., Reynolds, K., Go, W., and Spisak, S. (1994) Host fidelity is an effective premating barrier between sympatric races of the apple maggot fly. *Proc Natl Acad Sci U S A* **91**: 7990–4.

Fleming, T.H. and Holland, J.N. (1998) The evolution of obligate pollination mutualisms: senita cactus and senita moth. *Oecologia* **114**: 368–75.

Ganeshaiah, K.N., Kathuria, P., Shaanker, R.U., and Vasudeva, R. (1995) Evolution of style-length variability in figs and optimization of ovipositor length in their pollinator wasps: a coevolutionary model. *J Genet* **74**: 25–39.

Gorelick, R. (2001) Did insect pollination cause increased seed plant diversity? *Biol J Linn Soc Lond* **74**: 407–27.

Grant, V. (1949) Pollination systems as isolating mechanisms in Angiosperms. *Evolution* **3**: 82–97.

Grison-Pige, L., Bessiere, J., and Hoessaert-McKey, M. (2002) Specific attraction of fig-pollinating wasps: Role of volatile compounds released by tropical figs. *J Chem Ecol* **28**: 283–95.

Gross, B.L. and Rieseberg, L.H. (2005) The ecological genetics of homoploid hybrid speciatioN. *J Hered* **96**: 241–52.

Haine, E.R., Martin, J., and Cook, J.M. (2006) Deep mtDNA divergences indicate cryptic species in a fig-pollinating wasp. *BMC Evol Biol* **6**: 83.

Herre, E.A. (1989) Coevolution of reproductive characteristics in 12 species of New World figs and their pollinator wasps. *Experientia* **45**: 637–47.

Herre, E.A., Jander, K.C., and Machado, C.A. (2008) Evolutionary ecology of figs and their associates: recent progress and outstanding puzzles. *Annu Rev Ecol Evol Syst* **39**: 439–58.

Hodges, S.A., Fulton, M., Yang, J.Y., and Whittall, J.B. (2004) Verne Grant and evolutionary studies of Aquilegia. *New Phytol* **161**: 113–20.

Jackson, A.P., Machado, C.A., Robbins, N., and Herre, E.A. (2008) Multi-locus phylogenetic analysis of neotropical figs does not support co-speciation with the pollinators: The importance of systematic scale in fig/wasp cophylogenetic studies. *Symbiosis* **45**: 57–72.

Jesson, L.K. (2007) Ecological correlates of diversification in New Zealand angiosperm lineages. *N Z J Bot* **45**: 35–51.

Johnson, S.D., Linder, H.P., and Steiner, K.E. (1998) Phylogeny and radiation of pollination systems in Disa (Orchidaceae). *Am J Bot* **85**: 402–11.

Jousselin, E., van Noort, S., Berry, V., Rasplus, J.Y., Ronsted, N., Erasmus, J.C., et al. (2008) One fig to bind them all: Host conservatism in a fig wasp community unraveled by cospeciation analyses among pollinating and nonpollinating fig wasps. *Evolution* **62**: 1777–97.

Kato, M., Takimura, A., and Kawakita, A. (2003) An obligate pollination mutualism and reciprocal diversification in the tree genus Glochidion (Euphorbiaceae). *Proc Natl Acad Sci U S A* **100**: 5264–7.

Kay, K.M. (2006) Reproductive isolation between two closely related hummingbird-pollinated neotropical gingers. *Evolution* **60**: 538–52.

Kephart, S. and Theiss, K. (2004) Pollinator-mediated isolation in sympatric milkweeds (Asclepias): do floral morphology and insect behavior influence species boundaries? *New Phytol* **161**: 265–77.

Kiester, A.R., Lande, R., and Schemske, D.W. (1984) Models of coevolution and speciation in plants and their pollinators. *Am Nat* **124**: 220–43.

Ma, W.J., Peng, Y.Q., Yang, D.R., and Guan, J.M. (2009) Coevolution of reproductive characteristics in three dioecious fig species and their pollinator wasps. *Symbiosis* **49**: 87–94.

Machado, C.A., Robbins, N., Gilbert, T.P., and Herre, E.A. (2005) Critical review of host specificity and its coevolutionary implications in the fig/fig-wasp mutualism. *Proc Natl Acad Sci U S A* **102**: 6558–65.

Moe, A.M. (2011) *From pattern to process: ecology and evolution of host specificity in the fig-pollinator mutualism*. Minnesota, MN: Ecology, Evolution and Behavior, University of Minnesota.

Moe, A.M., Rossi, D.R., and Weiblen, G.D. (2011) Pollinator sharing in dioecious figs (Moraceae). *Biol J Linn Soc Lond* **103**: 546–58.

Moe, A.M. and Weiblen, G.D. (2010) Molecular divergence and host conservatism in Ceratosolen (Agaonidae) pollinators of geographically widespread Ficus species (Moraceae). *Ann Entomol Soc Am* **103**: 1025–37.

Molbo, D., Machado, C.A., Sevenster, J.G., Keller, L., and Herre, E.A. (2003) Cryptic species of fig-pollinating wasps: Implications for the evolution of the fig-wasp mutualism, sex allocation, and precision of adaptation. *Proc Natl Acad Sci U S A* **100**: 5867–72.

Mound, L.A. (2005) Thysanoptera: Diversity and interactions. *Annu Rev Entomol* **50**: 247–69.

Parrish, T.L., Koelewijn, H.P., van Dijk, P.J., and Kruijt, M. (2003) Genetic evidence for natural hybridization between species of dioecious Ficus on island populations. *Biotropica* **35**: 333–43.

Peccoud, J., Simon, J.C., von Dohlen, C., Coeur d'Acier, A., Plantegenest, M., Vanlerberghe-Masutti, F., et al. (2010) Evolutionary history of aphid-plant associations and their role in aphid diversification. *C R Biol* **333**: 474–87.

Pellmyr, O. (2003) Yuccas, yucca moths, and coevolution: a review. *Ann Missouri Bot Gard* **90**: 35–55.

Renoult, J.P., Kjellberg, F., Grout, C., Santoni, S., and Khadari, B. (2009) Cyto-nuclear discordance in the phylogeny of Ficus section Galoglychia and host shifts in plant-pollinator associations. *BMC Evol Biol* **9**: 248.

Sakai, S. (2001) Thrips pollination of androdioecious Castilla elastica (Moraceae) in a seasonal tropical forest. *Am J Bot* **88**: 1527–34.

Sanderson, M.J. and Donoghue, M.J. (1994) Shifts in diversification rate with the origin of angiosperms. *Science* **264**: 1590–3.

Sargent, R.D. (2004) Floral symmetry affects speciation rates in angiosperms. *Proc Roy Soc Lond B Biol Sci* **271**: 603–8.

Sun, H.Q., Huang, B.Q., Yu, X.H., Kou, y., An, D.J., Luo, Y.B., et al. (2011) Reproductive isolation and pollination success of rewarding Galearis diantha and non-rewarding Ponerorchis chusua (Orchidaceae). *Ann Bot* **107**: 39–47.

Vamosi, J.C. and Vamosi, S.M. (2010) Key innovations within a geographical context in flowering plants: towards resolving Darwin's abominable mystery. *Ecol Lett* **13**: 1270–9.

van Noort, S. and Compton, S.G. (1996) Convergent evolution of agaonine and sycoecine (Agaonidae, Chalcidoidea) head shape in response to the constraints of host fig morphology. *J Biogeogr* **23**: 415–24.

Waser, N.M. and Ollerton, J. (Eds) (2006) *Plant-Pollinator Interactions*. Chicago, IL: The University of Chicago Press.

Weiblen, G.D. (2004) Correlated evolution in fig pollination. *Syst Biol* **53**: 128–39.

Weiblen, G.D., and Bush, G.L. (2002) Speciation in fig pollinators and parasites. *Mol Ecol* **11**: 1573–8.

Weiblen, G.D., D.W. Yu, and West, S.A. (2001) Pollination and parasitism in functionally dioecious figs. *Proc Roy Soc Lond B Biol Sci* **268**: 651–9.

Whitfeld, T.J.S. and Weiblen, G.D. (2010) Five new Ficus species (Moraceae) from Melanesia. *Harv Paper Bot* **15**: 1–10.

Yang, C.F., Gituru, R.W., and Guo, Y.H. (2007) Reproductive isolation of two sympatric louseworts, Pedicularis rhinanthoides and Pedicularis longiflora (Orobanchaceae): how does the same pollinator type avoid interspecific pollen transfer? *Biol J Linn Soc Lond* **90**: 37–48.

Zerega, N.J.C., Clement, W.L., Datwyler, S.L., and Weiblen, G.D. (2005) Biogeography and divergence times in the mulberry family (Moraceae). *Mol Phylogenet Evol* **37**: 402–16.

Zerega, N.J.C., Mound, L.A., and Weiblen, G.D. (2004) Pollination in the New Guinea endemic Antiaropsis decipiens (Moraceae) is mediated by a new species of thrips, Thrips antiaropsidis sp. nov. (Thysanoptera: Thripidae). *Int J Plant Sci* **165**: 1017–26.

CHAPTER 27

Sexual system genomics and speciation

Rob J. Kulathinal and Rama S. Singh

27.1 In the beginning: Darwin and Wallace on sexual selection and speciation

The grand theory of natural selection that Charles Darwin and Alfred Wallace jointly proposed was meant to explain the origin and evolution of biological diversity among all living things. Both men shared a common vision on precisely how natural selection operated and, for the remainder of their lives, maintained a healthy connection that included a mutual appreciation and reciprocal endorsement of each other's work. However, Darwin and Wallace remained unconvinced of the other's selective explanations for two of nature's most visible and widespread biological features: the presence of spectacular secondary sexual dimorphic characters, and the phenomenon of hybrid sterility that commonly occurs when plant and animal breeders cross closely related species.

Darwin explained the evolution of secondary and seemingly maladaptive sexual traits such as the peacock's long tail (Fig. 27.1a) by sexual selection, i.e. sex-biased selection as the result of each sex possessing different strategies to increase their reproductive output (Darwin 1871). Darwin distinguished sexual from natural selection in order to explain how boldly dimorphic traits were the byproduct of selecting mates by females and outcompeting other males. In contrast, Wallace thought that the spectacular sexual dimorphisms found in nature could be best explained by natural selection for protective adaptation. For Wallace, sexual dimorphism in bird species with bright conspicuous male plumages evolved by gradually selecting for dull-coloration in females in response to predation pressure (Wallace 1890).

To explain why hybrid sterility occurs so frequently, Darwin and Wallace again evoked different models for selection to act upon. Since it would be counter-productive, in the context of a population's overall fitness, for sterility to evolve within a species by natural selection, Darwin explained it as merely an incidental outcome of natural selection on diverging populations (Darwin 1859). Wallace, on the contrary, held that natural selection could increase the degree of sterility in hybrids, ultimately proposing a mechanism that is the precursor of what is now known as sympatric speciation and reinforcement.

It is interesting to note that disagreements still exist over the explanations of these two phenomena among evolutionary biologists, more than a century after Darwin and Wallace's congenial correspondences. The genetic basis of hybrid sterility is a well-studied field of inquiry and a variety of explanations have been put forth (for a review, see Kulathinal and Singh 2008). Ecological and genetic studies of sexual selection now cover a full range of hypotheses from game theory to good-genes models (Shuster and Wade 2003). While it is most certainly true that neither Darwin nor Wallace comprehended the importance of hybrid sterility and sexual selection in the context of species formation, both these phenomena currently represent two of the most active areas of research in the field of speciation.

Rapidly Evolving Genes and Genetic Systems. First Edition. Edited by Rama S. Singh, Jianping Xu, and Rob J. Kulathinal.

Figure 27.1 Examples of morphological diversity that may have been driven by sexual selection. (a) Classic example of the formation of elaborate but maladaptive display feathers in the male peacock, *Pavo cristatus*. (b) Two stalk-eyed male flies (family Diopsidae) approach each other from opposite sides of a twig. (c) Male (striped) and female (plain) pair of Midas cichlids, *Cichlasoma citrinellum*, exhibiting extreme sexual dimorphism in size and coloration alongside their 1-day-old fry. (d) Morphological diversity in two 'picture-winged' species of Hawaiian Drosophila, *Drosophila nigribasis* and *D. macrothrix*, compared to the non-native Hawaiian, *D. suzukii* (from left to right). (e) Two species of Darwin's finches, *Geospiza magnirostris* (top), *Certhidea olivacea* (bottom). Beak sizes correlated to song used in mating. (f) Diversity of sex-specific reproductive tissues in flowers may be augmented by the coevolution in flower-pollinator systems. Images courtesy of A. Konings (c) and of K. Kaneshiro (d).

27.2 The Modern Synthesis and the development of speciation theory

After the Modern Evolutionary Synthesis of the 1930s and 1940s introduced species as reproductively isolated units (Dobzhansky 1937; Mayr 1942), post-Synthesis theories of speciation were based on two prevailing views in population genetics: that the genetic basis of most adaptive traits is complex and multigenic, and that gene flow is a powerful force of homogenization such that population differentiation cannot take place without geographical or temporal isolation (Dobzhansky 1951; Mayr 1963). Ernst Mayr's allopatric geographic theory of speciation was built on the basis of these two assumptions. Mayr emphasized the importance of geographic isolation and considered reproductive isolation as a byproduct of genetic divergence in isolation, much like Darwin did a century earlier (Mayr 1963). Theodosius Dobzhansky, on the other hand, focused on the evolution of reproductive isolating mechanisms proposing the role of natural selection in their perfection (Dobzhansky 1937, 1951). While Wallace had argued for the importance of natural selection under sympatric conditions, Dobzhansky extended the role of natural selection during secondary contact between species as an additional stage in the allopatric model of speciation that allowed for the reinforcement of premating isolation (Dobzhansky 1951).

Two of the most fruitful approaches in understanding the genetics of speciation include the Mendelian analysis of hybrid incompatibility genes and expanding the role of sexual selection from morphological to molecular levels. Dobzhansky first made use of genetic crosses to investigate the genetic basis of hybrid sterility between species, employing two closely related fruit fly species, *Drosophila pseudoobscura* and *D. persimilis* (Dobzhansky 1937). By backcrossing fertile F_1 hybrids to either parent, he showed that the X chromosome produced the largest effect on hybrid male sterility. Not much was done for 40 years until a new generation of speciation geneticists resurrected the technique *en force* (Coyne and Orr 1989) to study the genetic basis of Haldane's rule, i.e. the observation that hybrid inviability and sterility tend to occur more commonly in the heterogametic than homogametic sex (Haldane 1922). Observations from crosses between different species paved the way to new theoretical developments such as the faster-X (Charlesworth et al. 1987) and dominance (Orr and Turelli 1996) theories of hybrid incompatibilities. On the empirical side, a handful of genes affecting hybrid inviability and sterility have been identified (reviewed in Presgraves 2010). Indeed, the use of trans-species Mendelian genetics and its associated theoretical developments have taught us more about the genetic basis of speciation during the last 25 years than over the previous 150 years.

The second approach, pursued independently and in parallel, has allowed theories of speciation based on sexual selection and mating behavior to grow in scope (Fig. 27.1b,c). During the last three decades, formal mathematical models (e.g. Lande 1981; Kirkpatrick 1982) and empirical examples spanning a wide range of organisms (Arnqvist and Rowe 2005) have propelled the field of sexual selection into the forefront of evolutionary biology. An emphasis on classic male features can be seen in such early models as runaway selection which attempted to explain Darwin's original paradoxical observation that outwardly maladaptive traits are the evolutionary consequence of increasing male fitness (Fisher 1930; Lande 1981). Later theories of sexual selection emphasized the benefits of female discrimination through direct resource-based advantages (Maynard Smith 1991), the indirect appraisal of males by the good-genes model (Andersson 1994), and coevolutionary arms conflict between male and female strategies (Rice 1996).

With new molecular evidence, a new and encompassing view of sexual selection has begun to materialize (Coulthart and Singh 1988; Civetta and Singh 1999; Swanson and Vacquier 2002). This broadened perspective allows mechanisms of sexual selection to be extended beyond classical examples of female choice and exaggerated male phenotypes that have been limited to male secondary sexual traits and extreme male behavior (Darwin 1871). Foremost, the extension of sexual selection to the molecular level finally presents a direct link to the genetics of speciation by providing a common functional pool of genes and traits. This systems-based framework enables us to understand the origin of species-specific characters alongside species formation and divergence. In addition, microevolutionary processes that are rapidly evolving can be viewed to directly impact macroevolutionary patterns of diversity.

27.3 A new paradigm: the genomics of sexual systems and the origin of species

Our view of how species form has changed dramatically over the past 150 years. Early theories of speciation were shaped more by population dynamics, i.e. how populations split in both space and time and ultimately become species, than by the kind of genes or genetic mechanisms involved. A shift in emphasis to studying sex and reproduction-related systems holds the potential of providing us with greater insight on how populations diverge and ultimately become reproductively isolated. Using empirical evidence from the literature, we briefly outline the following framework for this shift in paradigm with highlights from the recent literature: (1) genomes can be broadly organized into sexual and non-sexual functional components, (2) sexual systems are generally more variable and possess higher rates of evolutionary change, (3) sexual selection is generally strong and repetitive, (4) sexual coevolutionary forces and interacting systems can drive rapid evolution, and (5) reproductive systems are generally the first to break down in interspecific hybrids.

27.3.1 Functional genomics: organization into sexual and non-sexual systems

Sexual systems are comprised of genes involved in male and female fertility in addition to morphological and behavioral traits involved in sexual selection. By being directly or indirectly involved in some aspect of reproductive function, sex genes are prone to sexual selective pressures. In contrast, non-sex genes primarily affect viability and are subject mostly to natural selection. This binary classification provides a heuristic framework to treat *reproduction* and *survival* as separate but coupled evolutionary systems open to independent modification at different evolutionary rates (Singh 2000). Ultimately, the basis of this dichotomy has its origin in the concept of individual fitness. As integrated components of fitness, fertility and viability are difficult to tease apart, yet historically, the viability component has received disproportionate attention. From the handful of studies that quantify both components of fitness, fertility appears more prominent. For example, in a meta-analysis of studies from the literature where selection was quantified in the wild, fertility fitness components were found to be more significant than viability and sexual selection stronger than natural selection (Kingsolver et al. 2001). It was also found that sex-specific variance in reproductive success produces a much larger effect than differences between males and females in survivability (Charlesworth 2001).

Modern genomic tools provide growing support for a functional treatment of the genomic landscape. Genomic approaches can quickly assay for sex-biased gene expression on a genome-wide scale for both male- and female-specific tissues in addition to somatic ones. Microarray, EST (expressed sequence tag), and now RNAseq studies reveal that a surprisingly large fraction of the genome is involved in reproductive function. In *Drosophila*, upwards to one half of all genes are expressed in the male testis (Ranz et al. 2003; Parisi et al. 2004). In addition, testes harbor a much greater fraction of tissue-specific genes than other tissues (Singh and Kulathinal 2005) suggesting the availability of a large number of targets for sexual selection to act upon.

Of course, any dichotomous perspective that classifies genes, traits, and systems as sex versus non-sex may be too simplistic due to the multifunctional and pleiotropic nature of most genes and traits. A more realistic representation pictures somatic non-sexual roles at one end of the functional spectrum and purely reproductive roles at the opposite end. This functional grouping allows us to better visualize how different selective processes act upon each genetic system with respect to mechanism (Fig. 27.2: sexual vs. natural selection, positive vs. purifying selection, rapid vs. slow evolution).

27.3.2 Higher variation among reproductive systems

Mounting evidence reveals that traits involved in pre- and postzygotic mechanisms of isolation are evolving rapidly, suggesting that sex and

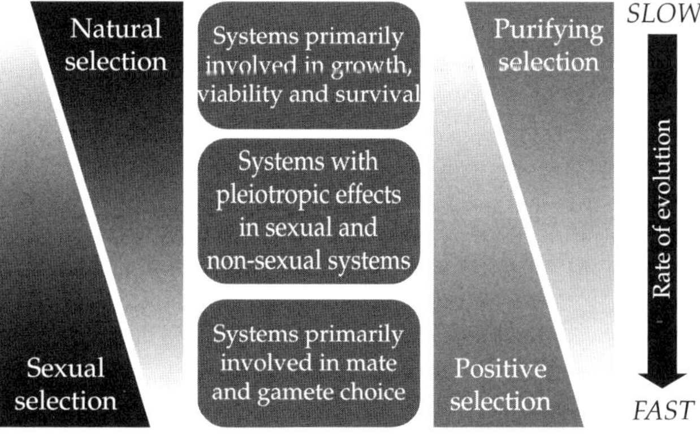

Figure 27.2 The relative contribution of natural versus sexual selection on various genetic systems. The effectiveness (right-hand side triangles) of purifying selection (negative selection on deleterious characters) and positive selection (eventual fixation of advantageous variants) differs between systems since genes related to sex and reproductive allow for more latitude on an organism's genetic endowment. Also indicated are expected correlated differences in evolutionary rates.

reproduction-related (SRR) genes are preferentially involved in speciation (Singh and Kulathinal 2000). Two pioneering studies of sexual systems set the stage on studies of reproductive characters and their association to speciation. William Eberhard's broad survey of animal taxa found male genitalia to be among the most disparate and diagnostic (Eberhard 1985). Of course, entomologists had distinguished related species on the basis of genitalia for centuries (for example, sexual characters were the basis for the Linnaean system of classification), however, Eberhard's comprehensive compilation of male gonadal morphology underscored the connection of high gonadal diversity to sexual selection. In the second pioneering study, Hampton Carson demonstrated that over a very short period of time (~6 million years), a founding population of fruit flies diverged to over 500 species across an expanding Hawaiian archipelago, and that much of the remarkable morphological diversity between species was driven by the evolution of behavioral preferences for conspecific mates in newly founded populations (Carson 1997; Fig. 27.1d). These two case studies point to the power of sexual selection driving the variability of both reproductive systems and species diversity.

Recent genomics studies are also finding that the genetic components of SRR systems are generally more liable to rapid evolutionary change. Various lines of evidence are summarized as follows: (1) among developmental programs, reproductive systems including sex determination (Hodgkin 1990), mating strategies (Shuster and Wade 2003), and floral pollination (Charlesworth et al. 2005; Cozzolino and Widmer 2005) are among the most rapidly evolving. (2) Reproductive genes show faster rates of sequence evolution (Civetta and Singh 1998; Wyckoff et al. 2000) and a more rapid loss of orthology (Haerty et al. 2007). Sex genes, by virtue of their non-ubiquitous expression and action, and not because of their dispensability (Torgerson et al. 2005), appear to be more open to mutational, selectional and neutral change than other genes. (3) Reproductive genes, particularly those expressed in the male testis, show higher rates of retention in the testis after evolving off the X chromosome onto an autosome via retroposition (Emerson et al. 2004). (4) Sexual systems are more likely to evolve novel genes (Dorus et al. 2008), thus, molecular diversity may be directly associated with morphological specialization in sexual systems. (5) Reproductive tissues exhibit greater variance in gene expression. Computational and microarray approaches using *Drosophila* point towards a complex set of transcripts (Telonis-Scott et al. 2009). Furthermore, in a large number of cases, ubiquitously expressed genes show significantly higher levels of expression in the testis when compared to ovary or somatic tissue (Parisi et al. 2004).

27.3.3 Strength of sexual selection

Why do SRR genes, traits, and systems tend to evolve faster? Male competition and female choice form the basis of Darwin's original hypothesis (Darwin 1871). Recent theories have focused on the antagonistic interaction (conflict) that arises between opposing male and female strategies. These mechanisms of conflict may be behavioral, morphological, or physiological. That females continually respond and coevolve to counter male strategies has been demonstrated in many taxa. In a now classic experimental evolution study, it was found that *Drosophila* females that were not given the opportunity to coevolve with males were less fit than females allowed to coevolve with those same males (Rice 1996). Such experiments reveal that what biologists actually observe in both nature and the laboratory is most likely the tip of the iceberg. Due to the coevolutionary and recurrent nature of sexual selection, the majority of directional change will most likely be cryptic. Furthermore, the cycle of inconspicuous change in the male's and female's sexual machinery does not cease to take place.

Empirically, the rate of speciation between groups of phylogenetically similar species of insects that differ in their capacity for sexual conflict has also been compared (Arnqvist et al. 2000). Species with multiple partners, allowing for sexually antagonistic strategies to develop, were compared to related species with monogamous mating systems. Results show that speciation rates are four times higher in species where conflict is present, providing evidence that sexual conflict is an important driver of evolutionary change, and speciation. While sexual conflict has become an important topic in the field of speciation, the rapid evolution of reproductive systems presents an alternative to the metaphor of 'fitness conflicts.' In this view, reproduction and fitness do not belong to males or females alone but to both sexes together, and battles (interactions) between males and males, females and females, and males and females are all part of a broader form of sexual selection (known as broad-sense sexual selection; Civetta and Singh 1999).

A more recent theory explains rapid sexual systems evolution as the direct result of male sexual drive (Singh and Kulathinal 2005; also known as male-driven sexual selection) whereby all aspects of males are under intense sexual selection due to their leading role in initiating mating and doing everything possible—from molecules to morphology—to increase their chances of mating with females, and thus increasing their overall fitness. An important consequence of male sexual drive theory is how the genome itself becomes masculinized by sexual selection. In general, genomes appear to maintain a disproportionate number of reproductive genes, an influx of new male-specific genes, and a biased distribution of harbored male versus female genes on sex chromosomes (Singh and Kulathinal 2005). Female reproductive systems often coevolve with these male genes into a rapidly evolving trajectory.

27.3.4 Sexual systems interaction, coevolution, and rapid change

Using an expanded role of sexual selection, it is highly probable that we will find natural and sexual selection to often interact and reinforce each other's effects. How often natural and sexual selection reinforce or oppose each other is an area of current interest. Selection on secondary sexual traits most likely affects survival. Similarly, natural selection on nonsexual traits, such as the shape and size of beaks in Darwin's finches, may affect mate choice and reproduction through their effect on song (Podos 2001) (Fig. 27.1e). Since sexual systems are more closely allied with fitness, and are more prone to be affected both by natural and sexual selection, they have the potential to evolve the fastest. Perpetual sexual interactions, coupled with inequalities in sex allocation, differences in selection intensity (due to males generally being more active and passionate), and sexual conflict in fitness interests, make sexual traits more liable to rapid, often exaggerated, and seemingly 'maladaptive' evolutionary change. The exaggerated tail length of peacocks and the wastage of reproductive gametes (such as the excessive production of pollen in plants and sperm in animals) most likely represent only a fraction of the maladaptive change with regard to reproduction in these organisms.

While pleiotropic effects across different genetic systems may yield greater selective pressure and constraint, intermolecular coevolution between male and female components can generate rapid change. These components may include signal-response systems such as pheromones and visual cues used in mating rituals as well as gametic interactions in fertilization. A growing number of intermolecular examples from the coevolution of sperm–egg proteins in marine invertebrates (Swanson and Vacquier 2002) to male-specific sperm accessory proteins and their targets in female *Drosophila* (Ram and Wolfner 2009) have been characterized.

The extension of sexual selection to beyond their original precopulatory scope also allows its inclusion in other taxa and processes that have not been part of traditional sexual selection literature. For example, reproductive mechanisms not aligned with the classic notion of female choice on male traits may provide a forum for sexual selection to prevail in plants. Because many angiosperms release pollen in massive amounts, pollination can be considered analogous to sperm competition (Delph and Havens 1998). Insect–plant pollination interactions may also be viewed to help orchestrate male–female coevolutionary interactions found in sexual selection models (for example, Fig. 27.1f). A growing number of examples between plants and animals are discovering such coevolving partners of rapid evolutionary change (e.g. Whittall and Hodges 2007). It is intriguing to consider that not only does strong and recurring selection maintain many of the dramatic sexual dimorphisms observed within species, but that it may also play a role in creating the complex web of species-to-species interactions in many ecosystems by evolving innovative new traits that are specific to a particular species and its partners.

27.3.5 Rapid breakdown of sexual systems in species hybrids

Darwin devoted an entire chapter in *Origins* to the evolution of hybrid sterility (Darwin 1859) and it is generally now observed that hybrid sterility evolves faster than hybrid inviability (Wu et al. 1996). This pattern contrasts the distribution of within-species mutations: in *Drosophila*, there are many more incidences of mutations affecting viability than sterility (Lindsley and Lifschytz 1972). This contrast suggests that the genetic basis of hybrid sterility is of a different nature than within-species sterility (Kulathinal and Singh 2008). Of course, the rapid evolution of reproductive genes may account for this pattern by generating a higher rate of deleterious interactions in the hybrid. In addition, sequence divergence is correlated to gene expression divergence as shown in *Drosophila* species hybrids (Artieri et al. 2007). However, the contrast of gene expression profiles between species and gene expression breakdown in their hybrids suggests that regulatory divergence in response to stabilizing selection alone may be capable of producing hybrid incompatibility. Indeed, the role of regulatory divergence as a result of stabilizing selection, at present, remains unappreciated.

27.4 Towards a post-genomics synthesis of speciation

Molecular studies of sexual systems are providing a general framework to explore genetic mechanisms of speciation among sexually reproducing organisms. By categorizing genes according to their function, the relative roles of natural versus sexual selection, as well as the importance of purifying versus positive selection and their associated evolutionary rates among functional classes, can be established (Fig. 27.2). As we extend our approach to studying speciation from a Mendelian one to a genomics one, it will be exciting to see how different genomes and transcriptomes from a range of species have evolved depending on such biological parameters as mating strategies, the strength of sexual isolation, as well as the genetic bases of evolved reproductive networks.

Applying a sexual/non-sexual gene pool dichotomy also allows species concepts and competing genetic theories of speciation to be related directly to one or more components of genetic systems. The result of this sexual systems approach is that speciation theories may not be as different to each other as they appear. For example, as an explanation to species formation on small isolated island populations—observations that inspired Darwin on his seminal voyage on the *Beagle*—Mayr proposed founder (also known as peripatric) speciation models. Various founder effect models prominently feature sexual traits and sexual selection (Carson 1989). Non-founder models include those that employ the 'Recognition' species concept' (Paterson 1985), which apply to only the subset of genes affecting species-specific sexual recognition signals and does not constrain the variation and evolution of genes affecting other functional components. Allopatric speciation with reinforcement and sympatric models also rely on a phase that involves homogametic mating to complete incipient reproductive isolation (Noor 1995). Table 27.1 summarizes the implications of a dichotomous gene pool concept to various speciation theories.

An understanding of sexual systems evolution may help resolve a host of controversies associated with the general problem of speciation. For example, the faster evolution of reproductive genes, along with the mechanics of dominance and hemizygosity, may provide a more unitary explanation of Haldane's rule (Haldane 1922) than the faster-male hypothesis by including cases in which the heterogametic sex is either male *or* female (Kulathinal and Singh 2008). The Dobzhansky–Muller model of reproductive incompatibility, representing the general framework used by speciation geneticists today, can be extended to include gene function (Fig. 27.2; Kulathinal and Singh 2008). For example, by introducing a simple binary parameter of sexual versus non-sexual function to loci in the Dobzhansky–Muller model, we may begin to envision a wider scope to the resolution of why hybrid sterility evolves faster than hybrid inviability. A focus on the study of sexual system genes may also offer new perspectives on other such phenomena as the evolution of sexual dimorphism, sex

Table 27.1 Sexual vs. non-sexual traits: effects on components of fitness and relevance to theories of speciation

Functional classification	Effect on fitness components		Relevance to theories of speciation
	Viability	Fertility	
Primary sexual traits (e.g. gametogenesis, fertilization, mating behavior)	Small, pleiotropic	Large, direct	[1]Organizational theory [2]Mate recognition [3]Genetic transilience
Secondary sexual traits (adapted for both sexual and non-sexual function)	Large, direct	Large, pleiotropic	[4]Runaway selection [5]Reinforcement
Non-sexual traits (e.g. development, metabolism, physiology)	Large, direct	Small, pleiotropic	[6]Allopatric speciation [7]Sympatric speciation

[1]Carson (1982), [2]Paterson (1985), [3]Templeton (1980), [4]Lande (1981), [5]Dobzhansky (1951), [6]Mayr (1940), [7]Bush (1969)

allocation, life-history traits, and the maintenance of sex (Singh and Artieri 2010).

27.5 Future prospects: sex as a major force in evolution

Molecular and evolutionary genomic studies of sexual systems are beginning to connect two of Darwin's greatest theories: sexual selection at the microevolutionary level, and species formation at the macroevolutionary level. Functional genomics provides a direct way to sort through which selective mechanism—natural or sexual—a gene or trait will be most prone to be acted upon. At present, genetic systems remain poorly characterized. For example, how pleiotropic are genes in sexual versus non-sexual systems? What proportion of genes from each genetic system has evolved *de novo*? What epigenetic landscape does a typical gene lie in? At a higher level of genetic organization, how different are each of the systems' networks in terms of size, robustness, redundancy, and degree of epistasis? Further experiments and exploration of these systems will inform us to whether sexual systems have indeed different intrinsic properties, and how these properties would impact the evolution of reproductive traits as well as incompatibilities in the hybrid.

In a historical context, the Evolutionary Synthesis that occurred over half a century ago furnished the first unification of biological principles (e.g. Dobzhansky 1937; Mayr 1940). Sexual system genomics and a molecular re-appraisal of sexual selection are offering a new perspective: while natural selection is responsible for survival and maintenance of multitude of characters, sexual selection is primarily responsible for some of the most spectacular aspects of organismal diversity, including sexual dimorphism and possibly the origin of species. Subsuming sexual selection under natural selection, although technically correct, tends to rob us of the rich view spanned by the evolutionary dynamics of sexual and reproductive systems. Studying the genomic consequences of rapidly evolving sexual systems at both the organismal and molecular levels holds much promise in our quest to understand one of the greatest of all mysteries: the origin of species.

References

Andersson, M. (1994) *Sexual selection*. Princeton, NJ: Princeton University Press

Arnqvist, G., and Rowe, L. (2005) *Sexual conflict*. Princeton, NJ: Princeton University Press.

Arnqvist, G., Edvardsson, M., Friberg, U., and Nilsson, T. (2000) Sexual conflict promotes speciation in insects. *Proc Natl Acad Sci U S A* **97**: 10460–4.

Artieri, C.G., Haerty, W., and Singh, R.S. (2007) Association between levels of coding sequence divergence and gene misregulation in Drosophila male hybrids. *J Mol Evol* **65**: 697–704.

Carson, H.L. (1997) Sexual selection: A driver of genetic change in Hawaiian Drosophila. *J Hered* **88**: 343–52.

Carson, H.L. (1989) Genetic imbalance, religned selection and origin of species. In L.V. Giddings, K.Y. Kaneshiro, and W.W. Anderson (Eds) *Genetics, speciation and the founder*, pp. 345–62. New York: Oxford University Press.

Charlesworth, B. (2001) The effect of life-history and mode of inheritance on neutral genetic variability. *Genet Res* **77**: 153–66.

Charlesworth, B., Coyne, J.A., and Barton, N.H. (1987) The relative rates of evolution of sex chromosomes and autosomes. *Am Nat* **130**: 113–46.

Charlesworth, D., Vekemans, X., Castric, V., and Glemin, S. (2005) Plant self- incompatibility systems: A molecular evolutionary perspective. *New Phytol* **168**: 61–9.

Civetta, A. and Singh, R.S. (1998) Sex-related genes, directional sexual selection, and speciation. *Mol Biol Evol* **15**: 901–9.

Civetta, A. and Singh, R.S. (1999) Broad-sense sexual selection, sex gene pool evolution, and speciation. *Genome* **42**: 1033–41.

Coulthart, M.B. and Singh, R.S. (1988) High level of divergence of male-reproductive-tract proteins between Drosophila melanogaster and its sibling species, D. simulans. *Mol Biol Evol* **5**: 182–91.

Coyne, J.A. and Orr, H.A. (1989) Two rules of speciation. In D. Otte and J.A. Endler (Eds) *Speciation and its consequences*, pp. 180–207. Sunderland, MA: Sinauer.

Cozzolino, S. and Widmer, A. (2005) Orchid diversity: an evolutionary consequence of deception? *Trends Ecol Evol* **20**: 487–94.

Darwin, C.R. (1859) *The origin of species by means of natural selection, or the preservation of favoured races in the struggle for life*. London: John Murray.

Darwin, C.R. (1871) *The descent of man, and selection in relation to sex*. London: John Murray.

Delph, L.F. and Havens, K. (1998) Pollen competition in flowering plants. In T.R. Birkhead and Møller (Eds) *Sperm competition and sexual selection*, pp. 149–73. San Diego, CA: Academic Press.

Dobzhansky, Th. (1937) *Genetics and the origin of species*. New York: Columbia University Press.

Dobzhansky, Th. (1951) *Genetics and the origin of species*, 2nd edition. New York: Columbia University Press.

Dorus, S., Freeman, Z.N., Parker, E.R., Heath, B.D., and Karr, T.L. (2008) Recent origins of sperm genes in Drosophila. *Mol Biol Evol* **25**: 2157–66.

Eberhard, W.G. (1985) *Sexual selection and animal genitalia*. Cambridge, MA: Harvard University Press.

Emerson, J.J., Kaessmann, H., Betran, E., and Long, M. (2004) Extensive gene traffic on the mammalian X chromosome. *Science* **303**: 537–40.

Fisher, R.A. (1930) *The genetical theory of natural selection*. Oxford: Clarendon Press.

Haerty, W., Jagadeeshan, S., Kulathinal, R.J., Wong, A., Ram, K.R., Sirot, L.K., et al. (2007) Evolution in the fast lane: rapidly evolving sex-related genes in Drosophila. *Genetics* **177**: 1321–35.

Haldane, J.B.S. (1922) Sex ratio and unisexual sterility in hybrid animals. *J Genet* **12**: 101–9.

Hodgkin, J. (1990) Sex determination compared in Drosophila and Caenorhabditis. *Nature* **344**: 721–8.

Kingsolver, J.G, Hoekstra, H. E., Hoekstra, J.M., Berrigan, D., Vignieri, S.N., Hill, C.E., et al. (2001) The strength of phenotypic selection in natural populations. *Am Nat* **157**: 245–61.

Kirkpatrick, M. (1982) Sexual selection and the evolution of female choice. *Evolution* **36**: 1–12.

Kulathinal, R.J. and Singh, R.S. (2008) The molecular basis of speciation: From patterns to processes, rules to mechanisms. *J Genet* **87**: 327–38.

Lande, R. (1981) Models of speciation by sexual selection on polygenic traits. *Proc Natl Acad Sci U S A* **78**: 3721–5.

Lindsley, D.L. and Lifschytz, E. (1972) The genetic control of spermatogenesis in Drosophila. In R.A. Beatty and S. Gluecksohn-Waelsch (Eds) *Proceedings of the International Symposium on 'The genetics of the spermatozoan'*, pp. 203–22. Bogtrykkeriet Forum, Copenhagen.

Maynard Smith, J. (1991) Theories of sexual selection. *Trends Ecol Evol* **6**: 146–51.

Mayr, E. (1942) *Systematics and the origin of species*. New York: Dover Publications.

Mayr, E. (1963) *Animal species and evolution*. Cambridge, MA: Harvard University Press.

Noor, M.A. (1995) Speciation driven by natural selection in Drosophila. *Nature* **375**: 674–75.

Orr, H.A. and Turelli, M. (1996) Dominance and Haldane's rule. *Genetics* **143**: 613–16.

Parisi, M., Nuttall, R., Edwards, P., Minor, J., Naiman, D., Lü, J., et al. (2004) A survey of ovary-, testis-, and soma-biased gene expression in Drosophila melanogaster adults. *Genome Biol* **5**: R40.

Paterson, H.E.H. (1985) The recognition concept of species. In E.S. Vrba (Ed.) *Species and speciation* (Transvaal Museum Monograph No. 4), pp. 21–9. Pretoria: Transvaal Museum.

Presgraves, D.C. (2010) The molecular evolutionary basis of species formation. *Nat Rev Genet* **11**: 175–80.

Podos, J. (2001) Correlated evolution of morphology and vocal signal structure in Darwin's finches. *Nature* **409**: 185–8.

Ram, K.R., and Wolfner, M.F. (2009) A network of interactions among seminal proteins underlies the long-term postmating response in Drosophila. *Proc Natl Acad Sci U S A* **106**: 15384–9.

Ranz, J.M., Castillo-Davis, C.I., Meiklejohn, C.D., Hartl, D.L. (2003) Sex-dependent gene expression and evolution of the Drosophila transcriptome. *Science* **300**: 1742–5.

Rice, W.R. (1996) Sexually antagonistic male adaptation triggered by experimental arrest of female evolution. *Nature* **381**: 232–4.

Shuster, S.M., and Wade, M.J. (2003) *Mating systems and strategies*. Princeton, NJ: Princeton University Press.

Singh, R.S. (2000) Toward a unified theory of speciation. In R.S. Singh and C. Krimbas (Eds) *Evolutionary genetics: From molecules to morphology*, pp. 570–604. Cambridge: Cambridge University Press.

Singh, R.S. and Artieri, C.G. (2010) Male sex drive and the maintenance of sex: Evidence from Drosophila. *J Hered* **101**: S100–S106.

Singh, R.S. and Kulathinal, R.J. (2000) Sex gene pool evolution and speciation: A new paradigm, *Genes Genet Syst* **75**: 119–30.

Singh, R.S. and Kulathinal, R.J. (2005) Male sex drive and the masculinization of the genome. *Bioessays* **27**: 518–25.

Swanson, W. and Vacquier, V. (2002) The rapid evolution of reproductive proteins. *Nat Rev Genet* **3**: 137–44.

Telonis-Scott, M., Kopp, A., Wayne, M.L., Nuzhdin, S.V., and McIntyre, L.M. (2009) Sex-specific splicing in Drosophila: Widespread occurrence, tissue specificity and evolutionary conservation. *Genetics* **181**: 421–34.

Torgerson, D.G., Whitty, B.R., and Singh, R.S. (2005) Sex-specific functional specialization and the evolutionary rates of essential fertility genes. *J Mol Evol* **61**: 650–8.

Wallace, A.R. (1890) *Darwinism: An exposition of the theory of natural selection with some of its applications*. London: Macmillan & Co.

Whittall, J.B. and Hodges, S.A. (2007) Pollinator shifts drive increasingly long nectar spurs in columbine flowers. *Nature* **447**: 706–9.

Wu, C.-I., Johnson, N.A., and Palopoli, M.F. (1996) Haldane's rule and its legacy: Why are there so many sterile males? *Trends Ecol Evol* **11**: 281–4.

Wyckoff, G.J., Wang, W., and Wu, C.-I. (2000) Rapid evolution of male reproductive genes in the descent of man. *Nature* **403**: 304–9.

Index